AF832346

Rebuilding Communities After Displacement

Mo Hamza · Dilanthi Amaratunga · Richard Haigh ·
Chamindi Malalgoda · Chathuranganee Jayakody ·
Anuradha Senanayake
Editors

Rebuilding Communities After Displacement

Sustainable and Resilience Approaches

Springer

Editors

Mo Hamza
Division of Risk Management and Societal Safety
Lund University
Lund, Sweden

Dilanthi Amaratunga
Global Disaster Resilience Centre, School of Applied Sciences
University of Huddersfield
Huddersfield, UK

Richard Haigh
Global Disaster Resilience Centre, School of Applied Sciences
University of Huddersfield
Huddersfield, UK

Chamindi Malalgoda
Global Disaster Resilience Centre, School of Arts and Humanities
University of Huddersfield
Huddersfield, UK

Chathuranganee Jayakody
Global Disaster Resilience Centre, School of Applied Sciences
University of Huddersfield
Huddersfield, UK

Anuradha Senanayake
Global Disaster Resilience Centre, School of Applied Sciences
University of Huddersfield
Huddersfield, UK

ISBN 978-3-031-21413-4 ISBN 978-3-031-21414-1 (eBook)
https://doi.org/10.1007/978-3-031-21414-1

© The Editor(s) (if applicable) and The Author(s), under exclusive license to Springer Nature Switzerland AG 2023
This work is subject to copyright. All rights are solely and exclusively licensed by the Publisher, whether the whole or part of the material is concerned, specifically the rights of translation, reprinting, reuse of illustrations, recitation, broadcasting, reproduction on microfilms or in any other physical way, and transmission or information storage and retrieval, electronic adaptation, computer software, or by similar or dissimilar methodology now known or hereafter developed.
The use of general descriptive names, registered names, trademarks, service marks, etc. in this publication does not imply, even in the absence of a specific statement, that such names are exempt from the relevant protective laws and regulations and therefore free for general use.
The publisher, the authors, and the editors are safe to assume that the advice and information in this book are believed to be true and accurate at the date of publication. Neither the publisher nor the authors or the editors give a warranty, expressed or implied, with respect to the material contained herein or for any errors or omissions that may have been made. The publisher remains neutral with regard to jurisdictional claims in published maps and institutional affiliations.

This Springer imprint is published by the registered company Springer Nature Switzerland AG
The registered company address is: Gewerbestrasse 11, 6330 Cham, Switzerland

International Scientific/Peer-Review Committee

Prof. Mo Hamza, Lund University, Sweden
Prof. Dilanthi Amaratunga, University of Huddersfield, UK
Prof. Richard Haigh, University of Huddersfield, UK
Dr. Chathuranganee Jayakody, University of Huddersfield, UK
Dr. Chamindi Malalgoda, University of Huddersfield, UK
Ms. Anuradha Senanayake, University of Huddersfield, UK
Prof. Irene Lill, Tallinn University of Technology, Estonia
Assoc. Prof. Emlyn Witt, Tallinn University of Technology, Estonia
Prof. Nishara Fernando, University of Colombo, Sri Lanka
Ms. Maduri Fernando, University of Colombo, Sri Lanka
Dr. Nuwan Dias, University of Huddersfield, UK
Dr. Kinkini Hemachandra, University of Huddersfield, UK
Ms. Georgina Clegg, University of Huddersfield, UK
Mr. Asitha De Silva, University of Huddersfield, UK
Mr. Malith Senevirathne, University of Huddersfield, UK
Ms. Shavindree Nissanka, University of Huddersfield, UK
Mr. Thushara Kamalrathne, University of Huddersfield, UK
Ms. Aravindi Samarakkody, University of Huddersfield, UK
Mr. Thisara Perera, University of Huddersfield, UK
Mr. Mahesh Senanayake, University of Huddersfield, UK
Prof. Champa Navaratne, University of Ruhuna, Sri Lanka
Dr. Gayan Wedawatta, Aston University, UK
Dr. Pournima Sridarran, University of Moratuwa, Sri Lanka
Dr. Filippo Dionigi, University of Bristol, UK
Ms. H. Unnathi S. Samaraweera, University of Colombo, Sri Lanka

Preface

The number of people worldwide forced to flee their homes due to conflict, violence, disaster risk, and environmental degradation driven by climate change is on the rise. These displacement situations are also becoming more protracted, with many lasting several years or slowly turning to a permanent form of relocation.

Forced displacement has become a significant development and humanitarian challenge. There is an urgent need to develop more sustainable and resilient approaches that can better support displaced communities and ensure a long-term prospect for such a population. But the picture is not complete without including the host communities and the kind of support they also need. Forcible displacement often results in resettle in areas that are already facing their own development challenges, which can lead to tensions between the two communities.

Sustainable and resilient approaches for rebuilding displaced and host communities go beyond addressing physical needs, including housing, safe drinking water, sanitary facilities, electricity, and transport infrastructure. The displaced have lost assets and livelihoods, and are unable to plan for the future. For some communities, such as those involved with fisheries or agriculture, land or place is both a livelihood and a strong sense of identity. No access to grazing land means indigenous people are deprived of a livelihood and a loss of local identity. No access to livelihood creates unemployment and fewer economic opportunities. This can lead to marginalisation, poverty, and further far reaching and more severe consequences such as human trafficking when vulnerable populations are exploited. Providing access to previous livelihoods or creating new economic opportunities give displaced people the opportunity for financial independence. This also helps to avoid tensions with host communities, who often view immigrants as a threat to their livelihoods or a drain on local resources.

Effective planning of settlements can help to ensure that displaced communities have access to education and training opportunities. It is crucial that new arrivals gain employability skills and match their vocational and professional qualifications with those of the host country. Access to quality language training programmes is also essential, along with orientation to help them integrate, such as financial support, housing, rights and responsibilities, employment, culture, politics, and available services and facilities.

There is an emerging discourse on the best approaches for rebuilding displaced and host communities. This recognises that sustainable and resilient approaches for rebuilding must go beyond physical aspects. Broader needs must be addressed, including sociocultural, institutional, human capital, livelihood, and economic perspectives. There is also a need to promote social cohesion between displaced and host communities.

This book contributes to this discourse. The chapters of this book are primarily drawn from contributions made to the online International Symposium on Rebuilding Communities After Displacement, held on 13 and 14 December 2021 (http://regardproject.com/portal/index.php/events/research-symposium). The symposium invited leading scientists around the world to share their knowledge and experience. The event was jointly organised by:

- Division of Risk Management and Societal Safety, Lund University, Sweden.
- Global Disaster Resilience Centre, University of Huddersfield, UK.
- Centre for Sustainable Transitions (CST), University of Central Lancashire, UK.
- Building Lifecycle Research Group, Tallinn University of Technology, Estonia.
- Social Policy Analysis and Research Centre (SPARC), University of Colombo, Sri Lanka.

The symposium encouraged debate and discussion on how a holistic approach to the built environment can help address the challenges associated with displacement. This included effective policies and methods for rebuilding communities and integrating newcomers after forced displacement. It also promoted knowledge-sharing among the research community and practitioners in the field and encouraged multi-disciplinary and multi-sectoral interaction.

Authors of selected high-quality research articles presented at the symposium were invited to contribute to this book volume, which addresses five themes associated with displacement.

- Understanding and managing for better response provision for displaced communities.
- Governance of displacement, and efforts to prevent and respond to internal displacement.
- Economic and policy interventions for sustainable resettlement.
- Resilience and environmental consideration of resettlement planning.
- Social cohesion between displaced and host communities.

In reflecting upon these themes, this book brings together a collection of empirical research, case studies, discussion, and insights on sustainable and resilient approaches to rebuilding displaced communities.

As Editors, we hope that this important collection of scientific results will provide a valuable point of reference, reflecting our progress to date in tackling the development challenges associated with forced displacement.

Lund, Sweden	Prof. Mo Hamza
Huddersfield, UK	Prof. Dilanthi Amaratunga
Huddersfield, UK	Prof. Richard Haigh
Huddersfield, UK	Dr. Chamindi Malalgoda
Huddersfield, UK	Dr. Chathuranganee Jayakody
Huddersfield, UK	Ms. Anuradha Senanayake
September 2022	

Acknowledgements The REGARD project has been co-funded by the Erasmus+ Programme of the European Commission. The European Commission support for the production of this publication does not constitute an endorsement of the contents which reflects the views only of the authors, and the Commission cannot be held responsible for any use which may be made of the information contained therein.

Contents

Part IV Socio-economic Interventions for Sustainable Resettlement

Editors and Contributors

About the Editors

Prof. Mo Hamza is Professor of Risk Management and Societal Safety at Lund University, Sweden. In his career spanning 36 years so far, he has worked with international development organisations including: World Bank, ADB, UNDP, UNISDR, USAID, DfID, IFRC, IUCN, Swedish Red Cross, and the Swedish Civil Contingencies Agency (MSB). His primary areas of professional expertise and research work are: disaster risk and vulnerability reduction, post-disaster recovery, climate change impact and adaptation in fragile and failed states, and environmental displacement. Previously he was Chair of Social Vulnerability Studies at the United Nations University, Bonn, Germany; a Senior Research Fellow at the Stockholm Environment Institute (SEI), Sweden; and an advisor to the MIT Climate CoLab. He has undertaken consultancy and research work in: Afghanistan, Bangladesh, Botswana, Egypt, Eritrea, Ethiopia, India, Indonesia, Jordan, Kyrgyzstan, Liberia, Nigeria, Pakistan, Saudi Arabia, Sierra Leone, Sri Lanka, Sudan, Uganda, UK, South Eastern Europe, and the Balkan States. He is currently an advisor to the Swedish Civil Contingencies Agency (MSB) where he is responsible for their capacity development processes. Among various and more recent publications, Mo Hamza was the co-guest editor of a double special issue in *Global Discourse Journal* (2022) "Critical Exploration of Crisis: Politics, Precariousness and Potentialities", and the author of "Refugees' Integration in the Built Environment—The Sweden Case" in *Sustainability Journal* 2021. He was also the lead author and editor of the World Disasters Report (2015) "Focus on local actors, the key to humanitarian effectiveness". He can be contacted on: mo.hamza@risk.lth.se.

Prof. Dilanthi Amaratunga is a leading international expert in disaster resilience with an extensive academic career that has a strong commitment to encouraging colleagues and students to fulfil their full potential. Currently she is leading University of Huddersfield, UK's Global Disaster Resilience Centre. She is recognised for her career-long impact up until 2020, and is placed among the global top 2% of influential scientists, according to the report of "Composite Citation Metrics" by Elsevier BV Netherlands and Stanford University, USA, released in August 2021. She has project managed to successful completion a large number of international research projects (over £ 20 million) generating significant research outputs and outcomes. She provides expert advice on disaster resilience to national and local governments and international agencies including the UNDRR. To date, she has produced over 500 publications, refereed papers, and reports and has made over 100 keynote speeches in around 40 countries. Among many leadership roles, she is the joint chief editor of the *International Journal of Disaster Resilience in the Built Environment*. She is a member of the European Commission and UNDRR's European Science and Technology Advisory Group (ESTAG) representing the UK, a Steering Committee member of the UK Frontiers of Development programme, a Steering Committee member of the UK Alliance for Disaster Research, motivation of which is to bring together the UK's rich and diverse disaster research community to facilitate collaboration and partnership. She is a Fellow of the Royal Institution of Chartered Surveyors (RICS), a Fellow of The Royal Geographical Society, and a Fellow and a Chartered Manager of the Chartered Management Institute, UK. Further details are viable here: www.dilanthiamaratunga.net.

Prof. Richard Haigh is Professor of Disaster Resilience and Co-director of the University of Huddersfield's Global Disaster Resilience Centre in the UK. His research interests include multi-hazard early warning, disaster risk governance and resilience in the built environment. He is the Editor-in-Chief of the *International Journal of Disaster Resilience in the Built Environment* and Co-Chair of the International Conferences on Building Resilience, which started in 2008. He is an expert member of Working Group 1 of Intergovernmental Coordination Group on the Indian Ocean Tsunami Warning and Mitigation System (ICG/IOTWMS). Richard and his team won the Newton Prize for Indonesia, celebrating best research innovation partnership for the project Mainstreaming Integrated Disaster Risk Reduction and Climate Change Adaption Strategies into Coastal Urban Agglomeration Policy. He has led and successfully delivered research grants funded by the UK Natural Environment Research Council, the European Commission's Horizon 2020, Erasmus+, Framework Seven Programme, Lifelong Learning, Asia Link and European Social Fund, the UK Newton Fund, the UK Foreign and Commonwealth Office, British Council, and UK Parliamentary under Secretary of State for Business, Energy and Industrial Strategy. Many of these projects were carried out in close collaboration with government, non-government organisations, and industry partners. He has published widely, including an edited book by Wiley Blackwell, 15 edited book chapters, and 45 peer-reviewed journal articles, the majority of which are in ISI or SCOPUS indexed journals. He has delivered over 80 invited speeches and keynote presentations for audiences in 25 countries across North America, South America, Europe, Asia, Africa, and Australasia.

Dr. Chamindi Malalgoda is a senior lecturer and the quantity surveying course leader at the Department of Design and Built Environment, School of Arts and Humanities, University of Huddersfield, UK. She is also a member of the Global Disaster Resilience Centre (GDRC) of the University. Chamindi's research interest includes disaster risk reduction and resilience in the built environment. Chamindi completed her Ph.D. in 2014 in the theme of empowering local governments in making cities resilient to disasters and is currently managing several European funded research projects related to disaster resilience and management. She has led and successfully delivered research grants funded by the Erasmus plus programme under Strategic Partnerships and International Credit Mobility including REGARD (Rebuilding after displacement) project which was completed in March 2022. She has contributed to over 40 publications, which include refereed journals, conference papers, and various high-impact reports. She is also a reviewer of a number of international journals in the fields of disaster resilience, construction, and property management and recently guest edited a special issue on "Rebuilding Communities Following Disasters and Conflict Induced Mass Displacements" in the *Sustainability Journal*. For further information, see https://pure.hud.ac.uk/en/persons/chamindi-malalgoda.

Dr. Chathuranganee Jayakody is a Senior Tutor in Planning and Environmental Management at the School of Environment Education and Development (SEED), Faculty of Humanities, The University of Manchester. Her role includes teaching in the areas of urban planning, neighbourhood planning, infrastructure planning and supervising master's dissertations in planning, urban design, global urban development, and urban regeneration and development. Before joining the University of Manchester, she was a post-doctoral researcher at the University of Huddersfield and a research member of the Global Disaster Resilience Centre. She is also a part-time tutor at School of Architecture, Manchester Metropolitan University. She has worked in the Planning Control and Building Control Department at Bolton Metropolitan Borough Council. She completed her Ph.D. in "Architecture and the Built Environment" at the University of Huddersfield, and she did her master's degree in urban design at the University of Salford. Her current research interest is in the areas of built environment perspective of displacement, planning and designing sustainable shared spaces and public open spaces, urban planning, and urban designing interventions for disaster resilience.

Ms. Anuradha Senanayake is a Ph.D. researcher at the Global Disaster Resilience Centre, University of Huddersfield, UK. By profession she is a lecturer attached to the Department of Sociology, University of Colombo, Sri Lanka, and an Attorney-at-Law at the Supreme Court of Sri Lanka. She is actively involved in youth advocacy and volunteerism in the Rotary District 3220 Sri Lanka and Maldives, where she held the position of Rotaract Country Representative from 2017 to 2018. She completed her M.Phil. under the theme "The Role of Governance in Relocation Disaster Risk Reduction Strategy" at the University of Colombo, Sri Lanka. She also holds an LLM in "International Human Rights" from Birmingham City University, UK. Her research interests lie in the areas of displacement and relocation, disaster risk governance, disaster justice, urban sociology, sociology of law, and sociology of medicine.

Contributors

Mokbul Morshed Ahmad Development Planning Management and Innovation (DPMI), Asian Institute of Technology, Khlong Luang District, Pathum Thani, Thailand

Bayes Ahmed Institute for Risk and Disaster Reduction (IRDR), University College London (UCL), London, UK

Dilanthi Amaratunga Global Disaster Resilience Centre, School of Applied Sciences, University of Huddersfield, Huddersfield, UK

Attia Atiq Department of Geography, Government College University, Lahore, Pakistan

Shalini Attri Department of English, BPS Women's University, Sonipat, Haryana, India

Cloe Barbera Institute of Advanced Development Studies (INESAD), La Paz, Bolivia

B. M. R. L. Basnayake Department of Agribusiness Management, Faculty of Agricultural Sciences, Sabaragamuwa University of Sri Lanka, Belihuloya, Sri Lanka

Anamaria Bukvic Department of Geography, Center for Coastal Studies, Virginia Tech, Blacksburg, VA, USA

Andrew Carmichael University of Central Lancashire, Preston, UK

Veronika Cummings Chair of Human Geography, Institute of Geography, Johannes Gutenberg University of Mainz, Mainz, Germany

Achini M. De Silva Department of Agribusiness Management, Faculty of Agricultural Sciences, Sabaragamuwa University of Sri Lanka, Belihuloya, Sri Lanka

D. A. M. De Silva Department of Agribusiness Management, Faculty of Agricultural Sciences, Sabaragamuwa University of Sri Lanka, Belihuloya, Sri Lanka

Janaranjana Ekanayake National Aquatic Resources Agency (NARA), Colombo, Sri Lanka

Alireza Fallahi Chairman of Post-Disaster Reconstruction Group, SBU, Tehran, Iran

Maduri Fernando Social Policy Analysis and Research Centre, University of Colombo, Colombo, Sri Lanka

Nishara Fernando Department of Sociology, University of Colombo, Colombo, Sri Lanka

Ximena Flores-Palacios La Paz, Bolivia

K. L. W. I. Gunathilake Department of Geography, University of Kelaniya, Kelaniya, Sri Lanka

S. K. Gunatilake Department of Natural Resources, Faculty of Applied Sciences, Sabaragamuwa University of Sri Lanka, Belihuloya, Sri Lanka

Richard Haigh Global Disaster Resilience Centre, School of Applied Sciences, University of Huddersfield, Huddersfield, UK

Chathuranganee Jayakody Global Disaster Resilience Centre, School of Applied Sciences, University of Huddersfield, Huddersfield, UK

Naduni Jayasinghe Faculty of Arts, Social Policy Analysis and Research Centre, University of Colombo, Colombo, Sri Lanka

P. P. G. Jayathilake Social Policy Analysis and Research Centre, University of Colombo, Colombo, Sri Lanka

Thushara Kamalrathne Global Disaster Resilience Centre, School of Applied Sciences, University of Huddersfield, Huddersfield, UK

Ahmad Raza Khan Department of Political Science, Government College University, Lahore, Pakistan

Anjali Korala Department of Political Science and Public Policy, University of Colombo, Colombo, Sri Lanka

Irene Lill Department of Civil Engineering and Architecture, Tallinn University of Technology, Tallinn, Estonia

Champika Lasanthi Liyanage University of Central Lancashire, Preston, UK

Shakeel Mahmood Department of Geography, Government College University, Lahore, Pakistan

Chamindi Malalgoda Global Disaster Resilience Centre, School of Arts and Humanities, University of Huddersfield, Huddersfield, UK

Masaaki Minami Faculty of Science and Engineering, Iwate University, Morioka, Japan

Anke Niehof Wageningen University, Wageningen, The Netherlands

Jane Liise Nurk Department of Civil Engineering and Architecture, Tallinn University of Technology, Tallinn, Estonia

Temi Oni-Jimoh University of Central Lancashire, Preston, UK

Watchara Pechdin Disaster Preparedness, Mitigation and Management (DPMM), Asian Institute of Technology, Khlong Luang District, Pathum Thani, Thailand

Melissa Quetulio-Navarra Ateneo de Manila University, Quezon City, Philippines

H. Unnathi S. Samaraweera University of Canterbury, Christchurch, New Zealand;
University of Colombo, Colombo, Sri Lanka

Samiullah Department of Geography, University of Peshawar, Peshawar, Pakistan

Minura Sandamith Department of Agribusiness Management, Faculty of Agricultural Sciences, Sabaragamuwa University of Sri Lanka, Belihuloya, Sri Lanka

Pathmanesan Sanmugeswaran The Open University of Sri Lanka, Colombo, Sri Lanka

K. B. Seenapatabendige Department of Architecture, Faculty of Built Environment and Spatial Science, General Sir John Kotelawala Defense University, Sooriyawewa, Sri Lanka

Anuradha Senanayake Global Disaster Resilience Centre, School of Applied Sciences, University of Huddersfield, Huddersfield, UK

Azadeh Shojaei Researcher at Post-Disaster Reconstruction Group, SBU, Tehran, Iran

Lilian Nwanyisonde Smart Global Disaster Resilience Centre, School of Applied Sciences, University of Huddersfield, Huddersfield, UK

Shinsuke Tanimoto Faculty of Science and Engineering, Iwate University, Morioka, Iwate, Japan

Felix Villalba-Romero University of Malaga, Malaga, Spain

Isuru Wicramarthne Department of Agribusiness Management, Faculty of Agricultural Sciences, Sabaragamuwa University of Sri Lanka, Belihuloya, Sri Lanka

Emlyn Witt Department of Civil Engineering and Architecture, Tallinn University of Technology, Tallinn, Estonia

Christopher W. Zobel Department of Business Information Technology, Center for Coastal Studies, Virginia Tech, Blacksburg, VA, USA

List of Figures

List of Tables

Part I
Response Management for Displaced Communities

Chapter 1
Rebuilding After Displacement: Identifying the Needs of Displaced Communities from the Perspective of the Built Environment

Andrew Carmichael, Champika Lasanthi Liyanage, Felix Villalba-Romero, Temi Oni-Jimoh, and Chathuranganee Jayakody

Abstract The United Nations Refugee Agency (UNHCR)'s submission on the inadequacies of policies imposed by certain international initiatives strongly confirms the need to focus on built environment programmes when rebuilding displaced communities. This chapter identifies the needs of internationally displaced communities in the United Kingdom (UK) within the past decade (2009–2019), in the context of the built environment. This will assist in their progress towards successful resettlement and integration. The chapter is based on the findings of qualitative analysis of interviews with experts working with the displaced in the UK. The interviewees were from different groups responsible for managing different aspects of the refugee experience. Research results reveal that the UK government's standards vary across the range of different categories of displaced persons. This creates a deprived group within the displaced communities with numerous needs. Some of these are built environment-related needs that should be met to enable successful integration into the UK.

Keywords Built environment (BE) · Displaced communities (DC) · Integration · Resettlement · UK

1.1 Introduction

In recent times, the world has witnessed an increase in mass displacement of communities from one geographical area to another, due to the rise in armed conflicts

A. Carmichael (✉) · C. L. Liyanage · T. Oni-Jimoh
University of Central Lancashire, Preston, UK
e-mail: acarmichael2@uclan.ac.uk

F. Villalba-Romero
University of Malaga, Malaga, Spain

C. Jayakody
Global Disaster Resilience Centre, School of Applied Sciences, University of Huddersfield, Huddersfield, UK

© The Author(s), under exclusive license to Springer Nature Switzerland AG 2023

M. Hamza et al. (eds.), *Rebuilding Communities After Displacement*,
https://doi.org/10.1007/978-3-031-21414-1_1

and disasters caused by climate change. The United Nations High Commissioner for Refugees (UNHCR) highlights that there is a dearth of built environment policies within international initiatives to help resettlement of displaced communities (UNHCR 2018). Thus, there is a need to include a mission of rebuilding communities after displacement with particular focus on the built environment. In the 1951 Refugee Convention (UNHCR 2010), the rights of the displaced were ratified by 145 states including the United Kingdom (UK). This included the obligation to protect refugees within their borders, in accordance with the terms of the convention, and ensure their resettlement and integration.

While the UK has hosted many immigrants during its history, it has supplied even more migrants to the world, only registering a higher net number of arrivals from the mid-1980s (DEMIG 2015). These included asylum seekers with a well-founded fear of persecution, and it was beholden on the country to be responsible for their welfare in line with their international commitments. Asylum seekers are those people who have left their country of origin and formally applied for asylum and protection, but whose application is yet to be concluded. Currently, most of these people are settled in London and the South-East by the government's dispersal programme. There is then a period of waiting for them until a decision has been made concerning their asylum claims. Those whose claims have been successful become refugees, a legal recognition of their status. Until such time, however, the lives of asylum seekers appear on hold in the UK, as they are not permitted to work, study, or volunteer until their claim has been granted (Mahamed 2016; Mayblin and James 2019).

1.1.1 Resettlement and Integration of the Displaced Communities in the UK

Resettlement of a displaced person in the UK depends on the path through which the displaced person enters the country. This study identifies and concurs with Mahamed (2016) that there are two routes (Fig. 1.1) to becoming refugees in the UK; the resettlement channel and the asylum channel. The key difference in the application process is the origin of the request, either external, before arriving in the UK (resettlement) or on arrival in the UK (asylum). This research observes this as a significant factor, with implications for the current experience of both types of displaced persons.

Notwithstanding the route chosen, there is a desire for integration of displaced persons with their host communities. Integration, enabling the displaced person to live harmoniously with their host communities, is considered a two-way process, thereby placing demands on both the incoming and host communities (ECRE 2005). In the year 2000, the UK's Labour government established an integration policy which expressed a wish to make refugees aware of the provisions available to them; assist them in gaining access to these provisions, such as housing, employment, and

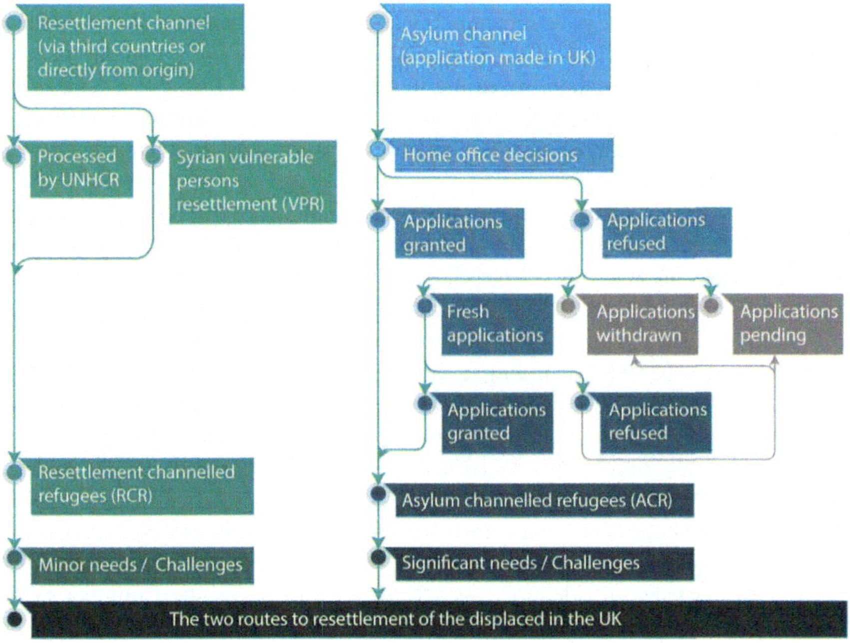

Fig. 1.1 Two routes to resettlement of displaced persons in the UK

English language services; enable them to realise their potential as equal members of the society; and motivate them for community participation (Broadhead and Spencer 2020). Yet two decades, and various changes of government, since this policy, there remain some attendant barriers to resettlement with most displaced persons in dire need of fundamental daily provision (Mayblin and James 2019; Fitzpatrick et al. 2016). Moreover, more recent work (Hamza 2021) notes that scholars question the success of such policy in the face of these challenges and highlights the trauma faced by displaced persons and the perception of them as either a threat, burden, or both, by the host population. This chapter in identifying the extent of the needs of displaced persons, and the experience of all parties engaged in the provision of those needs, advances the debate on the success of policy initiatives and where redress may be made.

1.2 Methodology

Data collection utilised both primary and secondary sources of data. Initially, secondary data collection, comprising of a broad literature review of existing publications, was undertaken. This was conducted using both searches of the Scopus database using relevant keywords to identify academic literature and more general web-based and citation searches for government publications, and publications of both national and international organisations. This provided an in-depth understanding of what is known and unknown about resettling the displaced and allowed for coding and thematic analysis in accordance with established techniques (Miles and Huberman 1994; Saldana 2013). The identification of themes assisted in structuring the primary sources of data collection.

Primary sources of data collection utilised the qualitative approach drawn from the constructivist paradigm (Allen 1994). Primary data collection was carried out using semi-structured interviews where participants were asked to respond to open-ended questions. This offered opportunity to examine how they interpreted the question and allowed them to be expansive with their provision of information. Participants were recruited using purposive sampling, a rational and effective method when wanting information-rich data (Patton 1990). The sampling focused on people in organisations that work with displaced communities because it is not only difficult in the UK to approach groups of displaced persons directly, but it may also be undesirable to subject them to additional interviews given their experiences and psychological status. Interviews were transcribed and the findings again analysed, coded, and themes identified.

Twelve interviewees (identified as UK INT followed by their number in the interviewing stage) were drawn from six different types of organisations: national government, local authorities, voluntary organisations/NGOs (international), voluntary organisations/NGOs (local), academia, and displaced persons representatives. Some of the organisations co-operate with each other towards outcomes, for example, voluntary organisations work with local authorities, which are ultimately funded by the national government. Hence, an interviewee might be recognised as a stakeholder with more than one organisation (Fig. 1.2). Of the twelve interviewees, seven are described as working with voluntary organisations (6 local and 1 international). Six interviewees work with local authorities, four of them being included in the national government category as well due to the funding of their activities, with the other two also belonging to the voluntary organisation category. Two interviewees are purely focused on research work, while two more are displaced persons themselves and additionally work with voluntary organisations.

The interviewees were acknowledged as working with effectively three (although officially four) different types of displaced persons. Initial asylum seekers arriving in the UK are one group. Those who have been processed through the resettlement channel a second, and the third is those persons who have been in the asylum channel (which includes unaccompanied asylum-seeking children). Five of the interviewees (UK INT 1, 3, 7, 9, and 12), work with all these different types of displaced persons, while another three (UK INT 2, 4, and 5) work with three of them (Fig. 1.3). Two further interviewees (UK INT 10 and 11) are dedicated to working with the resettlement group, while the final two (UK INT 6 and 8) work with internally displaced persons. This latter type of displaced person lies beyond the scope of this research, but awareness of which provides interviewees with an understanding of community

Fig. 1.2 Organisations within which interviewees are a stakeholder

needs. Consequently, these two interviewees were retained for their specialist knowledge, and perspective, they offered to what is a collective group with highly robust information about the state of displaced persons in the UK.

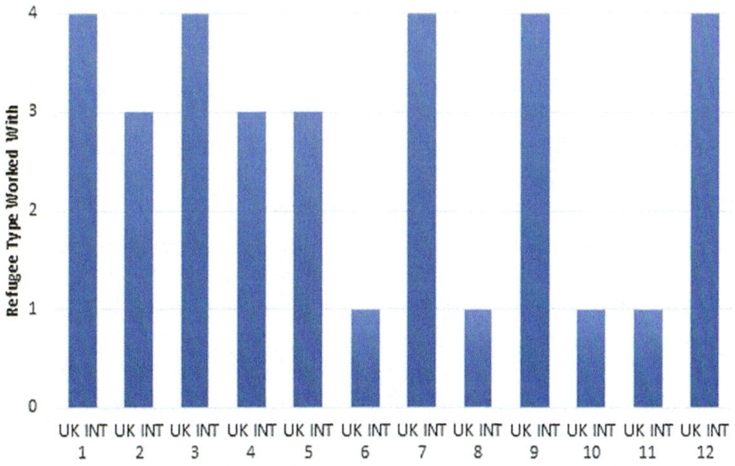

Fig. 1.3 Number of refugee types that interviewees work with

1.3 Literature Review

The review of literature identified both overarching themes that are relevant to consideration of all the needs of displaced communities (and indeed their hosts), and more specific requirements in terms of aspects of the built environment. There are four factors that are of particular importance when seeking to successfully resettle and integrate displaced persons: the individuality of the integration process; the two-way effort needed for integration; viewing integration initiatives from a long-term perspective; and that efforts towards the collection of data relating to integration are increased. There are a "plethora of challenges" that "resettlement and integration of a displaced population into a new society faces" (Hamza 2021, p. 1), and these four factors have been found to act across the range of needs, whether they be housing, health, economic, or socio-cultural requirements. Typifying this interconnectivity of requirements is the seemingly single issue of housing that Jayakody et al. (2022, p. 368) demonstrate as being of significantly broader import, the home being at once "social centre for family and friends, a source of pride and cultural identity and a resource that commands both political and economic importance".

Recognising the individuality of displaced peoples' experience is essential for positive integration (Baxter 2018; UNHCR Bureau for Europe 2013). Integration processes and activities towards successful integration in the host community work best when they include personal motivation and ambition (Baxter 2018; Fandrich et al. 2013). Moreover, consultation with displaced persons as to the services they would find of benefit and involvement of them in the design of such services bolsters self-reliance, dignity, and social interaction (Coley et al. 2019; Fandrich et al. 2013; Robila 2018; UNHCR 2013). However, the reality may be that such considerations are sacrificed to the needs of the host community, such as with the placement of displaced persons on vocational training courses that will have a positive impact on the labour market in their newly arrived at country, rather than reflecting their own best-suited and desired employment (Chaloff 2006).

This imbalance benefits neither host nor displaced community, the literature highlighting the need for involvement of both in successful integration (Coley et al. 2019; Corcoran 2006; Fandrich et al. 2013; Froy 2006). Acceptance by the existing population must be matched by adaptation by the incoming group to create a mutually advantageous position. One community should not be favoured over the other, and the implication of this is that in the same manner that it is advisable for displaced persons to be consulted about the integration process (Coley et al. 2019; Fandrich et al. 2013; Robila 2018; UNHCR 2013) similarly should the receiving population have chance to discuss the potential impacts upon them. This is a position encountered repeatedly within the literature, particularly regarding readying the host community for the arrival of displaced persons (Araya et al. 2019; Fandrich et al. 2013).

That there needs to be longer-term strategising on resettlement of displaced persons (Coley et al. 2019; Fandrich et al. 2013; Froy 2006) is clearly a point to be remembered when considering the interview responses in this study; whether sufficient time has elapsed for the interviewee to have experienced different outcomes, and if consideration has been given to such time-scales. Coley et al. (2019) note the need for patience in forging networks among the newly changed communities, but pressures exist in society for short-term measures to alleviate problems, or perceived issues. Allied to this is the need for more research and data. It may be self-serving for researchers to call for more research, yet it provides the means to greater understanding of longer-term trends and encouraging positive outcomes from resettlement.

Discussion of more specific aspects of the built environment can be summarised within the categories of; housing, socio-cultural, social infrastructure, economic, and physical infrastructure needs. Housing provides a significant role in the success or otherwise of displaced communities in their new built environments. A potential source of conflict with existing communities (Haigh et al. 2016), housing is likely both the most valuable asset of a displaced person and their community and the worst impacted by a displacement event (Wijegunarathnaa et al. 2017). Not only is the physical structure left behind, but an individual's sense of place may be lost. Evolving over time, note Jayakody et al. (2022), housing moves from an immediate need for shelter in the aftermath of an event, to a place that meets social and emotional requirements, providing a base for economic activity and the hub for belonging and cohesion.

The theme of individuality in the process of integrating displaced persons is particularly relevant to discussions of socio-cultural needs, important in language (and proficiency in the host community language) (Coley et al. 2019) and opportunity to practice and exchange culture (Congress of Local and Regional Authorities 2019). Linked to housing, there is a need to avoid segregation of populations and widespread mention of the value of mentors who can bridge the gap between host and incoming populations (Froy 2006; Nakeyar et al. 2018). Education plays a pivotal role in the development of integrated communities, not just for the influence it can have on language skills, but for numerous aspects of social infrastructure (Cassity and Gow 2005; Christie and Sidhu 2002; Taylor and Sidhu 2012). Similarly, the need for appropriate healthcare transcends different aspects of displaced persons' needs, whether it be physical or mental health considerations.

Finding employment is highly relevant to integration, both in terms of the economic and social benefit that work can provide, and perceptions within the wider community. This can, however, be challenging for displaced persons, with possibly unfamiliar language, the demand for existing skills, and the state of local job availability all relevant factors. Generally, there can be quite a disparity between the employment rate and earnings of displaced populations with that of their hosts

(Coley et al. 2019; Kone et al. 2019). Finally, physical infrastructure needs are typically already existent in developed countries and do not need to be specifically provided in response to displacement. However, it is clear the extent to which access to such infrastructure relates to the previously discussed considerations of language, economic status, and social networks. The value of physical infrastructure lies not in the building works themselves but in the services they provide (Araya et al. 2019). Even in a society with a high level of physical infrastructure such as the UK, displaced persons can be marginalised and potentially excluded by simple acts of not being able to afford transport fares or being resettled in isolated areas. These themes and aspects were used as reference points for the analysis of our interview responses.

1.4 Research Findings and Discussion

The research interview findings covered many topics and discussions, with identification of a wide range of needs and barriers to be overcome before the successful integration of displaced persons. This chapter focuses on these needs from the perspective of the built environment in the context of the existing literature.

1.4.1 Housing Provision

Article 25 of the Universal Declaration of Human Rights (UDHR) highlights the right of every human, displaced or not, to adequate housing. Within the UK, housing for displaced persons is provided through various channels; national government working via local authorities; the displaced persons themselves, through friends and families they can live with; and voluntary organisations (Fig. 1.4). The assistance offered by voluntary organisations in finding accommodation for displaced persons comes either through their private acts of charity, or through work in conjunction with local authorities. This would be in collaboration with housing associations and local authority social housing, who provide affordable housing to the general population (Evans 2003). And this is by no means the total extent of their assistance, as noted by one of our interviewees; "Yeah, yeah…The charity organisations, they do help…They are ready to facilitate people here in terms of many things. Not only the accommodation" (UK INT 9). A package known as Section 95 Support is available for asylum seekers during their immigration processing period. It was identified by one of our interviewees for the benefits it brings; "If you want housing, you should stay with the Home Office to have the package of housing, free legal aid and child support" (UK INT 2). Housing under this package is provided by government (under the National Asylum Support Service) through private contractors and subcontractors which are supervised by the local authorities; "X is a private company that runs contracts for the government…they have the contract to provide housing

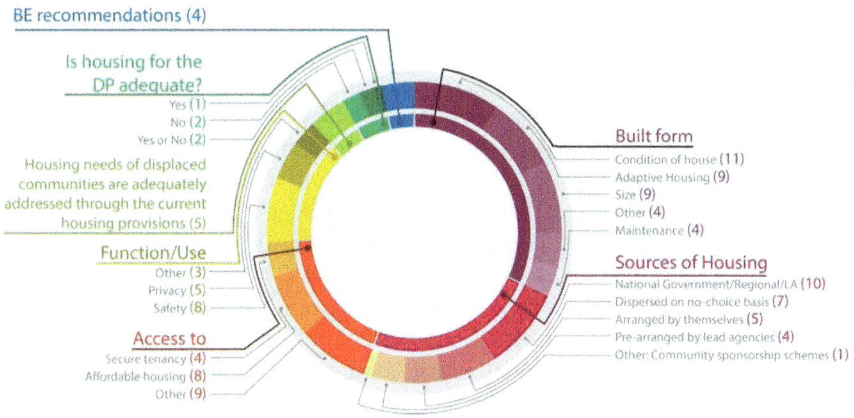

Fig. 1.4 Interviewee responses to housing needs/suitability of displaced persons

for asylum seekers" (UK INT 7). In what appears to be a vital point, housing for resettlement channel refugees is pre-arranged before their arrival in the UK.

1.4.1.1 Housing Suitability

Six of the interviewees (UK INT 1, 2, 3, 10, 11, and 12) submit that housing provided for resettlement channel refugees is suitable; "Yeah, the conditions and standards are in a good condition" (UK INT 10). This is consistent with literature which acknowledges that housing for resettled refugees has quality assured (Cebulla et al. 2010; Evans 2003). One respondent did contradict this claim arguing that there are still issues with some of the housing supplied to this group; "Yeah, even though there are conditions, we have standards for housing. So, there's a lot of issues that we have, broken appliances, structural issues, heating. Sometimes it's in ventilation and lighting" (UK INT 11).

In terms of the size of house offered to resettlement channel refugee families, interviewees talked in relatively positive terms about provision. One claimed that typically they receive larger houses; however, others agreed that the size of housing allocated is dependent on the size of the family, as informed by the UNCHR who have made a comprehensive assessment of the potential refugee family and everything that pertains to them. This, it is suggested, is proportional; "If you are a single person, they give you one single bedroom, full flats...It depends on the size of the family" (UK INT 5); "Obviously everything is based on the family sizes. You wouldn't put, like you wouldn't put them in an overcrowded house. So, there is a list I think depending on the age or who is or what needs to be in the room" (UK INT 10). Additionally, the resettlement channel refugees benefit from what one interviewee described as government subsidising the rent; "The Home Office, they pay what's called a top up, so they will pay something like an average of an extra £100 a month towards rent because strangely they recognised that the housing benefit rate will not cover a decent house" (UK INT 7). The overall response to this pre-arranged, externally initiated, resettlement channel refugee housing is one of positivity.

Initial asylum seekers and asylum channel refugees appear to receive different treatment in relation to housing. Interviewees highlighted the distribution of these latter groups; "They are dispersed by government on a no-choice basis. No, they cannot choose" (UK INT 9); "They don't have any agency in choosing where they are going to live" (UK INT 10). This reflects findings elsewhere that non-resettlement channel refugees are sent anywhere there is available housing (Mahamed 2016). Moreover, it appears that dispersed refugees are typically housed in deprived areas with multiple problems, and little or no experience of diverse communities, factors that can lead to social tension and racism towards the displaced persons (Jeffery 2010).

Interviewees expressed concerns about the condition of housing for these two groups (UK INT 1, 2, 3, 5 and 7); "So, one family with four children, they live just down here on the Quay, and they really hated the house. There were issues with the house, there were problems with the house" (UK INT 7). Some respondents claimed that some people do get good housing while others are not as lucky, that policies on the issue are "unpleasant and not pretty" (UK INT 2). Although there are housing standards to be satisfied by the government agencies it is felt that the standards are rarely met. Some asylum seekers houses are unfurnished and with no decorations, or in so poor condition as to receive public condemnation regarding their state.

1.4.1.2 Adaptive Housing for Disabled Persons and Children

Where adaptive housing is required by displaced persons, interviewees (UK INT 1, 3, 7, 10, 11, and 12) suggest it is forthcoming for resettlement channel refugees, and some of the initial asylum seekers (UK INT 7 and 9). In contrast, it appears difficult to access by their asylum channel refugee counterparts, although as one interviewee suggests "If they're coming for resettlement in the UK and the individual who is travelling, or a member of the household, has a disability that requires specific accommodation, where possible that need will be met" (UK INT 1). Additionally, some occupiers can adapt their houses according to their needs with permission from their landlords; "So, they're able to amend it but, again, they'd have to talk to the landlord about what they require" (UK INT 3). However, again it seems less co-ordinated, and more a matter of positive circumstances than planning.

1.4.1.3 Safety and Privacy

Safety of the displaced communities in their different neighbourhoods has been deemed a mixed experience across all groups; "I think there are good and bad parts in all neighbourhoods" (UK INT 1). Privacy might be an issue especially in shared accommodation, and rural areas where there might be just one new family in the host community could be isolating. One respondent, in what may be seen as quite a defensive message, asserted that resettled refugees and asylum seekers are very safe in their accommodation because the houses have met the housing standards set out by local government; "I would say, any housing the City Council is involved with for example by paying benefits or deposits or first month's rent, they will be very safe as well because of their exacting standards" (UK INT 7). However, not only have previous findings suggested these standards are not universally met but the same interviewee (UK INT 7) noted that people from the same country are not placed together, so they do not form groups which might be a detriment to their sense of community and feeling of security in their new locations.

1.4.1.4 Affordability and Tenure

As housing is provided for by the government, friends, and families or voluntary organisations, there is seemingly limited concern for housing affordability by displaced persons. Social housing provided by the local authorities through council houses and housing associations is affordable, particularly for asylum channel refugees who were noted as tending to be dispersed into less desirable areas that typically are supported by cheap rent rates; "They've been dispersing asylum seekers to the areas where rent is cheaper, so we don't have many asylum seekers in the more expensive areas like Manchester or Stockport in terms of Greater Manchester" (UK INT 4). However, as numbers expand, affordable types of housing are not always easily accessible because of long waiting lists; "So, the waiting list for a council house or social housing in Lancaster is something" (UK INT 7). This forces the requirement to move to more expensive housing. In this situation, displaced persons gravitate towards other people's houses as lodgers or try and find shared housing where they can combine benefits to cover the housing cost. For resettlement channel refugees, affordability is also an element to be considered because there is a certain budget for housing sourced for them. The way this is negotiated is laid out by one interviewee (UK INT 10); "Obviously the way housing is, it's sourced. You will have obviously a certain budget and you won't go to a local authority with the most expensive housing because we know that the rent won't be covered".

It was claimed by three respondents that everyone is supposed to be on a six-monthly introductory tenancy, following which their tenancy becomes secured. One of these interviewees said that only those who have access to housing through the local authorities and private housing have access to secured tenancy whereas those who are classed as lodgers in other people's houses do not. However, a second respondent stated that the tenancy is secured between 6 and 12 months depending on the landlords, while the third contradicted this further in saying that it depends on the different local authorities because their rules differ. Consequently, there remains some confusion over the issue.

In summary, there appears a lot of disparity between housing provided for the different status displaced persons, with resettlement channel refugees seemingly experiencing the best examples of housing provision, and asylum channel refugees finding their needs poorly responded to. Initial asylum seekers seem to be receiving of a more arbitrary outcome. This research suggests that there is some way to go before all displaced persons are being treated with equality and fairness in relation to housing. Moreover, while the physical need of housing may be met with simple provision of a dwelling, and individuality seems well-catered for in terms of family members and house size, there is little consideration of the social and emotional role of housing, and no discernible role for host community or displaced persons views when allocating displaced persons to an area. Cost and availability seem to be the main drivers in what are not perhaps longer-term solutions.

1.4.2 Socio-Cultural Needs

The socio-cultural needs of displaced persons in the UK, as identified by this research, are multi-faceted. While some are predicated on culture and religion, others are linked to safety, stability, emotional need, and social inclusivity (Fig. 1.5). These are all features that are impacted by built environment considerations.

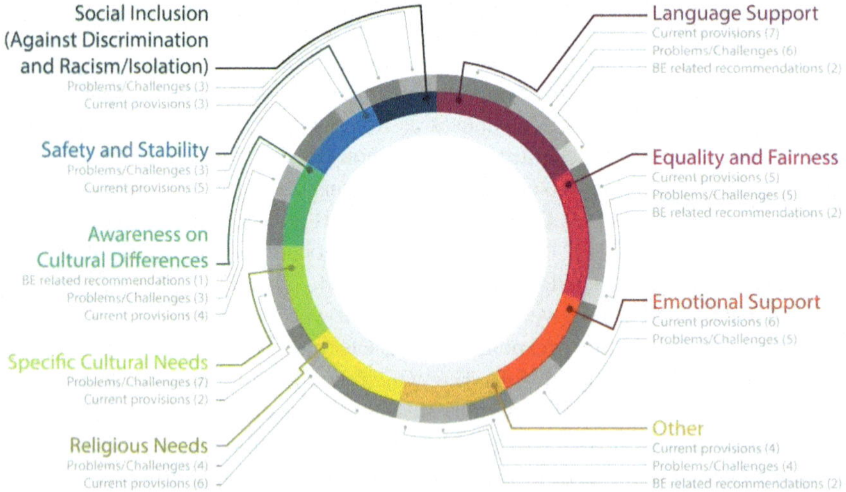

Fig. 1.5 Interviewee responses to socio-cultural needs of displaced persons

1.4.2.1 Language

This said, language is the most important area of socio-cultural concern, and integration is made more difficult in the absence of a common language. The displaced person simply needs to be able to communicate to access the support available to them. For which, in the UK, they need to learn the English language. This research found that all public bodies in the UK are supposed to provide interpretation for displaced persons. Some, such as the NHS, do, while others in allied provision such as dentistry, which can be a combination of both public (NHS) and private care, do not according to one interviewee (UK INT 7). Voluntary organisations can offer support by providing interpreters and the use of friends and families for interpretation is widespread, especially in health care settings. This is far from ideal, and sometimes not possible, so continued provision of information in languages other than English and the use of interpreters are major needs (Cebulla et al. 2010). Sensitivity in dealing with the needs of the displaced person with respect to English language and interpretation is required in certain cases, for example domestic violence. In such cases, there is a need to match the gender of the victim with that of the interpreter for ease of communication. There additionally exists provision of lessons for as many displaced people as would want to learn the English language for the first two years of their rehousing. In a sign that perhaps that provision, with its laudable long-term ambition, needs reinforcing, one interviewee highlights the challenge; "So, without English language, they won't be able to orientate themselves…In the beginning we provide them with bespoke lessons for the first two years and now we are trying to strengthen that provision with other sorts of projects and services" (UK INT 11).

1.4.2.2 Cultural/Religious

Refugees are often people with specific cultural backgrounds, they may even have been fleeing persecution because of that background. Their ways of life can be different in terms of culture, religion, and language. In this situation, a process of mutual restructuring by both the refugee community and their host communities is imperative (Manz and Panayi 2012). That everyone has a right to worship should be evident, but it appears that religion is not considered when relocating displaced people to their respective houses (UK INT 7). A lack of mosques in some areas poses a problem to Muslims, practising Christians would similarly like to be housed close to churches where they can worship. The strength of this need is reflected in efforts to attend services: "If their denomination is not near, they are ready to travel miles to attend church…I have known people travelling 10, 20 miles to go to a church" (UK INT 5).

Allied to religious needs can be the importance and consequently availability of suitable food. Some people cannot eat meat that is not halal for example. As with the general observations of housing provision, it was noted by our interviewees that the prospect of success with matching this requirement is greater with resettlement channel refugees than it is with initial asylum seekers and the asylum channel refugees.

There are also gender-related socio-cultural needs. Some cultures do not permit women to speak in public or relate with men who are not their husbands or members of their families. This can impair the quality of support they receive (Hutchinson and Baqi-Aziz 1994; Manz and Panayi 2012), a suggestion supported by one of our interviewees; "Probably the Muslims as well, the culture does not really permit a family woman to be outspoken or to come out in the open. So probably when they have issues, they might not really be able to share if the caseworker is male" (UK INT 11).

1.4.2.3 Emotional Stability

This research acknowledges that some displaced persons may already be dealing with mental health challenges prior coming to the UK. This could increase, especially in the case of asylum seekers because of the stress and anxiety of waiting for the decision of the Home Office concerning their applications. Many displaced people have suffered, and are potentially still suffering, from the trauma of whatever made them flee or leave their habitual place of residence. There is little or no official provision for emotional support aside from a nominated caseworker, and voluntary organisations seek to fill the gap. Mind Matters is one such body, alongside the House of Tara, an organisation that runs a wellbeing drop-in centre for displaced persons, rendering services such as massage and sound healing. Our interviewees could only offer anecdotal evidence, yet the feeling was there are considerable welfare needs; "Some people already have mental health issues when they come. So, I think probably statistically I don't know among asylum seekers; I don't know if there's a higher rate, but a lot of people are depressed" (UK INT 7).

1.4.2.4 Awareness of Cultural Differences

Lack of cultural awareness and understanding of refugee concerns by support professionals and the limited provision of culturally appropriate services for refugees and asylum seekers are examples of cultural differences. The Race Equality Cultural Awareness Programme (RECAP) is an in-depth training course aiming to improve cultural awareness among support workers, especially NHS staff. This is not solely focused on the host community. Training on social and cultural orientation is a requirement for displaced persons within 12 months of their arrival.

1.4.2.5 Social Inclusion

Although initiatives such as the training aim to assist social inclusion of displaced persons, exclusion remains a problem. This happens when displaced communities of asylum seekers and refugees are faced with a combination of difficulties such as unemployment, poor housing, a lack of skills, low incomes, discrimination, health concerns, and family breakdown. With exclusion from the outcomes and opportunities enjoyed by mainstream society (Zetter and Pearl 2000), most displaced persons have experienced isolation, especially during the early phase of their transitioning into the UK. Some experience isolation only because they have very little money, while for others, the fear of violence occurring is very real, something particularly exacerbated in areas that are less diversified (UK INT 7 and 10). This can be compounded by adverse reactions from their host communities; "They are having a difficult enough time transitioning to life in the UK. The last thing they want to do is be on the public face of the newspaper" (UK INT 1). Hate crimes towards them from the host communities can sadly be a feature. One interviewee suggested that the incidence of hate crime, violence, racism, and discrimination have been given impetus by recent political events; "But even more recently, because of Brexit, now the incidence of hate crime has increased dramatically in the UK" (UK INT 11).

1.4.2.6 Equality and Fairness

One interviewee felt displaced persons in the UK contend with racism and discrimination; "Well, like there's racism everywhere in this country, you can't pretend there isn't" (UK INT 10). Those seeking asylum are the most impacted as they experience discrimination from different angles, and they have no recourse to the resources available to refugees. Refugees dispersed across the host communities also face discrimination and racism in the workplace (Cebulla et al. 2010; Evans 2003; Harris 2003). This can be a shock for displaced persons, most especially resettlement channel refugees, who have high expectations from the UK, only to discover that the process does not work as they believed and there might be some considerable time delays relating to the severity of their circumstances. When placed on waiting lists for months, they may think they are being treated unfairly because they are refugees, yet this is not so as the system applies to everyone including local people; "They are being treated fairly; refugees are being treated like other people on housing benefit" (UK INT 7).

This research identified responses that clarify that not all communities in the UK are racist or discriminatory towards displaced persons. Interviews highlighted that some are very welcoming, and help displaced people integrate and settle (UK INT 7, 9 and 10). There is evidence of two-way effort in community integration. Displaced persons should not assume that the UK is a hostile environment but should try and harness as much support as is available for them. They should feel free to speak with their caseworkers about whatever they might be going through because they are there to help (UK INT 10). However, as with issues surrounding housing it appears, there is limited capacity within the system of resettlement to meet the individual socio-cultural needs of displaced persons.

1.4.3 Social Infrastructure Needs

1.4.3.1 Education

Approximately, 45% of displaced persons already have qualifications before coming to the UK, and these groups find it easier to get jobs, because, alongside their knowledge and skills, most can speak English to a satisfactory level (Cebulla et al. 2010; Chiswick and Miller 2002). However, they need to have their qualifications certified by the National Recognition Information Centre (NARIC) in the UK to get jobs commensurate to their level of qualification. As one interviewee relates; "It is requested to get statements from NARIC UK now in terms of… the official organisation in the UK accredited by the government to provide qualification equivalents" (UK INT 11). While this may be yet another process to endure, the more than half (Cebulla et al. 2010) of displaced persons who do not have any qualifications are extra-disadvantaged especially when seeking employment because of their poor level of English ability. The provision of English language skills and "English for Speakers of other Languages" (ESOL) training, basic numeracy, literacy, and IT skills was established for this purpose. However, there are concerns about the quality, appropriateness, and limited availability of the ESOL provision (Evans 2003). Children of displaced persons gain access to primary and secondary education quickly through the help of their support worker, voluntary organisations, and sometimes the NHS district nurse; "So, in terms of younger refugees and asylum seekers, or children of refugees and asylum seekers, they do get into school" (UK INT 7).

However, support is needed to access higher education in terms of understanding of language, culture, and the working environment, as evidenced by one of the interviewees (UK INT 10). Voluntary organisations mainly help in training people in these subjects. Furthermore, vocational training like NVQs; and training in different skills, are generally available to the different types of displaced persons as with any other UK citizen. Some voluntary organisations arrange certified courses from time to time such as food hygiene, health, and safety at work, for those who can understand the English language to help them become self-reliant. Opportunities such as this will enhance the rate of integration of both refugees and asylum seekers. Yet barriers to education are many; childcare for women with young children, lack of sufficient time for study, health problems, long waiting times for application decisions, most courses

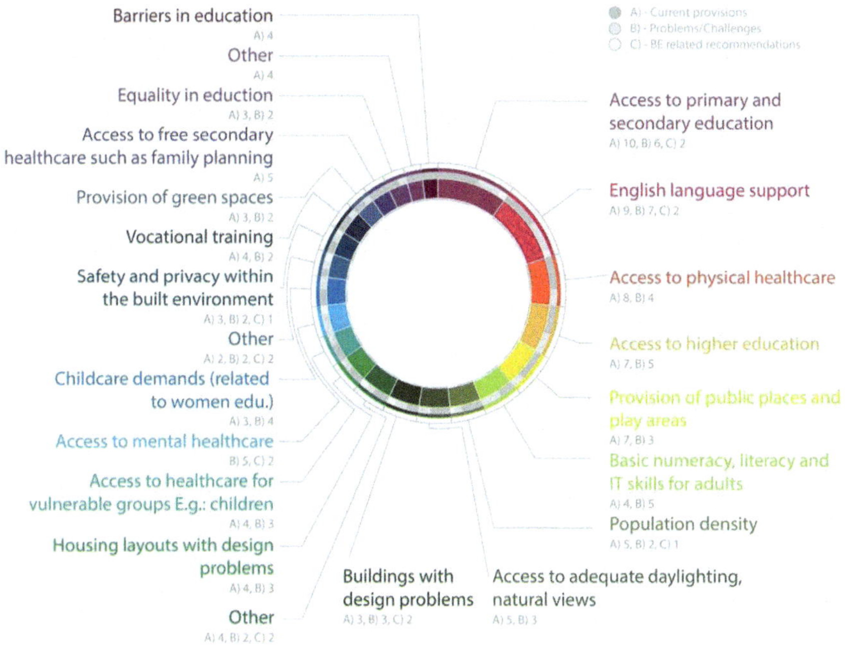

Fig. 1.6 Interviewee responses to social infrastructure needs of displaced persons

being full time, non-UK qualifications not being accepted, problems with funding and illiteracy, and placing low or no value on education. Hence, refugees need extra help with psychological re-orientation to understand the importance of education in making their lives better, or indeed to gain control of their lives to become an asset to the UK. As was evident in the discussion of cultural differences, the importance of learning the English language to integrate well into the UK is paramount (Fig. 1.6).

1.4.3.2 Health and Wellbeing

Refugees seem to have poorer mental health than the general UK population and these might have implications for integration (Mahamed 2016). They have many worries ranging from; uncertainty about the decision on their claims, anxiety over being removed from the country, concern about family members back home, adapting to a different culture and language, experiences of negotiating the immigration system, poverty, and homelessness, and ensuring basic necessities of life such as food, clothing, and shelter (UK INT 7). It is a general belief that everyone should have access to a general practitioner and hospital care. Refugees are also eligible for mental healthcare as confirmed by an interviewee; "So, you would have the common anxiety disorders treatments like, you know, … therapies and … CBT for a few weeks or group therapy. You can access them just as any other UK resident can access them" (UK INT 10). Yet, as previously mentioned, some barriers such as language and issues with interpreters impede this access. Secondary healthcare is available in the UK, but it is not for free, it must be paid for. This applies to everyone in the UK and not just asylum seekers or refugees, but clearly those who cannot work, or have limited funds, will be unable to use this.

1.4.3.3 Built Environment Factors that Affect Health and Wellbeing of Displaced Persons

There are some aspects of the built environment that could have profound effects on the health and wellbeing of displaced communities. Buildings with design problems such as emissions from gas, exposure to pollutants, smoke from wood fire used for cooking or heating and tobacco all have harmful effects on the health of the inhabitants (Cooper 2014; Hartig and Lawrence 2003). Some houses are reported to have bad lighting conditions with just one translucent window, subsequently impairing the daylight. Because displaced persons may already be facing mental health challenges, inability to access natural daylight could also increase their stress and affect their wellbeing. Apartment buildings reduce social networking, leading to loneliness, and restrict children from playing outside the residential unit (Evans 2003). The availability of public places and play areas is important because they are indicators of social wellbeing in both adults and children (Cooper 2014; Janahi et al. 2018). This research's findings concur with this thought but note that provisions for these do exist in the UK; "I think people have access to libraries and public places. I mean, it is good for their health and wellbeing. There are no restrictions if they want to go to the public places" (UK INT 9).

Refugee communities in high-density areas are prone to infections especially in pregnant women and unborn children and overcrowding and noise have been found to result in psychological distress for the inhabitants (Cooper 2014). This research found no cases of overcrowding in housing provided for displaced persons when it was sourced based on certain standards. However, for shared accommodation, where asylum channel refugees live with friends and families, or combine their housing benefits together to rent a house, there can be a problem with noise and too many occupants. Additional feelings of being unsafe exist in such situations.

1.4.4 Economic Needs

1.4.4.1 Employment

Asylum seekers are not permitted to work, do voluntary work or training in the UK except those whose professions fall within identified areas of shortage in the UK and have not received a decision within twelve months (Mahamed 2016; Mayblin and James 2019). "If you've come here registered as an asylum seeker, and the

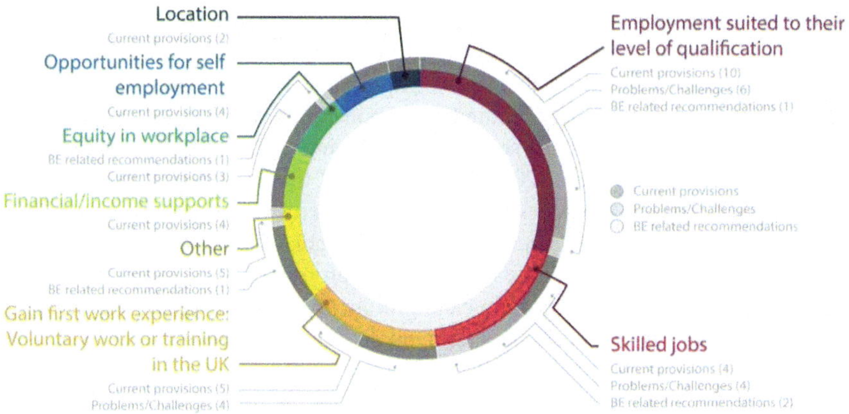

Fig. 1.7 Interviewee responses on employment needs of displaced persons

home office…has not communicated with you for a period of up to 12 months. Technically… You can apply for permission to work" (UK INT 12) (Fig. 1.7).

Once established as refugees, displaced people can work in the UK because they have the same rights as British citizens. Most gain employment through informal workplace networking, via mouth-to-mouth sources. Again though, there are challenges in securing employment commensurate with their qualifications because of language barriers and the need for NARIC; "No, because if their qualification is not recognized…we have lawyers working in a factory, you know, and that is quite common" (UK INT 10). Lack of such certification means they are required to take top-up courses and training. These are expensive and take a considerable time for completion, leaving them in low-skilled jobs with lower pay. In addition to these challenges, refugees can face the problem of racism and discrimination by employers. A government approach that actively promotes a hostile environment effectively turns the public services, employers, and landlords in the UK into immigration officers. Consequently, opportunities for self-employment are reportedly among the fastest avenues for refugees to gain access to the labour market in the UK. There are opportunities for refugees with realistic business plans and assisted through interest-free loans to set up their own businesses such as restaurants or tailor shops.

1.4.4.2 Financial/Income Supports Available from the Government/Voluntary Organisations

In the UK, those who have equal rights with UK citizens receive financial support in terms of benefits for the unemployed and job seekers. On the other hand, asylum seekers are not eligible to claim such benefits, but are instead supported directly by the Home Office through the National Asylum Support System (NASS). This includes housing and a stipend of £5.39 per day, sometimes increased to £38 per week depending on age (Mahamed 2016; Mayblin and James 2019). Subsequently, many asylum seekers are living in what can be classed as abject poverty, whereas they would have provided for themselves if gainfully employed (Bloch 2002). The UK government needs to bear in mind its responsibility according to human rights law, to ensure that the asylum seekers and refugees are not left impoverished or destitute (Mayblin and James 2019). In the light of this evidence, it perhaps needs to reconsider its stand and provide asylum seekers with access to the labour market so they can both help themselves and take the financial burden off the government and voluntary organisations. This seems more appropriate longer-term thinking that would give greater opportunity for meaningful involvement in the community while discouraging images of a host population indefinitely supporting displaced persons.

1.4.5 Physical Infrastructure Needs

All interviewees agree that physical infrastructure such as transportation, communication, water, sanitation, solid-waste, and energy are readily available in the UK, but at a fee. This, displaced persons pay for themselves, just like any other UK residents, although they may receive a discount based on their income. Moreover, there is no fixed sum for this in the UK as provision of such services is regional and often through private organisations. Similarly, there are variations in local authority council tax rates that contribute towards service provision (Fig. 1.8).

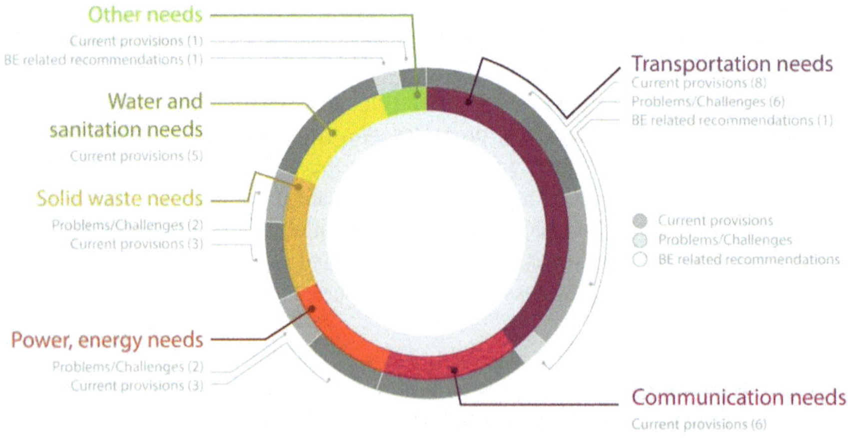

Fig. 1.8 Interviewee responses to physical infrastructure needs of displaced persons

1.4.6 Governance Needs

Research findings reveal that government policy relating to the resettlement scheme adequately addresses the needs of these refugees. However, for their counterparts, the asylum channel refugees and asylum seekers, their needs have been escalated by an increasing attitude of official hostility to their arrival. This has resulted in poverty, destitution, and social exclusion (Mayblin and James 2019), which further impedes their integration. The paradox was noted by one of our interviewees; "Well, it's quite contradictory. You have quite the welcoming programme; welcoming and supportive and well-funded resettlement programme. And then you have a hostile environment for Asylum Seekers" (UK INT 7). It is almost akin to state sponsored discrimination between the groups.

1.4.7 Communities with Special Needs

These include women, children/young people, the elderly and the disabled. They are the most vulnerable and their needs are many. Faced with extra needs with respect to immigration, adaptive housing, health, education, family, and social relationships issues, some, including children, have experienced detention. Women are exposed to gender-related persecution, domestic abuse and torture, female genital mutilation, forced persecution, and childcare issues (Bonnerjea 1994; Burnett and Peel 2001; Dransfield and Clark 2018; Girma et al. 2014; Hek 2005). Disabled asylum seekers can be highly deprived (Ward et al. 2008; Yeo 2017) due to unsanitary accommodation without disability adaptations; the disabled, being considered in some quarters as burdensome, are likely to experience negative effects on their mental health. There is recognition of this during the relocation of displaced people; "So, they've got extra needs and we sort of recognize that" (UK INT 10), yet actual provision may be in short supply; "You are not entitled to those provisions for disabled people" (UK INT 4) (Fig. 1.9).

1.5 Conclusion

This research submits that the UK accepts displaced persons into the system, and the country has enacted several policies to regulate the intake of the displaced. Review of the literature identified areas in which existing policy could be considered flawed and provision for displaced persons capable of improvement. Desk-based research and primary data sources identified two different groups of refugees; those who come into the UK via the resettlement channel, and those that arrive via the asylum channel. The experience of each under the general themes of displacement and resettlement, and at a more specific level of built environment provision, appear to differ quite

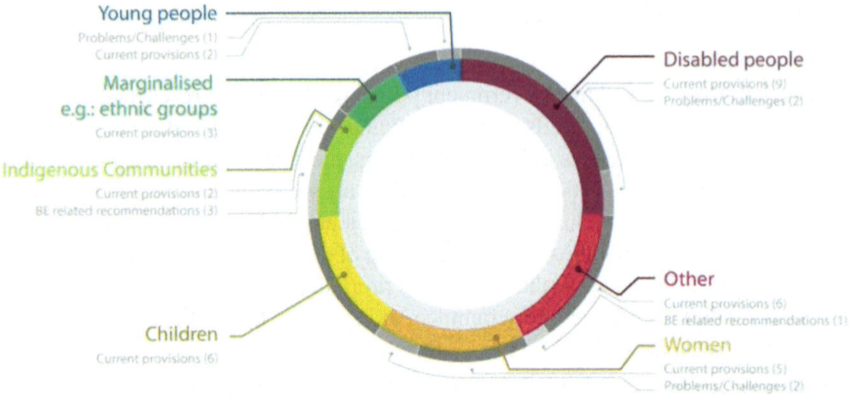

Fig. 1.9 Interviewee responses to communities with special needs among displaced persons

significantly. The UK has been continually accused by researchers, voluntary organisations, and some of the public of a double standard and high level of partiality and unfairness. The UK's official support towards the resettlement channel refugees appears to be warm, welcoming, and fully funded, whereas the attitudes towards the asylum channel refugees and asylum seekers are said to be hostile, damaging, demoralising and only partially funded. Consequently, the asylum seekers and asylum channel refugees have greater needs. This has resulted in a lot of challenges for these disadvantaged groups in their integration. Not that everything has been perfect for resettlement channelled refugees either, there is clearly work to be done on improving the response for all.

In terms of the built environment, this chapter highlights the huge importance of built environment initiatives towards fulfilling the needs of the displaced communities. The built environment goes beyond housing and encompasses the space and facilities required to live. This requires us to perceive living as the way one exists and copes with the physical, social, economic, and psychological environments on a day-to-day basis, reflected in daily activities, values, interests, attitudes, and opinions, whether at home, work or leisure and as influenced by family, culture, and social class. The built environment is the space, which comprises structures, facilities, and resources that accommodate and permit the individual to live a good and satisfactory life with a sense of wellbeing. Therefore, the built environment has a role to play in meeting the needs of the displaced communities in the UK (Fig. 1.10), ranging from; housing, socio-cultural, social infrastructural, economic, physical infrastructure, governance, and communities with special needs over and above those typical experienced by many.

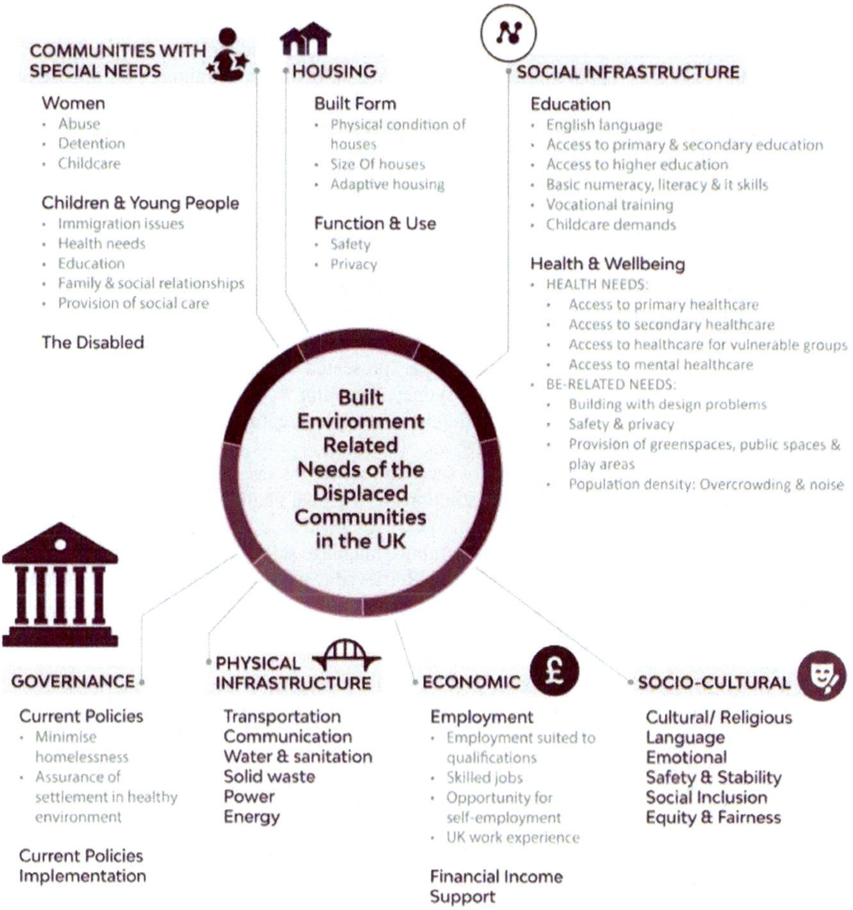

Fig. 1.10 Built environment-related needs of displaced communities in the UK

Acknowledgements The REGARD project has been co-funded by the Erasmus+ Programme of the European Commission. The European Commission support for the production of this publication does not constitute an endorsement of the contents which reflects the views only of the authors, and the Commission cannot be held responsible for any use which may be made of the information contained therein.

References

Allen MSW JA (1994) The constructivist paradigm. J Teach Soc Work 8(1–2):31–54. https://doi.org/10.1300/J067v08n01_03

Araya F, Faust KM, Kaminsky JA (2019) Public perceptions from hosting communities: the impact of displaced persons on critical infrastructure. Sustain Cities Soc 48:101508

Baxter M (2018) Bringing refugees from crisis to flourishing: the role of resettlement agencies and the church in facilitating integration and stability. Soc Work Christ 45(3):19–35
Bloch A (2002) Refugees' opportunities and barriers in employment and training (Vol. 179). Leeds: Corporate Document Services
Bonnerjea L (1994) Family tracing: a good practice guide. Save the Children. https://resourcecentre.savethechildren.net/node/13669/pdf/family_tracing.pdf
Broadhead J, Spencer S (2020) Policy primer: integration. The Migration Observatory, University of Oxford. https://migrationobservatory.ox.ac.uk/resources/primers/policy-primer-integration/#:~:text=Return%20to%20top-Integration%20policy%20in%20the%20UK,on%20community%20cohesion%20in%20England
Burnett A, Peel M (2001) Health needs of asylum seekers and refugees. BMJ (Clin Res Ed) 322(7285):544–547. https://doi.org/10.1136/bmj.322.7285.544
Cassity E, Gow G (2005) Shifting space and cultural place: the transition experiences of African young people in west Sydney schools. Paper presented at the Australian Association of Educational Research, annual conference, Sydney, Australia
Cebulla A, Daniel M, Zurawan A (2010) Spotlight on refugee integration: findings from the Survey of New Refugees in the United Kingdom
Chaloff J (2006) Innovating in the supply of services to meet the needs of immigrants in Italy. In: From immigration to integration: local solutions to a global challenge. OECD Publishing, pp 145–188
Chiswick BR, Miller PW (2002) Immigrant earnings: language skills, linguistic concentrations and the business cycle. J Popul Econ 15(1):31–57. Retrieved from https://link.springer.com/content/pdf/10.1007%2FPL00003838.pdf
Christie P, Sidhu R (2002) Responding to globalisation: refugees and the challenges facing Australian schools. Mots Pluriels 21
Coley J, Godin M, Morrice L, Phillimore J, Tah C (2019) Integrating refugees: what works? What can work? What does not work? A summary of the evidence, 2nd edn. Project report. Home Office
Congress of Local and Regional Authorities (2019) The role of local and regional governments in protecting internally displaced persons (IDPs). Retrieved from https://rm.coe.int/the-role-of-local-and-regional-governments-in-protecting-internally-di/168097f950
Cooper R (2014) Wellbeing and the environment. In: Wellbeing. Wiley. Retrieved from https://doi.org/10.1002/9781118539415.wbwell055
Corcoran MP (2006) Local responses to a new issue: integrating immigrants in Spain. In: From immigration to integration: local solutions to a global challenge. OECD Publishing, pp 239–284
DEMIG (2015) DEMIG POLICY, version 1.3, online edn. International Migration Institute, University of Oxford, Oxford. Retrieved from www.migrationdeterminants.eu
Dransfield R, Clark L (2018) An exploration of the mental health needs of asylum-seeking women and children in the UK: implications for health visiting practice. Community Practitioner. Retrieved from http://eprints.leedsbeckett.ac.uk/4877/1/AnExplorationoftheMentalHealthNeedsofAsylumSeekingWomenAM-CLARK.pdf
ECRE (2005) The way forward Europe's role in the global refugee protection system. http://www.ecre.org/files/exsum%20ECRE%20WF%20Integration_Jul05.pdf
Evans GW (2003) The built environment and mental health. J Urban Health Bull N Y Acad Med 80(4). Retrieved from https://doi.org/10.1093/jurban/jtg063
Fandrich CM, Moritz T, Papadopoulou A, Treviranus B (2013) Comparative study on the best practices for the integration of resettled refugees in the EU member states. Directorate General for Internal Policies Policy Department C: Citizens' Rights and Constitutional Affairs
Fitzpatrick S, Bramley G, Blenkinsopp J, Johnsen S, Littlewood M, Netto G, Sosenko F, Watts B (2016) Destitution in the UK. Joseph Rowntree Foundation, York
Froy F (2006) From immigration to integration: comparing local practices. In: From immigration to integration: local solutions to a global challenge. OECD Publishing, pp 31–100

Girma M, Radice S, Tsangarides N, Walter N (2014) Detained: women asylum seekers locked up in the UK. Women for Refugee Women, London

Haigh R, Hettige S, Sakalasuriya M, Vickneswaran G, Weerasena LN (2016) A study of housing reconstruction and social cohesion among conflict and tsunami affected communities in Sri Lanka. Disaster Prev Manag Int J 25(5)

Hamza M (2021) Refugees' integration in the built environment: the Sweden case. Sustainability 13:12812. https://doi.org/10.3390/su132212812

Harris J (2003) All doors are closed to us: A social model analysis of the experiences of disabled refugees and asylum seekers in Britain. Disabil Soc 18(4):395–410

Hartig T, Lawrence R (2003) The residential context of health. J Soc Issues 59(3):455–473. https://papers2://publication/uuid/EDDB0560-D3ED-4AE1-9F31-FDD0EB643471

Hek R (2005) The experiences and needs of refugee and asylum-seeking children in the UK: a literature review. Retrieved from https://dera.ioe.ac.uk/5398/1/RR635.pdf

Hutchinson MK, Baqi-Aziz M (1994) Nursing care of the childbearing Muslim family. J Obstet Gynecol Neonatal Nurs 23:767–771. Retrieved from https://doi.org/10.1111/j.1552-6909.1994.tb01951.x

Janahi H, Raman S, Zapata-Lancaster G (2018) Understanding the impact of the residential built environment design on inhabitants' wellbeing. In: ARCC conference repository

Jayakody C, Malalgoda CI, Amaratunga D, Haigh R, Liyanage C, Hamza M, Witt E, Fernando N (2022) Addressing housing needs of the displaced people promoting resilient and sustainable communities. Int J Disaster Resil Built Environ 13(3):368–385. https://doi.org/10.1108/IJDRBE-09-2021-0124

Jeffery L (2010) Forced displacement, onward migration and reformulations of 'home' by Chagossians in Crawley, UK. J Ethn Migr Stud 36(7):1099–1117

Kone Z, Ruiz I, Vargas-Silva C (2019) Refugees and the UK labour market. Econref working paper 04-2019. Centre on Migration, Policy and Society. https://www.compas.ox.ac.uk/wp-content/uploads/ECONREF-Refugees-and-the-UK-Labour-Market-report.pdf

Mahamed F (2016) A new perspective on the resettlement and integration of refugees and forced migrants. #rethinkingrefugee, Ashley Community Housing. https://www.ach.org.uk/sites/default/files/2018-04/ACH_Provocation_Mar17.pdf

Manz S, Panayi P (2012) Refugees and cultural transfer to Britain: An Introduction. Immigrants & Minorities 30(2–3):122–151

Mayblin L, James P (2019) Asylum and refugee support in the UK: civil society filling the gaps? J Ethn Migr Stud 45(3):375–394. Retrieved from https://doi.org/10.1080/1369183X.2018.1466695?needAccess=true

Miles MB, Huberman AM (1994) Qualitative data analysis: an expanded sourcebook

Nakeyar C, Esses V, Reid G (2018) The psychosocial needs of refugee children and youth and best practices for filling these needs: a systematic review. Clin Child Psychol Psychiatry 23(2):186–208

Patton M (1990) Qualitative evaluation and research methods. Sage, Beverly Hills, CA

Robila M (2018) Refugees and social integration in Europe. United Nations Department of Economic and Social Affairs (UNDESA), Division for Social Policy and Development, New York

Saldana J (2013) Coding manual for qualitative analysis

Taylor S, Sidhu RK (2012) Supporting refugee students in schools: what constitutes inclusive education? Int J Incl Educ 16(1):39–56. https://doi.org/10.1080/13603110903560085

UNHCR (2010) Convention and protocol relating to the status of refugees

UNHCR (2018) Protection and solutions in urban settings. High commissioner's dialogue on protection challenges. https://www.unhcr.org/5c0fd0954.pdf

UNHCR Bureau for Europe (2013) A new beginning: refugee integration in Europe, Sept 2013. Available at http://www.refworld.org/docid/522980604.html

Ward K, Amas N, Lagnado J (2008) Supporting disabled refugees and asylum seekers: opportunities for new approaches. Metropolitan Support Trust, London

Wijegunarathna EE, Wedawatta G, Prasanna LJ, Ingirige B (2017) Long term satisfaction of resettled communities: an assessment of physical performance of post-disaster housing. Procedia Eng 212:1147–1154

Yeo R (2017) The deprivation experienced by disabled asylum seekers in the United Kingdom: symptoms, causes, and possible solutions. Disabil Soc 32(5):657–677. https://doi.org/10.1080/09687599.2017.1320268

Zetter R, Pearl M (2000) The minority within the minority: refugee community-based organisations in the UK and the impact of restrictionism on asylum-seekers. J Ethn Migr Stud 26(4):675–697. https://doi.org/10.1080/713680501

Chapter 2
Micro-narratives on People's Perception of Climate Change and Its Impact on Their Livelihood and Migration: Voices from the Indigenous Aymara People in the Bolivian Andes

Ximena Flores-Palacios, Bayes Ahmed, and Cloe Barbera

Abstract Millions of people are being forced to migrate involuntarily due to extreme weather events and anthropogenic climate change each year. Climate-induced human migration is now contemplated as a colossal humanitarian challenge. Climate migration is an emerging issue that has long been neglected for political reasons. Most recent works in this field are primarily linked to defining or estimating climate migrants and constructing theoretical frameworks on the potential drivers of population mobility. However, the miseries of the marginalised people and climate migrants are still overlooked and insufficiently researched, especially in the front line indigenous communities residing in high-altitude regions of South America. To overcome such limitations, we studied Aymara, an indigenous people in the northern Bolivian Altiplano. We collected thirty micro-narratives from some selected case study areas through in-depth interviews and conducted two focus group discussions. The fieldwork was conducted between April and May 2021. We aimed at understanding the impact of climate change on livelihoods, local economy, and the migration pattern of the Aymara people and their adaptation strategies in the face of global warming and extreme weather events. The participants stressed that climate change adversely affects the Aymara community despite their inherited traditional knowledge. They face severe water scarcity, weather variability, glacier melting, changes in precipitation and farming seasons, loss of biodiversity, and environmental degradation. Climate change exacerbates pre-existing vulnerabilities associated with livelihoods, poverty, people's health, and food security. Consequently, young Aymara adults predominantly migrate to larger cities or sometimes even cross borders, searching

---

X. Flores-Palacios
La Paz, Bolivia

B. Ahmed (✉)
Institute for Risk and Disaster Reduction (IRDR), University College London (UCL), Gower Street, London WC1E 6BT, UK
e-mail: bayes.ahmed@ucl.ac.uk

C. Barbera
Institute of Advanced Development Studies (INESAD), La Paz, Bolivia

© The Author(s), under exclusive license to Springer Nature Switzerland AG 2023
M. Hamza et al. (eds.), *Rebuilding Communities After Displacement*,
https://doi.org/10.1007/978-3-031-21414-1_2

for alternative occupations and better living standards. We recommend that high-level intergovernmental and local policy-makers recognise climate migrants and take urgent steps to protect them and rebuild communities by safeguarding their human rights, cultural heritage, and dignity.

Keywords Indigenous people · Climate change · Vulnerability · Traditional knowledge · Environmental migration · Bolivia

2.1 Introduction

Human-induced climate change is the defining development challenge of our time. It is predicted that the adverse impact of climate change will intensify extreme weather events in all regions worldwide in the coming decades (IPCC 2021). Climate extreme (extreme weather or climate event) can be defined as the occurrence of a value of a weather or climate variable above (or below) a threshold value near the upper (or lower) ends of the range of observed values of the variable. For simplicity, extreme weather and climate events are collectively referred to as 'climate extremes' (IPCC 2012). Although climate change is a global issue, there is compelling evidence that the Andean Mountains in South America are highly vulnerable because of their unique geophysical features (Global Environment Facility 2021; Schoolmeester et al. 2016). Indigenous mountainous communities in the Bolivian Altiplano (high plateau) are already experiencing climate change's effects on their lives and livelihoods. Moreover, unpredictable climatic conditions challenge traditional knowledge and threaten people's livelihoods, the development and maintenance of their culture, and their survival (McDowell and Hess 2012; OXFAM 2020).

In the Bolivian Altiplano, Aymara, an indigenous people in the Andean region of South America, are particularly vulnerable because they live in a harsh environment, face pervasive poverty, have unequal access to land with high fragmentation, are exposed to the impacts of climate change and variability, and have limited access to technology, infrastructure, and financial resources to support their livelihoods and preserve their environment (Ministerio de Planificación del Desarrollo et al. 2015). Consistent evidence indicates that people in the Bolivian Altiplano have adapted to different climatic conditions over generations. People's capacity to adapt to environmental change has been based on an in-depth understanding of their land and ecosystems (Ayala et al. 2016; Ford et al. 2020). However, as climate change increasingly affects these populations, they respond and adapt in unique ways.

They combine their traditional knowledge with modern technical knowledge to adapt to climate change and develop other income-generating activities. Furthermore, migration is part of a broader livelihood portfolio and is an important adaptation strategy in the context of climate change. The IPCC Sixth Assessment Report (2021) states that there is now robust evidence that the impacts of climate change are driving rural communities to diversify their income sources, such as urban wage labour, migration, and remittances.

This study explores the Aymara people's perception of climate change and its impacts on their livelihoods and migration in five selected municipalities (Batallas, Pucarani, Puerto Pérez, Huatajata, and Taraco) in the Northern Altiplano in the department of La Paz, Bolivia.

The research questions are as follows:

(a) How do the Aymara people perceive the effects of climate change?
(b) What are the impacts of climate change on the Aymara people's livelihoods and well-being?
(c) How are the Aymara people responding to the impacts of climate change?
(d) How have climate change impacted Aymara people's decisions to migrate?

2.2 Conceptual Framework

The conceptualisation of critical issues such as vulnerability resilience, traditional knowledge, and human mobility in the context of the Bolivian Altiplano will give us the framework to understand the effects of climate change on people's lives and livelihoods and their coping mechanisms, including migration.

2.2.1 Vulnerability and Resilience

Extensive evidence indicates that climate change and variability are disproportionately distributed among vulnerable and marginalised population groups (IOM 2018; Thomas et al. 2019). Furthermore, the IPCC indicates that "climate change will exacerbate multidimensional poverty in most developing countries, including high mountain states [where Aymara population live and thrive]" (Olsson et al. 2014). Indigenous peoples are among the first to face the direct impacts of climate change and variability due to their strong dependence on their ecosystems and natural resources for their livelihoods and food security (United Nations 2022).

This is relevant for the Aymara people in Bolivia, who live in fragile areas that are particularly exposed to climate change and environmental degradation (Gonda 2020; Schoolmeester et al. 2016). Even though climate change disproportionally affects indigenous communities, they have demonstrated resilience in the context of slow- and sudden-onset events. A growing number of research provide evidence of indigenous resilience grounded in cultural aspects, history, and traditions that connect people to their communities and environments (Cuesta et al. 2012; Ford et al. 2020; IOM 2018; Nakashima et al. 2012).

While there is evidence of indigenous people's resilience, in most cases, they are still depicted as victims of climate change because their needs, problems, and challenges are generally ignored in policy debates. These narratives are incomplete, and they can delegitimise indigenous people's agency and portray them as unable to decide their future. Moreover, traditional knowledge is presented as romanticised and generally as a means of corroborating scientific knowledge (Belfer and Maillet 2017). Indigenous peoples' long history of resilience and adaptation has long been neglected in research and policy. They must be recognised as agents of change, and their voices need to be heard (Baird 2008; Etchart 2017; Oelz et al. 2017).

Resilience in Aymara communities is rooted in their traditional knowledge. The IPCC Assessment AR4 noted that traditional knowledge is "an invaluable basis for developing adaptation and natural resource management strategies in response to environmental and other forms of change" (IPCC 2007). Traditional knowledge and wisdom on how to live in harmony with nature have been accumulated over many generations by indigenous people in the Andean region. Since ancient times, they have developed a remarkable capacity to adapt to harsh environmental conditions, manage their ecosystem, conserve biodiversity, and domesticate a great diversity of food plants and animals. In addition, irrigation technologies and terraces have been built on the slopes to increase and stabilise cultivable areas (Fundación Tierra 2019; Meldrum et al. 2018). Although Andean communities' resilience is threatened by their vulnerability to the adverse impacts of climate change and variability, people still use their traditional knowledge to inform decision-making about fundamental aspects of day-to-day life.

2.2.2 *Migration*

Extensive literature has documented that Andean communities in Bolivia are characterised by high levels of mobility which is a traditional way to reduce vulnerabilities. Over the years, large numbers of Andean people have migrated internally and across borders to escape harsh environmental conditions, land scarcity, poverty, unemployment, lack of access to infrastructure, health, education, technology, and other agricultural support systems (Balderrama et al. 2011; Kollmair and Banerjee 2011; UDAPE and INE 2018). However, climate change is exacerbating existing vulnerabilities and causing new ones.

In the Andean region, those who usually migrate are the young adults who move to urban areas and abroad. Few youths aspire to remain in rural areas and make a living out of agriculture as they do not have access to assets, goods, and services or opportunities to acquire new skills. Too often, their only option is to migrate (Urioste 2012). These mobility patterns affect the social structures of communities as new redistributions of roles and responsibilities emerge among those who stay behind. Women, the elderly, and children are often left in the communities to look after plots of land and animals. Meanwhile, male family members migrate to urban areas to seek employment and diversify their family income (Gil Montero 2006).

In the Andean region, four types of migration have been identified: (i) long-term migration to another province in the country and abroad, mainly to Argentina and Brazil, (ii) temporary migration to urban areas or rural municipalities, mainly in the lowlands (rural–rural migration), (iii) circular migration and multi-residency to neighbouring cities, and (iv) return migration mostly of elders (Kaenzig and Piguet 2014; Schwilch et al. 2017).

2.2.3 Environmental Migration

According to IOM (2019), the working definition of climate migration refers to "the movement of a person or groups of persons who, predominantly for reasons of sudden or progressive change in the environment due to climate change, are obliged to leave their habitual place of residence, or choose to do so, either temporarily or permanently, within a state or across an international border". Climate migration is thus a subcategory of environmental migration; it defines a particular type of ecological migration, where the change in the environment is due to climate change (IOM 2019).

Environmental factors have always impacted human mobility; throughout history, people have left places with deteriorating or harsh conditions. However, the accelerated and combined impacts of climate change have become so severe that the scale of movement is increasing and has unparalleled implications on people's lives and livelihoods (Brookings 2014; IPCC 2007). Over the last decades, there has been a growing recognition that environmental and climate changes are important drivers of migration. Policy-makers and researchers worldwide have advocated incorporating migration into climate change debates (Foresight 2011; Flores-Palacios 2016).

Many studies in the Andean region of South America have anticipated that the impacts of climate change will become increasingly severe in the decades ahead, contributing to human mobility (Kaenzig and Piguet 2014). In the southern Ecuadorian Andes, Gray (2009) found that environmental factors were correlated to short-distance migration but not to international migration. Milan and Ho (2013) studied the relationship between increasing rainfall variability, livelihoods, and human mobility in the highlands of Peru. They found that rainfall variability and climatic conditions, in general, play an essential role in shaping local livelihoods and thereby influence migration decisions primarily driven by broader economic considerations.

In the highlands of Bolivia, migration in the context of climate change is a topic still underexplored. However, there is growing evidence that the adverse effects of climate change contribute directly and indirectly to both temporary and permanent migration and displacement within the country and across borders. Climate impacts add new layers of vulnerability to the Andean populations living in this region. Like other Andean countries, Bolivia has experienced an average increase in temperatures, and one of the most severe consequences is the acceleration of the melting of glaciers. This is severe because entire communities depend on glaciers for water, irrigation, power generation, and ecosystem conservation. In the last fifty years, many tropical glaciers in Bolivia have entirely disappeared, and projections indicate that many of the rest are likely to melt over the coming decades (Barnett et al. 2005; Ramirez 2006).

There are few studies on the impacts of melting glaciers on population movements in the Bolivian Altiplano. These have shown that glaciers retreat due to climate change, leading to water scarcity and migration. For instance, Alurralde et al. (2011) pointed out that if the temperatures continue to increase and the rainfall patterns keep changing, the Illimani glacier's melting rate will become faster and devastate the downstream communities depending on its water resources and could boost future migration rates in the area.

Kaenzig (2013) also studied the Illimani glacier and the communities situated on the hillsides of the mountain in the municipality of Palca. He found that climate change has no direct implications on migration patterns as glacier retreats cannot be considered a driver. Episodes of seasonal droughts or hazards cannot be solely attributed to glacier retreats. Still, they must consider other climatic factors, such as precipitation and temperatures, and non-climatic factors, such as the governance of water distribution (Brandt et al. 2016).

Balderrama et al. (2011) studied the Northern Potosi region, which experiences gradual environmental degradation, drought, and desertification. The authors found that ecological degradation coupled with a reduction in the amount of land available for farming, mining enclaves, desertification, and climate variability has led to the impoverishment of the indigenous population, triggering migration in the region to other rural areas or neighbouring cities.

2.3 Study Area Profile

The study area is in the West of the Department of La Paz and belongs to the Northern Bolivian Altiplano. This area is part of the ancestral territory and the cradle of ancient Andean civilisations. It is inhabited by the Aymara people, one of the largest ethnic groups in Bolivia. The fieldwork was conducted in the Municipalities of Batallas, Pucarani and Puerto Pérez in Los Andes Province, the Municipality of Huatajata in Omasuyos Province, and the Municipality of Taraco in Ingavi Province.

The region's altitude ranges from 3000 m above sea level in the inter-Andean valleys to almost 6500 m above sea level in the peaks of the Cordillera Real, which is home to most of Bolivia's glaciers. A large part of this region is influenced by the presence of Lake Titicaca (the largest lake in South America) and the glaciers of the Cordillera Real Mountain range. The estimated population for 2020 of the selected municipalities is 66,000 inhabitants. All municipalities are rural, except for Batallas, which has a small urban centre. Poverty in the study area is widespread. In Batallas, the proportion of people living in poverty is 76.4%; in Pucarani is 82.2%; in Puerto Pérez is 83.2%; in Huatajata is 29.6%, and in Taraco is 85.3%. Many households lack essential services such as water supply, electricity, and sanitation. Water consumed comes from streams, natural springs, and wells in unfavourable conditions.

Communities in the Northern Altiplano are highly vulnerable to climate change due to several factors: (i) agricultural production is dependent on climatic variables such as rainfall for sowing and cultivation, and it is vulnerable to frost and hailstorms,

(ii) water scarcity, (iii) small landholdings, (iv) loss of agrobiodiversity, and (v) low level of investment in households farming system (Devenish and Gianella 2012; Rangecroft et al. 2013).

The study area is characterised by small farms comprising various plots at different altitudes. Farmers grow potatoes, barley, oats, broad beans (*Vicia faba*), quinoa, oca (*Oxalis tuberosa*), and papaliza (*Ullucus tuberosus*). Livestock is also raised, and small-scale fishing is performed by the communities on Lake Titicaca's shores. Other incipient activities include trade, transport, and tourism. The Aymara population living in the study area presents high levels of mobility. Over the years, many people have migrated internally and abroad for employment, education, family reunification, and a better life.

2.4 Methodology

2.4.1 Rationale

This qualitative study explores the intersections between climate change, people's livelihoods, and current mobility patterns and trends in the Northern Bolivian Altiplano. The approach considers people's knowledge and experiences at the centre of the study as it is crucial to give prominence to people on the front line of climate change. As Paton and Fairbairn-Dunlop (2010) wrote, "the voices of academics, scientists, politicians and development practitioners dominate the climate change debate, yet local knowledge and beliefs, local realities, as well as local voices and actions are essential elements of navigating the way forward". While listening to the people's voices may serve a political purpose and be seen as a right to be engaged, axiomatic also is that the voices carry considerable knowledge (traditional and other), which must be counted, and which can valuably inform future initiatives (Flores-Palacios 2016). Ethical awareness and appropriate engagement with Aymara communities were the study's core.

2.4.2 Design

The fieldwork, including pilot interviews for the study, was carried out between 12 April and 28 May 2021 by maintaining all the local COVID-19 regulations. The data collection methods were in-depth semi-structured interviews and focus group discussions (FGDs). Thirty participants were selected for qualitative micro-narrative collections, and two FGDs were conducted. The participants' voices provided diverse and rich sources of information. Due to the spread of COVID-19 in the country, travelling to all municipalities was limited. Thus, the fieldwork was conducted in Batallas and Pucarani. Participants from Puerto Pérez, Huatajata, and Taraco participated in the fieldwork in Batallas.

As oral tradition is congruent with Aymara culture, we emphasised making people's knowledge noticeable through oral testimonies. Twenty-six open-ended questions were formulated for the in-depth semi-structured interviews (Appendix 1) and fourteen for the FGDs (Appendix 2). Both methods helped to collect crucial qualitative information. The questions were oriented towards knowledge creation, especially around people's perception of the impacts of climate change on families' livelihoods, natural resource endowments, and population movements.

In addition, a comprehensive literature review of different disciplinary insights relating to climate change and human mobility and the government policies and practices in the study area was undertaken to contextualise the study.

2.4.3 Participants' Profile

The respondents were selected with the support of a research assistant, an Aymara community leader fluent in Aymara, who identified the participants based on their knowledge and experience. A total of thirty people (twenty-three male and seven female) participated in the study, aged between 36 and 75 years old. There were more male than female respondents, and most were also in the middle-aged group, with the mean age of participants being 49 years old. Of the 30 participants, 18 were engaged in agricultural and livestock activities, nine were involved in farming and livestock activities, and were public servants. Three Aymara participants were associated with universities; one was a historian, a master's student, and a university lecturer.

2.4.4 Process of Gathering Information and Data Analysis

The first author and principal researcher for this study conducted in-depth semi-structured interviews and FGDs with the support of the local research assistants. Participants had time to elaborate on their comments based on their knowledge, experiences, values, and beliefs. In conversational methods, silence and other forms of non-verbal cues have meaning. Therefore, awareness of cultural issues is essential (Kovach 2010). Participants were not simply knowledge providers; their role was to bring their traditional knowledge and experiences to light. This study applied a qualitative thematic analysis to identify, analyse, and report patterns within the data.

2.5 Results and Discussion

2.5.1 People's Perception of Climate Change

2.5.1.1 People's Perception of Changing Weather Patterns

The participants' perceptions of climate change and variability correspond with meteorological information and scientific estimates made in the Andean region. Rising temperatures, glaciers retreat, and shifts in the frequency and intensity of precipitation and frost events have been documented in the Andean highlands over the last decades (Hock et al. 2019; Meldrum et al. 2018).

Male, 41 years (BATALLAS)

We have experienced temperatures, rain, and frost changes in the last few years. But before, everything had its season; we knew when it was going to rain and what time there was going to be frost. Ice has come very early and affects our production, same with hailstorms. Our seasonal calendar has changed.

When participants were asked about specific weather-changing patterns, they described many processes of change that have occurred in the last decades. They pointed to abnormal weather variability, increases in mean temperatures, changes in precipitation, and hailstorms and frosts during unusual times.

Male, 55 years (HUANCANÉ BATALLAS)

Talking about climate change hurts us. Now with these changes, everything is happening unexpectedly. Frost, hails come suddenly, and we do not know when it will rain.

Historical records show that the climate regime—temperature and precipitation—is changing in the Altiplano, and forecasts indicate that such variations will continue for several decades, as well as an increase in extreme events (Seth et al. 2013). Given climate's significant influence on people who depend on agriculture and livestock in the Altiplano, climate change and variability imply challenges to the traditional relationships between people and their environment and impose additional pressures on the agriculture-based economy and natural resources (Young and Lipton 2006).

A standard view among interviewees was the shifting seasons due to climate change. Some mentioned that the higher-elevation zones previously excessively cold for arable farming have become productive and now support the cultivation of hardy crop varieties. These perceptions coincide with well-documented *studies* in the Bolivian Altiplano that link shifting seasons to warmer global temperatures (Hoffmann and Requena 2012; Valdivia et al. 2010).

Male, 55 years (HUANCANÉ BATALLAS)

Everyone is worried. Now you cannot farm well anymore; nothing is in its season. Before, we had seasons when we planted different crops, but now we do not know when to cultivate what. As soon as the frost starts, there is no more production.

2.5.1.2 People's Perception of Glacier Retreat, Water Scarcity, and Loss of Biodiversity

Melting of the Glaciers

The most visible impact of climate change in high mountain regions is glacier retreat. The IPCC (2007) reports that Andean inter-tropical glaciers will likely disappear in the coming decades, negatively impacting water availability in the region. In addition to global increases in temperature, the greater frequency of El Niño in recent decades has contributed to rapid glacial retreats (Cook et al. 2016; Radford 2016). Participants said they depend on glacial melt for some of their water supply, and now they are facing seasonal water shortages, threatening their livelihood.

Male, 41 years (BATALLAS)

Glaciers are melting because of the heat. Now it is very sunny. You must imagine how impressive was the mountain range. Now the snow-capped mountains are in a hot frying pan. They are going to melt. The problem is that this trend will continue; the glaciers will melt, the snow will disappear, and it is all because of the heat.

Water Scarcity

In the study area, like in the many parts of the Bolivian highlands, water supply mainly originates from rainfall, glacial melt, and groundwater aquifers. Although there is no information on how much of their water supply comes from glacial melt, there is some evidence that it represents a significant percentage (OXFAM 2009). As highlighted by the participants, drinking water and irrigation schemes are urgently necessary for the area.

Male, 42 years (PAJCHIRI PUERTO PÉREZ)

I remember in the 90s that there was water. We are worried that there is no water, and it will worsen. We know that water is life. It is a fundamental element for humans and animals. If you do not have water, you will not produce anything. We depend on water for our food.

Some participants said that water scarcity could lead to conflict over access to and control over resources. These conflicts can escalate into violence if not addressed, disrupting the communities. Brandt et al. (2016) and Field (Rangecroft et al. 2013) mentioned that water scarcity is associated with lower water run-off due to climate change, increasing glacier melting, water overuse, technical irrigation problems, social conflicts, and the lack of water governance.

Male, 63 years (CHIRAPACA BATALLAS)

Water shortage due to melting glaciers will bring instability to our communities. That is the reason why today we must fight for water. My community is having problems with another community because they will not let us access the water. So, the other communities may not mention this, but they say they own the water.

Loss of Biodiversity

The Andean region is one of the most critical biodiversity hotspots in the world. Yet, they represent one of the regions most severely threatened by human activities and most vulnerable to climate change. This double pressure makes this area a key priority for global conservation efforts (Gonda 2020). In the study area, biodiversity loss is linked to environmental degradation, overexploitation of natural resources, and climate change. Moreover, the loss of agrobiodiversity is the primary concern for people in the Bolivian Altiplano. Andean families have accumulated knowledge of their ecosystems and natural resources, including developing and using plant genetic resources, soil conservation, and pest management. They also have specialised knowledge of wild plants used for food, medicinal plants, and fodder (FAO 2012). However, the temperature increase and precipitation variability adversely impact water supplies and crop yields. Native Andean crop varieties risk disappearing, and entire ecosystems are deteriorating with the drastic shift in weather patterns (Valdivia et al. 2010).

Participants reported that the number of varieties of some crops decreased due to the loss of seeds post-harvest and the adverse weather effects, notably the irregular rain and frost. Participants expressed concerns that potatoes and other edible tubers' genetic diversity are being lost at an alarming rate. They pointed out that they cannot afford to lose further agrobiodiversity when resilient varieties are needed in times of climate change. It is essential to mention that the area surrounding Lake Titicaca is an important centre of origin for *the* genetic diversity of potatoes and other edible tubers (Graves 2006).

2.5.1.3 People's Perception of Environmental Degradation

Besides the effects of climate change and variability, environmental degradation is critical in these fragile ecosystems. Land degradation and soil erosion in the study area are mainly due to intensive land use, reduced vegetation cover, inadequate livestock practices, reduced fallow periods, and use of chemical inputs and tractors, combined with insufficient rain (NCAP Project 2011; Orsag 2009).

Male, 55 years (HUANCANÉ BATALLAS)

When I was a boy, we had many animals, cows, and sheep, but now we only have up to 5 to 10 cows, and the same with sheep; families rarely have more. Decreased soil quality, limited grazing land, and climate change are the main reason for miserable livestock.

Another issue highlighted by participants was the pollution in Lake Titicaca; they said that this situation is adversely affecting them. High levels of pollution have been documented in Lake Titicaca resulting from mining activities and domestic and industrial waste from rivers of nearby cities of La Paz and El Alto that drain into the lake. Lake Titicaca is only about 281 m at its deepest, and climate change has dried up several areas near the shore, concentrating the pollutants dumped there by factories, mining activities, farming, and industry (Bloudoff-Indelicato 2015).

Male, 44 years (COHANA PUCARANI)

Cohana Bay is polluted. Totora (*Schoenoplectus californicus subsp. tatora*) was food for our livestock; now, it looks like cardboard. Moreover, birds like *chhuqha* (*Fulica ardesiaca*), tiki tiki (*Gallinula chloropus*) and *qiñu* (*Rollandia microptera*) have gradually disappeared, and fishing is not good. Besides damaging the environment, pollution is forcing young people to migrate to different destinations. We are abandoned, and there is no response from the state authorities.

2.5.2 The Impacts of Climate Change on Livelihood and Well-Being

2.5.2.1 Impact on People's Livelihoods

Families in the study area depend on natural resources for their livelihoods, and their main income-generation activities are carried out through traditional practices. They are mainly engaged in rainfed agriculture and livestock. Besides, off-farm income represents an essential source of livelihood for many farmers with part or full-time employment by some household members. As Urioste (2017) mentions, pluriactivity is a vital characteristic of rural economies in Bolivian highlands. The author states that while traditional peasants lived from agricultural and livestock work and with strong links to endogenous productive systems that connected agro-ecological regions more intensely through the exchange in-kind, today's pluriactive peasants are more related to urban markets, which in turn requires knowledge of new codes such as mercantile exchange or the use of technology. Farming is not the only occupation of rural people; non-farming employment is a complementary source of income.

Female, 36 years (IGACHI BATALLAS)

We are farmers; we produce potatoes, quinoa, and broad bean. We also make oats to feed the livestock. We have it mainly for our consumption, but if there is a surplus, it is for sale. In the case of potatoes, half is for consumption and half for sale. Cheese for consumption and milk for sale; we consume little cheese.

The fishery has been a traditional activity in Lake Titicaca that generates income and provides a source of animal protein. However, various local species have gone extinct due to overfishing and pollution.

Male, 55 years (HUANCANÉ BATALLAS)

My community is Huancané, on the shores of Lake Titicaca. We produce potatoes, broad beans, barley, oat, and alfalfa *[Medicago sativa L.]*. We also have seaweed in the lake, called chanco (*Myriophyllum elatinoides*), but it is nearly all gone. We use totora to feed our livestock, although there is not much totora. Fishing is not good now; there is pollution, maybe.

Climate change is seriously affecting people's livelihoods in the study area. Precipitations have become irregular and unpredictable. The dry season is now extended, causing delays in sowing due to a lack of rainfall. On the other hand, rain is often heavy during the wet season and causes flooding, especially around Lake Titicaca. When participants were asked about the effects of climate change on their livelihoods, community members mentioned the impacts on agriculture, livestock, and fishing. They said that the traditional agricultural calendar has changed and referred to losses in production because of frosts, droughts, and hailstorms, seriously affecting people's food security and income. In addition, environmental degradation associated with human activities was also mentioned.

Female, 36 years (CHIRAPACA BATALLAS)

The problem now is climate change. Ten, twenty years ago, you could produce quite well, there were not so many frosts, or at least they were not so harsh and abrupt. There was plenty of water and the rest.

Higher temperatures and reduced water availability are spreading pests and diseases, affecting crops, and prompting people to use more pesticides.

Male, 63 years (CHIRAPACA BATALLAS)

Twenty years ago, we didn't use fertilisers or pesticides; we used natural fertilisers. With that, our product was better. We need more chemicals and fertilisers, but production remains low and is sometimes wormy. This is due to climate change.

Diseases in livestock resulting from the increase in daily temperatures and decrease in nightly temperatures have also been reported. Participants said that limited irrigation access is an issue affecting all the communities. The communities benefiting from irrigation grow potatoes, broad beans, quinoa, alfalfa, and barley, which ensure a higher income than that earned by the communities which do not rely on irrigation water.

2.5.2.2 Impact of Climate Change on Well-Being

Climate change and variability impact depend on biophysical factors, people's resilience, and social and institutional support systems. Participants pointed out that the effects of climate change exacerbate the poverty trap, food insecurity and health. Moreover, it also affects people's traditional knowledge and social cohesion necessary to make a living from the land and preserve their cultures (Garcia-Alix 2008; Minority Rights Group International 2019).

Poverty

Poverty and climate change are deeply intertwined because climate change disproportionally affects the most vulnerable communities. People in poverty have a higher chance of experiencing the impacts of climate change due to increased exposure and vulnerability (Kronik and Verner 2010; OXFAM 2020). This is particularly relevant for the Andean communities because climate change is already disrupting livelihoods, forcing families to leave their communities, and pushing people further into poverty, as observed in the study area.

An additional problem affecting Aymara farming systems is *minifundio*, or smallholding predominates in the highlands. Extreme land fragmentation is prominent constraint farmers face in improving their living conditions in rural areas (Fundación TIERRA 2013; Regalsky et al. 2015). Land scarcity associated with *minifundio* prevents traditional crop diversification and rotation methods, which require some plots of land to rest fallow each year instead of planting irrigated parcels twice annually.

Male, 45 years (EL ALTO)

Family sizes are growing, people have more children, and the land is being parcelled; there is less land to cultivate.

Male, 50 years (PUCARANI)

Families do not have enough land, and no water comes down from the mountains, forcing people to migrate.

In a limiting context such as the *minifundio*, families in the study area are already experiencing the consequences of climate change. Still, they are not equipped to adapt to the present and future impacts. Climate change sometimes renders traditional practices obsolete and puts harvests at risk, affecting food security and incomes.

Male, 63 years (CHIRAPACA BATALLAS)

Climate change has affected my income. We look for means of work because sometimes life teaches us that. If production does not give us enough, we must find other work means. To cope with climate change's impacts, we must diversify and invest more money in our activities. Since agriculture and dairy farming are not going well, I decided to dedicate myself to farming cuniculture, giving me better results.

Food Security

Climate change has significant implications for food insecurity. It creates uncertainty regarding the availability of water and other resources needed for agriculture and shapes the frequency and intensity of extreme weather events that can adversely affect rural livelihoods (FAO 2015).

In the study area, it was observed that climate change is jeopardising nutrition and food security. With changes in the hydrological cycle and other extreme weather events, farmers produce less, have lower yields, and obtain less income from their produce. Decreases in food production are causing problems of malnutrition, which endangers particularly young children's health and compromises people's immune systems, making them more susceptible to infectious diseases (FAO 2015). Participants mentioned the reduced ability to produce staple food, such as *chuño*, because of shorter seasons and fewer viable locations.

Male, 55 years (HUANCANÉ BATALLAS)

Before, we had plenty of food, also fish, which is so rare now; it used to be food that the poor would eat; currently, only the rich can afford it.

Health

Climate change poses many threats to people's health, such as disorders linked to the availability and quality of water and food, conditions related to extreme heat events, respiratory disease, vector-borne infections, cancer, and other pathologies associated with climatic disasters and extreme temperatures (Feo et al. 2009). People in the study area are worried about climate change's threat to human health.

Female, 45 years (TARACO)

Our health is affected by the intense heat; you cannot do anything when it is scorching. Now at 6–7 o'clock, we start managing the cattle, but as soon as the sun comes out, it burns you, and you do not want to do anything. It takes away your strength and your energy. You will faint if you go on with that tremendous heat and sun. That is why I could tell you that if you wanted to harvest the potatoes in the sun, you would not be able to do it.

Narratives of impending climate change, apart from the actual impacts experienced, have generated anxiety about the future. IOM adverts that the adverse effects of climate change have a profoundly negative impact on indigenous communities' daily lives and psychological well-being (Traore Chazalnoel and Carini 2018).

2.5.3 Adaptation to Climate Change and Indigenous Knowledge

Climate change impacts are expected to increase the vulnerability of Aymara communities in the coming years. To face these challenges, they are mobilising their in-depth knowledge of the territories that have been the source of their livelihoods for generations. As in the rest of the Altiplano region, families in the study area have long passed on knowledge from generation to generation. However, climate change is threatening ancestral wisdom that may endanger the survival of their culture (Nakashima et al. 2012; Ramos-Castillo et al. 2017).

This study highlights six essential aspects of traditional knowledge in the context of climate change: weather forecasting and early warning systems; conservation of agrobiodiversity and crop diversification; food preservation; soil and water management; water harvesting; and a combination of traditional and modern knowledge.

2.5.3.1 Weather Forecasting and Early Warning Systems

Traditional knowledge has helped Aymara communities forecast local weather and extreme events by observing the stars, plants, animal behaviour, and natural phenomena at certain times and seasons. With the onset of climate change, the Aymara traditional knowledge system has been challenged. However, it is still a source of crucial information on weather forecasting, as mentioned by the participants in the study area.

Male, 63 years (CHIRAPACA BATALLAS)

Our parents and grandparents taught us how to predict the weather through animal behaviour. Animals gave us indications, for example, through differences in egg-laying behaviour or animals digging in the earth. All this taught us what stage we were in and when it would rain, but nowadays, even the animals do not know.

Male, 55 years (HUANCANÉ BATALLAS)

Before, our grandparents understood animal behaviour, but now most of us don't know what it means. Now the *leque leque* [*Vanellus resplendes*] lays its eggs anywhere. The *churi churi* used to put its nest on top of the totora when the rain was coming and when there was no rain, it would put lower down. And regarding foxes, they howl a lot in some seasons and others.

2.5.3.2 Conservation of Agrobiodiversity and Crop Diversification

Andean traditional farming systems have adapted to changes in agroecological conditions according to norms and practices to ensure sustainable management of resources for the use of present and future generations (Fundación Tierra 2019; Meldrum et al. 2018). In the study area, Aymara farmers have used their traditional knowledge to domesticate, improve, and conserve thousands of crop species and varieties. For instance, in the community of Corquemaya in Batallas, participants explained how they choose, at sowing time, the varieties that can best respond to the climatic challenges they foresee. They also spread the risk between different crops and types as a measure of adaptation to climate change.

Preserving crop diversity is vital for food security, the environment, and sustainable development (Esquinas-Alcázar 2005; Flores-Palacios 1997; Khoury et al. 2016). Moreover, the need to preserve crop diversity in climate change adaptation is crucial, as highlighted in various studies (de Haan 2020; Bedmar Villanueva et al. 2007). Crop diversification is a traditional risk management practice and may find renewed importance in spreading the risk of harvest loss, enabling long-term crop adaptation, preserving seeds, and conserving wild crop relatives (Walshe and Argumedo 2016).

2.5.3.3 Food Preservation

In the study area, it was observed that food preservation is another traditional form of adaptation to climate change. Aymara communities use an ancient technique to preserve foods, such as the transformation of potato into *chuño* [black sun-dried potato] and *tunta* [white sun-dried potato], and oca (*Oxalis tuberosa*) into *kaya* through processes of freezing drying, and dehydration. Once dried and with minimal care in storage, the products can be preserved for many years, even decades, making food available in times of scarcity (Guidi et al. 2002).

2.5.3.4 Soil and Water Management

Traditional practices to manage soil are still in place in Aymara communities, as observed in the study area. Participants mentioned that communities near Lake Titicaca use a traditional *camellones* or *waru-waru* (high beds) technique. The *camellones* are elevated fields surrounded by channels. During the rainy season, the gutters fill with water that would have otherwise flooded the land, and during the dry season, this same water is used for irrigation.

2.5.3.5 Water Harvesting in America

As observed in the study area, some ancient rainwater harvesting technologies are still used, such as the *wijiña* (conical deposit dug in the ground to accumulate rainwater) and the *kurmi khotas* (traps to harvest rainwater) (Loayza-Aguilar et al. 2020). People also use rainwater harvesting practices with traditional agroecological methods, *q'otañas* (a form of rainwater harvesting using clay layers of soil) and *q'ochas* (shallow depression that stores water). These practices improve the local water supply as it enhances soil infiltration and biomass production and is structured to trap sediments (Mwenge Kahinda et al. 2021).

2.5.3.6 Combination of Traditional and Modern Knowledge

Apart from using their traditional knowledge, Andean people are also incorporating modern technologies and practices. For instance, in the Lake Titicaca region, communities are implementing traditional and contemporary adaptation measures such as the conservation of water springs, water courses, wetlands (bofedales), pastures, and soil. Moreover, farmers are promoting ecological production and land use planning mechanisms to support the development of Municipal Land Use Plans and local prevention strategies to reduce the impacts of climate change (NCAP Project 2011).

2.5.4 Human Migration

2.5.4.1 Drivers of Migration

Several reports have shown that migration is time and place specific, and it is a multifaceted phenomenon in environmental migration (FAO et al. 2018). Economic, social, cultural, and political drivers still dominate migration decision-making. However, in the case of societies dependent on agriculture, like the Aymara communities in Bolivia, environmental factors have always played a significant role in migration decisions, as changes in agricultural productivity are associated with climate-related factors. Research in rural areas of Latin America shows that increased temperatures, rainfall variability, and extreme weather events can negatively affect crop yields in ways that may induce migration (De Sherbinin 2020; FAO 2018; Milan et al. 2016).

This study shows that migration in the study area is multifaceted and that migrants have not moved due to a single cause. People migrate because of better work opportunities, unemployment, environmental change, climate change and environmental degradation, food insecurity, and limited access to health and education in their places of origin. Moreover, participants reported that the scarcity of farming land is one of the main reasons rural people migrate, especially to the nearby urban centres and neighbouring countries.

Female, 36 years (IGACHI BATALLAS)

My brothers have already left; they are joining friends working in Argentina. They left because of poverty here and lack of opportunities in Batallas. We have small land holdings, and climate change affects our agriculture.

2.5.4.2 Migration Patterns

In the study area, four types of migration have been identified: (i) long-term migration, internal, and international, (ii) temporary migration to urban areas or rural municipalities, mainly in the lowlands (rural–rural migration), (iii) circular migration and multi-residency to neighbouring cities, and (iv) return migration of elders.

All the participants mentioned that they have family members who have moved internally and across borders. Temporary and permanent migration is a livelihood strategy to escape poverty and land scarcity and adapt to climate change and variability. Internal migration in the study area is both temporary and permanent. Migrants moving to urban areas work primarily in the construction, commerce, and transport sectors. The relative proximity of the communities to La Paz and El Alto allows for a form of multi-residence. One particularity of the population movements in the Andean region is that migrants do not cut-off ties with their place of origin.

Participants noted that those who move to urban areas, particularly El Alto, try to buy a house while maintaining their farmland.

Female, 38 years (EL ALTO)

In the 1970s, 80s and '90s, people migrated to La Paz and El Alto; some of them had their houses there, so they lived there and sometimes here. They work in commerce or as taxi or minibus drivers in the city.

Circular migration is also familiar with shorter movements to urban areas. This option allows migrants to maintain their land and family ties with their communities while working outside for short periods. Rural–rural migration also exists in the study area. Some people migrate to the lowlands, for instance, to Yungas, a low-lying ecoregion of the department of La Paz. One of the research findings was that the municipality of Batallas is receiving migrants. It seems that this municipality is creating bridges among rural communities, providing the population with the opportunity to access essential services and commerce.

Male, 60 years (CORQUEMAYA BATALLAS)

Here in Batallas, there are people from Batallas. We are receiving migrants from other provinces, not only returnees. Even people from Peru are coming.

In the study area, some young people are migrating abroad. Participants said they have family members living in Argentina, Brazil, and some others in Spain. Usually, migration abroad is long term; one participant noted that it is harder to come back often when you are far away.

Female, 36 years (IGACHI BATALLAS)

Some family members have migrated to Brazil, Argentina, and other foreign countries. They migrate mainly because there is not much work here. Besides, the young people are all leaving to search for opportunities; only the older people are left behind here.

Male, 41 years (CANTÓN HUANCANÉ BATALLAS)

People from Huancané migrate because our agriculture is not good. Young people go to countries like Brazil and Argentina. Those are the two main destinations for our migrants. The youth are the ones that are looking for a better life for their family and themselves.

There is evidence of return migration in the study area. Some people return to their communities to look after their elderly parents, and other adults return because life in urban areas has become difficult for them.

Male, 55 years (HUANCANÉ BATALLAS)

I used to live in La Paz, but now I am back in Huancané. Before, people used to migrate to the cities; now, some return to their communities. We are coming back because of the pandemic, and there are no jobs in the city. We are coming back, although we know that life in rural areas is very hard.

Male, 41 years (CANTÓN HUANCANÉ BATALLAS)

Older adults are returning to their communities. I think they have accomplished their goals in the city; their children are grown up. So, it is time for them to look after their land.

Male, 41 years (BATALLAS)

…at the moment, migration is the other way around in Batallas. People who migrated to the city for economic reasons are returning; even those who emigrated to Argentina are returning.

Some people prefer to stay in their communities because of the strong sense of belonging to their lands and communities and because they believe their traditional knowledge will help them adapt to climate change. In the case of returnees, particularly elders, they do not want to migrate again.

Female, 36 years (IGACHI BATALLAS)

Older adults are not migrating anymore. They are so used to living here. They do not go because they must look after their property and livestock.

Male, 74 years (PUERTO JAJACHI COHANA PUCARANI)

I am not thinking of migrating. I have already lived in other places; now it is my time to live here, in my house, in my area of origin. This is the place where I belong.

Some stay in their communities as they cannot migrate because they lack support such as capital and networks in potential destinations. However, those who remain cannot be considered "trapped populations", as defined by Warner et al. (2012), namely people who struggle to survive in their areas of origin and cannot easily use a migration to adapt to the adverse impacts of climate change. Families and communities usually play an essential role in buffering socio-economic and environmental risks because practices of solidarity are part of the Aymara culture.

As regards remittances, most participants pointed out that they are scarce. Several participants said that remittances are used to improve food security and nutrition, part is invested in agricultural improvements, and there is not much left for other purposes.

Male, 36 years (PUERTO JAJACHI COHANA PUCARANI)

Generally, when family members leave, like the husbands or young men, they send money back to support their families, their mothers, and their fathers who are struggling. They know the situation here; loved ones are being helped.

Although families in the study area receive some remittances, migrants often buy urban plots or houses in their destination places or invest their earnings in their children's higher education.

In the case of the study area, migration is not only common for men but also for women. Our results are consistent with the findings of Balderrama et al. (2011) and Kaenzig (2013), who conducted research in the Altiplano region. People who are migrating are primarily young men and women. Participants reported that usually, the head of the household, especially men, moves to urban areas, mainly La Paz and El Alto, to seek employment and then the rest of the family follows them. Another group comprises young people who move to urban centres after school to look for jobs or pursue technical and university studies. These migration patterns affect the communities' social structures as there is a redistribution of roles and responsibilities among those who stay, mainly women and older people. The outflow of the younger age group is creating imbalances in the population's age and skillsets that remain in their communities.

Female, 37 years (COHANA PUCARANI)

Families in Corquemaya are having trouble growing food and maintaining their livestock; there is now no food for them. So, some family members are thinking about leaving to find jobs elsewhere.

Female, 55 years (CORQUEMAYA BATALLAS)

Primarily young people migrate from Batallas. They move out because there is not much work here. They migrate for economic reasons, which is why they go to the cities; some even go to neighbouring countries.

2.5.4.3 Linkages Between Climate Change and Migration

The nexus of climate change and migration is emerging in research, policy, and practice. It is increasingly being recognised that mobility can be used to improve adaptive capacity and further development agendas and outcomes, particularly in developing countries (Foresight 2011). Environmental migration, like general migration, usually occurs within national boundaries, and when international migration is reported, it is generally to neighbouring countries. Social networks are essential for destination migration as people tend to choose places where they have contacts and people who can help them settle in and find work (Obokata et al. 2014).

The pattern of climate migration in the study area is internal, but cross-border migration is also to neighbouring countries and Europe.

Male, 45 years (BATALLAS)

Climate change is what causes migration. How will we sustain the increased population in the communities if there is no production? We need rain for agriculture and livestock. If it does not rain, how will we do it?

Families may follow a mixed strategy in which some household members move while others stay behind.

Male, 48 years (HUATAJATA)

Climatic changes affect us; unpredictable rain and frequent hailstorms reduce our crop productivity. With no help from our local and national governments, we have to do everything to survive; one strategy is moving to the cities to look for jobs, and young people are migrating.

Although there are cases in which it is possible to disentangle migration drivers, such as sudden-onset events, environmental migration must be recognised as a multicausal and multidimensional phenomenon. Ecological and climate factors often act in combination with economic, social, political, and demographic aspects, as pointed out by IOM (2014). Regarding the study area, both climatic variability and water scarcity alone may not directly drive migration, but they are usually combined with other stressors impacting agricultural production and livelihood security.

Male, 63 years (CHIRAPACA BATALLAS)

I am not sure if climate change is the only reason to migrate, I would say more for work because there is not much work in rural areas. But our livelihoods depend on climatic conditions, so I think climate change is influencing migration. How are we going to deal with this situation? We do not receive support from the government; we do not have access to technology, and we receive very little help when we lose our crops because of frost or hail. We live independently, do what we can, and move away.

Female, 36 years (IGACHI BATALLAS)

I think climate change could be a contributing factor to migration. Weather conditions were extreme this year, there has not been any good harvest this year, and our livestock got sick. When this situation happens, people look to go to other places. In Yungas, for example, you can diversify your crops, plant fruits and coca, and increase your income.

Male, 41 years (CANTÓN HUANCANÉ BATALLAS)

I think that a combination of factors is driving migration. The first one is that we have small plots. Secondly, climate change has affected the productive capacity and economic incomes. The third is water scarcity. These are causing migration. It all adds up.

2.6 Conclusion and Policy Implications

Over centuries, the Aymara people have been protectors and custodians of biological diversity, landscapes, and natural resources. They have accumulated knowledge of their ecosystems, including soil conservation, management of pests, and the development, use, and preservation of plant and animal genetic resources. Their resilience is rooted in traditional knowledge, as their capacity to adapt is based on an in-depth understanding of their environments.

The study aimed to explore the linkages between climate change, livelihoods, and migration in selected municipalities of the Northern Bolivian Altiplano to understand how they respond to human-caused climate change and other multiple challenges in their rural communities. The study reveals that families are still marginalised due to poverty, unequal access to land with high fragmentation, and limited access to health, education, technology, credit, and infrastructure.

The study establishes that climate change exacerbates people's pre-existing vulnerabilities affecting their livelihood patterns as they greatly rely on natural resources in fragile environments. The risks associated with climate change increase with poverty, land scarcity, and limited institutional support. Vulnerability to these risks often depends on specific local contexts and public policies, which in the case of the Bolivian Altiplano have, in many cases, failed to address the struggles of the local communities.

Despite the enormous challenges facing the Aymara communities in the study area, participants said they had developed various mechanisms to strengthen their adaptive capacity to cope with both slow- and sudden-onset events. They are combining their traditional knowledge and modern technology to respond to climate change. They are also building on their family and community social support systems and using population movements to enhance their adaptive capacity. How families respond to climate change and environmental hazards depends on the severity of the event, levels of vulnerability, and the resources and strategies available to the households.

In the selected communities, human mobility in the context of climate change is a risk management and livelihood diversification strategy to reduce vulnerability to environmental and non-environmental risks. In this case, migration generates additional income with off-farm activities. Migration appears very strongly as a multi-causal phenomenon shaped by economic, social, demographic, environmental, cultural, institutional, and climate change drivers. Evidence of circular, temporary, and permanent migration has been a traditional household strategy in the studied communities. In recent years, involuntary migration has increased due to low agricultural productivity, poor economic growth, limited access to land, off-farm work opportunities, and the impacts of anthropogenic climate change.

Migrants from the study area are primarily young adults who move to pursue a better living standard, find better jobs, diversify income, and respond to climate change. The destinations are typically large urban areas, mainly La Paz and El Alto, and other urban centres and rural municipalities in the lowlands. International migration is also directed to Brazil, Argentina, and Spain, where migrant networks often facilitate people's social and economic integration in the destination places.

Four implications for policy design have been identified in this study. First, people's perceptions and experiences must be incorporated into research and policy. Their knowledge of struggles and proposed solutions should be central in understanding and addressing climate change-related challenges. Second, environmental migration must be studied within specific social and cultural contexts to implement tailored interventions. Third, the design of any climate change adaptation initiative must be conducted in consultation with local communities to ensure that their priorities are considered. Finally, data production and research are necessary to contribute to a broader understanding of the linkages between climate change, people's livelihood, and migration.

Acknowledgements This study was funded by UCL Global Engagement Funds (2020–2021). We are indebted to the local people for their active participation and support during the fieldwork. We thank Alejandro Mamani for assisting us in conducting surveys and Cecilia Skarne for translating the micro-narratives into English.

Appendix 1: Guiding Questions for In-Depth Interviews

Demographics

Age (years):

Gender:

Marital status:

Number of children:

Source of income:

Place of residence:

Background Information

1. How long have you/your family lived in this community?
2. What crops do you grow? Are you also involved in livestock and fishing?

3. How have agriculture and livestock changed in the last ten years? Do you produce more crops today than in the past? Compare how it is today and how it was ten years ago.
4. How has the weather changed in the last decade, e.g. temperatures, rainfall, frost? Are droughts more frequent? Compare how it is today and how it was ten years ago.
5. Have you noticed that the glaciers are melting? Why? How does it affect you?
6. Do you think these changes in climate and melting glaciers will continue in the future?

Impacts of Climate Change and Disasters

7. Are you concerned about climate change?
8. Do you think your family has been affected by climate change?
9. How are the changes in climate affecting your health?
10. How are the changes in climate affecting your work? Agriculture, livestock, and fishing?
11. How are changes in climate affecting your food consumption? Are you producing less food?
12. What do you (or your family) do with your crops? For your consumption, for sale?
13. How are changes in climate affecting your income? Do you have to look for other jobs besides agriculture, livestock, or fishing?
14. Does your family have ways of dealing with climate change? Could you give examples? Do they move to other places?
15. Do you think climate change has also affected other families in your community?
16. Is there anything else you would like to say about climate change?

Migration

In the past ten years

17. Have any of your family members migrated to other places? Where (internal movements, abroad)? Why?
18. What is migration like in your community? Do people migrate for short periods or stay permanently in other places? Or do they come and go? How often?
19. Who leaves? Young people (men and women)? Adult men/women? The elderly? Who stays?
20. How many people have migrated from this community to another place? What percentage, more or less? Where do they go?
21. Are changes in the climate a reason to migrate to other places? Since when have people been migrating because of changes in the climate?
22. Do you plan to migrate because of climate change or for other reasons?
23. Do you think more people will migrate in the future? Why?
24. Do you or your family receive financial support from your family or friends who have migrated?

COVID-19

25. How has COVID affected the community?
26. Have people from the community migrated because of COVID? Or have people returned to the community?

Appendix 2: Guiding Questions for Focus Group Discussions (FGDs)

Climate Change

1. Have you heard the words climate change? What do these words mean to you?
2. Why do you think climate change is happening and is becoming "worse" today?
3. Do you think your family/community has been affected by climate change? How?
4. Do you have ways of dealing with climate change (examples)?
5. Anything else you would like to say about climate change?

Migration

In the last ten years

6. Have any family members moved out of the village (any idea of the number)? Where have they gone? What would be the main reasons they moved (school, work, other reasons)? Who is leaving? (Men/women)? (young/adult)?
7. Do you think they have moved because of climate change (Yes, No)? Reasons (% related to climate change?)
8. Do they keep in contact with you? Do they come back to the community on occasion? Do they support the community?
9. Do you think family members who have moved influence family and community life? How?
10. What are the effects of this emigration on the community?

Future

11. Do you think you will move away from here shortly? Why?
12. Where would you go?
13. Would you say that climate change would be a reason to leave? Or would it force you to move?
14. If you migrate, what would you miss most about the community?

References

Alurralde J et al (2011) Living with glaciers, adapting to change the experience of the Illimani project in Bolivia. In: XIVth IWRA world water congress

Ayala G, Torrez E, Veizaga A, Limachi O (2016) Las ciencias ancestrales como mecanismo de adaptación al cambio climático. Plural

Baird R (2008) The impact of climate change on minorities and indigenous peoples. Minority Rights Group International

Balderrama C et al (2011) Migración rural en Bolivia: El impacto del cambio climático, la crisis económica y las políticas estatales. IIED. Human settlements working paper series. In: Rural-urban interactions and livelihood strategies—31

Barnett TP, Adam JC, Lettenmaier DP (2005) Potential impacts of a warming climate on water availability in snow-dominated regions. Nature 438(17):303–309

Bedmar Villanueva A, Halewood M, López Noriega I (2007) Agricultural biodiversity in climate change adaptation planning. Eur J Sustain Dev 6(2):1–8

Belfer EFJ, Maillet M (2017) Representation of indigenous peoples in climate change reporting. Clim Change 145:57–70

Bloudoff-Indelicato M (2015) What are North American trout doing in Lake Titicaca? Smithsonian Magazine, 9 Dec

Brandt R, Kaenzig R, Lachmuth S (2016) Migration as a risk management strategy in the context of climate change: evidence from the bolivian andes. In: Migration, risk management and climate change: evidence and policy responses. Springer, pp 43–61

Brookings (2014) LSE project on internal displacement: climate change and internal displacement. Retrieved from http://www.brookings.edu/

Cook S et al (2016) Glacier change and glacial lake outburst flood risk in the Bolivian Andes. Cryosphere 10(5):2399–2413

Cuesta F et al (2012) Panorama Andino sobre cambio climático: vulnerabilidad y adaptación en los Andes Tropicales. CONDESAN, SGCAN, Lima, Peru

de Haan S (2020) 10 ways to enhance Andean agrobiodiversity during COVID-19 and climate change. Available at: https://cipotato.org/

De Sherbinin A (2020) Climate impacts as drivers of migration. https://www.migrationpolicy.org/article/climate-impacts-drivers-migration

Devenish C, Gianella C (2012) Sustainable mountain development in the Andes: from Rio 1992 to Rio 2012 and beyond. Consorcio para el Desarrollo Sostenible de la Ecorregión Andina (CONDESAN), Lima

Esquinas-Alcázar J (2005) Protecting crop genetic diversity for food security: political, ethical and technical challenges. Nat Rev Genet 6:946–953

Etchart L (2017) The role of indigenous peoples in combating climate change. Palgrave Commun 3:17085

FAO (2012) Conservation and sustainable use under the International Treaty. Rome

FAO (2015) Climate change and food security: risks and responses. Rome

FAO (2018) The state of food and agriculture: migration, agriculture and rural development. Rome

FAO, IFAD, IOM, WFP (2018) The linkages between migration, agriculture, food security and rural development. Rome

Feo O et al (2009) Cambio climático y salud en la región andina. Rev Per Med Exp Salud Pública 26(1)

Flores-Palacios X (1997) Contribution to the estimation of countries' inter-dependence in the area of plant genetic resources, s.l. Commission on genetic resources for food and agriculture, background study paper no. 7

Flores-Palacios X (2016) Samoa: exploring the linkages between climate change and population movements. Doctoral dissertation, Auckland University of Technology

Ford J et al (2020) The resilience of indigenous peoples to environmental change. One Earth 2(6):532–543

Foresight (2011) Migration and global environmental change. Future challenges and opportunities. Final project report. The Government Office for Science

Fundación TIERRA (2013) Gestión comunal de los derechos de propiedad de la tierra. TIERRA, La Paz

Fundación Tierra (2019) La agricultura familiar campesina: una guía para comprender sus valores alimentarios, ecológicos y sociales

Garcia-Alix L (2008) The United Nations permanent forum on indigenous issues discusses climate change. Indigenous Aff 1(2):16–23

Gil Montero R (2006) Despoblamiento diferencial en los Andes Meridionales: Sud Chichas y la Puna de Jujuy en el siglo XIX. Bull Inst Fr Études Andin 35(1):55–73

Global Environment Facility (2021) Andean neighbors working together on climate resilience

Gonda C (2020) Cambio climático y biodiversidad en los Andes tropicals. FARN

Graves C (2006) La papa, tesoro de los Andes, 2da edn. CIP, Lima, Perú

Gray C (2009) Environment, land, and rural out-migration in the Southern Ecuadorian Andes. World Dev 37(2):457–468

Guidi A, Esprella R, Aguilera J, Devaux A (2002) Características de la cadena agroalimentaria de chuño y tunta para el Altiplano Central de Bolivia. Fundación PROINPA, Cochabamba, Bolivia

Hock RRG et al (2019) High mountain areas supplementary material. In: IPCC special report on the ocean and cryosphere in a changing climate. IPCC

Hoffmann D, Requena C (2012) Bolivia en un mundo 4 grados más caliente: escenarios sociopolíticos ante el cambio climático para los años 2030 y 2060 en el altiplano norte. Insituto Boliviano de la Montaña and PIEB

IOM (2014) IOM outlook on migration, environment and climate change

IOM (2018) Environmental migration and indigenous peoples: what is at stake?

IOM (2019) Glossary on migration. Int Migr Law 34

IPCC (2007) Climate change 2007: synthesis report. In: Contribution of working groups I, II and III to the fourth assessment report of the Intergovernmental Panel on Climate Change (IPCC). Geneva, Switzerland

IPCC (2012) Summary for policymakers. In: Managing the risks of extreme events and disasters to advance. A special report of working groups I and II of the IPCC. Cambridge University Press, UK

IPCC (2021) Summary for policymakers. In: Climate change 2021: the physical science basis. Working group I to the sixth assessment report of the IPCC. Cambridge University Press

Kaenzig R (2013) The circulation of people. A critical discussion on the impact of glacier shrinkage upon population mobility in the Bolivian Andes. Working paper 7—2013. MAPS. University of Neuchâtel

Kaenzig R, Piguet E (2014) Migration and climate change in Latin America and the Caribbean. In: People on the move in a changing climate: the regional impact of environmental change on migration. Springer

Khoury CK et al (2016) Origins of food crops connect countries worldwide. Proc R Soc B Biol Sci 283(1832)

Kollmair M, Banerjee S (2011) Drivers of migration in mountainous regions of the developing world: a review. International Centre for Integrated Mountain Development, Kathmandu, Nepal

Kovach M (2010) Conversational method in indigenous research. First Peoples Child Fam Rev 5(1):40–48

Kronik J, Verner D (2010) The role of indigenous knowledge in crafting adaptation and mitigation: strategies for climate change in Latin America. In: Social dimensions of climate change: equity and vulnerability in a warming world. World Bank, pp 145–169

Loayza-Aguilar J, Blanco-Capia LE, Bernabé-Uño A, Ayala-Flores G (2020) Saberes locales sobre tecnologías y estrategias de producción agropecuaria para la resiliencia climática. J Selva Andina Biophere 8(1):32–41

McDowell JZ, Hess JJ (2012) Accessing adaptation: multiple stressors on livelihoods in the Bolivian highlands under a changing climate. Glob Environ Change 2:342–352

Meldrum G et al (2018) Climate change and crop diversity: farmers' perceptions and adaptation on the Bolivian Altiplano. Environ Dev Sustain 20:703–730
Milan A, Ho R (2013) Livelihood and migration patterns at different altitudes in the Central Highlands of Peru. Clim Dev 6(1):69–76
Milan A, Schraven B, Warner K, Cascone N (2016) Migration, risk management and climate change: evidence and policy responses
Ministerio de Planificación del Desarrollo, UNFPA, UDAPE (2015) Vulnerabilidad poblacional al riesgo de desastres en Bolivia. Editorial Greco, La Paz
Minority Rights Group International (2019) Minority and indigenous trends 2019
Mwenge Kahinda J, Bahal'okwibale PM, Budaza N, Mavundla S, Nohayi NN, Nortje K, Boroto RJ (2021) Compendium of community and indigenous strategies for climate change adaptation—focus on addressing water scarcity in agriculture. FAO, Rome
Nakashima DJ et al (2012) Weathering uncertainty: traditional knowledge for climate change assessment and adaptation. UNESCO-UNU
NCAP Project (2011) Enhancing adaptive capacity in semi-arid mountainous
Obokata R, Veronis L, McLeman R (2014) Empirical research on international environmental migration: a systematic review. Popul Environ 36(1):111–135
Oelz M, Kumar Dhir R, Harsdorff M (2017) Indigenous peoples and climate change: from victims to change agents through decent work. ILO
Olsson L et al (2014) Livelihoods and poverty. In: Climate change 2014: impacts, adaptation, and vulnerability. Part A: global and sectoral aspects. Contribution of working group II to the fifth assessment report of the IPCC. Cambridge University Press, Cambridge, United Kingdom, pp 793–832
Orsag V (2009) Degradación de suelos en el Altiplano Boliviano causas y medidas de mitigación. Rev Anal 1(3)
OXFAM (2020) Bolivia: climate change, inequality and resilience
OXFAM International (2009) Bolivia. Climate change, poverty and adaptation
Paton K, Fairbairn-Dunlop P (2010) Listening to local voices: Tuvaluans respond to climate change. Local Environ 15(7):687–698
Radford T (2016) Bolivian glaciers melt at alarming rate. Climate News Network, 22 Oct
Ramirez E (2006) Impacto del cambio climático sobre la disponibilidad de recursos hídricos. In: Retroceso de los glaciares y recursos hídricos en Bolivia—De la investigación a la acción. Instituto Boliviano de la Montana—UMSA, La Paz, Bolivia, pp 19–31
Ramos-Castillo A, Castellanos E, Galloway McLean K (2017) Indigenous peoples, local communities and climate change mitigation. Clim Change 140:1–4
Rangecroft S et al (2013) Climate change and water resources in arid mountains: an example from the Bolivian Andes. Ambio 42(7):852–863
Regalsky P, Núñez del Prado J, Vásquez Rojas S, Chumacero J (2015) La problemática de la tierra luego de 18 años de titulación: Territoros, minifundio, individualización. TIERRA, La Paz
Schoolmeester T et al (2016) Outlook on climate change adaptation in the tropical Andes Mountains. In: Mountain adaptation outlook series. United Nations Environment Programme, GRIDArendal and CONDESAN, Nairobi
Schwilch G et al (2017) Impacts of out-migration on land management in mountain areas of Nepal and Bolivia
Seth A, Thibeault J, Garcia M, Valdivia C (2013) Making sense of twenty-first-century climate change in the Altiplano: observed trends and CMIP3 projections. Routledge
Thomas K et al (2019) Explaining differential vulnerability to climate change: a social science review. Wiley Interdiscip Rev Clim Change 10(2)
Traore Chazalnoel M, Carini F (2018) Environmental migration and indigenous peoples: what is at stake? IOM
UDAPE, INE (2018) Migración interna en Bolivia
United Nations (2022) Indigenous peoples. Department of Economic and Social Affairs

Urioste M (2012) La Reforma Agraria abandonada: valles y altiplano. In: Proceso agrario en Bolivia y América Latina. Plural editores, La Paz, Bolivia, pp 19–51

Urioste M (2017) Pluriactividad campesina en tierras altas: "con un solo trabajo no hay caso de vivir". Foro Andino Amazónico de Desarrollo Rural, La Paz, Bolivia

Valdivia C et al (2010) Adapting to climate change in Andean ecosystems: landscapes, capitals, and perceptions shaping rural livelihood strategies and linking knowledge systems. Ann Assoc Am Geogr 100(4):818–834

Walshe R, Argumedo A (2016) Ayni, ayllu, yanantin and chanincha: the cultural values are enabling adaptation to climate change in communities of the Potato Park, in the Peruvian Andes. GAIA Ecol Perspect Sci Soc 25(3):166–173

Warner K et al (2012) Where the rain falls: climate change, food and livelihood security, and migration. CARE France and UNU-EHS

Young K, Lipton J (2006) Adaptive governance and climate change in the tropical highlands of western South America. Clim Change 78:63–102

Chapter 3
From Zero to Hero? Changes in the Estonian Context for Refugees

Emlyn Witt, Jane Liise Nurk, and Irene Lill

Abstract Until the Russian invasion of Ukraine in February 2022, there were very few displaced people in Estonia—372 or approximately 0.03% of Estonia's population as of January 2020. For earlier waves (1997–2019) of asylum seekers and refugees in Europe, Estonia was not a popular destination, and, by 2020, 30% of displaced people who had been granted protection in Estonia had left for other countries. However, since February 2022, more than 30,000 Ukrainians displaced by the war have sought refuge in Estonia so that now, by mid-April 2022, more than 2% of Estonia's population comprises displaced people. This research investigates the context and problems faced by displaced people in Estonia. It reports the findings of a desk study of existing literature and media reports and a series of nine interviews conducted with 14 key informants from relevant organisations that deal with refugees, beneficiaries of subsidiary protection and asylum seekers. The findings show that the difficulties faced by displaced people in Estonia are linked to their residence permit durations and Estonian language skills for accessing necessary services and navigating the complexities of daily life. Government policies and structures for dealing with displaced people are relatively new and subject to change, and the wide array of organisations with unclear and overlapping remits poses a challenge. The attitudes of Estonians towards displaced people have also been diverse, subject to rapid changes and dependent upon the specific group of immigrants. Whereas the latest wave of Ukrainian war refugees has been warmly welcomed into Estonia, the historical constraints and limitations of the Estonian system remain largely unchanged and adequately providing for them will be a considerable challenge.

Keywords Mass displacement · Refugees · Migration · Estonia · Invasion of Ukraine

---

E. Witt (✉) · J. L. Nurk · I. Lill
Department of Civil Engineering and Architecture, Tallinn University of Technology, Tallinn, Estonia
e-mail: emlyn.witt@taltech.ee

© The Author(s), under exclusive license to Springer Nature Switzerland AG 2023

M. Hamza et al. (eds.), *Rebuilding Communities After Displacement*,
https://doi.org/10.1007/978-3-031-21414-1_3

3.1 Introduction

Immigration has been a problematic issue in Estonia. Until 1945, the Estonian population was almost exclusively (97%) comprised of ethnic Estonians, but, during the Soviet occupation, a policy of mass immigration saw the proportion of Russian immigrants in Estonia reach 30.3% of the total population by 1989 (Zabrodskaja 2009). Since regaining its independence in 1990, the immigration discourse has thus been closely bound with national identity, and language politics and security concerns which are aggravated by fears that Estonia could become a transit country for asylum seekers travelling between Russia and the Scandinavian countries. These fears are founded on past experiences from the early 1990s when approximately 400 asylum seekers were arrested as they attempted to transit through Estonia (Veebel 2015) and are reinforced by recent (2021) events on the Belarusian border with Lithuania, Latvia and Poland where migrants have been encouraged to travel to Belarus and then to cross illegally into those countries as a form of hybrid attack (ERR 2022m). A national regulation, valid since 1993, limits the number of immigrants (from outside the EU and now also excluding Ukrainian war refugees) into Estonia to 0.1% of the permanent population of Estonia (i.e. to a total of approximately 1300 people) and the Convention Relating to the Status of Refugees (a.k.a. the Geneva Convention of 28 July 1951) has only relatively recently (1997) been ratified by Estonia. Prior to that, all asylum seekers and economic migrants were considered to be illegal immigrants (Veebel 2015).

Since there are no internally displaced people in Estonia currently, all displaced people are cross border in nature and they fall into three categories:

1. Asylum seekers (whose asylum applications are in process)
   Then those who have already been awarded international protection—these are either:
2. Refugees (whose asylum applications have been successful and resulted in them being recognised as refugees) or
3. Beneficiaries of subsidiary protection (who have not qualified as refugees but have been recognised as being at serious risk if they were to return to their home countries).

3.1.1 Prior to February 2022

Until the Russian invasion of Ukraine began on 24 February 2022, the number of displaced persons in Estonia was small both in absolute terms and in relation to the size of the Estonian host population. According to Eurostat database information, the number of asylum applications (per capita) in Estonia was considerably below the European Union (EU) average for the years 2009–2021 (0.00–0.02% for Estonia, 0.04–0.25% for the EU). The European Migration Network's Annual Policy Report on Migration and Asylum Estonia 2016 (Lauren 2016) shows the number of

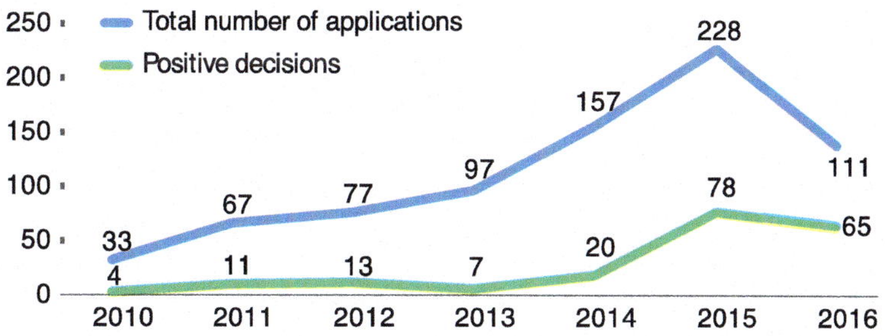

Fig. 3.1 Asylum decisions according to Lauren (2016). *Source* Estonian Police and Borber Guard Board

successful asylum applications 2010 to mid-2016 (Fig. 3.1). The low rate of recognition as well as a low level of guarantees for asylum seekers led to criticism of Estonia both internationally and locally (Luuk 2016; Veebel 2015). It is notable that the actual processing times of applications have historically been relatively quick—a Ministry of the Interior representative confirmed to the authors that average asylum application processing times in Estonia were 60 days in 2018 and 70 days in 2019 compared to the permitted 6 months.

During the 2015–2017 wave of refugees into Europe (mainly from Syria, Afghanistan and Iraq), there was active local opposition to the European Commission's temporary measures for relocating displaced people to Estonia based on an agreed quota. For example, in 2015, the Estonian Prime Minister argued that the numbers should be far lower than the quota even though the Estonian government had agreed to relocate 550 refugees (Veebel 2015). A poll conducted in June 2015 showed 48% of approximately 13,000 respondents (responding in both Estonian and Russian languages) considered that accepting refugees would influence Estonia in a very negative way, while more than 80% were negative about refugees coming to Estonia (Vaitmaa 2015). This coincided with a rise in national populism, anti-refugee sentiment being used for political gain and repeated attacks on the Vao accommodation centre for asylum seekers (Luuk 2016).

Following the 2015 agreement to resettle a maximum of 550 people in Estonia and the subsequent wave of resettlement which began in 2016, the EU leadership called again on member states to voluntarily resettle a further 50,000 persons. This led to an Estonian national programme of resettlement being launched in 2017. This 2-year programme was designed to resettle 80 persons. However, by October 2019, only seven persons had actually been relocated under it and, under a new Estonian government formed in April 2019 (and openly opposed to further resettlement), no further relocations were expected according to a Ministry of the Interior representative interviewed by the authors at the time—Interview I#1.

As well as being reluctant to accept refugees, Estonia was also an unpopular destination for them—according to an article in the Estonian business daily newspaper, Äripäev, refugees (and/or asylum seekers) themselves were resistant to being relocated from camps in Greece to Estonia

(Äripäev 2016; Cooney 2017). In correspondence with the authors in January 2020, a representative of the Estonian Ministry of the Interior provided figures for the period 1997–2019 as follows: total number of applicants = 1202 (persons); total number granted protection = 531 (280 as refugees + 251 subsidiary protection status); 213 out of the 531 came via relocation and resettlement during 2016–2019. As of 24 January 2020, there were 372 persons in total (refugees + subsidiary protection + their family members) living in Estonia with valid residence permits based on their protection status.

From the summer of 2021, the Belarusian authorities were accused of using migrants to exert pressure on the neighbouring states of Lithuania, Latvia and Poland in a form of hybrid attack on those countries in apparent retaliation for sanctions against Belarus and their support for the Belarusian opposition. Estonian Defence Forces were sent to support these countries' efforts to reinforce their borders, and, in April 2022, the Estonian government approved a bill to amend the State Borders Act and associated legislation in order to better allow the state to "contain mass immigration in an emergency if foreigners cross the Estonian border illegally and submit unfounded applications for international protection" (ERR 2022m). With this new migrant crisis on the western border of Belarus, it seems reasonable to deduce that Estonian public opinion towards immigrants (of all types) was particularly low by early 2022.

3.1.2 *Following the Russian Invasion of Ukraine in February 2022*

Within days of the full-scale invasion of Ukrainian territory by Russian forces on 24 February 2022, the fastest growing refugee crisis since World War II was underway (UNHCR, Ukraine Situation Flash Update 9, April 2022) and Estonia began to receive refugees in considerable numbers—more than 1000 per day in the first weeks of March. By mid-April, the number received by Estonia exceeded 30,000 (ERR 2022a) with more than 20,000 applications for temporary protection having been registered with the Estonian authorities (see Fig. 3.2). While these numbers are very small in comparison with the overall crisis in which, according to UNHCR figures, more than 5 million Ukrainians have fled Ukraine and a further 8 million are internally displaced within the country, they are substantial in relation to the increase in the numbers of displaced people in Estonia and in terms of Estonia's own population of only 1.3 million people (UNHCR, Ukraine Situation Flash Update 9, April 2022).

Prior to the 2022 invasion of Ukraine, the small numbers of displaced people in Estonia and their dispersion across the country made it difficult to consider them in terms of "communities". Even though Ukrainians have made up a substantial proportion of (the few) asylum seekers in recent years, there has been a Ukrainian community resident in Estonia, and (until early 2022) this had remained largely constant at

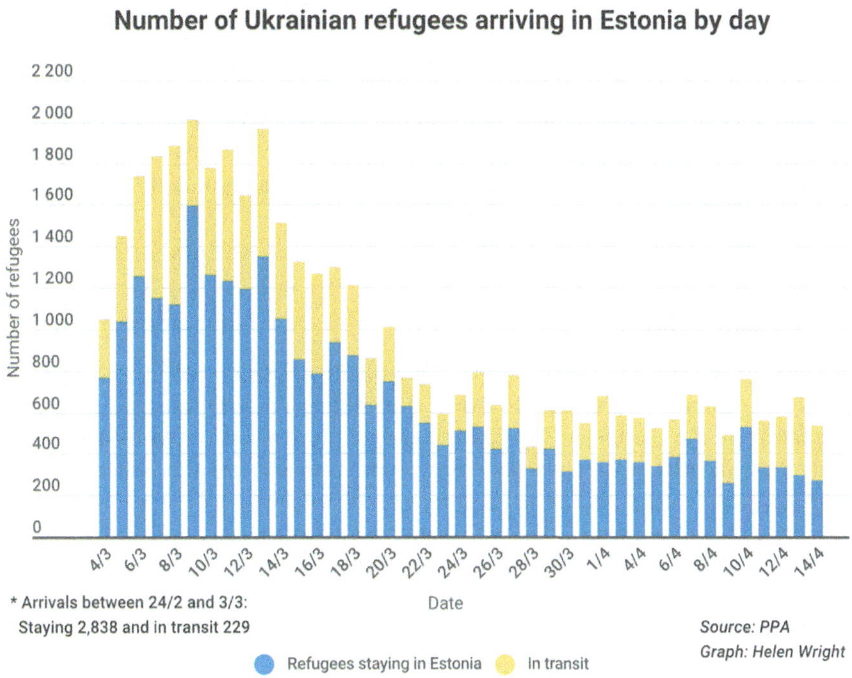

Fig. 3.2 Daily Ukrainian war refugee arrivals to mid-April 2022. *Source* ERR (2022a)

about 23,000 people from 2012 to 2017 despite a jump (from 395 Ukrainian immigrants in 2014 to 1200 in 2015 and increasing to 1785 in 2019) after the hostilities in Crimea and Eastern Ukraine in 2014 (Eurostat 2022; Statistics Estonia 2022). More recently, Ukrainians have made up a substantial proportion of the migrant workers in the Estonian agricultural and construction sectors, and, while statistics for early 2022 are not yet available from the Statistics Estonia and Eurostat databases, a representative of the Estonian Refugee Council suggested that almost 50,000 Ukrainians were living and working in Estonia prior to the current wave of refugees arriving (Cole 2022). Historically, the Ukrainian community had been somewhat assimilated into the large ethnic Russian population in Estonia because of their close ethnic and linguistic ties. In the context of the current conflict, however, the Ukrainian community, particularly its recent arrivals, have been more actively welcomed by and are associating rather with the Estonian community. A recent poll showed that 92% of Estonian-speakers support the reception of refugees from Ukraine compared with 51% of other nationalities (which, in practice, equates to Russian-speakers) (ERR 2022b). The feelings of solidarity with the Ukrainians are very high and 51% of families in Estonia have given money to charities or causes supporting Ukraine (ERR 2022c, d), and volunteers coordinated by the Estonian Refugee Council, an NGO, have directly evacuated more than 8000 Ukrainian refugees and their accompanying pets to Estonia (ERR 2022e).

This chapter outlines the Estonian context and the issues faced by displaced people within it. The context description has been derived on the basis of an interview survey undertaken before the large-scale invasion of Ukraine in February 2022, and, while the findings remain relevant and robust, it has been necessary to supplement them with a fully up-to-date commentary (based on national news media reports) to reflect the almost 100-fold increase in the number of displaced people in Estonia since February 2022.

3.2 Methodology

Within the framework of the REGARD (Rebuilding After Displacement) project co-funded by the European Commission and led by University of Huddersfield, a study was undertaken to understand the country context and the issues faced by displaced people in Estonia. A series of nine semi-structured interviews were carried out in October–November 2019 with fourteen key informants. Interviewees were purposively selected on the basis that they were identified as representatives of the most important organisations involved with displaced people.

The interview schedules covered:

- The needs of displaced communities—including housing needs, sociocultural needs, social infrastructure needs, economic needs, physical infrastructure needs, governance needs and communities with special needs.
- The roles of the host community and the built environment in social cohesion—including needs of the host community, perceptions, impacts on facilities, services and resources, economy, housing cost, labour market, social structure, sociocultural differences, fairness and equity, integration of the displaced and the role of the built environment.

In addition to the interviews which were carried out (Table 3.1), further interview requests were made to the local representative of the UNHCR and the Estonian Police and Border Guard Board, but no interviews could be arranged. An interview was also agreed with a representative of the Ministry of Culture, but the interviewee did not take place and no alternative interview was set up.

A representative of the Tartu Welcome Centre was also included as an interviewee although this NGO does not work specifically with refugees and asylum seekers but rather with all foreign nationals who come to live in Tartu (Estonia's second city).

A qualitative, thematic analysis of the nine interview transcripts was conducted using NVivo (v.12) according to both predetermined themes (to enable country comparisons) and emergent themes. This provided an understanding of the Estonian situation that prevailed prior to February 2022. These findings are described in the next section together with updated information as reported in the national (state) media. Media articles were selected on the basis of their relevance from a single source—the Estonian Public Broadcasting service (Eesti Rahvusringhääling, ERR).

Table 3.1 Interview details

Interview ref.	Organisation	Organisation type	Interview date	No. of interviewees
I#1	Estonian Ministry of the Interior	National Government Ministry	14.10.2019	1
I#2	Estonian Ministry of Social Affairs	National Government Ministry	14.10.2019	1
I#3	Estonian Refugee Council	NGO	14.10.2019	1
I#4	Johannes Mihkelson Centre	NGO	16.10.2019	1
I#5	Estonian Human Rights Centre	NGO	21.10.2019	1
I#6	Tartu Welcome Centre	NGO	25.10.2019	1
I#7	Töötukassa (Estonian Unemployment Insurance Fund)	National Government Agency	31.10.2019	2
I#8	AS Hoolekandeteenused (Welfare Services PLC)	Public Limited Company	8.11.2019	2
I#9	Mustamäe District Administration	Local Government	12.11.2019	4

3.3 Findings

3.3.1 Governance and Organisational Context

The Estonian system for dealing with displaced people is led by national government but relies on close collaboration both among the many national government ministries and agencies involved and also with local governments, NGOs and private companies as well. Figure 3.3 illustrates the overall arrangement.

According to the Ministry of the Interior (Interview I#1), the key organisations in this arrangement are the Ministry of the Interior itself, the Ministry of Social Affairs and the Estonian Refugee Council (Pagulasabi, an NGO). The Ministry of the Interior formulates immigration policy which is executed through its agency, the Police and Border Guard Board (Interview I#1). The Ministry of Social Affairs provides housing, services and social security (Interview I#2). It contracts with a publicly limited company (AS Hoolekandeteenused—which approximately translates as Welfare Services plc) to provide accommodation and initial integration services through two reception centres and in helping beneficiaries of international protection to find permanent accommodation (Interview I#8). The Estonian Refugee Council provides beneficiaries of international protection with support

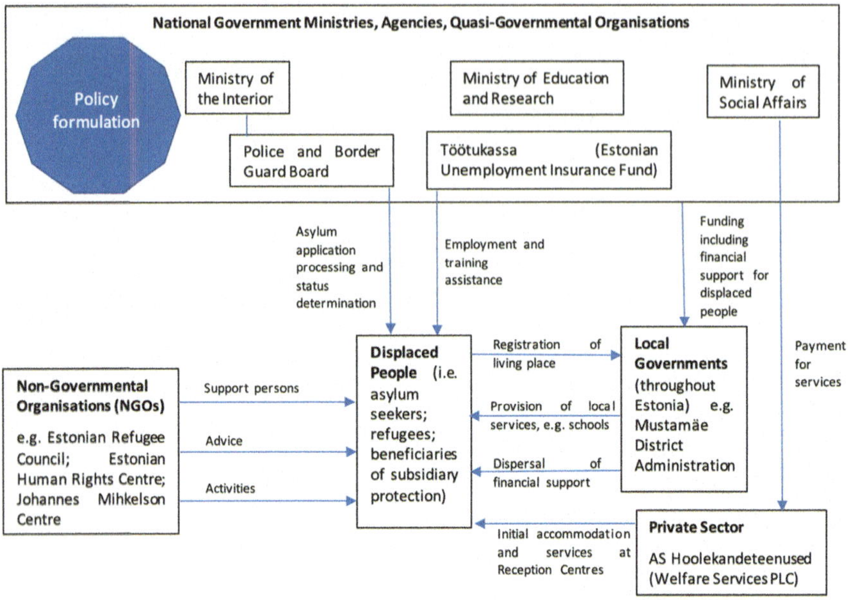

Fig. 3.3 Overall organisational arrangement of the Estonian system for displaced people

services, advice and activities aimed at integration and assistance. Support services are primarily delivered through support persons who help beneficiaries to understand and navigate the systems in Estonia and become self-sufficient as quickly as possible (Interview I#3).

In addition to the Ministry of the Interior and the Ministry of Social Affairs, other national government ministries, e.g. the Ministry of Education and Research and the Ministry of Culture, have important roles to play in the resettlement of displaced people as do other national level agencies such as Töötukassa (the Estonian Unemployment Insurance Fund) and Haigekassa (the Estonian Health Insurance Fund).

According to the local government officials interviewed (Interview I#9), there are separate funds allocated by national government to local governments to support beneficiaries of international protection. However, according to the Ministry of the Interior, there is apparently no special subsidy or monetary amount actually dispersed to the beneficiaries of international protection—this is the same amount as for everyone else receiving income support (Interview I#1).

Besides the Estonian Refugee Council which is focused on beneficiaries of international protection as its name suggests, other NGOs which deal with a wider range of clients also play an important role in assisting displaced people, for example, the Estonian Human Rights Centre which provides legal advice to asylum seekers and those who already receive protection (Interview I#5) and Johannes Mihkelson Centre which has programmes aimed at integrating refugees in society (Interview I#4).

There is active coordination between the various organisations and agencies involved in dealing with displaced people in Estonia. One coordinating body brings together all the government bodies and another brings together all the partners and both meet at least twice a year (Interview I#1). There appears to be considerable overlap in the remits and service provision by the different organisations, so there is scope for the refugee services to be greatly rationalised for more efficiency and better service delivery. This need for better integration has been raised in various contexts, e.g. with respect to the provision of support persons by different organisations (UNHCR 2016).

The political situation in 2019 led to an effective moratorium on further resettlement of displaced people in Estonia and on the development of high-level policy, and, under the resulting low demand for their services, the existing governance structures appeared adequate until recently. Since February 2022, and specifically for war refugees from Ukraine, all relevant information has been organised on a government website (https://kriis.ee/) including in Ukrainian language, and additional reception centres have been set up.

3.3.2 National Systems and Programmes

Prior to the 2015 European migration crisis, Estonia did not have any specific systems or programmes in place to deal with relocation and resettlement of displaced people. Asylum seekers and refugees primarily came from Russian-speaking cultures and were typically individuals and, in particular, single men.

Following the 2015 agreement to resettle people from Southern Europe, the first Estonian national relocation and resettlement programme was put in place which lasted until 2017. The intention was to make this "the best programme in the world"; however, it was designed against a backdrop of very little experience and the peculiarities of the Estonian context (Interview #1). In addition, the new wave of displaced people comprised primarily family groups and represented cultures and languages which were less familiar. The selection of people for resettlement was made by selection missions sent out to the camps in Southern Europe and, when the selected people arrived to Estonia, everything was organised and ready for them (asylum decisions, residence permits, a place to live, a support person to assist them in adapting to their new life, etc.). The general lack of available accommodation opportunities in local governments coupled with an intention to spur integration with local communities and to avoid creating relatively large groups of foreigners led to the dispersion of the displaced people throughout Estonia in single family groups (Interview I#1). This policy of dispersion was later found to be a mistake as it placed resettled people in suboptimal locations, led to their relative isolation and made integration activities and services much more difficult to organise. Subsequently, newly arrived asylum seekers and refugees have been accommodated at the reception centres in Vao and

Vägeva for some months first before being assisted to find more permanent accommodation according to their own needs and preferences and, typically, in the more populated urban centres of Tallinn and Tartu (Interviews I#1, I#3, I#5).

With the 2017–2019 national resettlement programme nearing its end with well below the initially envisaged number of displaced people having been resettled under it (7 out of an anticipated 80), with the significant exodus of people resettled under the initial, 2015–2017, programme and with an effective moratorium on further resettlement, the existing systems looked set to remain in place with little development.

Following the February 2022 Russian invasion of Ukraine, the EU's interior ministers adopted (on 3 March 2022) a temporary protection directive to provide Ukrainian refugees with residence permits in a streamlined process. On 8 March 2022, the Estonian government decided to implement this directive and, at the same time, approved the amendment of a number of laws to clarify the legal basis for Ukrainian refugees to reside and work in Estonia. With these amendments, the temporary protection/residence permit application process for Ukrainian war refugees was reduced to 1 day (Government of Estonia 2022).

3.3.3 Housing

Housing is a major problem for displaced people in Estonia. The lack of social housing in Estonia in general and particularly in the capital, Tallinn, is a matter of concern for disadvantaged and marginalised groups among the Estonian population as well as for displaced people (UNHCR 2016). The slightly better availability of social housing in rural municipalities was linked to the earlier government strategy to resettle refugees and other beneficiaries of protection in rural areas throughout the country where subsidised housing exists (UNHCR 2016). By 2017, this strategy of dispersion had been accepted by the authorities as having failed and, since then, displaced people have been resettled in urban centres (Interview I#1).

3.3.3.1 Temporary Accommodation

A legacy of that strategy has been the location of some of the housing in which displaced people have been accommodated being quite isolated (Interview I#2), and it is unclear whether this contributed to the large number of refugees (from the 2015 to 2017 wave) who subsequently left Estonia (Interview I#1). Both the current accommodation centres in Vao and Vägeva are located in quite remote rural areas, but this is a considerable improvement on the arrangement prior to 2014 when the accommodation centre was located in Illuka near the border with Russia and was even more isolated (Interviews I#5, I#8). Refugees were expected to stay in one of the accommodation centres on their arrival and, after about four months, to find accommodation on the rentals market. In doing so, they were assisted by support staff and the accommodation centre (ERR News 2018).

With the influx of Ukrainian war refugees in large numbers (more than 1000 per day in March 2022), the existing accommodation centres in Vao (with a maximum capacity of 70 people) and Vägeva (maximum capacity 50 people) have no significance and the authorities have resorted to using available hotel accommodation which has largely been filled. A passenger ferry, the MS Isabelle, with capacity to accommodate 2100 people has been chartered by the Estonian state to provide additional temporary accommodation. The Secretary General of the Ministry of Social Affairs was quoted in early April 2022 as stating that: "We can offer dignified accommodation to 25,000–30,000 people without refugee camps" (ERR 2022f). Since the number of refugees received has already exceeded the upper limit quoted (ERR 2022a), it is clear that the current arrangements are already at their limits.

3.3.3.2 Permanent Housing

Refugees have faced considerable challenges when attempting to secure accommodation in the private market. Some of these challenges are bureaucratic—refugees require a formal rental contract whereas many landlords avoid these to avoid paying tax on rentals. Landlords have also been resistant to allowing refugees to use the rental accommodation address when registering in the local population register—which is a further requirement for accessing any social services. Others are discriminatory—refugees are often simply refused as tenants on the grounds that they are refugees. In other cases, landlords are hesitant to rent to tenants who have relatively short-term (one year) residence permits. Buying an apartment is also usually out of the question for refugees as banks will not extend loans to buy apartments to holders of short-term residence permits, and, generally, refugees do not have the required funds to put down a deposit (UNHCR 2016). In addition, most refugees envisage returning home "soon", so long-term financial commitments are unattractive to them.

Consequently, a social housing placement organised by the Ministry of Social Affairs in coordination with a local municipality is often the only available option, and, although, as noted above, refugees may participate in these decisions, the consequences of declining an offer include being allowed only to stay a further 2 months in the temporary accommodation and may result in the refugee having to find their own accommodation and paying all the related costs (UNHCR 2016).

Finally, it is notable that, as of 2016, there were no current plans for the state to build or renovate any further social housing (UNHCR 2016).

Discrimination against displaced people, as well as foreigners in general, by landlords was confirmed in the interviews (Interviews I#2, I#6), both from the point of view that much of the rental market is a black market activity where landlords avoid paying taxes and fear they would be caught if their address is being used as the official residence of a displaced person (Interview I#5) as well as simply not trusting foreigners with their properties (Interview I#2).

The general shortage of social housing was confirmed in the interviews and that this was a factor in the initial strategy of dispersing displaced people throughout Estonia (Interview I#1). However, it was also noted that social housing is often of low quality and may be poorly maintained and that larger blocks of social housing would be likely to also accommodate socially problematic tenants (Interviews I#5 and I#6).

One of the major challenges in terms of finding housing in the 2015–2017 wave which comprised refugees from Syria, Iraq and Afghanistan was the size of displaced people's families which tended to be much larger (up to seven children) than local, Estonian families (usually 1–3 children), and this is reflected in the housing stock (Interviews I#1, I#2, I#5, I#8). For example, the interview with officials of a local government revealed that both the families that had resettled in that municipality had eight family members and, in both cases, have been housed in four-room apartments which are rather small for them (Interview I#9).

The current wave of Ukrainian war refugees is predominantly small family groups of women and children many of whom have come to stay with friends or relatives (Government of Estonia 2022). In addition, the support for refugees among the Estonian population is very high, and additional accommodation is being made available to them by private individuals. While this has eased the overall accommodation problem, it has still strongly driven demand for lower-end rental apartments in Tallinn and other cities so that, by the end of March 2022, rents had increased by 15–20% in just one month (ERR 2022g).

The housing which is available to displaced people does, in some cases, has problems in terms of its quality—thermal insulation and state of repair—but this is not specific to displaced people as the same problems afflict other groups in Estonia to the same extent (Interview I#4).

3.3.4 Sociocultural Context

3.3.4.1 Social Infrastructure

Displaced communities need improved support to access health care (including psychological counselling), school, social care, language tuition, education funding (grants or loans) and childcare for women with children. Until 2022, existing school and healthcare facilities and capacities were adequate for the small numbers of displaced people, but with the large numbers of Ukrainian refugees (35% of them children), this is no longer the case (Cole 2022; ERR 2022a, h).

3.3.4.2 Education

Education is compulsory in Estonia and should be made available for all children. Children of the displaced communities have the same rights to access education as the children of their host communities (UNHCR 2016). Yet, despite considerable efforts to expand schools and absorb children (as well as teachers from the among the displaced communities) into the school and kindergarten systems, applications for places greatly outstrip current capacities and even expected increases in capacity which could be achieved by the start of the new school year in September (ERR 2022a, h).

3.3.4.3 Health

For earlier waves of refugees from further away, there was a need for the provision of necessary diagnostic and further training on knowledge regarding the exotic diseases of foreign regions as admitted by the Estonian Physician Family Physician Association (Estonian National Audit Office 2016). For Ukrainian war refugees, an issue has been that only a third of them are vaccinated against COVID-19 and many have already accessed the healthcare system with general medical concerns and to continue courses of treatment which were left off in Ukraine. However, it is recognised that a major challenge will be to address their mental health needs, and, according to the Head of the Department of Health Management and Continuity at Estonia's Health Board (in March 2022), no plan was yet in place (ERR 2022j). On a positive note, a school psychologist hotline offering free and anonymous support from qualified psychologists in Ukrainian language has already been set up (ERR 2022k).

3.3.4.4 Language

Estonian language training is made available to displaced persons. However, the network of Estonian language learning opportunities and support accessible has been complicated (Interviews I#1, I#6, I#7). In addition, Estonian is considered a difficult language to learn, and, with no utility outside Estonia, there have always been problems with the motivation to learn it. This is particularly true among Russian and English speakers who can quite easily get by without Estonian language as these languages are widely spoken in Estonia. The temporary nature of the residence permits granted to refugees also influences their attitudes and intentions to settle in Estonia, and many refugees simply do not envisage a long-term stay. However, without Estonian language, employment, education and social integration opportunities and access to services are, indeed, limited (Statistics Estonia 2009; UNHCR 2016). The UNHCR (2016) reported a severe shortage of translation services, particularly for less widely spoken languages, leading to the use of online translation services such as Google Translate by service providers and refugees. The Ukrainians and Belarussians are the biggest groups of displaced people in Estonia, and these have no problems with language since they have been able to communicate in Russian which is also widely spoken in Estonia.

Children are usually quick to learn and understand the language once they enrol in schools and adults are supported in learning the Estonian language by NGOs, Töötukassa (Estonian Unemployment Insurance Fund) and other organisations (Interviews I#1, I#6, I#7).

According to press reports, some recently arrived Ukrainian war refugees have already started formal language learning, and the government is organising summer (2022) schools for those wishing to learn Estonian and 10,000 places will be made available (ERR 2022i).

3.3.4.5 Social Services and Support

Access to social services and support as well as information about these is closely tied to Estonian language proficiency, and this has been noted as a barrier for refugees. The role of support persons has been essential in helping refugees become aware of and access social services, but there have been problems in coordinating numerous different support persons from different agencies and NGOs without clear role definitions (UNHCR 2016).

The interviews indicated that many of the organisations involved are offering programmes of support and actively promoting social inclusion and access to social services. However, as well as the language barrier, there are constraints to participation such as where displaced people have been housed in more remote locations and in cases where their work commitments prevent their participation through lack of time (Interview I#4).

3.3.4.6 Religion

While the Estonian constitution protects the freedom to practice one's own religion and prohibits the incitement of religious hatred, violence or discrimination, there are reports of anti-Jewish and anti-Muslim incidents. The Conservative People's Party of Estonia (EKRE) have organised events to protest against immigration and the "Islamization of Europe" (US State Department 2016). Veebel (2015) noted that the then Minister of Social Affairs, Margus Tsahkna, suggested that Estonia should learn from the mistakes of other countries and give priority to Syrian Christians when relocating refugees to Estonia and also argued that wearing the burqa in public should be banned.

Generally, the interviews carried out suggested that religion is not a major factor affecting displaced people in Estonia. There have been a few cases where religion has been cited as a barrier to certain employment opportunities, for example, a need to pray regularly (during working hours) and wear religious garments (a burka) rather than standard work clothes (Interview I#7).

3.3.4.7 Violence and Discrimination

However Estonian law protects against any form of discrimination in the workplace, in terms of other forms of intolerance, it requires that concrete harm to life, health or property has been caused in order to classify and act as incitement to hatred. The Estonian Penal Code does not address hate crimes (Estonian Human Rights Centre 2017; UNHCR 2016). Interview I#5 noted that, in this respect, Estonia was failing to implement an EU framework decision to criminalise racist and xenophobic hate speech and hate crime. The police and prosecutors therefore have an inadequate toolbox to react to such incidents, and very few cases are even considered for prosecution. In addition, if convictions are achieved, these only result in small fine penalties. It was also opined that this was unlikely to change in the near future for political reasons (Interviews I#5).

While draft legislation in relation to amendments to the penal code to address hate speech remains under discussion and public debate (ERR 2021), a related bill which bans symbols of aggression (specifically intended to ban the symbols associated with the Russian invasion of Ukraine) and prescribes punishments for joining an aggressor's armed forces is currently before parliament. The original draft of this bill included a section relating to hate speech, but this section was removed in order to expedite the bill through parliament (ERR 2022l).

The interviews suggested that racism and discrimination continue to be experienced in everyday life. The status of "displaced person" does not seem to make a difference in this case so that all people (visitors, students, refugees, etc.) who appear to be different in appearance to a typical, white, Estonian face similar issues (Interview I#2). Incidents of racism and discrimination in the work place have also been noted—both by employers and also by fellow workers (Interview I#7), but there are also positive developments in that some employers have signed up to a diversity charter in order to combat this and promote a multicultural society (Interview I#5).

3.3.5 Economic Context

3.3.5.1 Employment

Asylum seekers cannot work until they have been granted international protection which takes up to 6 months (Interviews I#1, I#8). Once they have been granted protection, they have the right to work, but there is a general mismatch between their education, qualifications and expectations and the employment that is available to them in Estonia (Interviews I#1, I#2, I#5, I#9). When they find employment, it is typically in relatively menial, low paid work (Interviews I#4, I#5) and the need to put in many hours of work to earn enough has a knock-on effect to their ability to participate in other activities including language learning and training (Interview I#4). There is also the possibility of exploitation by employers where employees feel they will not be able to find another job (Interview I#2).

The general policy on the employment of refugees is to enable them to work as quickly as possible (Valitsus.ee 2019), and, historically, most refugees have found permanent work within nine months from receiving their residence permits (UNHCR 2016). However, statistically, it has been found that immigrants are more likely than natives to be employed in jobs that require a lower level of education than their actual qualifications (Statistics Estonia 2009), but the Estonian Integration Strategy 2008–2013 stated that "*employment and salary differences can be explained through regular aspects not related to ethnicity (education, working in town or in the country, gender) and through the knowledge of the language*" (Estonian Ministry of Culture 2008). However, some refugees have expressed disappointment that the jobs offered to them were not in line with their qualifications and that the salaries offered were too low to meet their financial obligations (UNHCR 2016). It is also notable that the unemployment rate for Estonian citizens in 2015 was 5.6% compared to 9.3% for foreign citizens and that migrant workers are often not aware of their rights and do not know how to access help and support (Estonian Human Rights Centre 2017).

Cultural differences and gender roles also have an effect on employment—it was suggested that often it is the men who go to work first and the women take care of the family. Even if childcare possibilities are available, these are not necessarily taken up (Interviews I#3, I#4).

Although the application process for protection for refugees who fled Ukraine since 24 February 2022 is extremely quick (1 day), it is slower for those who left earlier and the same, historical issues relating to the employment of refugees remain despite many well-intentioned Estonian employers making a special effort to offer work opportunities to Ukrainian refugees. As of the end of April 2022, more than 2000 Ukrainian refugees had already found work in Estonia, but, without stable accommodation and childcare, challenges remain for many of them (ERR 2022n).

There are support programmes in place to encourage employers to employ displaced people such as the "My First Job in Estonia" which provides regular support and subsidises the wages of employees who are beneficiaries of international protection (Interviews I#5, I#9).

Self-employment is also a possibility, and there are examples of refugee-led enterprises in the catering industry in Tallinn and Tartu (Interview I#2). There are no restrictions in terms of owning or managing businesses and refugees have equal access to self-employment.

3.3.5.2 Financial/Income Support

Asylum seekers receive a small allowance from which they must meet all their basic needs, and this has led to some complaints that they struggle to afford food items such as fresh fruit for children and shampoo (Kaldur and Kallas 2011). Beneficiaries of international protection are eligible for the same social benefits and allowances as Estonian nationals and have full access to employment (Estonian National Audit Office 2016).

The level of financial support (as per 2019) is intended to ensure that, after paying all the taxes, housing and so on, a minimum monthly household income of 160€ for the first adult plus 120€ for every further adult and 180€ for every under-aged child would be achieved. In addition, if the family has more than three underage children, then they receive an additional 500€. There are further benefits available, for instance, if a child is less than 1 year and 6 months, then there is parental support and then there is specific family support money for every child in the family (Interview I#3). In the case of a large family, this can disincentivise employment as it gives rise to a situation where it does not make financial sense to go to work as this would simply reduce the support to the extent of the earned salary (Interview I#3).

There appears to be a discrepancy with the financial support granted to refugees as compared with ordinary Estonians. Some respondents reported that it is the same while local government officials indicated that it was considerably and unnecessarily higher (Interview I#9).

Access to banking was cited as a problem, particularly for asylum seekers who are unable to open bank accounts (Interview I#8). Banks are also unwilling to lend to displaced people, for example, to enable them to buy property, since their residence permits are temporary and limited to 3 years (Interview I#1). In this context, it should be noted that Ukrainian war refugees are being issued with 1-year resident permits in the first instance (Government of Estonia 2022), so it can be assumed that these restrictions also affect them.

3.4 Conclusions

Until Russia's large-scale invasion of Ukraine began on 24 February 2022, Estonia was not a popular destination for displaced people, and, even when granted protection, a large proportion then chose to leave Estonia for other countries. However, for Ukrainian war refugees, Estonia has proved to be a popular destination thus far. An established, existing community of Ukrainians including some who had relocated to Estonia after the initial Russian hostilities in 2014, the large number of Ukrainians engaged in seasonal work in the Estonian construction and agricultural sectors, and the genuinely sympathetic and welcoming attitude of most Estonians towards the displaced Ukrainians (including volunteers transporting more than 8000 refugees from the Ukrainian border to Estonia), has seen more than 30,000 Ukrainian war refugees arrive in Estonia so far.

The interview survey conducted in late 2019 defined the Estonian systems, programmes and context with regard to displaced people. The sudden, recent and very large influx of refugees from Ukraine has fundamentally changed Estonia in 1 month from being a country with almost no refugees to one where refugees comprise more than 2% of the total population. Even though the attitudes of most Estonians are far more positive towards Ukrainian war refugees than they have historically been towards refugees from elsewhere, the basic elements of the Estonian system remain unchanged and greatly challenged by the large numbers of refugees. Major challenges are currently faced in finding adequate housing, school, kindergarten and childcare places, employment opportunities, etc., and it is unclear how the social and health (including mental health) service systems will perform.

Acknowledgements This research was supported by the Rebuilding After Disaster (REGARD) project and the Build Resilience in Tropical Agro-Ecosystems (BRITAE) project both co-funded by the Erasmus+ Programme of the European Union. The European Commission support for the production of this publication does not constitute an endorsement of the contents which reflects the views only of the authors, and the commission cannot be held responsible for any use which may be made of the information contained therein.

References

Äripäev (2016) Newspaper article: Eestisse saatmine põgeniku jaoks liiga hirmus, Available at: https://www.aripaev.ee/uudised/2016/05/17/eestisse-saatmine-pogeniku-jaoks-liiga-hirmus. Accessed 20 Apr 2019

Cole M (2022) A friend is better than a phrasebook. Feature article. Estonian Public Broadcasting News Service. Available at: https://news.err.ee/1608547582/feature-a-friend-is-better-than-a-phrasebook. Accessed 20 Apr 2022

Cooney S (2017) 10 things you should know about refugees in Estonia. The Borgen Project blog. Available at: https://borgenproject.org/estonian-refugees/. Accessed 20 Apr 2022

ERR News (2018) Refugees in Estonia given a freer rein in finding their own accommodation. Estonian Public Broadcasting News Service. Available at: https://news.err.ee/834045/refugees-in-estonia-given-a-freer-rein-in-finding-their-own-accommodation. Accessed 20 Apr 2022

ERR News (2021) Rait Maruste, endine riigikohtu esimees ja Euroopa Inimõiguste Kohtu kohtunik (Reformierakond). Estonian Public Broadcasting News Service. Available at: https://www.err.ee/1608126385/rait-maruste-sonavabadus-vaenukone-ja-vihakuriteod. Accessed 20 Apr 2022

ERR News (2022a) PPA: 536 refugees from Ukraine arrive in Estonia on Thursday. Estonian Public Broadcasting News Service. Available at: https://news.err.ee/1608566278/ppa-536-refugees-from-ukraine-arrive-in-estonia-on-thursday. Accessed 20 Apr 2022

ERR News (2022b) Public support for hosting Ukrainian war refugees remains high. Estonian Public Broadcasting News Service. Available at: https://news.err.ee/1608541219/public-support-for-hosting-ukrainian-war-refugees-remains-high. Accessed 20 Apr 2022

ERR News (2022c) 22 half of households in Estonia have donated money to Ukraine. Estonian Public Broadcasting News Service. Available at: https://news.err.ee/1608563503/half-of-households-in-estonia-have-donated-money-to-ukraine. Accessed 20 Apr 2022

ERR News (2022d) Survey: most people support Aliens Act amendments for refugees only. Estonian Public Broadcasting News Service. Available at: https://news.err.ee/1608561163/survey-most-people-support-aliens-act-amendments-for-refugees-only. Accessed 20 Apr 2022

ERR News (2022e) Refugee Council wraps up scheduled evacuation buses from Ukrainian border. Estonian Public Broadcasting News Service. Available at: https://news.err.ee/1608551947/refugee-council-wraps-up-scheduled-evacuation-buses-from-ukrainian-border. Accessed 20 Apr 2022

ERR News (2022f) Estonia far off from needing to build refugee camps, says ministry official. Estonian Public Broadcasting News Service. Available at: https://news.err.ee/1608544177/estonia-far-off-from-needing-to-build-refugee-camps-says-ministry-official. Accessed 20 Apr 2022

ERR News (2022g) Real estate expert: rental prices surge due to Ukrainian refugee demand. Estonian Public Broadcasting News Service. Available at: https://news.err.ee/1608546586/real-estate-expert-rental-prices-surge-due-to-ukrainian-refugee-demand. Accessed 20 Apr 2022

ERR News (2022h) Tallinn to open a school for refugees on Räägu tänav. Estonian Public Broadcasting News Service. Available at: https://news.err.ee/1608556399/tallinn-to-open-a-school-for-refugees-on-raagu-tanav. Accessed 20 Apr 2022

ERR News (2022i) Estonia to open 10,000 summer language school places for Ukrainians. Estonian Public Broadcasting News Service. Available at: https://news.err.ee/1608552637/estonia-to-open-10-000-summer-language-school-places-for-ukrainians. Accessed 20 Apr 2022

ERR News (2022j) Health official: no set plan for handling refugees' mental health needs. Estonian Public Broadcasting News Service. Available at: https://news.err.ee/1608541585/health-official-no-set-plan-for-handling-refugees-mental-health-needs. Accessed 20 Apr 2022

ERR News (2022k) School psychologist hotline offering support in Ukrainian on Wednesdays. Estonian Public Broadcasting News Service. Available at: https://news.err.ee/1608541078/school-psychologist-hotline-offering-support-in-ukrainian-on-wednesdays. Accessed 20 Apr 2022

ERR News (2022l) Hate speech section taken out of symbols of aggression bill. Estonian Public Broadcasting News Service. Available at: https://news.err.ee/1608558940/hate-speech-section-taken-out-of-symbols-of-aggression-bill. Accessed 20 Apr 2022

ERR News (2022m) Estonia to amend State Borders Act to better contain illegal migration. Estonian Public Broadcasting News Service. Available at: https://news.err.ee/1608555880/estonia-to-amend-state-borders-act-to-better-contain-illegal-migration. Accessed 20 Apr 2022

ERR News (2022n) Ukrainian refugees struggling to find housing, childcare in Estonia. Estonian Public Broadcasting News Service. Available at: https://news.err.ee/1608579902/ukrainian-refugees-struggling-to-find-housing-childcare-in-estonia. Accessed 20 Apr 2022

Estonian Human Rights Centre (2017) Racism, racial discrimination and migration in Estonia 2015–2016. Estonian Human Rights Centre. Available at: https://humanrights.ee/en/2017/07/rassism-rassiline-diskrimineerimine-ja-migratsioon-eestis-2015-2016/. Accessed 20 Apr 2022

Estonian Ministry of Culture (2008) Estonian integration strategy 2008–2013. Available at: https://www.google.com/url?sa=t&rct=j&q=&esrc=s&source=web&cd=1&cad=rja&uact=8&ved=2ahUKEwi1-MPOzOXhAhXxxMQBHSoABI0QFjAAegQIABAC&url=https%3A%2F%2Fec.europa.eu%2Fmigrant-integration%2F%3Faction%3Dmedia.download%26uuid%3D29E25CE3-E42E-6E12-E0D82B731A093A9E&usg=AOvVaw0DH6s79U8tInh8ZbjQmoeF. Accessed 20 Apr 2022

Estonian National Audit Office (2016) Capability of the state and local governments to accept people who are requesting or have received international protection—is the state capable of meeting the commitments it has imposed on itself with legal instruments? Available at: https://www.riigikontroll.ee/DesktopModules/DigiDetail/FileDownloader.aspx?AuditId=2379&FileId=13589. Accessed 20 Apr 2022

Eurostat (2022) Eurostat database. Available at: https://ec.europa.eu/eurostat/data/database. Accessed 20 Apr 2022

Government of Estonia (2022) Estonia to simplify process of taking in Ukrainian refugees. Government of the Republic of Estonia press release of 08.03.2022. Available at: https://www.valitsus.ee/en/news/estonia-simplify-process-taking-ukrainian-refugees. Accessed 20 Apr 2022

Kaldur K, Kallas K (2011) Varjupaigataotlejate rahulolu ja ootused neile pakutavate teenuste ja vastuvõtutingimuste osas. Küsitluse raport. Institute of Baltic Studies, Tartu, Estonia. Available at: https://www.ibs.ee/wp-content/uploads/IBS_VPT_intervjuude_raport.pdf. Accessed 20 Apr 2022

Lauren A (2016) Annual policy report on migration and asylum Estonia 2016. European Migration Network. Available at: http://emn.ee/wp-content/uploads/2016/10/apr.sisu_.pdf. Accessed 20 Apr 2022

Luuk L (2016) Migration crisis in the EU: a new threat Estonia? Master's thesis, Tallinn University of Technology

Statistics Estonia (2009) Immigrant population in Estonia. Statistics Estonia, Tallinn. Available at: https://www.stat.ee/publication-download-pdf?publication_id=18391. Accessed 20 Apr 2022

Statistics Estonia (2022) Estonian national statistics database. Available at: https://www.stat.ee. Accessed 20 Apr 2022

UNHCR (2016) Integration of refugees in Estonia—participation and empowerment. UNHCR Regional Representation for Northern Europe, Stockholm. Available at: https://www.valitsus.ee/sites/default/files/content-editors/failid/unhcr-print_version_estonia-integration_mapping.pdf. Accessed 20 Apr 2022

US State Department (2016) Estonia 2016 International Religious Freedom Report, United States Federal Government State Department, Available at: https://www.state.gov/documents/organization/269054.pdf. Accessed 20 Apr 2019

Vaitmaa E (2015) Veebiuuring. Üle 80% vastanutest ei taha pagulasi Eestisse. – Eesti Päevaleht. 26.06.2015. Available at: https://epl.delfi.ee/news/eesti/hirm-pagulaste-ees?id=71780185. Accessed 20 Apr 2019

Valitsus.ee (2019) Pagulasküsimus, Estonian government website - webpages on refugee information, last updated 9 April 2019, Available at: https://www.valitsus.ee/et/pagulased. Accessed 20 Apr 2019

Veebel V (2015) Balancing Between Solidarity and Responsibility: Estonia in the EU Refugee Crisis. Journal on Baltic Security 1(2):28–61. https://doi.org/10.1515/jobs-2016-0020

Zabrodskaja A (2009) Language testing in the context of citizenship and asylum: The case of Estonia. Language Assessment Quarterly, 6(1):61–70

Chapter 4
Challenges of Resilience Building Among Traditional Agricultural Communities Displaced by the Landslides

B. M. R. L. Basnayake, D. A. M. De Silva, and S. K. Gunatilake

Abstract Landslides are controversial issues worldwide and cause wide range of socioeconomic issues. The present study was conducted to examine the approaches and challenges faced by traditional agricultural communities in re-building landslide-depredated mountain landscape. Primary data were obtained through field observations, in-depth interviews with key stakeholders of rehabilitation program, and storytelling exercise with the participation of disaster victims of Aranayaka mountain landslide, Sri Lanka. Land use maps of the area highlighted that more than 50% of land utilized for agricultural purposes. Paddy, tea, spices, and Kandyan home gardens were identified as prominent livelihood options. Agriculture operations assured household food and income security. Landslide of 2016 destroyed the entire system, and generations-old farming communities were displaced without any assurance for living. Improper resettlement plans were unable to address the issues of housing, livelihood, and food and income security of displaced people. Unrest, conflicts, trauma, and poverty shock made communities more vulnerable. Employment options shifted from owners of farmlands to daily wage laborers, and most of the females migrated to Middle East as domestic helpers. Land access and resettlement programs fall short of community expectations, with negative impacts on livelihoods of displaced farming community and absence of meaningful involvement by communities in decision-making.

Keywords Farming communities · Landslides · Resilience building · Sri Lanka

B. M. R. L. Basnayake (✉) · D. A. M. De Silva
Department of Agribusiness Management, Faculty of Agricultural Sciences, Sabaragamuwa University of Sri Lanka, P.O. Box 02, Belihuloya, Sri Lanka
e-mail: ruwinibasnayake@agri.sab.ac.lk

S. K. Gunatilake
Department of Natural Resources, Faculty of Applied Sciences, Sabaragamuwa University of Sri Lanka, P.O. Box 02, Belihuloya, Sri Lanka

© The Author(s), under exclusive license to Springer Nature Switzerland AG 2023
M. Hamza et al. (eds.), *Rebuilding Communities After Displacement*,
https://doi.org/10.1007/978-3-031-21414-1_4

4.1 Introduction

A disaster can be defined as a result of the combination of hazard, vulnerability, emergency, and risk which impacting both individually and communally (Moe and Pathranarakul 2006). Among natural disasters, landslide is the most frequent type of natural disasters mainly caused due to soil cover, slope stability, hydrology, drainage, precipitation, and landform as well as anthropogenically induced factors such as land use and improper management (Cruickshank 1995).

Displacement can be defined as community who have been forced to leave their own built homes or places and lands of habitual residence, in order to avoid the effects of disasters or environmental degradation (Ocha 1998). Based on the Global Report on Internal Displacement (2021), there were 7 million internally displaced people across the world at the end of 2020 as a result of natural disasters. Commonly occurring natural disasters are floods, storms, earthquakes, volcanic eruptions, wildfires, landslides, and extreme temperatures. Landslide-induced displacement is an unexpected tragedy as well as a major development obstacle which only can be overcome by a well-established proper functional mechanism. Since the displacement is a life-changing event, the attention should be focused to address all aspects of restoration of lives such as building houses, establishing livelihood opportunities, rehabilitation, and reintegration in the process of re-location (Sangasumana 2018). Commonly displaced people are often from rural areas where their primary income source and ways of living were agricultural-related (George and Adelaja 2021) and political factors, and socioeconomic factors determine the vulnerability and resilience of people to the effects of climate change and displacement (Thomas et al. 2019). Managing displaced agricultural communities are quite challengeable when their agricultural land is lost for further cultivation, and they did not receive enough natural resources. Further, it will cause disintegration of social support networks, increased risk of structural discrimination and social, economic, and natural shocks (Opitz Stapleton et al. 2017).

In the recent past, Sri Lanka has experienced several significant natural disasters which completely crippled the rural livelihoods and regional economy. Among the recorded natural disasters, landslides were the most significant natural hazards in Sri Lanka that seconds only to floods and cyclones (Ratnayake and Herath 2005). As shown in Fig. 4.1, the frequency of occurrence and consequences of landslides were in very high level comparing to the other types of natural hazards. Essential relief works, managing conflicts, settle back to livelihood, and need of repairs to services are the major socioeconomic effects of these disasters (Institutional Disaster Management Plan 2021).

Moreover, central highlands of the country, and most of the districts Badulla, Nuwara Eliya, Kegalle, Ratnapura, Kandy, and Matale are highly susceptible for landslide hazards and recent data shows that during the period from 2003 to 2017 there was a sudden increase in the occurrence of landslides mainly due to unplanned development programs, environmental degradation, human intervention, and effects of climate change (Sangasumana 2018). As shown in Fig. 4.2 which is developed

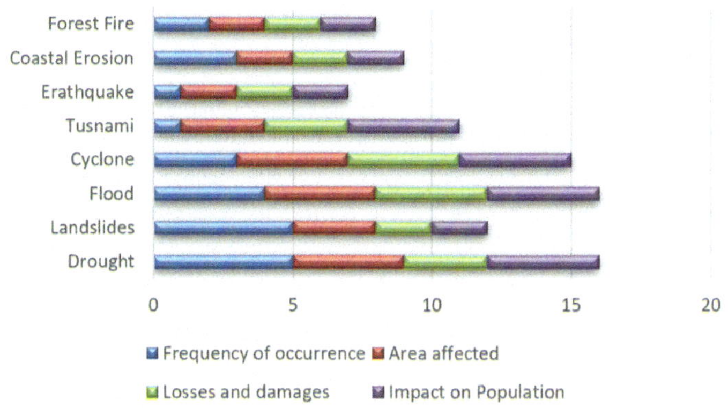

Fig. 4.1 Frequency and consequence of natural hazards in Sri Lanka. *Source* Desinventar, NDMP—2013–2017 (5 = extremely high, 4 = very high, 3 = high, 2 = significant, 1 = rare)

from the secondary data of Disaster Management Centre (2022) from 1977 to 2020, the damage of landslide incidents in Kegalle district is having rapid changes due to improper land use pattern.

In May 2016, a calamitous landslide occurred in Aranayaka divisional secretariat in Kegalle district which ultimately affected 340,150 inhabitants and 84 deaths were recorded (Institutional Disaster Management Plan 2021). Aranayaka, the location of landslide, was rich in natural resources, traditional agricultural landscapes enriched with Kandyan home gardens, and agroforestry system. Rich agro-biodiversity of

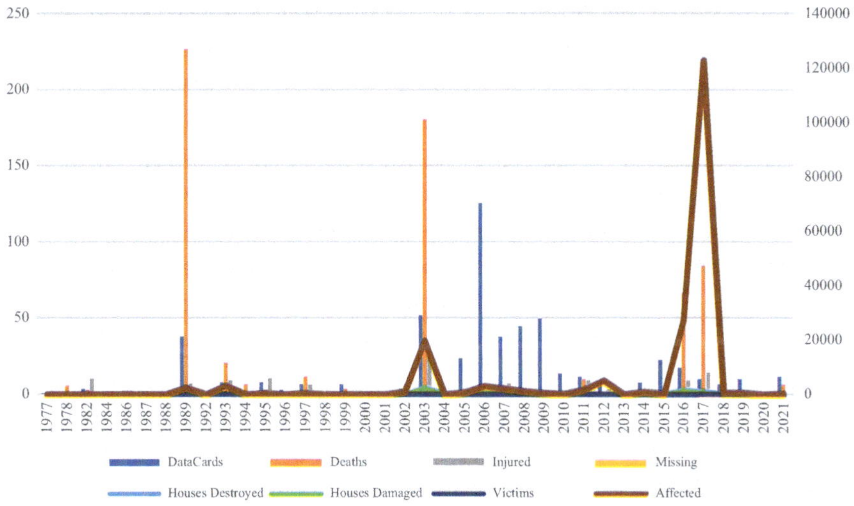

Fig. 4.2 Timeline of landslide incidents Kegalle district, Sri Lanka. *Source* Disaster Management Centre, Sri Lanka

Table 4.1 Crop diversification of typical Kandyan home garden system

Crop type	Example
Vegetables	Tomato, brinjal, leafy vegetables, bitter gourd, snake gourd, luffa, cucumber, capsicum, beans, pumpkin, wing beans
Fruits	Banana, mango, papaya, pineapple, guava, avocado, water melon Durian, passion fruit, pomegranate, mangosteen, jack fruit, mandarin Bael, ambarella
Spices	Cardamom, black pepper, clove, turmeric, ginger, mace, nutmeg, coriander, curry leaf, rampe
Roots and tuber crops	Cassava, sweet potato, Kiriala, Innala
Timber	Mahogany, teak, jackfruit, Alstonia, Albizia, berry, Filicium
Cash crops	Coffee, cocoa, vanilla, arecanut, coconut
Medicinal plants	Ranawara, aloe vera, thebu, *Cyperus rotundus*, thippili, pawatta, kothala himbutu, ginger

Source Bélair et al. (2010)

the area significantly contributed to rural food and income security, culture, traditions, and well-being of the people. This traditional Kandyan home garden, also called as forest gardens, consists of agroforestry ecosystems which fully covering the environmental, economic, and social livelihood needs.

The population of Aranayaka is approximately 68,464, and most of them were engaged in home gardens and agriculture plantations. Although recorded household income level was low, non-accounted but naturally provided resources contribute much on food security of the people assuring income security. Backyard production, home gardens, wild collected food, etc., were hidden income sources to the people. Further, typical Kandyan home garden system was there that consists of annual and perennial crops, such as tea, rubber, paddy, cardamom, black paper, jackfruit, fruit crops, vegetables, coconut, coffee, and cocoa as shown in Table 4.1.

Most Kandyan home garden production is for their own consumption; however, any marketable surplus can provide a safeguard against future crop failures and also additional extra income. Kandyan home gardens provide a considerable proportion of household income (30–50%), mostly through the production of timber, minor-export crops, and cash crops (Siriwardana et al. 2018). The crops are not grown according to any specific pattern and appear to be in a random, intimately mixed pattern (Perera et al. 2018). Moreover, as per Thamilini et al. (2019) study mainly because of the higher agro-biodiversity in these home garden systems, several essential nutrients including most of micronutrients are provided by the production in the daily diet of the household, enhancing the dietary diversity of the rural household. Figure 4.3 illustrates the land use pattern of traditional Kandyan home garden system that clearly shows the arrangements and layering of cultivation around the home.

The Aranayaka landslide incident collapsed the whole traditional agricultural system, their alternative income source, and their sustainable livelihood career. Because of improper resettlement planning, majority of them have to resettle in

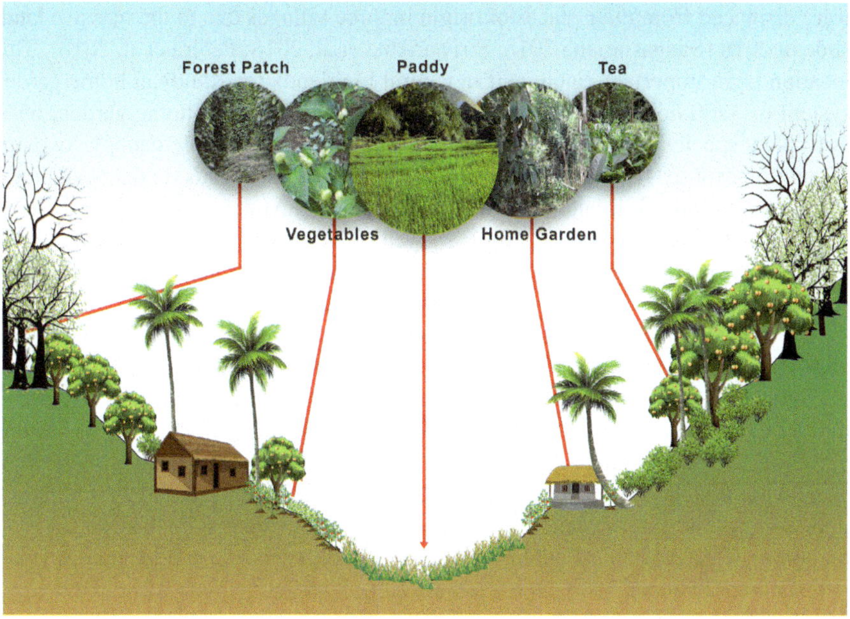

Fig. 4.3 Graphical representation of the typical land use pattern in the study area

duplex flats and in limited land space which is more challengeable to the rural traditional agricultural communities.

Even though landslides have been predominantly studied in Sri Lanka in the category of natural phenomena, highlighting geological, geomorphologic, and meteorological attributes, there is little or no attempt on seeking out the damage that occurred to the traditional agricultural systems and challenges faced by traditional agricultural communities in landslide-induced internal displacement. Therefore, the study mainly aimed to examine the approaches and challenges faced by traditional agricultural communities in the re-building landslide-depredated mountain landscape.

4.2 Method

Study was conducted in Aranayaka divisional secretariat, Kegalle district of Sabaragamuwa province, Sri Lanka. (Elevation: 200 m up to 1000 m altitude; Annual Rainfall: 2000–4000 mm; Mean annual temperature: 24 and 26 °C, RH: 65–80 (day) and 75–90 (night).) Further, the Aranayaka area is located in the wet zone of the country, and it receives heavy rains during the rainy periods (May–September, southwest monsoon; October–November, inter-monsoon) and bright sunshine during the dry season (March–December) (Thomas et al. 2019). Approximately 200 families were displaced from their places of origin in three villages due to the massive landslide in 2016 (Sangasumana 2018; Siriwardana et al. 2018; Perera et al. 2018). This location is an important catchment in central highlands for Kandyan home garden system of 3260 home gardens in 4795 ha, 37–143 species per home garden, 654–5663 plant species density per ha (including annuals), and 45–98% canopy coverage (Thamilini et al. 2019; Gunarathna et al. 2018) and provided lands for tea plantations, rubber plantations, farmlands, and human settlements. Moreover, the study location was selected purposely and priority has given to the severity of the incident and the damage for the traditional agricultural lands. Figure 4.4 land use map illustrates the study location.

The study was conducted as two-way approach. First, conceptualization initiated with the expert consultation with academics, researchers, responsible administrative officers, and community leaders to identify the problems and issues in agricultural communities. Then based on the information collected, field survey guided by questionnaires was prepared with key informant interviews and focus group discussions were conducted with vulnerable parties. Several exploratory field investigations were done in relocated areas to grasp the overall view. The sample size was randomly selected from 45 affected farm households who were relocated in Panapurewaththa, Pallepanguwa, Narangammana, Bakulwala, Dippitiya, and Ussapitiya areas of Kegalle district. Five focus group discussions were conducted with the participation of re-settled families, and 03 key informant interviews were conducted with grass root level government officers; Grama Niladhari, Agricultural Research and production Assistant, Samurdhi development officer, and Economic development officer in the study area to obtain the data on impact of the landslide on agricultural lands, involvement of the institutions and resettlement planning priorities. Secondary data was also obtained to estimate the damage, mainly from relevant documents published by the Disaster Management Centre (DMC), National Building and Research Organization (NBRO), Kegalle District Secretariat, and Aranayaka Divisional Secretariat.

The study used both qualitative and quantitative data. The quantitative data was processed through MS Excel 2019 and SPSS 26.0 version. Regression analysis and frequency analysis were used to analyze the collected data. The area map was developed to highlight the landslide area, commercial crop growing areas, agricultural lands, forest reserves, water bodies, and houses using ArcGIS (version 10.4). Land use maps of Sabaragamuwa province obtained from Department of Land Use Policy Planning were the primary sources for developing the maps.

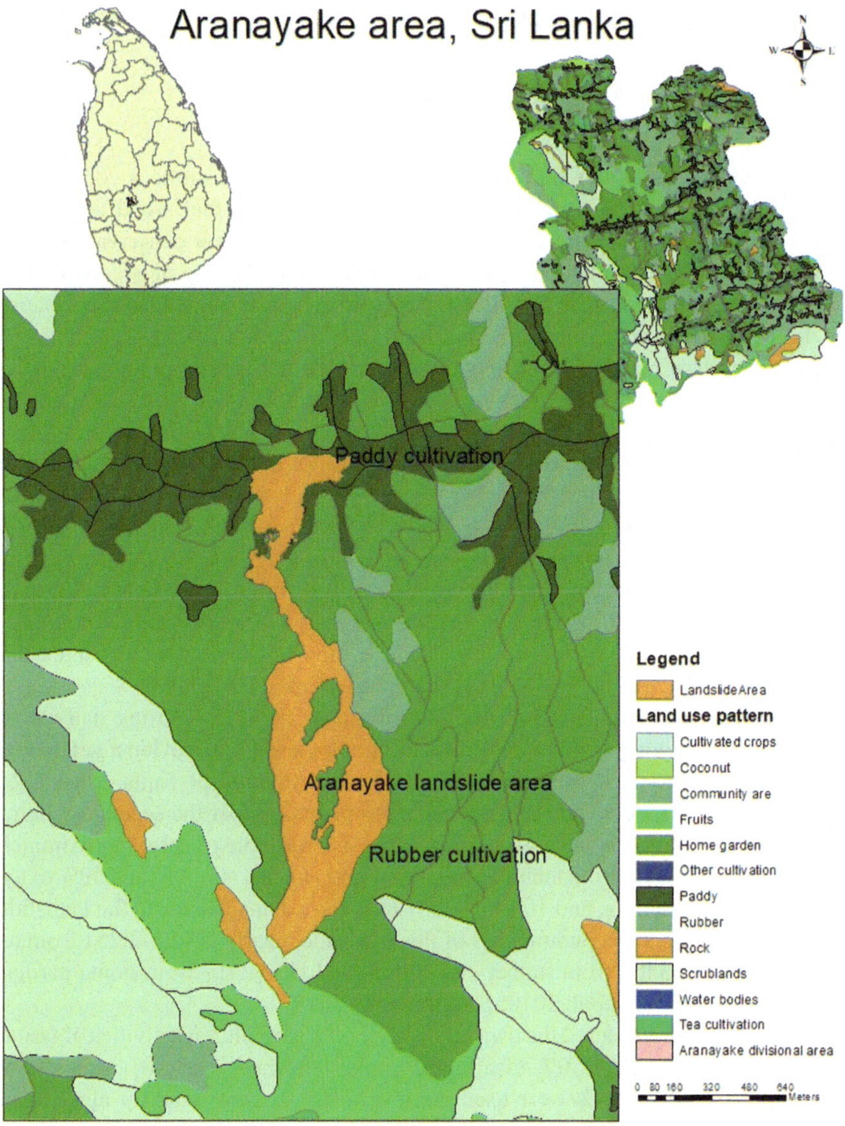

Fig. 4.4 Agricultural land usage map of the study area after the landslide incident

4.3 Results and Discussion

4.3.1 Impact of Agricultural Practices

In Sri Lanka, smallholder farms such as home gardens constitute a considerable amount of country's total annual crop, herbal medicines, fuel wood and timber production, spices and condiments, shade, natural beauty, soil conservation, meal diversity, income, raw materials for cottage industry, and many more. Despite Sri Lankan home gardens being considered desirable and sustainable land use systems, by directly contributing to food and nutrition security and indirectly contributing to improve the rural household income and thereby the purchasing power that makes people live in these home gardens, they have more access to food, and financial resources (Bélair et al. 2010; Gunatilleke 1991).

The study revealed that as per Table 4.2, 100% of the sample were engaged in agricultural activities before the landslide and everybody had their own home garden to fulfill the daily food consumption as well as earn extra income. Majority of the poor communities around the world live in rural areas and a large share of them depend on agricultural practices as their main income source (Jayathilake and Hennayake 2018; Mattsson et al. 2018; Promper et al. 2014). Aliening with global rural community, majority of rural people who live in Aranayaka, landslide area, was also dependent on their backyard or Kandyan home garden production.

These are mainly managed through family labor rather than using daily wage labors with locally appropriate indigenous technologies. They provide a year-round wide range of products such as staple foods, many kinds of fruits, vegetables, spices, firewood, fodder, timber, and medicinal plants. As per the secondary literature sources, this area which is shown in Table 4.3 was able to earn approximately US $160,000 annually from their home gardens and plantations crops and 59% of tea, 13% of home gardening, and 10% of paddy land were damaged due to the landslide. When considering the economic loss of this landslide, nearly US $142,251 from tea cultivation, US $3500 from rubber cultivation, and US $2417 from home gardens were recorded (Perera et al. 2018; Sangasumana 2018).

And when we considered the own land size for the agricultural activities of farmer communities, more than 50% of affected community had nearly 01 acre of land and same as nearly 44.4% were used more than 0.5–01 acres land for agricultural practices. Figures 4.5 and 4.6 illustrate the own land size and utilized agricultural land size before the landslide incident. However, after establishment of resettlements, as per Fig. 4.7 they only received 0.06–0.09-acres land size for both living

Table 4.2 Types of crops grown before the landslide

Home garden	Plantation	Paddy	Spices	Vegetables and fruits
100%	61.10%	41%	100%	44.40%

Source Field Survey, 2021

Table 4.3 Economic value of affected agricultural land area

Crop type	Affected area (acres)	Economic value (LKR)
Tea	83.2	44,941,481
Rubber	25.2	1,135,324
Home garden	17.8	763,634
Paddy	14.3	501,623
Total	140.68	47,342,061.55

Source Perera et al. (2018)

and agricultural purposes. It is clear that even the majority of them had more than 30 years of experiences on agricultural activities, they have to give up generations-old farming practices overnight. Moreover, limited public land availability for establishing resettlements negatively affected relocating farming communities. In general, re-settlers never receive similar piece of land which they can practice their agriculture. Limited availability of land and lack of other resources badly affected the sustainability of resettlements and unrest among the people hinder the success of re-building communities.

Descriptive statistics revealed that the average income level of Aranayaka farmer community was between 160$ and 300$ and their average expenditure was 103$. And it shows that they had very lower expenditure rate because of sustainability of their farming and they could manage to find nutritious daily food intake easily from their backyard production. However, after the landslide incident they had to spend 16-month period in temporary settled camps by compromising their needs and give up or limit their usual agricultural practices. Thus, there should be proper socioeconomic development planning specially for overcome or at least to manage this issue.

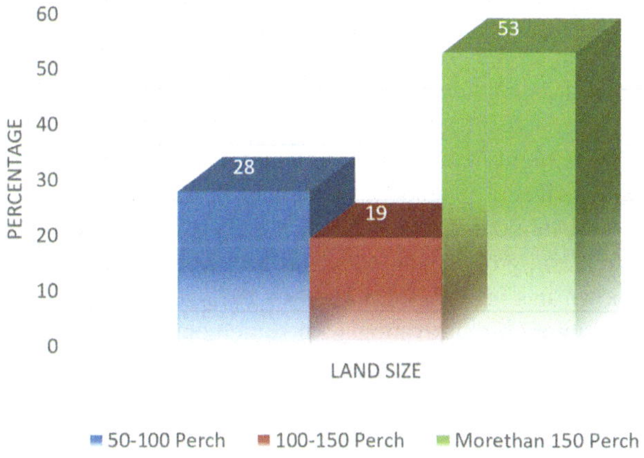

Fig. 4.5 Own land size before the landslide incident. *Source* Field Survey, 2021

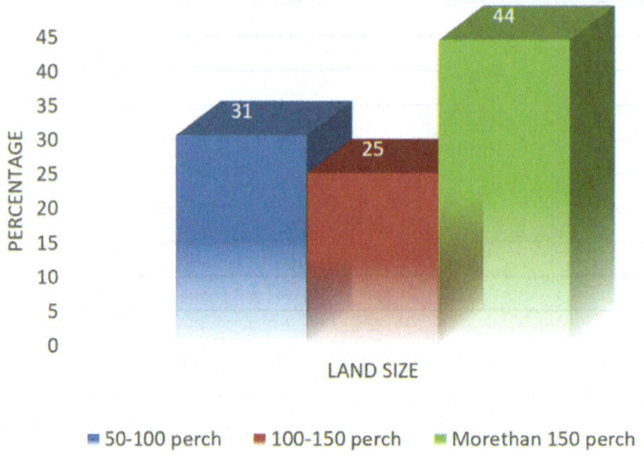

Fig. 4.6 Utilized agricultural land size before the landslide incident. *Source* Field Survey, 2021

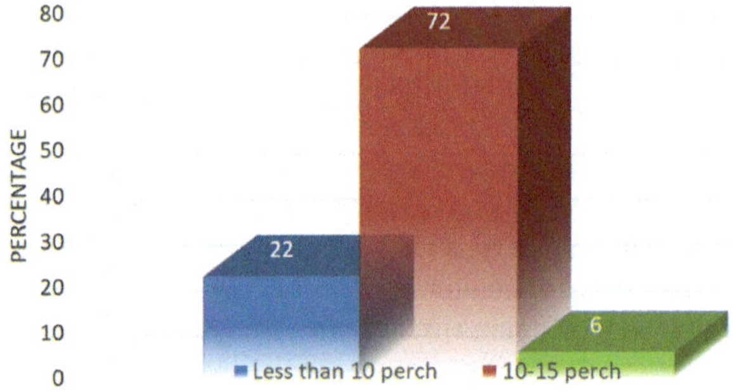

Fig. 4.7 Received land size after the landslide's incident. *Source* Field Survey, 2021

As per the regression model ($Y = 3987 + 849x$), their monthly average income earned from Kandyan home garden (KHG) has a significant contribution from the land size of KHG ($R^2 = 0.586$, $p < 0.01$) (Fig. 4.8), which means the affected farmer community was highly engaged in agricultural activities which had higher productivity and they had self-dependent sustainable system. Further, this could be caused to activate landslide due to intensive farming without having awareness on soil type, drainage pattern, ground water, and surface water levels.

Moreover, as shown in Fig. 4.9, from 1943 to 2012 period drastically the land use was changed for intensive cultivation with settlement expansion. Surrounding natural forests and shrub areas were gradually converted into small-medium scale tea plots, rubber cultivation, and home gardens by the people (Abeysinghe et al. 2004). Specially, the conversion of forest into tea lands was highly caused for this unfortunate landslide incident, these medium scale tea holders were not much considered on managing proper drainage systems and soil conservation practices, and they only

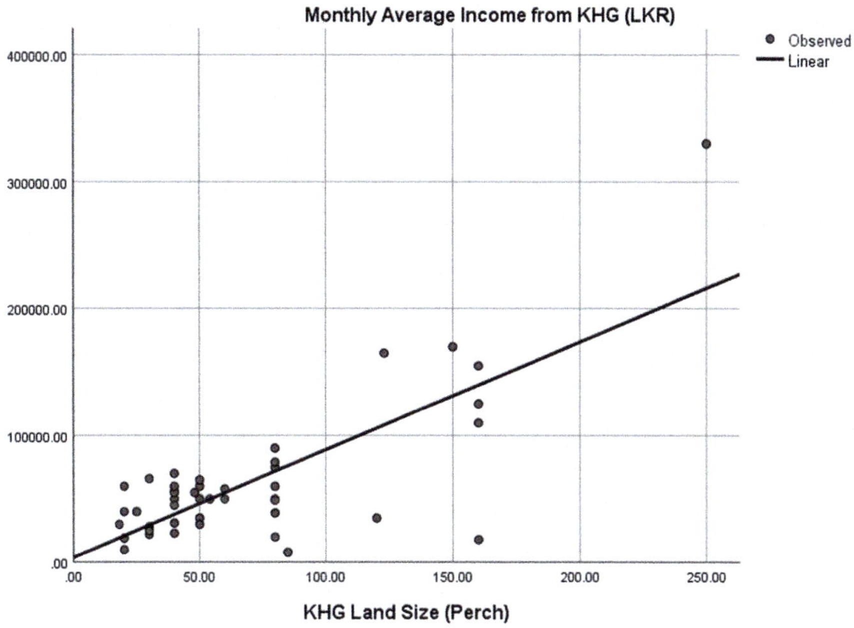

Fig. 4.8 Liner relationship between monthly average income and KHG land size. *Source* Field Survey, 2021

focused on final harvest again caused for triggering the landslide and resulted in losing their traditional agricultural system. Here, the government officers and extension officers also have a big role to play by educating and aware of this rural community.

Further, according to field survey, majority of respondents are having only up to primary-O/L education and 39% are engaging in agricultural activities more than 30 years in agriculture sector which is shown in Table 4.4. It could be caused for less awareness on prevailing measures and more focusing on economic gain

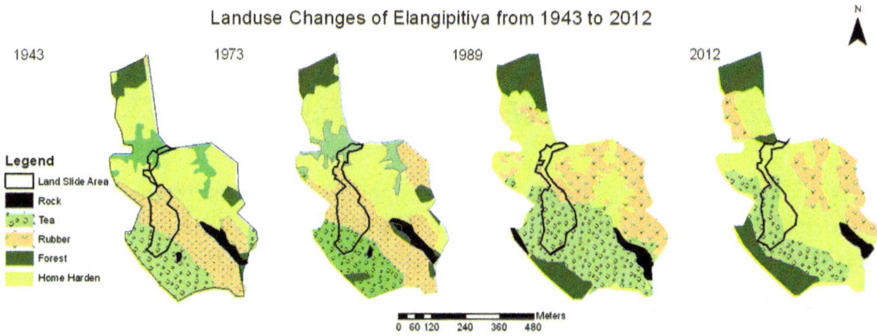

Fig. 4.9 Land use changes in Aranyaka landslide area from 1943 to 2012 period

Table 4.4 Education level and experience in agriculture sector of the respondents

Education level	Percentage	Year of experience	Percentage
Primary	31.1	10–15 years	27.8
Up to O/L	33.3	15–20 years	13.9
Up to A/L	28.9	20–25 years	8.3
Higher education	6.7	25–30 years	11.1
		More than 30 years	38.9

Source Field Survey, 2021

Table 4.5 Reasons for give-up from agriculture

Reasons	Respond (%)
Loss of the agricultural land due to landslide	34.30
Economic issues	8.60
Since you did not receive land for agricultural activities after the landslides	37.10
Due to distance for the agricultural land	40
Other	17.10

Source Field Survey, 2021

from it. Higher focus on the profit maximizing caused a neglect to the managing of sustainability of the environment and holistic land management eventually increasing the slope instability which made higher landslide vulnerability of Aranayaka area. Change in land cover due to human activities is one of the crucial factors impacting not only the natural process "landslide" and thus the hazard, but also the spatial distribution of elements at risk (Rathnaweera and Nawagamuwa 2013).

As per Table 4.5, at the current scenario, even after 05 years of the incident, 57% are not engaging in the agriculture, mainly because of loss of land due to landslide and distance to the agricultural lands from resettlements. Further, majority of them are engaging in non-agricultural-based jobs (40%), daily wage labors (31.4%), and own business (17%) after the landslide, and females are working as migrant workers.

4.3.2 Recognized Approaches and Challenges

As per collected data, resettlements were designed without concerning the lifestyle of the rural agricultural communities and livelihood programs were mismatched with their skills, talents, and experiences. As shown in the following Fig. 4.10, participants of the resettlement planning consist of government institutes, private sector, and NGOs except the vulnerable parties and villagers from resettlement areas. They have identified that homelessness, joblessness, and landlessness are main key issues of the internally displaced people (IDP). Further with the higher demand of food with no income, food prices were high, and due to the landslide incident, the value of the land and productivity of agriculture land were less.

As immediate solutions, the government was provided up to USD $4443 as subsidies to construct the place of residence and USD $1111 to purchase suitable land. Further, some of the vulnerable parties are residents in an apartment which was fully constructed by government. However, in this plan no lands were officially allocated for agricultural practices. Disaster recovery is crippled by improper recovery

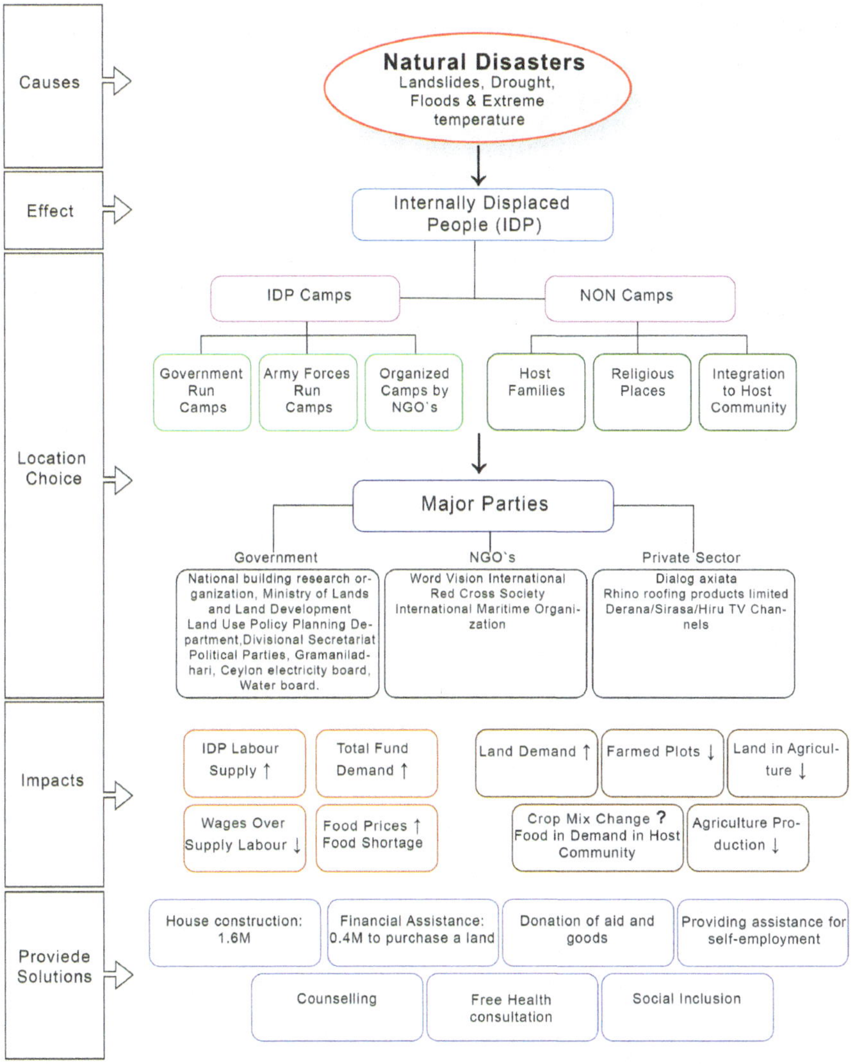

Fig. 4.10 Impact of displacement on agriculture host communities (adopted by George and Adelaja 2021)

plans, perceptions and actions of host communities, poor financing mechanism to establish resettlements as well as desires of the affected community. Trauma, social negligence, heavy dependency on government support, and abandoned generations-old agricultural livelihood made communities violate away from resilience-building programs.

As per Table 4.6, they are not in high satisfied level due to improper resettlement programs. Before the landslide incident, they had enough space, adequate land size, and self-determined house designing according to their willingness. They were complained that lot of issues in land allocation. They said that Grama Niladhari and officers from divisional secretariat register all the displaced people who were in temporary camps and searching for land, house designing and preparing the list of re-locations were made only with the participation of government officers, and then, they have granted lands/substitutes based on the preferences of the vulnerable parties. Further, the land distribution process was carried out by government officers as follows: ask vulnerable parties to select the re-location from the list as per their preferences, and then, from the location, the specific land plot was selected by a draw lots without considering the needs of vulnerable parties/living conditions/opportunities for farming activities. One of the main socioeconomic issues commonly they faced is inability to adapt to the new environment, and it has led to increase the anxiety and hopelessness in seeking new livelihood opportunities.

In the relocated areas, they are having issues such as proper garbage disposing, not having enough space to re-build their backyard production, water leaks due to poor construction, in some relocated areas again announced as vulnerable areas for landslides, etc. Currently, the government has given the approval for engaging in the agricultural activities in the affected area, and due to long distance, stealing the harvest by thieves is the main barrier to re-build their farming life. Further, people

Table 4.6 Satisfactory level of vulnerable agricultural community

Criteria	Satisfactory level	Reasons
Land selection	I	Long distance and again in high-risk areas
Land location	I	Long distance and issues in ownership
House design	II	Lack of space
Land extent	I	Lack of space, no community involvement
Reputation and social relationships	I	Unemployability and separation from neighbors
Education opportunities and facilities	II	Transport issues
Livelihood support	II	Support to re-start their usual life, lacking with durable solutions

High-III, moderate-II, and low-I

who are living in apartment houses also use the free space area for cultivating essential food crops to manage their daily consumption.

Proper awareness and continuous observation on soil conservation practices specially in tea lands, educating the farmers on impact of intensive farming, applying participatory resettlement planning method with the participation of all the possible stakeholders including vulnerable parties, villagers, government officers, farmer societies, private sector, NGOs, etc., and providing equal rights in decision-making and come up with common and sustainable decisions by considering and assessing actual damage for agricultural lands are important in future resettlement planning of displaced communities.

4.4 Conclusion

Typical Kandyan home gardens, Paddy, small holder tea, and rubber were the backbone of the rural economy which disaster devastated, and they were sustained with backyard production and agroforestry system which provided main income source of majority of villagers while ensuring the food and nutritious security. They earned average income level 160$–300$ with very lower expenditure rate further, and their monthly average income earned from Kandyan home garden (KHG) has a significant contribution from the land size of KHG ($R^2 = 0.586$, $p < 0.01$). Intensive farming and drastic changes of land use induced the tragedy incident in 2016, and the unexpected landslide buried houses and properties which resulting in displacement from their own lands. And they had to spend 16-month period in temporary settled camps by compromising their needs. Even though the majority of them had more than 30 years of experiences in agricultural activities, they gave up on it mainly due to they do not receive proper land area to continue their agricultural activities. They have changed their employability as daily wage workers, migrants, or non-agricultural-based jobs. It badly affected their income level as well as social well-being. Trauma, social negligence, heavy dependency on government support, and abandoned generations-old agricultural livelihood made communities violate away from resilience-building programs. As per collected data, resettlements were designed without concerning the lifestyle of the rural agricultural communities and livelihood programs were mismatch with their skills, talents, and experiences. Providing equal rights in decision-making and coming up with common and sustainable decisions by considering and assessing actual damage for agricultural lands, restoration, and regeneration through participatory community-based approaches with the participation of accountable officers and relief efforts should emphasize on low-cost appropriate technology which is essential to mitigate the challenges in resilience building among traditional agricultural communities who are displaced.

References

Abeysinghe B, Iwao Y, Bandara S (2004) Development of a risk assessment methodology for landslides in Sri Lanka. J Jpn Landslide Soc 41(2):131–141

Bélair C, Ichikawa K, Wong BYL, Mulongoy KJ (2010) Sustainable use of biological diversity in socio-ecological production landscapes. Background to the 'Satoyama Initiative for the benefit of biodiversity and human well-being'

Cruickshank R (1995) Manual on field mapping for landslide hazard zonation. National Building Research Organization, Colombo

Disaster Management Centre (2021) Institutional disaster management plan. State Ministry of National Security & Disaster Management, 09.12.2021. www.dmc.gov.lk

Disaster Management Centre (2022) Disaster information management system in Sri Lanka from 1974 to upto date. Disaster Management Center. Disinventor: http://www.desinventar.lk

George J, Adelaja A (2021) Forced displacement and agriculture: implications for host communities. Sustainability 13(10):5728

Global Report on Internal Displacement (2021) Available at: https://www.internal-displacement.org/global-report/grid2021. Accessed 06.01.2022

Gunarathna BWGID, Jayasinghe AB, Abenayake CC, Mahanama PKS (2018) Assessing the impact of land use on triggering landslides: a case of Sabaragamuwa Province, Sri Lanka. Bhumi Plann Res J 6(2)

Gunatilleke I (1991) Towards an era of environmentally sustainable and community-friendly forestry in Sri Lanka. In: Environment and economic development proceedings. Seminar organized by the University of Peradeniya and the Institute of Fundamental Studies, Kandy, pp 20–25

Jayathilake DMSC, Hennayake N (2018) Social construction of landslide; case of Aranayake-political ecological glances in Sri Lanka

Mattsson E, Ostwald M, Nissanka SP (2018) What is good about Sri Lankan homegardens with regards to food security? A synthesis of the current scientific knowledge of a multifunctional land-use system. Agrofor Syst 92(6):1469–1484

Moe TL, Pathranarakul P (2006) An integrated approach to natural disaster management: public project management and its critical success factors. Disaster Prev Manag Int J

Ocha U (1998) Guiding principles on internal displacement. United Nations Office for the Coordination of Humanitarian Affairs, Geneva

Opitz Stapleton S, Nadin R, Watson C, Kellett J (2017) Climate change, migration, and displacement. The need for a risk-informed and coherent research. ODI/UNDP report

Perera ENC, Jayawardana DT, Jayasinghe P, Bandara RMS, Alahakoon N (2018) Direct impacts of landslides on socio-economic systems: a case study from Aranayake, Sri Lanka. Geoenviron Disasters 5(1):1–12

Promper C, Puissant A, Malet JP, Glade T (2014) Analysis of land cover changes in the past and the future as contribution to landslide risk scenarios. Appl Geogr 53:11–19

Rathnaweera ATD, Nawagamuwa UP (2013) Study of the impact of rainfall trends on landslide frequencies; Sri Lanka overview. J Inst Eng Sri Lanka 46:35–42

Ratnayake U, Herath S (2005) Changing rainfall and its impact on landslides in Sri Lanka. J Mt Sci 2(3):218–224

Sangasumana P (2018) Post disaster relocation issues: a case study of Samasarakanda landslide in Sri Lanka. Eur Sci J

Siriwardana CS, Jayasiri GP, Hettiarachchi SSL (2018) Investigation of efficiency and effectiveness of the existing disaster management frameworks in Sri Lanka. Procedia Eng 212:1091–1098

Thamilini J, Wekumbura C, Mohotti AJ, Kumara AP, Kudagammana ST, Silva KD, Frossard E (2019) Organized homegardens contribute to micronutrient intakes and dietary diversity of rural households in Sri Lanka. Front Sustain Food Syst 94

Thomas K, Hardy RD, Lazrus H, Mendez M, Orlove B, Rivera-Collazo I, Roberts JT, Rockman M, Warner BP, Winthrop R (2019) Explaining differential vulnerability to climate change: a social science review. Wiley Interdiscip Rev Clim Change 10(2):e565

Chapter 5
Displaced Fishermen Off the Coast: Impact of Multiple Hazards on Life Above the Water

Achini M. De Silva, Minura Sandamith, Isuru Wicramarthne, B. M. R. L. Basnayake, Janaranjana Ekanayake, K. B. Seenapatabendige, and S. K. Gunatilake

Abstract The small-scale fisheries (SSF) sector tends to be firmly rooted in local communities, traditions, and values, and SSF contributes about half of fish catches. Study aimed to identify natural versus man-made hazards on displaced fishermen off the coast, perform vulnerability assessment, develop timeline on multiple hazards on coastal livelihood, and develop resilient building mechanism through GIS-based integrated coastal zone management plan. Storytelling exercise performed in 15 FI divisions to collect qualitative data along the multi-functional coast of the Galle district of Sri Lanka. Vulnerability assessment, case study analysis of multiple hazards, GIS mapping, change detection, and spatial modeling are used as decision support resources to develop integrated coastal resource management plan. Fishermen serve as starting point of network of collectors, traders, processors, and institutions while feeding thousands of families. Indian Ocean tsunami displaced fishing communities first and both natural and man-made hazards limit the coastal space for fishers while threatening their livelihood. Blooming coastal tourism added extra pressure on land values, shrinking the coastal land for fishing communities and infrastructure and logistic developments made fishers away from coast. GIS mapping of coastal zone, sourcing local knowledge through stakeholder consultations, was applied to develop participatory integrated coastal zone management plan to secure the coastal space for fishing communities.

A. M. De Silva (✉) · M. Sandamith · I. Wicramarthne · B. M. R. L. Basnayake
Department of Agribusiness Management, Faculty of Agricultural Sciences, Sabaragamuwa University of Sri Lanka, P.O. Box 02, Belihuloya, Sri Lanka
e-mail: achini@agri.sab.ac.lk

J. Ekanayake
National Aquatic Resources Agency (NARA), Craw Island, Mattakkuliya, Colombo, Sri Lanka

K. B. Seenapatabendige
Department of Architecture, Faculty of Built Environment and Spatial Science, General Sir John Kotelawala Defense University, Southern Campus, Sooriyawewa, Sri Lanka

S. K. Gunatilake
Department of Natural Resources, Faculty of Applied Sciences, Sabaragamuwa University of Sri Lanka, P.O. Box 02, Belihuloya, Sri Lanka

© The Author(s), under exclusive license to Springer Nature Switzerland AG 2023
M. Hamza et al. (eds.), *Rebuilding Communities After Displacement*,
https://doi.org/10.1007/978-3-031-21414-1_5

Keywords Fishing communities · Displacement · Multiple hazards · Sri Lanka

5.1 Introduction

Coastal zone of Galle district of, Sri Lanka, is one of the best beaches in the world and hottest tourist destinations in Asia. Coastal zone of south is very important due to their abundant natural resources and wide variety of interconnected ecosystems, such as coral reefs, sea grass beds, mangroves, sand dunes, coastal vegetation, lagoons, estuaries, and coastal wetlands (Alesheikh et al. 2007; Beatley et al. 2002). Coastal zone of Galle district is a remarkably 79.3 km length landscape, that plays a vital role in the country's economy. The development activities covering the fisheries sector, tourism and transport, and logistic development provide unlimited benefits to the coastal dwellers over a wide area. Coastal zone is a key element of the region's economy and culture (Beatley et al. 2002; Clark 2018). Coastal resources are shared among multiple users; small-scale fishers, lifeline of the coast, use the common property for hundreds of years, golden sandy beaches are home for thousands of travelers around the globe, and star-class hotels, restaurants, and other allied industries cater to global consumers. Coast is shared by many others: harbors, fish landing sites, boat yards, ice plants, fish processing centers, etc.

Fisheries industry, both small-scale, and commercial fisheries sectors are considered as socially and economically viable industry of the Galle district. Figure 5.1 shows the dynamic nature of fishing households and active fishermen over 30 years time. Fisheries cooperatives represent fishing communities in policy dialogs and welfare arrangements.

There were 47 community organizations with 3244 memberships to strengthen the local fisheries administration. Influence of other industries, especially the tourism, has made direct impact on fishing communities, and fisheries industry of the Galle district. Figure 5.2 shows the diversity and resource strength of fisheries sector, number of fishing fleets operating in Galle coast.

Each fleet type is a business unit, and it represents unique fishery, traditions, as well as economic status. Further, small-scale fisheries sector is considered as lifeline of the coastal belt ensuring food, nutrition, and economic security. Commercial fisheries are economically prominent sectors catering to both local and export markets. Catch basket of Galle represent all export-oriented high value species. Figure 5.3 brings the diversity of fish catch of the Galle district. Composition of fish catch plays a great role in main protein supplier to the local diet as well as export commodity.

Tourism considered as promising industry for the south coast where blooming tourism sector and established fisheries sector places huge demand on coastal natural resources. Golden sandy beaches of South coast are considered as one of the most attractive tourist destinations of the country. Both foreign and local guests spent 38% and 37% of guest nights, respectively, in South coast (SLTDA 2019). Resort region of the Galle district begins from world-famous Bentota beach and ended up in Galle. Highest number of guest accommodation facilities in Sri Lanka was 35% available in

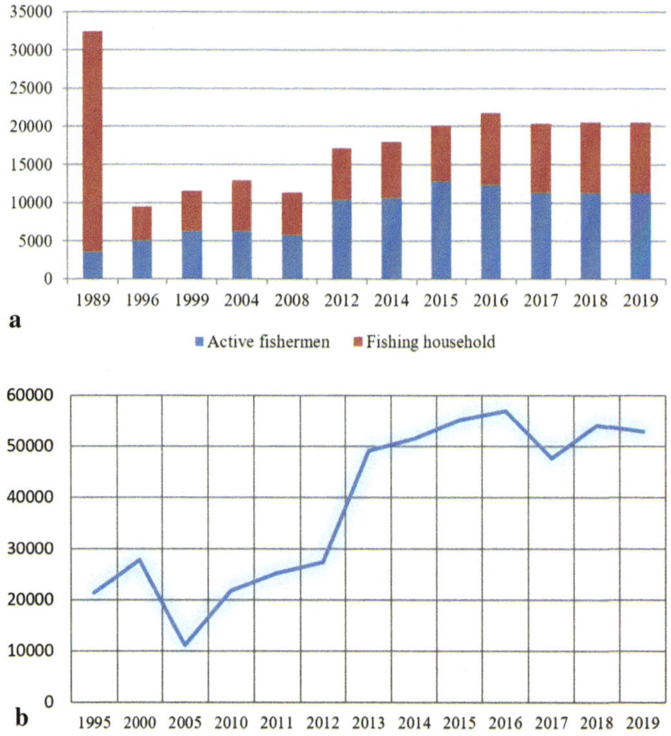

Fig. 5.1 Status of the fisheries industry of Galle district, Sri Lanka: **a** fishing households and active fisherman, **b** trend of total fish catch. *Source* Ministry of Fisheries (2020)

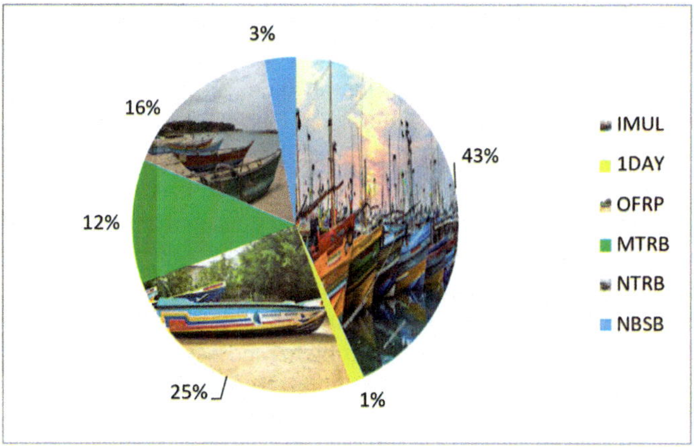

Fig. 5.2 Diversity of operating fishing fleets in Galle district, Sri Lanka. IMUL—Inboard Multiday Boats, IDAY—Inboard Single-day Boats, OFRP—Out-board Engine Fiberglass Reinforced Plastic Boats, MTRB—Motorized Traditional Boats, NTRB—Non-motorized Traditional Boats. *Source* Ministry of Fisheries (2020)

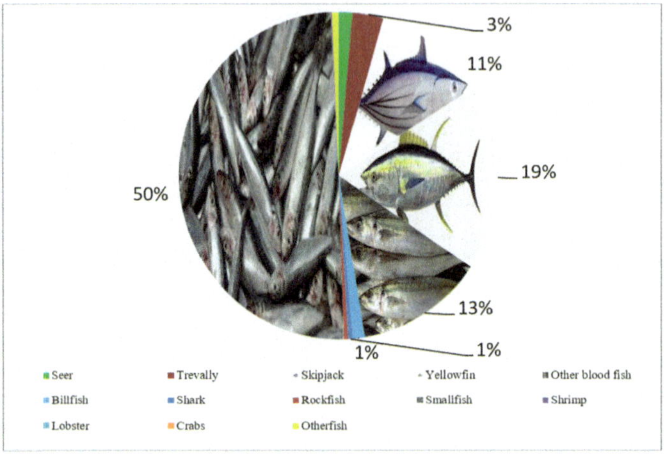

Fig. 5.3 Diversity of commercial fish production in Galle district, Sri Lanka. *Source* Ministry of Fisheries (2020)

South coast, and its room occupancy rate was 60% in 2019 (SLTDA 2019). Evidence from fisheries and tourism industries highlighted the demand on human as well as natural resources. Therefore, careful coastal zone management is essential to avoid conflicts among user groups while ensuring better returns.

Indian Ocean tsunami devastated the South and East coastline of Sri Lanka in 2004. Huge losses to human, physical, social, environmental, and financial resources crippled the economy. Forced displacement of coastal communities was the first experience of such kind which country and the people faced. Rehabilitation, resettlements, and relocation along with vulnerability of affected people were prime concerns of the post-disaster. Multi-functional as well as economically important South coast of the country received priority in rehabilitation and resettlement plans.

This study aimed to identify the natural and man-made hazards on displaced fishermen by the 2004 tsunami, perform vulnerability assessment, find out the timeline of multiple hazards on coastal livelihood, and develop resilient building mechanism through GIS-based integrated coastal zone management plan. Nearly, the two decades after the Indian Ocean tsunami, situation audit of the South coast, livelihood, economic activities, as well as natural and man-made hazards appeal to sustainable resource management initiative to overcome current issues. Moreover, large number of research publications highlighted the problems and issues of displacement, rehabilitation, relocation, resettlement, abandoned resettlements, and return to the original land. Therefore, scientific evidences, findings, policy recommendations, and living experiences of people and GIS systems will provide better answers to mitigate socioeconomic consequences. Table 5.1 provides the highlights of the published research on displacement, rehabilitation, relocation, and resettlements of the tsunami disaster.

Figure 5.4 brings the flow chart of the conceptualization of the study with key concerns of disaster, tsunami and its short-term impacts on people, communities, and industries (Birkmann et al. 2013; Cernea 2000; Clark 2018; Davidson et al. 2007; Dias et al. 2016; Kennedy et al. 2008; Kirsch et al. 2012; Ophiyandri 2011; Ruwanpura 2009). Long-term impacts on resettlements, livelihood, abandoned resettlements, and return to the original land have created dilemma on resettlement planning and sustainability of resettlements (Beatley et al. 2002; SLTDA 2019; Dias et al.

Table 5.1 Research highlights of the involuntary displacement by Indian Ocean tsunami, relocation, resettlement, and the vulnerabilities of the fishing communities

Source	Findings
Dias et al. (2016)	A resettlement program should re-establish the socioeconomic and cultural life of people. Reconstruction of a house does not solve the housing issue; it is vital to look into the indicators which can convert a house into a home and the surrounding into a neighborhood
Gunewardena (2008)	Government declared a no-build zone (buffer zone) of 100 m in the Southern and Western provinces and 200 m in the Northern and Eastern provinces. According to this policy, households who lived within the buffer zone limit were not allowed to rebuild their damaged or destroyed houses. Buffer zone policy considered only the construction of house but not the tourist hotels
Kennedy et al. (2008)	Satisfaction of the local community in resettlement programs is vital in order to ensure the long-term sustainability and usage of the housing resettlement programs
Davidson et al. (2007)	Community satisfaction is important in post-disaster resettlements in order to obtain the long-term involvement of the communities in rebuilding their socioeconomic lives
Ophiyandri (2011)	Community-based resettlement programs create community satisfaction and community satisfaction leads to a sense of ownership which is beneficial in creating disaster-resilient communities
Steinberg (2007)	Long-term satisfaction of people in housing resettlements is associated with community empowerment
Ruwanpura (2009)	Houses of resettlements should be embedded within local social relations
Buckle and Marsh (2002)	Agencies and aid workers must recognize that the unique infrastructure of villages and towns represents centuries of complex social, spatial, and architectural thought. Such places represent a cultural treasure that cannot be replaced with simple reconstruction of new homes and buildings
Shaw and Pardasani (2006)	Tsunami reconstruction and redevelopment in the Maldives: a case study of community participation and social action After a disaster, the social capital is destroyed and that social capital should be restored in the post-disaster reconstruction
Kirsch et al. (2012)	Return and rebuild: if the original site of residence is deemed safe for occupation after disasters, many people will opt to return and rebuild

(continued)

Table 5.1 (continued)

Source	Findings
Cernea (2000)	Relocation or resettlement has long been part of reconstruction efforts at the community level, such projects were done on an ad hoc basis, with little policy input, few responsible agencies, and minimal to poor capacities and resources As such, the outcomes of the vast majority of planned resettlement projects for disaster victims, in particular, as well as those displaced by other causes, have been very poor, resulting in the impoverishment of the affected populations
Birkmann et al. (2013)	The lack of provision of a functioning common social infrastructure (e.g., schools, public transport, etc.) before settling beneficiaries in their new locations, also contributed to new risks of livelihood disruption and additional costs for transport to work or social infrastructure such as schools or hospitals. Lack of access to employment or means of livelihood also led many to return to form locations or migrate
Josef (2009)	Displacement and varying degree of community vulnerability significantly affected rehabilitation, relocation, and resettlement success
De Silva and Yamao (2007)	Experienced fishermen with better educational backgrounds had a significant influence on the capacity building of livelihood assets. Relocation and resettlement plans brought persistent uncertainty to fishermen and threatened to disrupt their community bonds and social networks
Ibrahim (2020)	The wake of new threats and opportunities created by post-war development, policy changes and geopolitical constraints, and greater attention must be paid to SSFs. SSFs should be meaningfully incorporated into the post-war development agenda and "blue economy" policy reforms

2016; Ophiyandri 2011; Kirsch et al. 2012). The nexus between disaster, management, and sustainability of the system essentially requires participatory mechanism (Gunewardena 2008; Davidson et al. 2007; Buckle and Marsh 2002; Kirsch et al. 2012). Vulnerability analysis was performed to identify the exposure, sensitive variables, adaptive capacity, and impact of 4 main stages of recovery (Displacement, Rehabilitation, Relocation, and Resettlement).

Timeline of the remarkable events that occurred during last 3 decades (Fig. 5.5), especially the economic liberalization move toward to open economy and changes in traditional small-scale fisheries industry into commercial fisheries, open up new avenues in tourist industry with the onset of star-class hotels along the coast, coastal zone degradation based on coral mining, Indian Ocean tsunami, displacement of coastal communities, resettlement and rehabilitation efforts, establishment of coastal infrastructure, bloom in commercial fisheries industry through infrastructure development and resources allocation (huge foreign donations of fishing vessels), increase coastal erosion and establishment of physical structures to manage the coastal erosion, bloom in tourism industry and other allied coastal enterprises, and increased

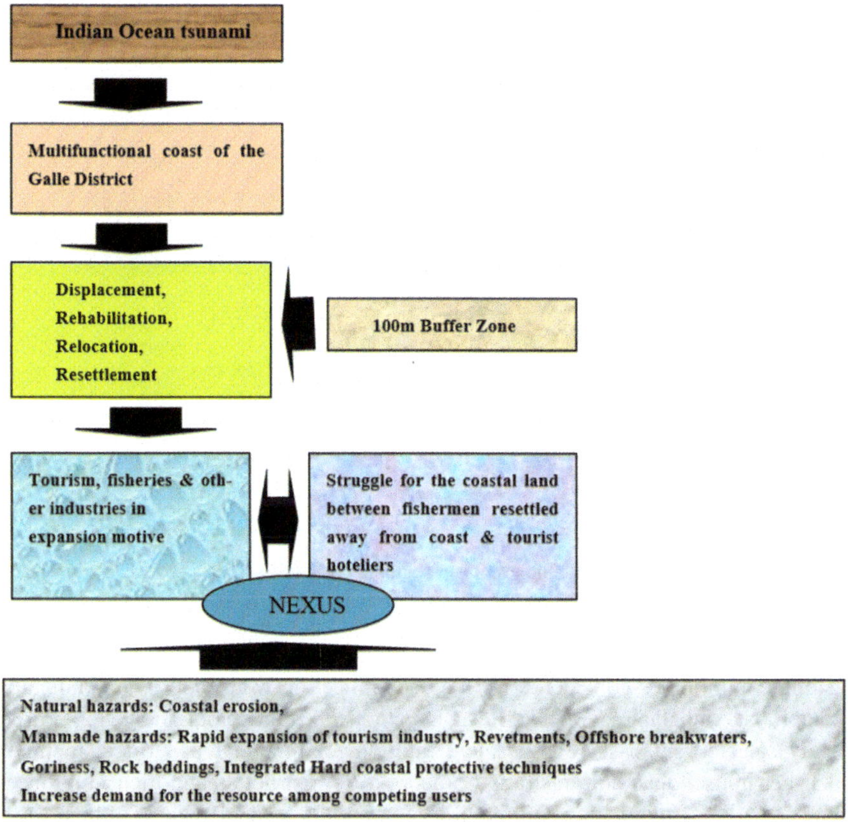

Fig. 5.4 Conceptualization process: disaster, displacement, relocation, resettlement, and associated issues of the coastal zone of the Galle district of Sri Lanka

pressure on coastal resources places demand on resource management efforts. List of events made various effects on coastal communities and our efforts were to explore the impact on displaced communities.

5.2 Methodology

Mixed method approach was instrumental to explore the impact of multiple hazards on displaced coastal communities of Galle district of Sri Lanka. Both quantitative and qualitative data were used along with both primary and secondary data sources. Key data collection tools were filed observations, in-depth interviews with key community stakeholders, government offices, members of NGOs, members of tourist hotel association, etc. Sampling procedure begun with cluster sampling and the coastline of Galle district, from Bentota to Gintota, 79.3 km, was considered for the study.

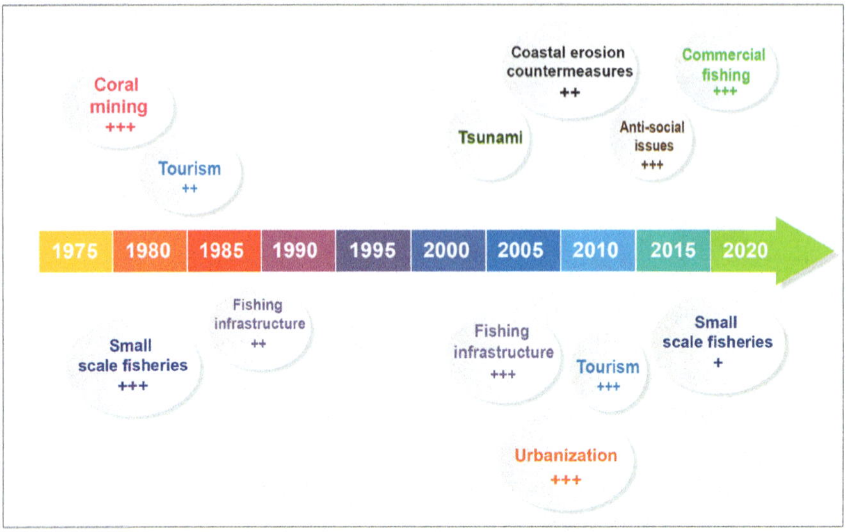

Fig. 5.5 Timeline of the remarkable events on southern coast of Sri Lanka

Both ends, Bentota and Gintota, are world-famous resort destinations. Clusters of fishing communities considered for the data collection were Induruwa, Ahungalla, Balapitiya, Ambalangoda, Hikkaduwa, Dodanduwa, and Gintota (Table 5.2). Purposively selected small-scale fishermen, multiday/commercial fishermen, members of fisher households, administrative officers of tourist hotels and their employees, representatives from fishing communities/community leaders, owners of boat yards, ice plants, fisheries inspectors, officers of coast conservation department, residents of resettlements, government officials/disaster management, local politicians, and scientists.

Qualitative data collection tools were mainly used to obtain the views of respondents. Respondents were purposively selected to represent prominent industries of each location. Sample was composed of fishermen, fish mongers, hoteliers, administrators of local government, officers of Department of coast conservation, fisherwomen, community leaders, boat yard owners, fish processors, ice plant owners, etc. Data collection was initiated with the participatory observations of the selected fishing villages along the coast of Galle district. Informal discussions with fishermen, families, and government officers were helpful to draw the outline of planned discussions. Storytelling exercise was planned with 4 steps:

1. Situation analysis: What is happening in coastal belt of the Galle district during last 16 years? Key themes were post-tsunami problems and issues, impact of natural hazards and man-made on displaced people during the recovery period.
2. Obstacle: What is causing a problem in the displaced fishing communities?
3. Action: How this problem/s going to address?
4. Result: What is the outcome?

Table 5.2 Sample profile

Location	No. of participants	Type of business	Role	Job experience
Induruwa	08	Fisheries Tourism Public service	Fishermen, fish mongers, hoteliers, local government administrators, coast conservation, fisherwomen	Fishers and hoteliers > 15 years Tourism industry > 10 years Government officers > 5 years NGOs > 10 years
Ahungalla	12	Fisheries Tourism Public service NGO	Fishermen, fish mongers, hoteliers, local government administrators, coast conservation, fisherwomen, community leader, NGO officer	
Balapitiya	10	Fisheries Tourism Public service Boat building	Fishermen, fish mongers, hoteliers, local government administrators, coast conservation, fisherwomen, community leader, boat yard owner	
Ambalangoda	10	Fisheries Tourism Public service Coast conservation	Fishermen, fish mongers, hoteliers, local government administrators, coast conservation, fisherwomen, community leader, boat yard owner, fish processor, ice plant owner, politician	
Hikkaduwa	08	Fisheries Tourism Public service Coast conservation	Fishermen, fish mongers, hoteliers, local government administrators, coast conservation, fisherwomen, community leader, boat yard owner, fish processor, ice plant owner, scientist	

(continued)

Table 5.2 (continued)

Location	No. of participants	Type of business	Role	Job experience
Dodanduwa	08	Fisheries Tourism Public service NGO	Fishermen, fish mongers, hoteliers, local government administrators, coast conservation, fisherwomen, community leader, NGO officer, scientist	
Gintota	10	Fisheries Tourism Public service Coast conservation	Fishermen, fish mongers, hoteliers, local government administrators, coast conservation, fisherwomen, community leader, boat yard owner, fish processor, ice plant owner, politician	

Source Field survey, 2021

Storytelling exercise was performed with the members of each selected cluster or fishing village. Respondents were selected purposively (Table 5.2) with the participation of community members. Date and time for the exercise were informed to the respondents. Research team visited the respondent's households and conducted the storytelling exercise. Open discussion on multiple hazards on multi-functional coast was initiated from the researchers and invited participants to express their views. All discussions were audio recorded with their prior permission and 10-min time slot was allocated to each participant. Each recorded discussion was uploaded and displayed to the team of researchers. First, key words were identified and listed along with the frequencies. Secondly, key problems, issues, etc., were sorted out and grouped into subthemes to analyze further.

Content analysis was performed to find out the multiple hazards on internally displaced people and socioeconomic consequences of natural and man-made hazards on communities. Word clouds were developed to present the results of content analysis. Case ordered meta-matrix was developed based on the stories of each research location and highlights were bundled into Table 5.3. Issues and problems were grouped into 4 categories based on the frequency of appearance in recorded stories. Color chart was used to indicate the status of the issue in each location. Red is used for critical impact, pink for severe impact, yellow for moderate and green for compatible impact (Table 5.3). Further, GPS locations of the fish landing sites, both traditional landing sites and harbors, and locations of physical structures such as wave breakers and coastal erosion were collected and area maps were developed using ArcGIS (version 10.4). The area map was developed to highlighting the areas of natural hazards, especially the coastal erosion. Land use maps of Southern province were obtained from the Department of Land Use Policy Planning which was the primary source for developing the maps. Finally, focus group discussions were performed to design the integrated participatory coastal resource management plan. First focus group was performed with key industry players of the southern coast; they were from fisheries small scale and commercial, tourism, boat building, ice production, cottage industries, and officers of local government. Focus group discussion aimed to explore the possibilities of the establishment of participatory approach on coastal zone management. Second and third focus groups were with community-level organizations and private and public partners to finalize the structure of community-based participatory coastal zone management plan (Fig. 5.15).

5.3 Results and Discussion

The present study focused to explore the issues and problems of internally displaced fishermen by the Indian Ocean tsunami during the post-tsunami period. Special focus was to find out the impact of natural and man-made hazards on recovering communities, explore the socioeconomic consequences, and develop GIS-based integrated participatory coastal zone management plan. Storytelling exercises were useful to perform the situation audit and identify the research gaps. Case analysis was useful to identify the status of the issues and the problems in each research location. Further, GIS modeling and mapping exercises were useful to find out the actual damages of coastal erosion.

The tsunami swept thousands of people away, destroyed infrastructure, and flooded large areas of coastal lands of Galle district. Both fisheries and tourism industries were devastated along with forced displacement of fishing communities, damaging to the fishing vessels, equipment, and houses. Loss account composed of 7032 completely damaged houses, 7680 partially damaged houses, and considerable damage to infrastructure including roads and bridges where 70% of buildings immediately located on the coastline have been completely destroyed and at least 30% of structures have suffered damage up to 1 km inland.

General context of the post-tsunami situation explains the consequences of displacement, relocation, resettlement, and abandoned resettlements and returned to the original lands. Buffer zone regulation was one of the key policy decisions that affected both fisheries and tourism industries of the Galle district (Clark 2018). The buffer zone regulation was re-imposed in 2005 which forced more than 70,000 tsunami-affected families to relocate into inland areas away from their land of origin (Alesheikh et al. 2007; Clark 2018; De Silva and Yamao 2007). The buffer zone regulations were re-imposed with the intention of safeguarding coastal communities from future tsunamis or other coastal hazards. Coastline of Galle district is home for economically important industries, fisheries, boat building, tourism, transport, and logistics, which provide unlimited benefits to the coastal dwellers over generations.

Table 5.3 Post-tsunami impacts on coastal communities

Impact—post-tsunami	I	A	B	AB	H	D	G
Relocation	Moderate	Moderate	Severe	Critical	Critical	Critical	Severe
Resettlement	Moderate	Compatible	Severe	Critical	Severe	Critical	Severe
Issues of social inclusion—none acceptance	Severe	Severe	Severe	Critical	Severe	Critical	Severe
Incompatible housing facilities	Moderate	Moderate	Severe	Severe	Severe	Severe	Moderate
Loss of neighborhood, and social spaces	Severe	Severe	Critical	Critical	Critical	Critical	Severe
Loss of income-generating activities—women	Moderate	Moderate	Severe	Critical	Critical	Critical	Severe
Extra expenses—transport, education, health care	Critical	Severe	Severe	Critical	Critical	Critical	Severe
Abandoned resettlements	Moderate	Moderate	Severe	Critical	Severe	Severe	Severe
Returned to original land on coast	Moderate	Compatible	Moderate	Critical	Severe	Critical	Severe
Lost or sold-out coastal land	Critical	Critical	Critical	Moderate	Severe	Severe	Severe
Sold-out traditional landing site property	Critical	Critical	Severe	Moderate	Moderate	Moderate	Moderate
Shrink coastal space for bots and store fishing gear	Critical	Critical	Severe	Severe	Severe	Severe	Severe
Limited space for fish drying	Moderate	Moderate	Moderate	Severe	Critical	Critical	Severe
Off the fisheries and migrated to other sectors	Severe	Critical	Severe	Moderate	Severe	Moderate	Severe
Expansion of tourist hotels/rebuilding	Critical	Critical	Severe	Severe	Critical	Critical	Critical
Establishment of five-star hotels	Critical	Critical	Critical	Moderate	Severe	Severe	Critical
Establishment of beach restaurants and entertainment	Moderate	Moderate	Moderate	Moderate	Severe	Severe	Moderate
Expansion of tourism-allied businesses—men	Moderate	Compatible	Moderate	Compatible	Compatible	Compatible	Compatible
Expansion of tourism-allied businesses—women	Moderate	Compatible	Moderate	Compatible	Compatible	Compatible	Moderate
Establishment of boat building	Moderate	Moderate	Compatible	Moderate	Compatible	Moderate	Compatible

Source Field survey, 2011

I Induruwa, *A* Ahungalla, *B* Balapitiya, *AB* Ambalangoda, *H* Hikkaduwa, *D* Dodanduwa, *G* Gintota

Critical impact

Severe impact

Moderate impact

Compatible

Buffer zone policy (100 m) forced tsunami-affected fishermen, coastal businessmen, and enterprises to abundant their properties, but measures were not applicable to star-class hotels along the coast (Clark 2018).

Face of the coastal zone, key element of economy and culture of the region, was changed, and most of the properties were abandoned. Whether this was an effective policy decision? and consequences of it discuss separately. In general, most of the resettlements were established beyond the buffer zone boundary and most of the housing schemes were 5–10 km away from the coast (Alesheikh et al. 2007; Clark 2018). Resettlements away from the coast were welcomed by most of the dwellers at the beginning of rehabilitation process and unforgettable experience of tsunami promotes livelihoods away from the coast. Situation has changed with the returning to usual fishing operations. Where lack of other livelihood options available in inland areas, social issues with neighboring inland communities, inconvenience, and practical difficulties were badly affected on daily life of the displaced people (Beatley et al. 2002; SLTDA 2019). Most of them abandoned their home in resettlements and established temporary and semi-permanent houses on their abandoned coastal lands. In addition to that, some people were permanently moved to the land of origin and rented out their houses of resettlement schemes. Moreover, some families who already sold out their land of origin had to face difficulties in returning to the traditional villages. Case-ordered meta-matrix developed from the findings of the storytelling exercise explains the location-specific problems and issues of the post-tsunami era (Table 5.3). Of the sample, Balapitiya, Ambalangoda, Hikkaduwa, and Dodanduwa are considered as main fish landing sites of Galle district where fishing considered as key economic activity. In contrast, Induruwa and Ahungalla can be considered as more agriculture-based locations and converted into main tourist destinations of the Galle district.

Results of the content analysis explain in Figs. 5.6 and 5.7. Storytelling exercise captured the key concerns of the respondents over natural hazards that popped up in post-tsunami period. Coastal erosion, especially between Ambalangoda and Hikkaduwa, reached to alarming levels and threading to collapse the main land route from Colombo-Galle. Severe coastal erosion involuntarily displaced the tsunami and displaced small-scale fishermen. Large number of seasonal fish landing sites from Ambalangoda to Dodanduwa were damaged and abandoned. Seasonal small-scale fish harvests were able to cater the fresh fish demand of the tourist sector, local diet as well as small-scale local fish drying and salting provided extra income to fisherwomen. Sand deposition, strong winds, heavy rainfall, increased rainfall intensity, and changes of tidal pattern were identified as an important natural hazard of the post-tsunami period.

Relocation of coastal fishing communities, encroaching tourism. and other businesses toward the coastal belt has created various man-made hazards on social and spaces. Amplified anthropogenic activities, rapid urbanization, increased intensity of construction, coastal pollution, and man-made spaces significantly affected on fishers' movements, especially the small-scale fishers, on coast (Fig. 5.7).

Coastal belt of Galle district shows the highest diversity in income-generating activities or livelihood options. Coastal line was divided into 6 hypothetical zones

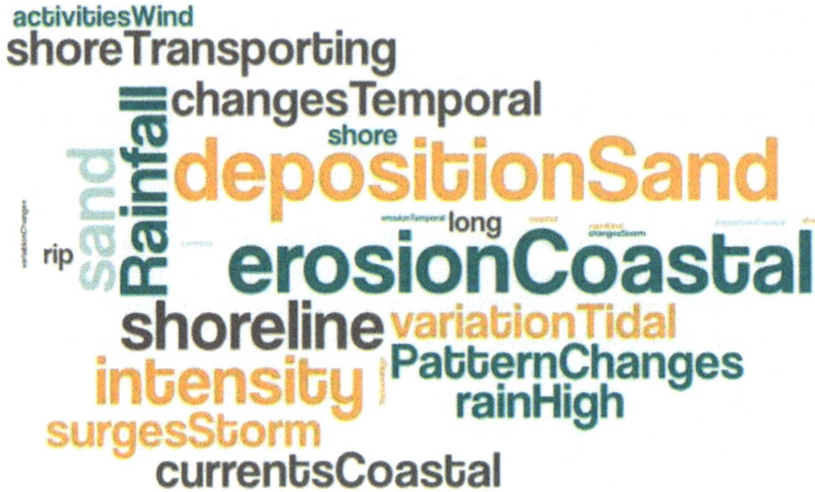

Fig. 5.6 Results of the content analysis of the storytelling activity on impact of natural hazards on southern coast

Fig. 5.7 Results of the content analysis of the storytelling activity on man-made hazards on southern coast

to identify its multi-functionality. Four clusters were highly diversified compared to others. Most diversified clusters were Ahungalla–Balapitiya, Balapitiya–Ambalangoda, Bentota–Kosgoda and Ambalangoda–Hikkaduwa, respectively (Fig. 5.9). Tourism-related businesses, star-class hotels, restaurants and fishing related businesses, fish landing sites, fish processing factories, boat-building yards, and ice plants

(a) (b)

Fig. 5.8 Ahungalla beach, living examples for coastal invasion by star-class hotels. *Source* Tripadvisor (2021a, b)

were prominent. Land tenure issues in small-scale fisheries (SSF) are emerging as serious threats to the livelihood of fishers. On the one hand, fishing communities were displaced and resettled away from the land of their origin. Land left was acquired by the hoteliers or added into the buffer zone as public good (Fig. 5.8). Tourism has shifted from sun and sea tourism in large cities (Hikkaduwa, Bentota), to remote coastal villages, combining sun and sea tourism with nature, eco, cultural, agro and other tourism activities. Rural coastal areas of Galle district, which offered the means of livelihood to SSF and molded the fishing communities into a particular way of life, have now become locations that produce colossal recreational values. For investors, this could form one important avenue of investment guaranteeing high returns to investment.

Land tenure of SSF is a more serious issue. Beach anchorage of SSF crafts has been a right enjoyed by fisher for generations, which is now been threatened by fencing of land adjoining beaches by those who now claim ownership to this age-old neglected land. While fishers used to leave their fishing equipment in temporary huts on the beach when they leave for home after fishing operations, this is no longer possible to the presence of diverse stakeholders who have no close links with SSF communities. The other major issue is the beach seine "padu" where seine nets are laid. Some hoteliers are maintaining the "padu" as their own beach area, threatening seine operations (Image 1). Present generation of the small-scale fishing community who used the beach for seasonal fishing gear operation was sold out their right on use of coast for the star-class hotel (Image 1). Unethical coastal encroaching through political power for the construction of star-class tourist hotels was common along the South coast. The retention of homesteads and forest areas which provided livelihood assistance is also highly uncertain and rights to these lands are likely to be lost for lack of clear titles.

It is quite evident that the above conditions would threaten the livelihoods of fisher families. Sustaining livelihoods in artisanal and small-scale fisheries therefore requires protected and priority access to fishing areas on the one hand, and secure land settlement rights along the coastal strip on the other. However, the land settlement rights of fishing communities in areas adjacent to the coast are generally not recognized. Insecure settlement on land makes their livelihoods highly precarious and vulnerable, particularly when faced with increasing competition for space and resources in the coastal zone. Multi-functionality of the Coastal zone of the Galle district is highlighted in Fig. 5.9. Ahungalla to Ambalangoda, about 15 km zone was identified as most diversified among other zones. Coastal zone is fulfilling the demands of competing users from tourism mainly and various subsectors of fisheries industry.

Coastal protective structures were established to mitigate the coastal erosion and facilitate the fisheries industry of the Galle district. Unfortunately, respondents were named those artificial structures as man-made hazard which negatively affected on livelihood of the people. Coastal erosion and the shoreline changes are considered as crucial issues especially for Sri Lanka. The erosion of sandy shorelines creates a considerable threat to life and property (OCHA 2004; Jayasuriya et al. 2006). In general, Coastal long-shore currents, coastal rip currents, waves, and wind activities, which carry sand off the shore and deposit it someplace else, have all been discovered to be some of the crucial natural factors in coastal erosion. While coastal erosion might intensify as a result of natural occurrences such as storm surges, sea-level increases, tidal fluctuations hurricanes, coastal floods, and tsunamis (OCHA 2004; Jayasuriya et al. 2006; Perera 1990; Godage 1992). Furthermore, increased

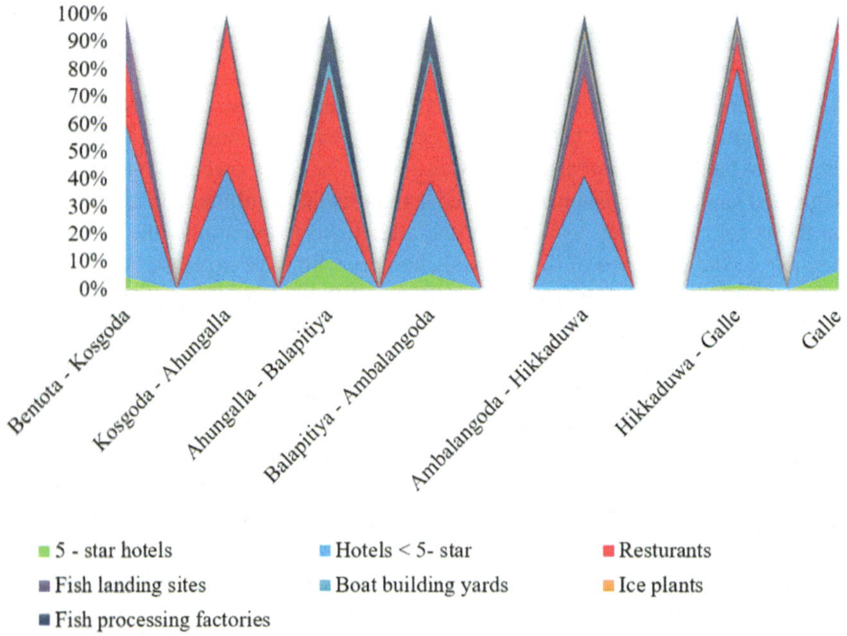

Fig. 5.9 Multi-functionality of the coast, Galle district, Sri Lanka. *Source* Field survey, 2021

environmental degradations such as sand mining, damage to the coral reef, coastal deforestation, and artificial modification through constructions, dredging, and filling are major contributors to coastal erosion (Ibrahim 2020; Perera 1990; Godage 1992). Whichever the course behind the generation of coastal erosion, it inhibits economic growth and provokes social conflict both directly and indirectly (Zhang et al. 2004).

Even though Sri Lanka has established a variety of coastline protective measures, the erosion processes are poorly understood, as a developing country, the efficacy of coastal protection structures has not been well evaluated (Perera 1990; Godage 1992; Zhang et al. 2004; Nayak 2017). Revetments, offshore breakwaters, goriness, coves, gabion walls, and immediate rock beddings like hard coastal protection techniques, and beach nourishment and dune construction like soft coastal protection techniques are the major strategies it has been used to protect Sri Lankan coastal areas (Image 2) (Zhang et al. 2004; Nayak 2017; Van Heuverswyn 2011; Prasad and Kumar 2014; Abeykoon et al. 2021; Ndour et al. 2018). However, the soft coastal protection techniques allow the shore to behave naturally without any constructive protection method (Fig. 5.10).

Construction of hard coastal protective structures such as revetments and groynes has been significantly increased in the last few years (Figs. 5.11 and 5.12). However, the man-made hard coastal protective structures are not sustainable solutions for coastal erosion. These structures pass the trouble into the neighboring sites. Besides, these hard protective coastal covers have been caused to generate other socioeconomic conflicts especially in the coastal small-scale fisheries community. Impacts of the coastal protective structures on the small-scale coastal fishing sector can be listed as disturbance for the beach accessibility and usage, damage to fishing crafts and fishing gears, traditional fishing ground losses—beach seine and stilt fishing, and life threats due to the alteration of the natural coastal current system mainly by the Groynes and Breakwaters.

Appropriate identification of the dynamic character of the coastline caused by wave patterns and coastal currents, mandatory subsidies for coastal zone management and preservation, buffering capacity from natural coastal ecosystems, and coastal-based enterprises are critical. Most importantly, the coastline management solutions could potentially influence both socioeconomic and environmental processes (Perera 1990; Godage 1992; Nayak 2017; Van Heuverswyn 2011). Therefore, the integrated planning approaches for erosive beaches are needed to be often used. To enable successful adaptive shoreline management, chronological shoreline and bathymetric data collection are necessary from feedback mechanisms through field surveys. Study findings revealed that natural coastal buffering systems such as sand dunes, dune vegetations, mangroves, and other coastal vegetations and corals would be more environmentally and economically effective long-standing solutions for the coastal erosion and we have to protect and strengthen these systems (Fig. 5.13).

Study identified several natural hazards that make significant impact of coastal livelihood, especially on fishing operations, fish processing, as well as tourism activities. Figure 5.14 explains the timeline of the physical changes of the coastal line of the Galle district. Coastal erosion is one of the key natural hazards which badly damaged the coastal line (Van Heuverswyn 2011; Prasad and Kumar 2014; Abeykoon et al.

Fig. 5.10 Images of hard coastal protective techniques constructed in coastal belt of Galle district, Sri Lanka. **a** Revetments, **b**, **c** offshore breakwaters, **d** goriness, **e** rock beddings, **f** integrated hard coastal protective techniques. *Source* Ekanayake (2021)

2021; Ndour et al. 2018). Ambalangoda to Hikkaduwa region was the worst affected region compared to the other regions of the Gale coast. Coral mining intensively to feed the traditional lime-making industry was the prime cause for the coastal erosion. Further, same region was the worst hit region compared to others during 2004, Indian Ocean tsunami accounted for huge losses. Moreover, Mangrove covered areas of river mouths of Bentota, Madu, Hikkaduwa, and Gin rivers were protected naturally from both tsunami and coastal erosion (Fig. 5.14a). Heavy uses of coastal resources are threatening to the existing mangrove forest cover of the district. Traditional fish

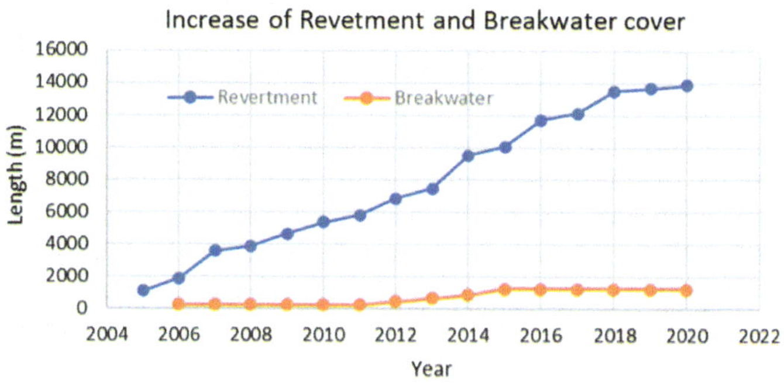

Fig. 5.11 Increase in revetment and breakwater cover in southern coastal area in Sri Lanka. *Source* Coast Conservation and Coastal Resource Management Department (2019)

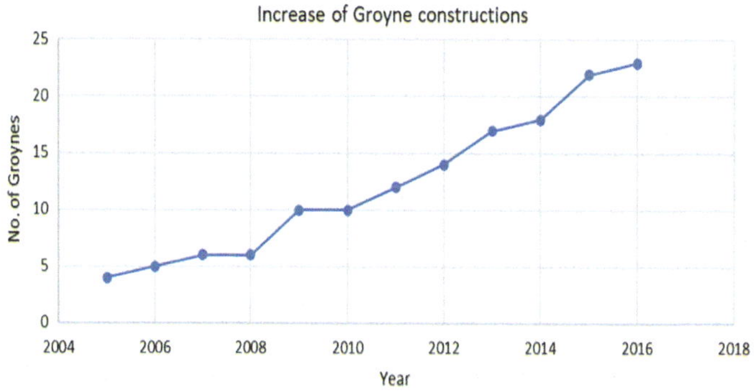

Fig. 5.12 Increase in number of groyne contractions in southern coastal area in Sri Lanka. *Source* Coast Conservation and Coastal Resource Management Department (2019)

Fig. 5.13 Images of dune vegetations and dune-forming plants. *Source* Ekanayake (2021)

landing sites (Fig. 5.14b) were one of the main economic activities from Bentota to Balapitiya (Bentota, Induruwa, Kosgoda, Ahungalla, and Balapitiya) region of the Galle district which is now considered as one of the most diversified coastal region of the Gale district. Large number of small-scale fishermen was engaged in fishing, especially during the months of September–April, main fishing season. Unfortunately, most of the landing sites, beaches were encroached by the tourist hotels. Timeline of the changes of coastline and its adverse effect on costal livelihood are explained by the GIS maps (Fig. 5.14c). Research evidence appeals sustainable coastal zone management plan for the economically, socially as well as naturally important coastal line of the Galle district.

Integrated Coastal Zone Management (ICZM) is a process of governing multi-faceted resource base and functions of its users with the participation of institutions and stakeholders (Roebeling et al. 2011). Coastal belt of Galle district is a victim of various natural and man-made hazards in different scales and appeals for sustainable coastal zone management to protect the resource base for future use. Focus group discussions were organized to design-integrated coastal zone management plan. Evidence of the research used as foundation for this work and 3 focus group discussions (FG 1—with key industry participants, FG 2—with community organizations, and FG 3—with public and private administrative officials) were instrumental to identify the concerns of the stakeholders and design coastal zone management plan to empower stakeholders to manage their resource base. Figure 5.15 explains the results of focus group discussions and scores allocated by the stakeholders.

One of the important aims of the study was to design participatory integrated coastal zone management mechanism (Fig. 5.16). Design phase involved two main steps; spatial mapping, developing area maps and visualizing the changes of the coastal belt based on both disasters. The process involves with integration of remote sensing data, aerial photographs, as well as point data provided by fieldwork. Second step is based on series of focus group discussions with various stakeholders of the resource base. Four different actor levels were considered for the discussions: first with individual participants from various sectors, second with community-level organizations, third with private sector companies, and fourth with both public and private parties. Area base participatory coastal zone management plan involves with resource audit, identifies gaps, designs resource use plans, and introduces management measures and recovery. Area-specific culture, traditions, as well as socioeconomic system will lead to develop unique management approach which all can agree and participate. Research intervention is essential to investigate the smooth functioning of the system within various sub-systems.

5.4 Conclusions

Indian Ocean tsunami of 2004 and threat of post-tsunami events largely changed the size and shape of the affected communities, especially the fishing communities. On the other hand, government, local administration authorities, donors and NGOs

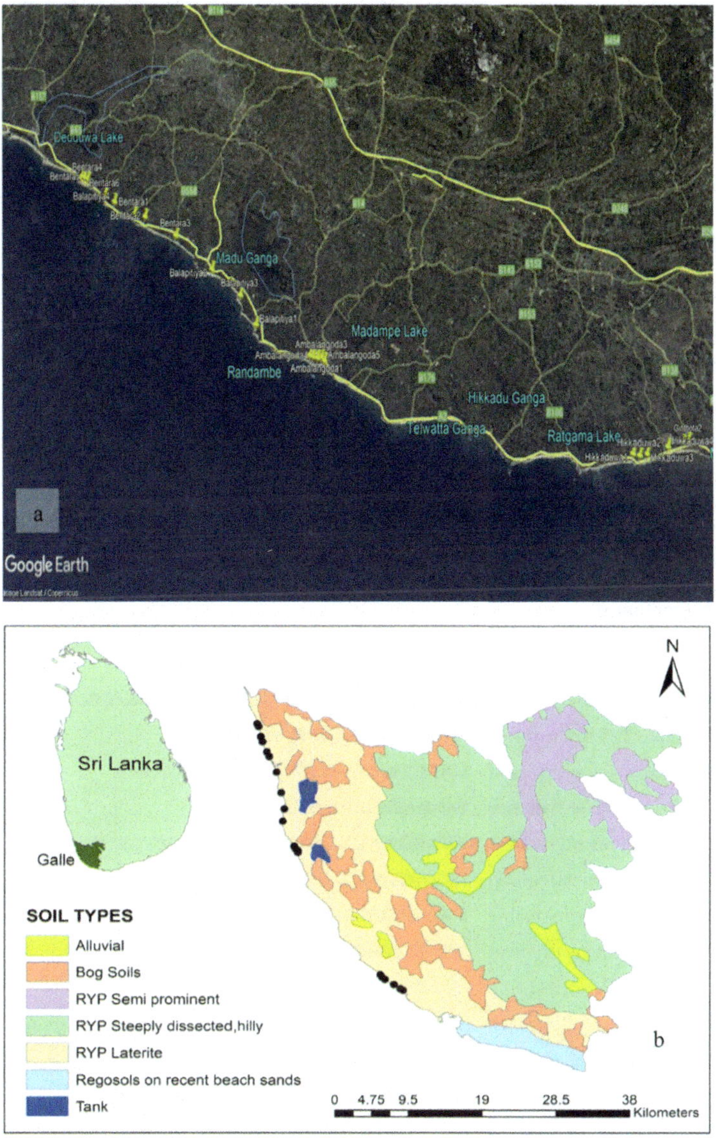

Fig. 5.14 Spatial analysis of the coastal line of Galle district and the timeline of the changes along the coast. **a** Main fish landing sites of the coastal line of Galle district, **b** traditional fish landing sites along the coast of Galle district and **c** timeline of the changes along the coastal line of the Galle district due to both natural and man-made hazards. *Source* Google maps and Department of Land Use and Policy Planning

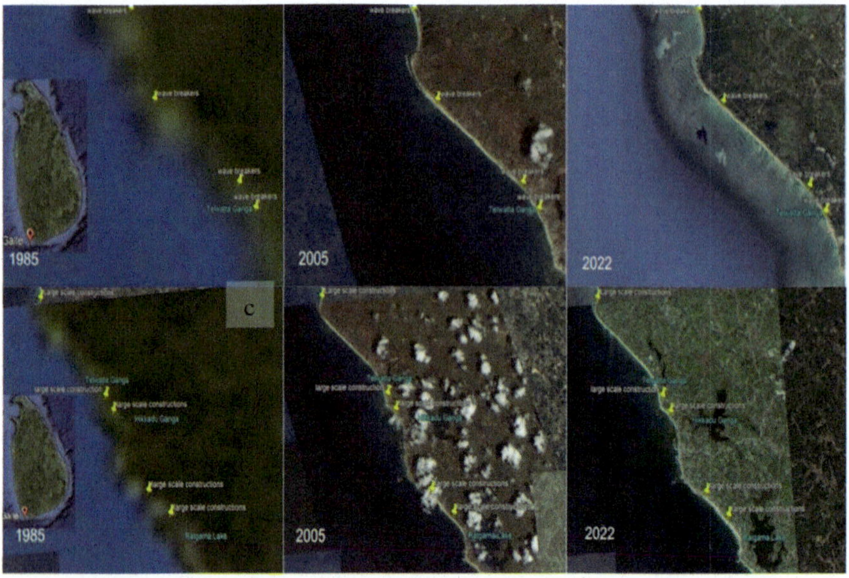

Fig. 5.14 (continued)

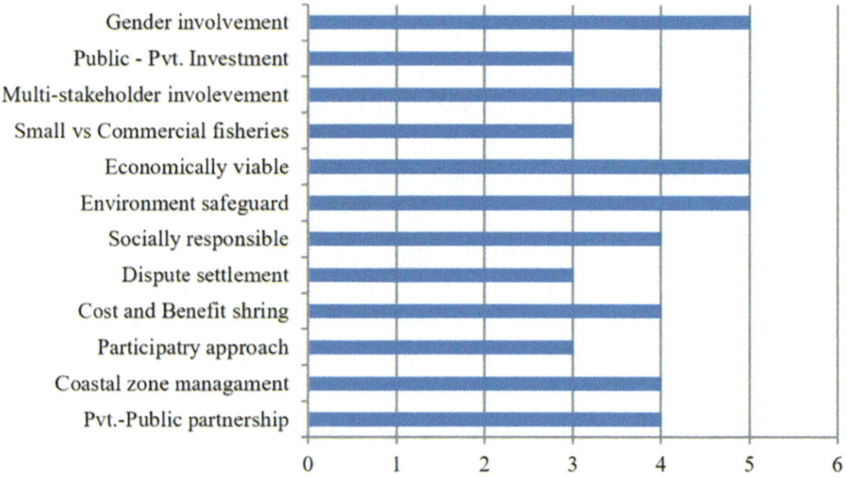

Fig. 5.15 Key concerns of the focus group discussions

were rushed on rehabilitation work. Relocating and establishing resettlements for displaced communities in safe distance from sea were priority and 100 m buffer zone, regulatory red tape made displaced people away from original land and settled in different locations where public lands were available. This research aimed to explore impact of relocation and resettlements on the livelihoods of displaced fishing

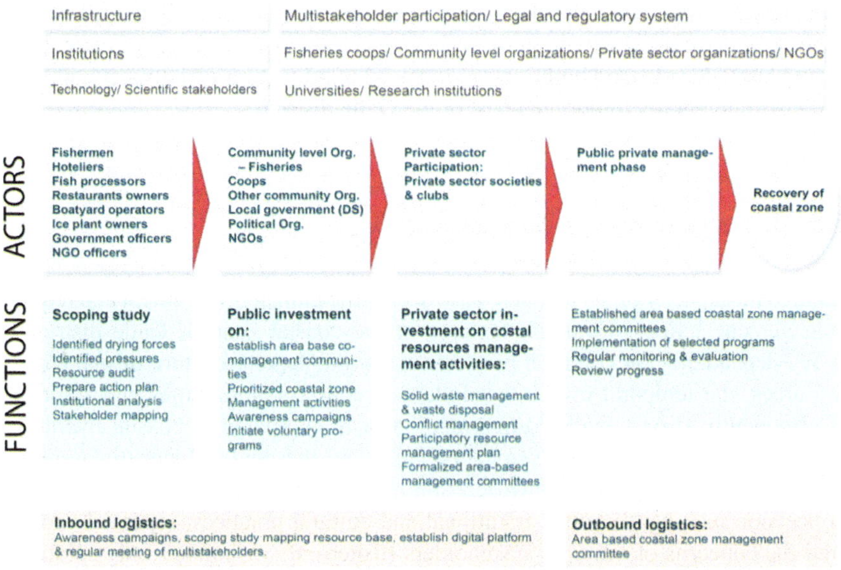

Fig. 5.16 Process flow and stakeholders of the area-based participatory coastal zone management plan

communities. Further, research eyeing to find out how do spatial factors, distance to sea, neighborhood environment, access to services, school, markets, etc., affect the coping strategies of the people. Post-tsunami impact was more on abandoned resettlements, returned to the original land, rented or sold out the properties at resettlement, shifting livelihood strategies away from fishery, if pace available establish temporary or semi-permanent housing in their original land. This was more common among main fish landing sites and adjacent villages compared to agriculture-based villages. Findings revealed that natural and man-made hazards in post-tsunami period accelerate fisher's departure.

Key natural hazards which highly affected on small-scale fisheries were tsunami, Coastal erosion, and effects of varied climates (storms and high intensity rains). Main man-made hazards were increased demand for coastal resources, competing uses and users; urbanization, industrialization, social consequences; and anti-social behaviors, narcotic drugs, and alcoholism. Both natural and man-made hazards make fishers away from coast, forced displacement of fisheries communities. Socioeconomic consequences: Fishers off the coast lead to decline fish production, impact on main protein supply to the local people's diet, threat to thousands of fisheries-related employment, fisher migration to other industries, move toward deep sea fishing, imbalance in harvesting coastal resources, fishers left unemployed or underemployment, etc. Man-made hard coastal protective structures are not sustainable solutions for the coastal erosion. They just pass the trouble in the neighboring sites. Finally, research aimed to develop integrated participatory coastal zone management plan with the participation of all stakeholders. Findings of the study laid the foundation to the coastal zone management plan. GIS maps identified the changes of coastal zone during last 3 decades and highlighted vital need of protecting socially, economically, and environmentally important coast of the Galle district.

5.5 Directions for Future Research

Nature and capacity of the resource base decide the human involvement and required management actions. Common features of the coastal zone of Galle district are heavy dependency on common pool resources, the depleted nature of many of the resources, the temporal cycles that influence coastal communities and resources, the frequently strong gender differentiation in roles related to coastal livelihoods (i.e., fishers, tourism service providers, etc.), and the dual influence of terrestrial and coastal livelihoods. Top-down decision-making ignored community participation in decision-making, forgotten traditional and cultural practices, separately considered the concerns of multiple stakeholders (fisher-only or hotelier-only solutions), and deleted gender involvement yielded negative consequences on resource base as well as the life above the water. Therefore, multi-disciplinary approaches explore the capacity of resource base as well as stakeholders will demarcate the management style. Identifying interests, values, and attitudes, aspirations of multiple stakeholders through social mapping would be advantageous to initiate self-governing management mechanism.

References

Abeykoon LCK, Thilakarathne EPDN, Abeygunawardana AP, Warnasuriya TWS, Egodauyana KPUT (2021) Are coastal protective hard structures still applicable with respect to shoreline changes in Sri Lanka?

Alesheikh AA, Ghorbanali A, Nouri N (2007) Coastline change detection using remote sensing. Int J Environ Sci Technol 4(1):61–66

Beatley T, Brower D, Schwab AK (2002) An introduction to coastal zone management. Island Press

Birkmann J, Garschagen M, Fernando N, Tuan V, Oliver-Smith A, Hettige S (2013) Dynamics of vulnerability: relocation in the context of natural hazards and disasters. In: Birkmann J (ed) Measuring vulnerability to natural hazards, 2nd edn. United Nations University Press, New York, pp 505–550

Buckle P, Marsh G (2002) Local assessment of disaster vulnerability and resilience: reframing risk. International Sociological Association (ISA), Brisbane

Cernea M (2000) Risks, safeguards, and reconstruction: a model for population displacement and resettlement. In: Cernea M, McDowell C (eds) Risks and reconstruction: experiences of resettlers and refugees. World Bank, Washington, DC, pp 11–55

Clark JR (2018) Coastal zone management handbook. CRC Press

Coast Conservation and Coastal Resource Management Department (2019) Performance report, coast conservation and coastal resource management department. https://www.parliament.lk/uploads/documents/paperspresented/performance-report-coast-conservation-department-2019.pdf. Accessed 15 July 2021

Davidson CH, Johnson C, Lizarralde G, Dikmen N, Sliwinski A (2007) Truths and myths about community participation in post-disaster housing projects. Habitat Int 31(1):100–115

De Silva DAM, Yamao M (2007) Effect of tsunami on fisheries and coastal livelihood: a case study of tsunami ravaged southern Sri Lanka. Disasters 31(4):19. https://doi.org/10.1111/j.0361-3666.2007.01015.x

Dias NT, Keraminiyage K, De Silva KK (2016) Long-term satisfaction of post disaster resettled communities: the case of post tsunami—Sri Lanka. Disaster Prev Manag. https://doi.org/10.1108/DPM-11-2015-0264

Ekanayake J (2021) Hard coastal protective techniques constructed in coastal belt of Galle district, Sri Lanka (images)

Godage D (1992) Coast erosion management plan and it's implementation. In: Seminar on causes of coastal erosion in Sri Lanka. CCD/GTZ coast conservation project, Colombo, Sri Lanka, pp 323–330

Gunewardena A (2008) Peddling paradise, rebuilding Serendib: the 100-meter refugees versus the tourism industry in post-tsunami Sri Lanka. In: Capitalizing on catastrophe: neoliberal strategies in disaster reconstruction. AltaMira, Lanham, MD, pp 69–92

Ibrahim FA (2020) Between the sea and the land: small-scale fishers and multiple vulnerabilities in Sri Lanka. Sri Lanka J Soc Sci 43(1):05–20. https://doi.org/10.4038/sljss.v43i1.7641

Jayasuriya S, Steele P, Weerakoon D (2006) Post-tsunami recovery: issues and challenges in Sri Lanka

Josef RK (2009) Facts on the seashore: conflict, population displacement, and coastal vulnerability on the eve of the Sri Lankan tsunami. In: Gamburd MR, McGilvray DB (eds) Tsunami recovery in Sri Lanka: ethnic and regional dimensions. Taylor and Francis. https://paa2011.princeton.edu/papers/112636

Kennedy J, Ashmore J, Babister E, Kelman I (2008) The meaning of 'build back better': evidence from post-tsunami Aceh and Sri Lanka. J Contingencies Crisis Manag 16(1):24–36

Kirsch TD, Wadhwani C, Sauer L, Doocy S, Catlett C (2012) Impact of the 2010 Pakistan floods on rural and urban populations at six months. PLoS Curr Disasters

Ministry of Fisheries (2020) Annual statistics report. Ministry of Fisheries, Colombo, Sri Lanka

Nayak S (2017) Coastal zone management in India—present status and future needs. Geo-Spat Inf Sci 20(2):174–183

Ndour A, Laïbi RA, Sadio M, Degbe CG, Diaw AT, Oyédé LM, Anthony EJ, Dussouillez P, Sambou H (2018) Management strategies for coastal erosion problems in West Africa: analysis, issues, and constraints drawn from the examples of Senegal and Benin. Ocean Coast Manag 156:92–106. https://invenio.unidep.org/invenio/record/17081/files/janpost.pdf

OCHA (2004) Tsunami disaster-Sri Lanka summary report. UN Office for the Coordination of Humanitarian Affairs (OCHA). https://reliefweb.int/report/sri-lanka/tsunami-disaster-sri-lanka-summary-situation-report-26-31-dec-2004. Accessed 15 Aug 2021

Ophiyandri T (2011) Community-based post-disaster housing reconstruction: examples from Indonesia. In: Amarathunga D, Haigh R (eds) Post-disaster reconstruction of the built environment: rebuilding for resilience. https://doi.org/10.1002/9781444344943

Perera HN (1990) Need for review and upgrading of master plan for coast erosion management. In: Seminar on causes of coastal erosion in Sri Lanka, pp 331–348

Prasad DH, Kumar ND (2014) Coastal erosion studies—a review. Int J Geosci 2014

Roebeling PC, Coelho CD, Reis EM (2011) Coastal erosion and coastal defense interventions: a cost-benefit analysis. J Coast Res 1415–1419

Ruwanpura KN (2009) Putting houses in place: rebuilding communities in post-tsunami Sri Lanka. Disasters 33(3):436–456

Shaw R, Pardasani M (2006) Tsunami reconstruction and redevelopment in the Maldives: a case study of community participation and social action. Disaster Prev Manag Int J 15(1):79–91

Sri Lanka Tourism Development Authority (SLTDA) (2019) Annual statistical report. Research and International Relations Division, Sri Lanka Tourism Development Authority. www.sltda.gov.lk

Steinberg F (2007) Housing reconstruction and rehabilitation in Aceh and Nias, Indonesia—rebuilding lives. Habitat Int 31(1):150–166

Tripadvisor (2021a) Hotel RIU Sri Lanka. https://www.tripadvisor.co.uk/Hotel_Review-g608521-d10039170-Reviews-Hotel_Riu_Sri_Lanka-Ahungalla_Galle_District_Southern_Province.html

Tripadvisor (2021b) White villa resort, Ahungalla. https://www.tripadvisor.com/Hotel_Review-g608521-d8006887-Reviews-White_Villa_Resort-Ahungalla_Galle_District_Southern_Province.html

Van Heuverswyn K (2011) Inleiding: van prestatiemetingen naar integraal management van performance. Die Keure, Brugge

Zhang K, Douglas BC, Leatherman SP (2004) Global warming and coastal erosion. Clim Change 64(1):41–58

Chapter 6
Drivers of Slow-Onset Displacement in the Coastal Mid-Atlantic Region and Preferences for Receiving Locations

Anamaria Bukvic and Christopher W. Zobel

> *"If we stick to former paradigms we are bound to be defeated in every battle. The point is not to prepare plans and tools to avoid surprise, but to be prepared to be surprised."*
> *Patrick Lagadec, 2009*

Abstract Coastal areas in the U.S. Mid-Atlantic region are experiencing profound changes due to sea level rise and more frequent and severe storm surges. Many coastal residents face the risk of displacement from areas of repetitive flooding due to direct flood impacts or indirect outcomes such as reduced accessibility to workplaces, schools, and services. Displacement of residents from such places can further exacerbate their adaptation and coping capacity, leading to the deterioration of socioeconomic conditions. Even though there are indications that people are already spontaneously moving out due to coastal flooding, it is unclear which concerns are driving this decision-making and how they can be utilized to improve coastal resilience and long-term sustainability. This chapter explores attitudes towards relocation among rural and urban coastal residents in flood-prone coastal locations in the Mid-Atlantic. It compares the drivers of willingness to consider permanent relocation in response to coastal flooding from five household surveys. The results show that residents in rural and urban settings are aware of their flood risk and are willing to consider moving elsewhere primarily due to concerns about future flood exposure, related financial considerations, and flood impacts on the overall quality of life.

Keywords Displacement · Relocation · Coastal · Flooding · Sea level rise · Surveys

---

A. Bukvic (✉)
Department of Geography, Center for Coastal Studies, Virginia Tech, Blacksburg, VA, USA
e-mail: ana.bukvic@vt.edu

C. W. Zobel
Department of Business Information Technology, Center for Coastal Studies, Virginia Tech, Blacksburg, VA, USA

© The Author(s), under exclusive license to Springer Nature Switzerland AG 2023
M. Hamza et al. (eds.), *Rebuilding Communities After Displacement*,
https://doi.org/10.1007/978-3-031-21414-1_6

6.1 Introduction

Coastal areas are rapidly changing due to accelerated sea level rise, more frequent and severe storms, and related geomorphological changes, profoundly impacting the communities and supporting systems on the shore. A growing body of risk information suggests that almost all human settlements in coastal regions will be affected to some extent by climate change and intensified coastal hazards (Alexander et al. 2012; Frey et al. 2010; Nicholls and Cazenave 2010; Sallenger et al. 2012). When these impacts exceed the households' coping capacity, individually or in combination, they will displace coastal residents, especially in areas with repetitive occurrences. The resulting coastal mobility, which is inherently a complex spatial and temporal phenomenon, will reshape spaces and places and redefine cultural and societal values in communities along the coast.

The early migration theories and more recent conceptual frameworks of human mobility in the coastal zone accounting for the complexity and emerging challenges like sea level rise offer useful insights into elements that shape relocation decision-making and process. Early research on flood-induced displacement and relocation was almost entirely risk-driven and often simplified to focus on preventing death or injury and reducing the direct costs of repetitive damages (Perry and Lindell 1997). Furthermore, such efforts were primarily undertaken in response to riverine flooding in the U.S. (Davidson 2005; Buss 2005) and internationally (Tierney and Nigg 1993).

More recently, however, the Hurricane Katrina and Sandy disasters catalyzed the research on displacement and relocation via acquisition/buyout programs in coastal areas that have been previously considered too valuable to be easily abandoned, like the river floodplains (Groen and Polivka 2010; Landry et al. 2007; Elliott and Pais 2006; Smith and McCarty 1996). Similarly, the dialogue shifted from discussing population mobility solely as hazard mitigation or disaster risk reduction strategy to envisioning it as a part of more comprehensive, forward-looking adaptation efforts. Further, such movement as an adaptation strategy has been associated with potential benefits for sending and receiving locations (Barnett and Webber 2009; Laczko and Aghazarm 2009; Leighton et al. 2011). More recently, population mobility in coastal areas has also been discussed in the context of community or disaster resilience (Binder et al. 2015).

Resilience is "the ability to prepare and plan for, absorb, recover from, or more successfully adapt to actual or potential adverse events" (National Research Council 2012, p. 16). It refers to the ability to "bounce back" from a disaster event and the related capabilities of preparing for future events and absorbing its impacts, including adapting to changing circumstances to support more effective recovery. Resilience can be considered both a capacity and a process, affected not only by the extent, frequency, and magnitude of the disruptive event, but also by the length of time over which the disruption occurs. In the case of flooding, for example, affected individuals or businesses will be impacted by the extent of the flood's impact on the local infrastructure and the length of time over which those impacts persist. They will also be affected by the recovery of their employees, the customer base, and other internal and external considerations, all of which will influence the decision to stay and cope with future flooding or relocate elsewhere.

The resulting need for a household, a business, or a community to be resilient to each experienced disruptive event can thus also be framed in terms of the sustainability of being able to maintain the overall functionality of the underlying socioeconomic systems regardless of the stressors. Sustainability can be defined as "meeting the needs of the present without compromising the ability of future generations to meet their own needs" (Brundtland and Khalid 1987). The concepts of adaptation, resilience, and sustainability provide an important lens through which the longer-term risk of displacement and preventive relocation can be assessed within a community.

Given the broader context of displacement risk in the coastal zone, it is critical to recognize the difference between short-term large-scale disruptive events like a hurricane, in which the impacts can be catastrophic in their extent and can persist for many days, weeks, or even months, and chronic, minor events such as nuisance or recurrent flooding to which a community has more time to adapt and become resilient. These latter flood events can also increase the risk of displacement due to their repetitive nature and compound impacts, particularly if they accelerate in strength and duration from one event to the next. Such escalation of impacts could deteriorate the capacity to cope with impacts or even the resolve to keep bouncing back, leading to a gradual onset of permanent displacement from coastal locations.

The awareness of climate change-driven shifts in the profile of coastal hazards and more frequent and severe chronic and episodic flooding spurred the discourse about the long-term livability in established communities along the coast—particularly concerning people's decisions to relocate. However, empirical understanding of people's ability to continue thriving under frequent-flooding conditions and coevolve with such changes is limited. It is also not well understood at which point households and whole communities in different coastal contexts would decide to relocate and what are the key components of this decision-making. The existing literature mostly studies the displacement and relocation preferences post-major disasters or presents specific case studies with unique circumstances, such as the relocation of indigenous communities due to land loss. These findings have a marginal application to other coastal places facing smaller-scale chronic or episodic flooding but still have significant implications on their livelihoods.

To address the aforementioned research gap, this chapter explores considerations driving the decision to move from flood-prone coastal locations using primary survey data. First, the chapter begins with a discussion of displacement risk and drivers of coastal displacement and relocation. Then, it compares and contrasts results from multiple household surveys about willingness to consider permanent relocation in response to coastal flooding. It also examines the associated evidence that can help shift the paradigm from reactive forced displacement to proactive planner relocation. Finally, it considers the concept of participatory relocation as an approach to enabling community self-determination, and we discuss other new opportunities for policy and planning innovation.

6.2 Literature

Relocation via property buyouts is not a new strategy. It has been successfully employed in the past, mostly in response to disastrous riverine floods, such as the Missouri buyouts after the 1993 and 1995 floods (Davidson 2005) and U.S. Army Corps projects in Arlington, Texas; Fairfax, Virginia; Hot Springs, South Dakota; and Glendive, Montana (Buss 2005). The recent coastal disasters Andrew, Katrina, and Sandy shifted the focus on the applicability of this response strategy to coastal settings (Groen and Polivka 2010; Landry et al. 2007; Elliott and Pais 2006). In each of these cases, relocation was introduced as a reactive policy rather than a preventive measure that may be necessary to address accelerated risks of recurrent coastal flooding. Very little is known about the proactive relocation in response to chronic coastal flooding when the risk and long-term community-level implications are still not fully visible. The difficulty in planning and implementing proactive relocation stems from the lack of empirical information about the key concerns that would prompt people to support this option, support needs, and the outcomes of this process on the households.

Families will still relocate in response to chronic minor flooding once they cannot cope with its impacts, such as the cost of repairs, replacement of damaged property, and significant interruption to their daily lives (Cole 2008). Regardless of the type of flooding, preventive relocation would save resources, time, and lives and allow for more careful consideration of sociodemographic, cultural, and economic determinants vital for re-establishing people's livelihoods in a new location (López-Carr and Marter-Kenyon 2015). Residents displaced by Hurricane Katrina primarily moved to nearby destinations within the economically robust urbanized counties instead of targeting less financially viable rural locations, regardless of flood exposure (Eyer et al. 2018).

Flooding can have compounding impacts on environmental and human systems and push and pull forces within them by changing their strength and direction in sending and receiving locations. Flooding and decreased natural shoreline protection (e.g., wetlands submergence and beach and dune loss) can prompt localities to prioritize coast armoring, further exacerbating coastal erosion and ecosystem degradation and amplifying impacts on the human system. Flooding can also, directly and indirectly, impact many community attributes that shape the quality of life of a place. The continued pressure of push forces and declining appeal of pull forces in sending locations would eventually lead to displacement, especially if pull forces in the receiving areas are appealing enough to facilitate this process. When living conditions exacerbate to the point of no return, a household or a community will reach a tipping point where displacement and relocation may represent the best and only viable option to solve the flood problem.

Tipping points in the social system are reached when minor changes lead to a more profound permanent change from a stable to an unstable state (Gemenne 2019). They have been explored in the literature (Gao et al. 2016) but not in the coastal context or a community scale. However, as the knowledge of cascading events and indirect flood impacts on coastal systems is expanding, it is clear that they can, in a complex interplay of interactions, simultaneously undermine the ability of households and communities to thrive. For example, sea level rise will gradually stress coastal wetlands through salinization. Eventually, if those wetlands do not manage to migrate upland or inland, they will become submerged and permanently lost, resulting in more flooding, decreased water quality and biodiversity, and shoreline erosion (Anderson et al. 2009). When these tipping points are reached, people will inevitably face the prospect of forced displacement.

The majority of the literature on relocation is focused on disaster-induced displacement. Even though disasters are likely one of the important relocation drivers, many coastal communities are being forced to move not because of experience with major disasters but due to chronic slow-onset hazard events. For example, some communities, such as Villages Newtok and Kivalina in Alaska, and Isle de Jean Charles in Louisiana, are actively pursuing permanent relocation due to coastal changes such as erosion and repetitive flooding (Bronen 2015; Lowlander Center 2015; Freudenberg et al. 2016). Current institutional and cultural preferences for coastal living still place a higher value on coastal real estate. In combination with the lack of financial support for relocation, homeowners typically opt to put their homes back on the market or rent them out.

Even in locations where in situ adaptation options are present, such as protection using shoreline hardening and natural barriers and accommodations like home elevation, homeowners may decide to move because of the uncertainty surrounding the changing conditions (Nicholls and Tol 2006). Although finding buyers or renters for a property in locations that offer some visible flood protection may be easier than in areas without any adaptation interventions, the risk of disruption can affect the business sector and economic prosperity. This is particularly true if local businesses and infrastructure are not also protected from the changing conditions. Thus, even in the presence of coastal protection features, relocation may be considered a viable adaptation strategy for some flood-prone communities (López-Carr and Marter-Kenyon 2015; King et al. 2014).

Everett Lee's Theory of Migration states that people need to be pushed from their original location or sending location and pulled by perceived benefits in receiving destination to decide to move (Fig. 6.1) (Lee 1966; Borjas 1989; Widgren and Martin 2002). Environmental problems such as extreme weather events, erosion, prolonged drought and water scarcity, and sea level rise can be important push factors forcing people to move regardless of the presence and strength of pull factors (Reuveny 2007). Push factors (or negative influences) include flooding, high cost of living, and limited access to work and services due to flooded roads and parking lots, while pull factors include positive community features such as amenities, perceived safety, and vicinity to health facilities and family. These factors differ in importance between different populations and geographic locations. For example, Wiseman (1980) notes that for older adults, factors such as availability of medical care for declining health, social support networks, cost of living, and issues with the deteriorating house may outweigh other considerations.

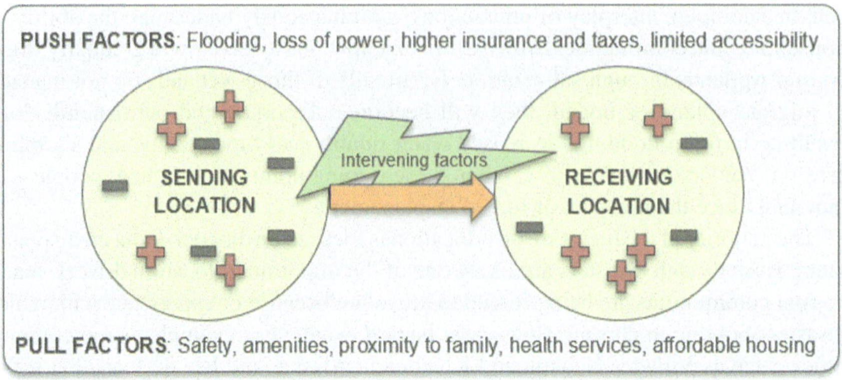

Fig. 6.1 The Everett Lee's theory of migration. Adapted from Lee (1966) and similar

Applying this theory to study the relocation of retirees to rural villages in Australia revealed that lifestyle change, home maintenance costs, social isolation, health problems, and reduced physical mobility were the key push forces, while built environment, affordability, small-community lifestyle, and desire to maintain the same life quality served as the pull migration factors (Stimson and McCrea 2004). Erickson et al. (2006) found that older adults may be pushed by a crisis to relocate but are more likely to be pulled by desirable residential arrangements and the vicinity of the family. However, such pull toward facilities that support older populations may inadvertently put them in harm's way if they are located in high-risk coastal locations. One such example is the Chamberlin retirement community in Hampton, Virginia, located on the Fort Monroe peninsula with only two access roads that frequently flood and no support amenities nearby to assist during emergencies.

Accelerated sea level rise in the Mid-Atlantic region will reshape sandy beaches, coastal headlands, spits, and barrier islands through erosion, migration, and splitting (Anderson et al. 2009). Many of these coastal landforms are also physically vulnerable to storm surge and tidal flooding while at the same time being densely populated and having an important role in local economies. Any processes that would undermine the livability, personal safety, and accessibility in such locations would serve as a strong and tenacious push force, making some of these areas priority locations for forced displacement. Even though there are many options available for shoreline protection, they may become too costly and disruptive over time. In such a case, although the most difficult and socially taxing of all adaption and resilience options, relocation may become a more viable strategy (Titus and Gill 2009).

If abandonment occurs in response to a major disaster as the ultimate push force, residents would not have an opportunity to carefully assess other push, pull, and intervening factors. Considerations that would otherwise serve as pull factors would be replaced by the necessity, availability of housing, proximity to places of importance such as work or dependent family members, and assistance stipulations. Others may witness the escalation of conditions, issues with recovery, and progressing impacts and choose to relocate proactively before the situation worsens. In many instances, this decision will be driven by concerns about property value and the ability to sell it.

Thus, slow-onset events would allow households to transition from living there to not living there on their terms and carefully assess push and pull forces in sending and receiving locations under different risk scenarios. For example, people may be willing to cope with some level of flooding if they can safely commute to work and the neighborhood still provides their favorite amenities and services. However, if the quality of life is adversely affected in any way, then the pull considerations like good schools suddenly become a push factor due to their frequent flood-related delays and closures. These pull and push forces differ between rural and urban locations and reflect the unique sociodemographic, historical, and cultural character of these settings.

For example, disaster resilience varies between urban and rural typologies and is generally higher in urban areas (Cutter et al. 2010). Rural locations have several disadvantages that shape their disaster preparedness and response capabilities, such as a lack of resources, access to training opportunities, the organizational capacity to manage resources, reliance on agriculture as the economic base, and ineffective communication between emergency responders and residents (Kapucu et al. 2013). The disaster impacts and recovery process also vary between rural and urban settings, especially in the post-disaster redistribution of vulnerabilities and length of recovery (Elliott and Pais 2010).

Rural residents are generally more self-reliant and can better identify their immediate recovery needs, organize and coordinate activities, and contribute to the response process, but they often have a lower socioeconomic status and educational attainment that may limit the level of preparedness (Brennan and Flint 2007). Such areas also tend to have higher rates of poverty, more racial and health inequalities, inadequate training opportunities, an aging population, and high out-migration rates (Cutter et al. 2010). Further, rural communities have more dispersed residential patterns, fewer resources, and limited access to contemporary equipment, technology, and training (Janssen 2006). Rural populations may also feel isolated, have reduced family and social support, and experience more challenging physical and social environments (Wainer and Chesters 2000).

In contrast, urban areas typically have higher residential densities, an aging-built environment, interdependent infrastructure and services, and overall higher complexity and potential for cascading events (Godschalk 2003). Other characteristics that shape urban profiles include a more fragmented and diversified social fabric; economic activities in manufacturing, services, and administration; higher education; easy access to amenities, information, and training opportunities; a lower sense of community; lower fertility and mortality; a more liberal political environment; more ethnical and cultural diversity; and higher in-and out-migration (Scott et al. 2007).

The available mechanisms to support relocation may influence the extent to which different rural and urban populations endorse this adaptation option. For example, buyout or acquisition programs in which property is acquired and converted into green space may be more challenging to implement in urban areas with continual development pressures and the need for housing. Particularly considering that participation in such programs is voluntary and that different neighbors may choose to participate at other times, this would result in a patchwork of green spaces without meaningful linear capability or benefits for the community. Moving a house inland on the property may not be an option in urban areas with smaller lots and densely built neighborhoods since only a few owners would have sufficient space to relocate a structure on the same lot. The same barriers hold for the Transfer of Development (TDR), a community planning tool that facilitates and incentivizes property transfer from a high-risk to a low-risk area with benefits to participating property owners and developers (Lung and Killius 2016). These options would work better in less populated settings, such as rural-urban interfaces and rural areas.

It is important to note that the interface between rural and urban areas is blurring in many locations and that migration from rural to urban areas is now based on a more comprehensive set of factors rather than just push forces in rural areas. For example, a case study in Uttarakhand, India, shows that most migrants moved for education, employment, income, and amenities that served as both push (shortage in sending location) and pull (abundance in receiving location) factors, whereas environmental degradation and declining agriculture were less important considerations (Hoffmann et al. 2019). Looking at climate change and rural-to-urban migration in Malawi, Suckall et al. (2015) observed that climate change may also affect the livelihoods of urban residents, prompting some to return to their rural locations and thus preventing or reversing rural-to-urban migration.

6.3 Methodology

This analysis presented in this chapter is focused on the U.S. East Coast Mid-Atlantic region comprising seven coastal states: New York, New Jersey, Pennsylvania, Delaware, Maryland, and Virginia (Fig. 6.2). This area has a very complex spatial configuration with rural-to-urban landscapes with some of the most populated metropolitan coastal areas. At the northern end of the region is the New York-Newark-Jersey City, NY-NJ-PA Metropolitan Statistical Area (MSA), the most heavily populated metropolitan area, with 19,261,570 residents and 2881.6 people per square mile (U.S. Census Bureau 2020). This area is also an epicenter of cultural, historical, geopolitical, and socioeconomic significance, with the concentration of financial and diplomatic capital such as corporate headquarters, international organizations, think tanks, and entrepreneurship. Near the region's center is the Washington-Arlington-Alexandria, DC-VA-MD-WV MSA metro area, which includes the seat of government for the United States.

In the southern part of the region is the Virginia Beach-Norfolk-Newport News, VA-NC MSA, which includes Naval Station Norfolk, the largest naval complex globally (CNIC 2021), and Newport News Shipbuilding, the largest industrial employer in Virginia and the nation's only builder of nuclear-powered aircraft carriers (Huntington Ingalls 2022). Beyond its cities and industry, this area is well known for its

Fig. 6.2 The U.S. Mid-Atlantic coastal region

beaches and tourism, with Virginia Beach and the historical sites of Jamestown, Yorktown, and Williamsburg, and the fishing and crabbing industry well-established along the coast of the Chesapeake Bay. The area has a rich cultural history, a long-standing Native American presence, and a continuous historical tie to some of the first European settlements in the Americas. The coastline in this region tends to be relatively rural, particularly along the Chesapeake Bay in Virginia and Maryland.

Although some portions of the Mid-Atlantic region are increasing in population, others are declining (U.S. Bureau of Labor Statistics 2022). The impacts of 2012 Hurricane Sandy on the northern parts of the region, and the subsequent recovery efforts, together with the ensuing pandemic and political changes, likely resulted in the population decline in 2020–21 compared to the previous year, especially in some coastal areas, with New York experiencing the most significant population loss of 327,955 people (Frey 2022).

This particular geographic region has moderate-to-high sea level rise risk that augments tidal and storm surge flooding and impacts densely populated urban areas, environmentally sensitive habitats, and valuable infrastructure and assets (Anderson et al. 2009). The Atlantic Coast has higher rates of relative sea level rise of 2–4 mm per year than the current global average of 1.7 mm per year, with the hot spot of physical vulnerability in the Mid-Atlantic regions between northern New Jersey and southern Virginia (Williams et al. 2009). This augmented SLR risk can be attributed to isostatic rebound, land subsidence driven by aquifer depletion, and overall low elevation and flat topography in this region (Ezer et al. 2013). The Mid-Atlantic coast also has high rates of storm-induced erosion, where even minor events can lead to beach loss, breach of barrier islands, repositioning of dunes and inlets, and other changes in coastal geomorphological features (Doran et al. 2013).

Considering that much of the coastal fringe is populated, especially on barrier islands, such storm surge and wind-driven land loss can lead to the incremental permanent displacement of residents and infrastructure. Even though this area has a lower hurricane risk than some other locations along the U.S. East Coast, it experienced three major disasters, hurricanes Isabel in 2003, Irene in 2011, and Sandy in 2012, that not only led to extensive damage in coastal communities but also served as events of social significance or benchmarks for personal risk assessment (Saunders and Senkbeil 2017). Hurricane Sandy was especially destructive, leading to widespread property damage, disruption in transportation and commercial activities, prolonged loss of power, a shortage of food, medicine, and gasoline, sewage overflows, and population displacement (White House 2012). As such, it instigated a more critical discussion on effective disaster risk reduction and adaptation strategies, including options that would help people move away from the hazard.

Figure 6.3 provides an overview of the urban and rural counties within the coastal Mid-Atlantic states. Most of the counties along the Atlantic coast are classified as urban rather than rural, except those along Virginia's Eastern Shore. It is also significant that there are several long stretches of rural counties on both Virginia and Maryland's sides of the Chesapeake Bay. This pattern partly reflects the geography of this region, and the relative lack of major highways, which make these counties less easily accessible than those in other coastal areas.

It is important to recognize that the population centers in many urban counties displayed in Fig. 6.3 are not distributed uniformly, and some contain rural places. These circumstances are particularly true further out from the major metropolitan areas. Local relocations may thus allow for transitioning from an urban to a rural location, even within the same county. As the map shows, however, relocating from

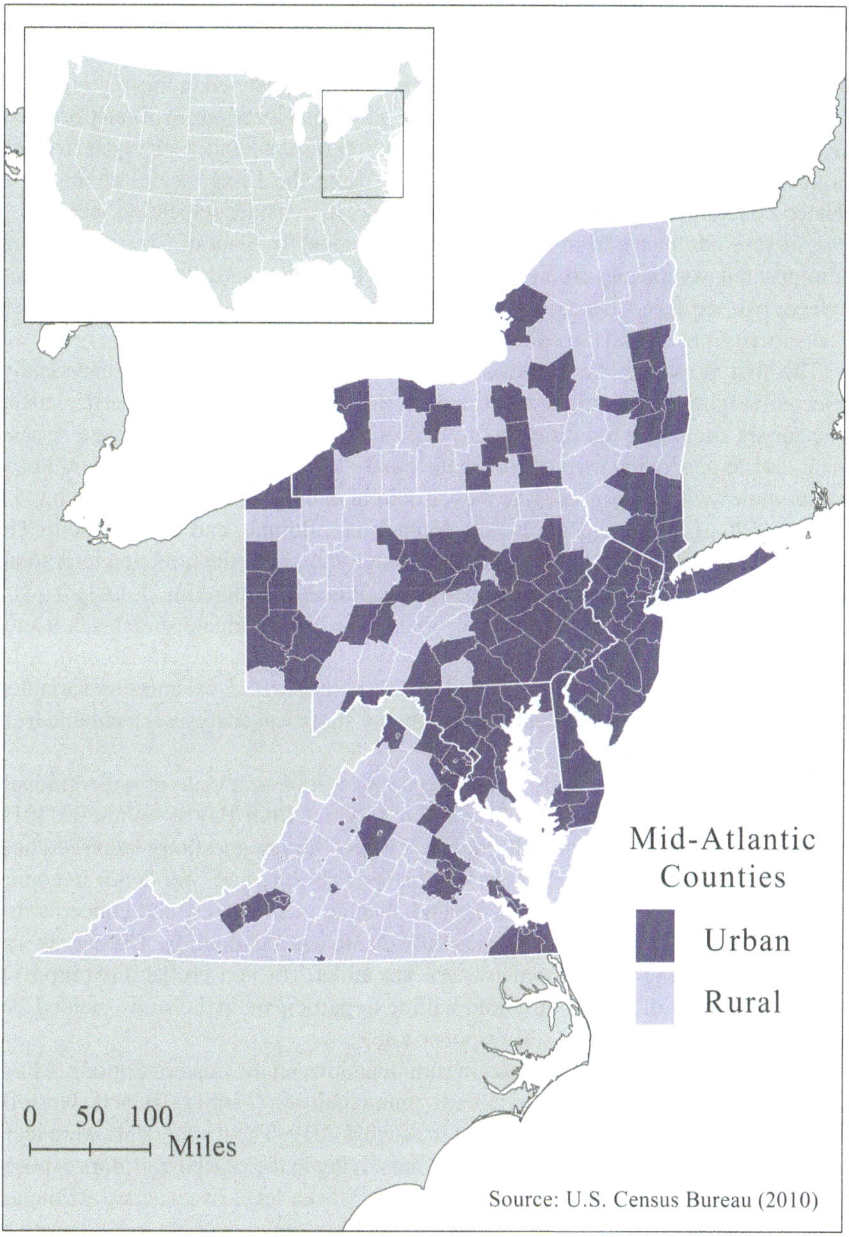

Fig. 6.3 Rural and urban counties in the Mid-Atlantic states

a primarily urban county to a primarily rural county or in the opposite way would require a relatively long-distance move in this particular region.

All surveys discussed in this chapter have been administered in this region except the last one, which was implemented in a few additional locations in urban Carolinas, Georgia, and Florida. The analysis compared the results from a survey instrument (a question with up to 16 offered responses using the Likert scale) about flood-related concerns that would prompt respondents to consider relocation included in five surveys administered in the Mid-Atlantic region. Two out of five surveys were administered in rural and urban areas to discern how attitudes differ among residents in these two settings. All surveys were administered in locations with high flood risk. The superscript responds to columns in Table 6.1.

The first presented door-to-door survey[1] was collected in Hurricane Sandy-affected neighborhoods eight months after the disaster (May 28–June 05, 2013). The survey sites were identified using FEMA's Remotely-Sensed Damage Assessment data to generate a map of Hurricane Sandy's high-impact areas in New Jersey communities with the highest levels of structural damages (Ventnor City, Longport, Margate City, Lavallette, Pine Beach, Manasquan, Belmar, and Long Beach). The survey was conducted in randomly selected high-density residential neighborhoods with single-family homes within a one-mile radius of the shoreline. It had 20 quantitative multiple-choice or ranking-type questions. Two graduate students collected 125 surveys during daylight hours, each session typically lasting 15–20 min. After data cleaning, the final sample included 118 responses with 5.6% missing variables. Additional information about survey design and statistical analysis is available from Bukvic et al. (2015).

The second survey[2] was also collected door-to-door by a team of undergraduate and graduate students in the City of Hampton, Virginia, from May 30 to June 09, 2019, in flood-prone residential neighborhoods. Twenty survey questions inquired about the respondent's socioeconomic characteristics, exposure and confidence in coping with flooding, preferred and implemented adaptation strategies, and concerns that would prompt them to relocate permanently due to coastal flooding. The survey was administered among coastal homeowners on a tablet. The non-probability purposive sampling strategy of every household willing to participate in the study yielded 200 completed surveys with a 14.5% response rate.

The third survey[3] was conducted in rural locations on the Eastern Shore in Maryland and in the urban Hampton Roads municipalities (Virginia Beach, Norfolk, Portsmouth, and Hampton) in Virginia in August 2019. The respondents were identified among a random sample of homeowners living in the coastal corridors exposed to the storm surge adjusted for this region's 2090 sea level rise scenario (Mitchell et al. 2022). The survey had 18 qualitative questions and two open-ended questions exploring past exposure to flooding, reasons and willingness to relocate, and factors important for selecting the receiving location (Bukvic and Zobel 2022).

The fourth survey[4] was conducted in the same rural and urban coastal locations prone to flooding using a phone modality in March 2021. The respondents were identified from the phone directory of households within designated flood-prone areas and asked to complete the same survey with 20 quantitative questions (Bukvic

Table 6.1 Synthesis of the key considerations that would prompt homeowners to consider relocating due to coastal flooding from five surveys administered in the Mid-Atlantic region

Surveys' results					
Place	New Jersey[1]	Hampton[2]	Multi-locations[3]	Multi-locations[4]	Multi-locations[5]
State	New Jersey	Virginia	Maryland and Virginia	Maryland and Virginia	East Coast: New York to Florida
Date	June 2013	June 2019	August 2020	March 2021	May 2021
Main risk	After Hurricane Sandy	Nuisance flooding	Nuisance flooding	Nuisance and storm surge flooding	Nuisance and storm surge flooding
Type	Door-to-door	Door-to-door	Mail	Phone	Online
Setting	Urban	Urban	Rural and urban	Rural and urban	Urban
Sample	$N = 118$	$N = 200$	$N = 102$	$N = 325$	$N = 1450$
I would consider permanently relocating from this community to a less flood-prone location if					
Most important considerations (Strongly agree and agree categories)					
1	Crime becomes worse	Insurance cannot cover flood damage	You experienced disastrous flooding	I cannot access services and amenities due to flooding	Crime becomes worse
2	>2 floods in next few years	Crime becomes worse	Insurance cannot cover flood damage	Crime becomes worse	I cannot access services and amenities due to flooding
3	Insurance cannot cover full reconstruction	I cannot access services and amenities due to flooding	Crime became worse	Insurance cannot cover flood damage	Insurance cannot cover the flood damage
4	Services and amenities do not restore their full function	I am offered with financial compensation (buyout)	You experienced damage to my home or vehicle	I am offered with financial compensation (buyout)	I cannot afford to pay taxes for a coastal property
5	We have one more flood in the next few years	I cannot afford to pay taxes for a coastal property	You were offered with financial compensation (buyout)	I cannot afford to pay taxes for a coastal property	I am offered with a comparable housing in a similar comm elsewhere

(continued)

Table 6.1 (continued)

Least important considerations (Strongly agree and agree categories)					
1	I can move together with my neighbors	I can move together with my neighbors	You could have a land/sea-tied job (e.g., fishing, farming, hunting, crabbing, etc.)	I can move together with my neighbors	I can move together with my neighbors
2	I receive assistance with finding a new job elsewhere	I experience one more flood in the next year	You experienced sporadic minor flooding (one per year)	I experience one more flood in the next year	My neighbors, friends, and family move out
3	I am provided with free legal advice on my options	Neighbors, friends, and family move out	You can move together with your neighbors	I receive assistance with finding a new job elsewhere	I experience one more flood in the next year
4	I am offered with comparable housing in similar community elsewhere	I receive assistance with finding a new job elsewhere	You experienced frequent minor flooding (two or more times per year)	Neighbors, friends, and family move out	I experience two or more floods in the next year
5	Neighbors, friends, and family move out	Businesses are closed	You could have a non-land-tied job (health industry, administration, education, tourism)	I am provided with free legal advice on my options	I receive assistance with finding a new job elsewhere

and Barnett 2022a). The final fifth survey[5] was the most extensive and targeted urban coastal residents in multiple locations on the U.S. East Coast, from New York to Florida. It was administered using Qualtrics online platform and research panels from May 25 to June 17, 2021. Respondents living in urban coastal zip codes bordering open water or a tidal estuary and having high flood exposure were invited to complete the survey with 20 quantitative questions using a smartphone or another digital device (Bukvic and Barnett 2022b).

6.4 Findings

The synthesis of survey instrument probing reasons for considering permanent relocation due to flooding shows that regardless of the prevalent coastal hazard and recent flood exposure (e.g., experience with minor nuisance flooding or a major hurricane disaster), respondents in flood-prone areas think similarly about the relocation drivers (Table 6.1). The surveys were administered in different geographic locations using various modalities but mostly the same questions. This approach allowed us to compare the most relevant response about concerns that prompted respondents to say yes to the possibility of relocation due to the actual or perceived risk of coastal flooding. These responses were similar across different time scales, from 2013 to more recent 2021, indicating that priority concerns did not notably change over time. Respondents would consider relocating in the future primarily due to financial considerations such as insurance rates and tax increases, frequent flooding, and increased crime.

According to the results, the most critical factors that would lead the respondents to consider permanently relocating to a less flood-prone location were worsening crime and insufficient insurance to cover flood damages. Although these are very different issues, they are related to a decreasing quality of life, and each is ranked among the three most important considerations in all five surveys. Four out of the five surveys also rank an inability to access available services and amenities due to flooding as one of the top five issues. As with crime and the cost of flood damages, this concern can persist over a more extended period and lead to cascading events that would further affect the quality of life in a coastal place. For example, flooded roads would prompt affected residents to seek alternative routes to the workplace or stores, potentially leading to traffic congestion, long commutes, and public health issues.

Three of five surveys indicated that the financial compensation (buyout) would be a significant factor in considering relocation. Three surveys emphasized the concern with paying taxes for coastal property. None of these considerations are related to a specific type of flooding or disaster event. Instead, they are much more closely tied to the issue of long-term livability and the feasibility of living in a coastal location, given the aggregate costs of ongoing flooding. Interestingly, only the homeowners exposed to Hurricane Sandy listed flood risk in their top five considerations. The respondents to the mail survey chose disastrous flooding as their top priority. This observation suggests that the experience with flood exposure (e.g., potential property damage, loss of service, and impact on finances) is a more relevant push factor than the overall concern with future flooding.

In contrast to the most important factors, those least influential were associated with the relocation behavior of the respondents' friends, family, and neighbors, not financial assistance. All five surveys listed the opportunity to move together with their neighbors as unimportant, with four indicating that this was the least detrimental factor. In four surveys, having friends, family, and neighbors move out was also considered a less important issue. In contrast to the relative importance of receiving financial compensation to relocate discussed above, assistance with finding a new job in the host location was one of the least important considerations in all five surveys. Free legal advice was also considered less important in several surveys, as was an option to provide comparable housing elsewhere. This finding suggests that, from a relocation management perspective, providing amenities directly related to relocation efforts is less critical than providing direct financial compensation to those who may have to move away from a coastal area.

In terms of the actual impacts of flooding, all four surveys unrelated to the Sandy disaster were conducted in areas prone to nuisance flooding. In each case, the flooding itself was not considered very important. The flooding was identified as an important consideration only in New Jersey's survey conducted after Hurricane Sandy. In Hampton, Virginia, respondents were explicitly unconcerned about closed businesses due to flooding. This outcome is likely because of an overall lack of commercial zones in the Hampton Roads area and the population's reliance on shopping and services in other parts of the metro region that are generally on higher ground.

6.5 Discussion

This chapter demonstrates that many coastal residents living in flood-prone locations are willing to engage in discourse on relocation. The results build upon the growing body of literature aiming to discern the fundamental concerns driving mobility in coastal areas. For example, Bukvic and Owen (2017) found that 49% of respondents would consider relocation/buyout after the Hurricane Sandy disaster or in the future primarily due to concerns with storm surge flooding and permanent sea level rise. This result indicates elevated awareness of episodic and chronic hazards in their coastal location. In the same survey, most respondents ranked unreliable transportation, blackouts, and unsafe drinking water as the second most important concern, suggesting that respondents are almost equally concerned with disaster-induced disruption in services and quality of life as with the actual coastal hazards. They also did not think that their neighbors, friends, and family's decision to leave the community (61%) was important factor in staying and recovering after the disaster or relocating (Bukvic and Owen 2017).

More recently, semi-structured interviews with local stakeholders engaged in coastal adaptation in rural communities on the Eastern Shore in Maryland and urban locations in Hampton Roads found there was a striking divergence in risk perceptions and confidence in staying in place between different localities due to their unique historical, economic, and sociocultural context (Bukvic and Harrald 2019). The same study identified other challenges and trends in local jurisdictions that significantly affect their adaptive capacity and ability to respond to coastal flooding (e.g., outmigration of younger people, problems with aging infrastructure, discontinued public services, and economic downturn). For example, many older coastal residents who require medication/regular medical assistance also reside in primary housing easily rendered inaccessible for days due to nuisance flooding. At the same time, the same residents are often inadequately prepared for the possibility of prolonged flooding due to a lack of resources for preventive measures and misaligned risk perceptions.

The empirical evidence presented in this chapter indicates which specific considerations can be explored to guide policy actions and programmatic incentives/disincentives for proactive relocation planning. This information is vital to ensure that all coastal populations have access to appropriate support needs to relocate if this option becomes necessary. The socially vulnerable coastal areas face a high risk of sea level rise-driven displacement due to their financial inability to afford structural protection, difficulty justifying public expenses, and the lack of social and political support for more orderly relocation (USGCRP 2014). There are many reasons why residents may delay relocating until they are forced to leave due to catastrophic or recurrent flooding. Many will simply be unable to move due to complex personal sociodemographic and economic circumstances, place-based dependencies, and external influences such as the inability to sell the home, legal predicaments, and proximity to the workplace. Such populations are often called "trapped" due to their inability to move away from the risk and improve their human condition (Ayeb-Karlsson et al. 2018).

Relocation should be implemented as a preventive measure to comprehensively assess sociocultural, political, and economic dimensions to improve relocation outcomes (López-Carr and Marter-Kenyon 2015). The research on coastal risks is already well-defined on both short- and long-term horizons but not on the implications for the human systems and adaptation pathways, especially those involving the significant reorganization of the social system and built environment to accommodate those changes proactively. The preference for maintaining the status quo stemming from this spatiotemporal gap in risk management will postpone action until people are forced to move by a major disaster or recurrent hazard events.

Early discussions on the relocation process should focus on whether it should be encouraged, required, and paid for and over what time it should be implemented (Blanco et al. 2009). For example, some communities have local ordinances that allow for the condemnation and acquisition of substantially damaged homes with structural damages exceeding 50% or carrying health and safety risks to occupants (Tate et al. 2016). Relocation is not a novel response and can be advanced on the vast knowledge about what works and what does not from examples of a successful relocation due to riverine flooding where flood losses were significantly reduced (Pinter and Rees 2021).

Our multi-survey results suggest that incentives (e.g., financial compensation and comparable housing in a similar community elsewhere) and disincentives (e.g., higher insurance policies and taxes) can play an important role in influencing coastal mobility and designing relocation programs. For example, New York State Governor Andrew Cuomo established a buyout program to incentivize relocation by compensating participating homeowners with the pre-storm market value of their home and offering an additional five percent bonus to those who stay locally or 10% to groups who sell collectively (Kaplan 2013). However, these programs generally have low participation rates (Braidwood 2013), perhaps due to their simplistic design solely based on financial exchange, disregarding other household needs. Another reason for limited buyouts participation may be the unsuccessful effort to engage stakeholders in participatory dialogue on the actual risks and realistic response options. Considering individual households have different needs and preferences for relocation assistance, the ideal relocation programs would include flexible, incentivized, and customized features rather than standardized assistance packages, strict deadlines, and equal participation requirements.

Barnett and Webber (2009) state that the fewer choices people have for moving, the more likely the outcomes of that movement will be adverse. Strategies that allow people to select between diverse options, such as financial compensation, replacement housing, new employment, or any combination of these and other provisions, may represent a more attractive incentive to potential relocates. The psychological dimensions of disaster experience may also contribute to relocation decisions and are often overlooked in assessing relocation drivers. For example, resource loss can be a valuable predictor of willingness to relocate by measuring how individuals personally and objectively value their resources and tend to experience psychosocial deterioration due to their loss (Hobfoll 1991).

Decisions made today will pave the road for what can be done and what problems will emerge in the future. Even though investing in protection may be more appealing for coastal locations today, solely relying on this approach can foster a sense of false security, encourage further development, and make significant financial investments in unsuitable long-term options. Most risk information about coastal hazards and climate change impacts is forward-looking, providing models and simulations of future flooding that reflect accelerated sea level rise and other anticipated morphological changes. Planning for population mobility aligned with this information will be crucial to prevent spontaneous displacement and support anticipatory relocation that may lead to better outcomes. This is not always easy to achieve on a local level, considering that a decision to support relocation may signal to some that other options are not possible and send a community on a downward trajectory.

Displacement and unstructured relocation can have ramifications, such as social and cultural costs, disruption of place attachment, abandonment of the built environment, impacts on infrastructure, and hollowed economies. One factor contributing to the level of "entrapment" of socially vulnerable coastal populations is community-based coping and adaptive responses that can create or limit conditions conducive to mobility (Ayeb-Karlsson et al. 2018). Recognition of comprehensive factors that shape mobility decisions, such as proximity to places of work, schools, family, and other aspects that may anchor families in place is equally important. For example, if a family postpones moving due to proximity to a dependent older adult, then mobility needs to be assessed to account for adverse impacts displacement would have on the elderly-care obligations.

6.6 Conclusions

Coastal residents are already being displaced from many coastal communities, albeit incrementally and spontaneously, reflecting deeply personal considerations related to direct and indirect flood impacts. Considering the localized nature of such displacement and the fact that most homeowners are forced to leave, resell or rent their homes, it is difficult to empirically discern the actual patterns and implications of this flood-driven outcome. The mass displacement from coastal areas is usually associated with disaster evacuation that often morphs into a temporary or permanent relocation. However, piecemeal displacement due to deteriorating post-disaster conditions resulting from slow recovery and recurrent flooding is also widespread but less evident. It is expected to accelerate with accelerated sea level rise and related augmentation of compound tidal, storm surge, and rainfall flooding. Understanding how to shift the discourse on relocation to prevent some of the adverse impacts of sudden displacement will be critical to coastal adaptation and resilience.

This chapter examined attitudes toward permanent relocation due to coastal flooding among rural and urban residents living in flood-prone locations. It synthesized concerns that would prompt such respondents to relocate from five surveys administered over different spatiotemporal scales in the Mid-Atlantic coastal region. Respondents in all surveys converge on some key factors that would prompt them to consider moving out, namely financial concerns with owning a coastal property (insurance coverage for flood damages and taxes), increase in crime, and future flood risk. Respondents also expressed interest in relocation of offered with fair compensation (buyout) or alternative housing elsewhere. These priorities observed among over 2000 respondents indicate which flood-related interventions are essential to coastal residents living in flood-prone corridors.

Financial instruments can be used as mobility incentives or disincentives, prompting people to move once they cannot pay high taxes or insurance policies that cover all flood losses. They can also signal flood risk, but these are not economically and politically popular interventions, especially in communities where revenues from coastal development are still exceeding the cost of dealing with flood damages. At the same time, local governments can justify the cost of living in a coastal location by investing more in flood risk reduction, increasing homeowners' confidence in staying and coping with flooding.

Most importantly, these findings can help localities develop more comprehensive adaptation and resilience strategies that are not narrowly focused on hazard mitigation but rather take a holistic perspective on all factors necessary for mobility management areas such as crime and affordable housing. Acknowledgment of needs and preferences among coastal relocatees can further spur innovation and new opportunities for sustainable urban and rural development in locations that can accommodate the incremental influx of residents from the coastal periphery.

Acknowledgements Authors are thankful to Margaret Crawford for the preparation of Figs. 6.2 and 6.3. This material is partly based upon work supported by the National Science Foundation under Grant No. 1920478.

References

Alexander KS, Ryan A, Measham TG (2012) Managed retreat of coastal communities: understanding responses to projected sea level rise. J Environ Plan Manag 55(4):409–433

Anderson KE, Cahoon DR, Gill SK, Gutierrez BT, Thieler ER, Titus JG, Williams SJ (2009) Executive summary. In: Coastal sensitivity to sea-level rise: a focus on the Mid-Atlantic region. A report by the U.S. climate change science program and the subcommittee on global change research. U.S. Environmental Protection Agency, Washington, DC, pp 1–8

Ayeb-Karlsson S, Smith CD, Kniveton D (2018) A discursive review of the textual use of 'trapped' in environmental migration studies: the conceptual birth and troubled teenage years of trapped populations. Ambio 47(5):557–573

Barnett J, Webber M (2009) Accommodating migration to promote adaptation to climate change, commission on climate change and development. Policy brief prepared for the Secretariat of the Swedish Commission on Climate Change and Development and the World Bank Report, 2010

Binder SB, Baker CK, Barile JP (2015) Rebuild or relocate? Resilience and postdisaster decision-making after Hurricane Sandy. Am J Community Psychol 56(1):180–196

Blanco H, Alberti M, Forsyth A, Krizek KJ, Rodriguez DA, Talen E, Ellis C (2009) Hot, congested, crowded and diverse: emerging research agendas in planning. Prog Plan 71(4):153–205

Borjas GJ (1989) Economic theory and international migration. Int Migr Rev 23(3):457–485

Braidwood R (2013) Telephone correspondence, 26 Sept 2013. Office of Emergency Management, Chesapeake, Virginia

Brennan MA, Flint CG (2007) Uncovering the hidden dimensions of rural disaster mitigation: capacity building through community emergency response teams. J Rural Soc Sci 22(2):7

Bronen R (2015) Climate-induced community relocations: using integrated social-ecological assessments to foster adaptation and resilience. Ecol Soc 20(3)

Brundtland GH, Khalid M (1987) Our common future. UN General Assembly document A/42/427. Oxford University Press, Oxford, pp 1–374

Bukvic A, Barnett S (2022a) Role of place attachment in relocation decision-making: perspectives of rural and urban coastal residents (Manuscript in preparation)

Bukvic A, Barnett S (2022b) What drives urban coastal residents to consider relocation? Insights from the U.S. East Coast. J Environ Manag 325(116429)

Bukvic A, Harrald J (2019) Rural versus urban perspective on coastal flooding: the insights from the US Mid-Atlantic communities. Clim Risk Manag 23:7–18

Bukvic A, Owen G (2017) Attitudes towards relocation following Hurricane Sandy: should we stay or should we go? Disasters 41(1):101–123

Bukvic A, Zobel C (2022) Attitudes toward relocation in rural and urban flood-prone jurisdictions on the U.S. East Coast (manuscript in preparation)

Bukvic A, Smith A, Zhang A (2015) Evaluating drivers of coastal relocation in Hurricane Sandy affected communities. Int J Disaster Risk Reduct 13:215–228

Buss LS (2005) Nonstructural flood damage reduction within the US Army Corps of Engineers. J Contemp Water Res Educ 130(1):26–30

CNIC (2021) Naval station Norfolk. [Online]. Available at: https://www.cnic.navy.mil/regions/cnrma/installations/ns_norfolk.html. Accessed 24 Apr 2022

Cole DW (2008) Sea level rise: technical guidance for Dorchester County. Coastal Zone Management Division, Department of Natural Resources, Annapolis, MD

Cutter SL, Burton CG, Emrich CT (2010) Disaster resilience indicators for benchmarking baseline conditions. J Homel Secur Emerg Manag 7(1)

Davidson B (2005) How quickly we forget: the National Flood Insurance Program and floodplain development in Missouri. Wash Univ J Law Policy 19:365

Doran KS, Stockdon HF, Sopkin KL, Thompson DM, Plant NG (2013) National assessment of hurricane-induced coastal erosion hazards: Mid-Atlantic coast. U.S. geological survey open-file report 2013-1131, p 28

Elliott JR, Pais J (2006) Race, class, and Hurricane Katrina: social differences in human responses to disaster. Soc Sci Res 35(2):295–321

Elliott JR, Pais J (2010) When nature pushes back: environmental impact and the spatial redistribution of socially vulnerable populations. Soc Sci Quart 91(5):1187–1202

Erickson MA, Krout J, Ewen H, Robison J (2006) Should I stay or should I go? Moving plans of older adults. J Hous Elderly 20(3):5–22

Eyer J, Dinterman R, Miller N, Rose A (2018) The effect of disasters on migration destinations: evidence from Hurricane Katrina. Econ Disasters Clim Change 2(1):91–106

Ezer T, Atkinson LP, Corlett WB, Blanco JL (2013) Gulf Stream's induced sea level rise and variability along the US Mid-Atlantic coast. J Geophys Res Oceans 118(2):685–697

Freudenberg R, Calvin E, Tolkoff L, Brawley D (2016) Buy-in for buyouts: the case for managed retreat from flood zones. Lincoln Institute of Land Policy, Cambridge, MA

Frey AE, Olivera F, Irish JL, Dunkin LM, Kaihatu JM, Ferreira CM, Edge BL (2010) Potential impact of climate change on hurricane flooding inundation, population affected and property damages in Corpus Christi. J Am Water Resour Assoc 46(5):1049–1059

Frey WH (2022) New census data shows a huge spike in movement out of big metro areas during the pandemic. Brookings The Avenue. Published April 14, 2022

Gao J, Barzel B, Barabási AL (2016) Universal resilience patterns in complex networks. Nature 530(7590):307–312

Gemenne F (2019) How climate-related tipping points can trigger mass migration and social chaos. Essay for Perry World House, in the Bulleting of the Atomic Scientists. Available at: https://thebulletin.org/2019/11/how-climate-related-tipping-points-can-trigger-mass-migration-and-social-chaos/. Accessed 29 Apr 2022.

Godschalk DR (2003) Urban hazard mitigation: creating resilient cities. Nat Hazards Rev 4(3):136–143

Groen JA, Polivka AE (2010) Going home after Hurricane Katrina: determinants of return migration and changes in affected areas. Demography 47(4):821–844

Hobfoll SE (1991) Traumatic stress: a theory based on rapid loss of resources. Anxiety Res 4(3):187–197

Hoffmann EM, Konerding V, Nautiyal S, Buerkert A (2019) Is the push-pull paradigm useful to explain rural-urban migration? A case study in Uttarakhand, India. PLoS ONE 14(4):e0214511

Huntington Ingalls (2022) Newport news shipbuilding. [Online]. Available at: https://newsroom.hii.com/presskits/newport-news-shipbuilding-facts. Accessed 24 Apr 2022

Janssen D (2006) Disaster planning in rural America. Public Manager 35(3):40

Kaplan T (2013) Cuomo seeking home buyouts in flood zones. The New York Times, 3 Feb 2013. Available at: http://www.nytimes.com/2013/02/04/nyregion/cuomo-seeking-home-buyouts-in-flood-zones.html?_r=0. Accessed 16 Dec 2015

Kapucu N, Hawkins CV, Rivera FI (2013) Disaster preparedness and resilience for rural communities. Risk Hazards Crisis Public Policy 4(4):215–233

King D, Bird D, Haynes K, Boon H, Cottrell A, Millar J, Okada T, Box P, Keogh D, Thomas M (2014) Voluntary relocation as an adaptation strategy to extreme weather events. Int J Disaster Risk Reduct 8:83–90

Laczko F, Aghazarm C (eds) (2009) Migration, environment and climate change: assessing the evidence. International Organization for Migration, Geneva, Switzerland, Geneva

Lagadec P (2009) A new cosmology of risks and crises: time for a radical shift in paradigm and practice. Rev Policy Res 26(4):473–486

Landry CE, Bin O, Hindsley P, Whitehead JC, Wilson K (2007) Going home: evacuation-migration decisions of Hurricane Katrina survivors. South Econ J 74(2):326–343

Lee ES (1966) A theory of migration. Demography 3(1):47–57

Leighton M, Shen X, Warner K (eds) (2011) Climate change and migration: rethinking policies for adaptation and disaster risk reduction. Source, no. 15. UNU-EHS, Bonn, Germany

López-Carr D, Marter-Kenyon J (2015) Human adaptation: manage climate-induced resettlement. Nature 517(7534):265–267

Lowlander Center (2015) Resettlement as a resilience strategy and the case of Isle de Jean Charles. Version 1.0, Oct 2015, Louisiana

Lung J, Killius M (2016) Tools for a resilient Virginia coast: designing a successful TDR program for Virginia's middle peninsula. Virginia Coastal Policy Center, 25, Virginia, US

Mitchell A, Bukvic A, Shao Y, McLaughlin D, Irish J (2022) Assessing the impacts of coastal flooding on a watershed scale in Hampton Roads, Virginia (manuscript under review)

National Research Council (2012) Disaster resilience: a national imperative. The National Academies Press, Washington, DC

Nicholls RJ, Cazenave A (2010) Sea-level rise and its impact on coastal zones. Science 328(5985):1517–1520

Nicholls RJ, Tol RS (2006) Impacts and responses to sea-level rise: a global analysis of the SRES scenarios over the twenty-first century. Philos Trans R Soc A Math Phys Eng Sci 364(1841):1073–1095

Perry RW, Lindell MK (1997) Principles for managing community relocation as a hazard mitigation measure. J Contingencies Crisis Manag 5(1):49–59

Pinter N, Rees JC (2021) Assessing managed flood retreat and community relocation in the Midwest USA. Nat Hazards 107(1):497–518

Reuveny R (2007) Climate change-induced migration and violent conflict. Polit Geogr 26(6):656–673

Sallenger AH, Doran KS, Howd PA (2012) Hotspot of accelerated sea-level rise on the Atlantic coast of North America. Nat Clim Change 2(12):884–888

Saunders ME, Senkbeil JC (2017) Perceptions of hurricane hazards in the Mid-Atlantic region. Meteorol Appl 24(1):120–134

Scott A, Gilbert A, Gelan A (2007) The urban-rural divide: myth or reality? Macaulay Institute, Aberdeen

Smith SK, McCarty C (1996) Demographic effects of natural disasters: a case study of Hurricane Andrew. Demography 33(2):265–275

Stimson RJ, McCrea R (2004) A push–pull framework for modelling the relocation of retirees to a retirement village: the Australian experience. Environ Plan A 36(8):1451–1470

Suckall N, Fraser E, Forster P, Mkwambisi D (2015) Using a migration systems approach to understand the link between climate change and urbanization in Malawi. Appl Geogr 63:244–252

Tate E, Strong A, Kraus T, Xiong H (2016) Flood recovery and property acquisition in Cedar Rapids, Iowa. Nat Hazards 80(3):2055–2079

Tierney KJ, Nigg JM (1993) Disasters and social change: consequences for community construct and affect. Disaster Research Center, Delaware

Titus JG, Gill SK (2009) Part II overview: societal impacts and implications. In: Titus JG (coordinating lead author), Anderson KE, Cahoon DR, Gesch DB, Gill SK, Gutierrez BT, Thieler ER, Williams SJ (lead authors) Coastal sensitivity to sea-level rise: a focus on the Mid-Atlantic region. A report by the U.S. Climate Change Science Program and the Subcommittee on Global Change Research. U.S. Environmental Protection Agency, Washington, DC, pp 85–86

U.S. Census Bureau (2020) American community survey 5-year estimates. Retrieved from Census Reporter Profile page for New York-Newark-Jersey City, NY-NJ-PA Metro Area. http://censusreporter.org/profiles/31000US35620-new-york-newark-jersey-city-ny-nj-pa-metro-area/. Accessed 15 Apr 2022

U.S. Bureau of Labor Statistics (2022) Mid-Atlantic information office. Available at: https://www.bls.gov/regions/midatlantic. Accessed August 22, 2022

USGCRP United States Global Change Research Program (2014) 3rd national climate assessment report. Washington, DC

Wainer J, Chesters J (2000) Rural mental health: neither romanticism nor despair. Aust J Rural Health 8(3):141–147

White House (2012) The Hurricane Sandy funding needs. Available at: https://www.esf.edu/glrc/library/documents/OMBLettertoHouseSpeakeronFundingNeedspostSandy_2012.pdf. Accessed 29 Apr 2022

Widgren J, Martin P (2002) Managing migration: the role of economic instruments. Int Migr 40(5):213–229

Williams SJ, Gutierrez BT, Titus JG, Gill SK, Cahoon DR, Thieler ER, Anderson KE, FitzGerald D, Burkett V, Samenow J (2009) Sea-level rise and its effects on the coast. In: Coastal sensitivity to sea-level rise: a focus on the Mid-Atlantic region. A report by the U.S. Climate Change Science Program and the Subcommittee on Global Change Research. U.S. Environmental Protection Agency, Washington, DC, pp 1–8

Wiseman RF (1980) Why older people move: theoretical issues. Res Aging 2(2):141–154

Chapter 7
Living with Landslide Risks: A Case of Resistance to Relocation Among Vulnerable Households Residing in the Kegalle District of Sri Lanka

K. L. W. I. Gunathilake, P. P. G. Jayathilake, Nishara Fernando, Naduni Jayasinghe, Dilanthi Amaratunga, and Richard Haigh

Abstract Landslides are fatal climate-induced natural disasters that leave more people homeless than any other calamity in Sri Lanka. Relocation is a frequently adopted policy initiative led by the government and related authorities in post-disaster reconstruction in Sri Lanka. The hesitancy of most vulnerable communities to relocate to new places has rendered relocation efforts difficult to execute and unsatisfactory. The current study was carried out using a purposively drawn sample of 21 residents who have refused to move away from their current place of residence following the Samasarakanda landslide which struck the area of Aranayaka in 2016. The main objective of this study is to provide an in-depth overview of their reluctance to relocate and suggest measures to strengthen policy planning measures associated with the relocation of vulnerable communities. Primary data collection was carried out using a questionnaire survey, and data were analyzed through the integrative use of Microsoft Excel, SPSS, and Arc GIS 10.5 analytical tools. The study revealed that 33% of the dwellers have wholly refused to relocate, while 66% have a partial willingness to cooperate with the process in the future. The respondents who ultimately declined to relocate have highlighted their lack of trust in government provisions (53%), insufficient financial support (73.7%), and the barriers to income gains through self-employment (38%) as the main reasons to refuse the process. The study further revealed lack of education among the dwellers, where the majority (80%) have not completed their secondary education, as one of the main reasons for a lack of understanding about the long-term benefits of relocation. Further, 24% of the informants had a clear idea about disaster risk reduction

K. L. W. I. Gunathilake (✉)
Department of Geography, University of Kelaniya, Kelaniya, Sri Lanka
e-mail: wathsalaindeevarie@gmail.com; wathsalag@kln.ac.lk

P. P. G. Jayathilake · N. Jayasinghe
Social Policy Analysis and Research Centre, University of Colombo, Colombo, Sri Lanka

N. Fernando
Department of Sociology, University of Colombo, Colombo, Sri Lanka

D. Amaratunga · R. Haigh
Global Disaster Resilience Centre, University of Huddersfield, Huddersfield, UK

© The Author(s), under exclusive license to Springer Nature Switzerland AG 2023

M. Hamza et al. (eds.), *Rebuilding Communities After Displacement*,
https://doi.org/10.1007/978-3-031-21414-1_7

pre-preparation plans, whereas most (76%) had no idea about disaster management plans. According to the findings of the study, policy planning associated with relocation should be aligned more with better coordination of socioeconomic and cultural requirements of the dwellers. A balanced framework of rational decision-making on resettlement alternatives and options, safety assessment, accessibility to existing public and social infrastructure and livelihood opportunities, community involvement in house designing, and consideration of socio-cultural values of the affected communities are urgent necessities in Sri Lanka with regard to the design and execution of natural disaster-induced relocation.

Keywords Resistance · Livelihoods

7.1 Background

Climate change has increased the frequency and intensity of extreme weather events and environmental deterioration (GCS 2018). Natural disasters have caused direct losses of 35 billion US dollars per year in developing nations during the last decade. These costs are growing, and they are now more than eight times higher than the losses caused by natural disasters in the 1960s (EM-DAT 2009). Floods, earthquakes, and tropical cyclones are the three primary kinds of natural catastrophes that account for 90% of the world's direct losses. People in developing countries bear a disproportionate share of the deaths and damage caused by such environmental shocks (Kahn 2005). Natural catastrophes do not always make developing countries vulnerable, but their effects are often more severe due to poor construction techniques and insufficient infrastructure. Given the magnitude of these losses, attention has been drawn to how natural disasters might jeopardize a country's long-term efforts to achieve and sustain economic growth (Freeman 2000).

Each year, natural disasters cause new internal displacements—18.8 million in 2017, compared to 11.8 million displaced by violence or war (IDMC 2017). Some governments are relocating or considering relocating the people from high-risk areas, either as a proactive measure before a disaster strikes or as a reactive response in the aftermath of a natural disaster.

The international community increasingly acknowledges that relocation may be necessary to protect people from disasters or climate change damage (IASC 2011). The Sendai Framework for Disaster Risk Reduction (SFDRR) encourages countries to develop public policies "to relocate" human settlements in disaster-prone areas and consider relocating public facilities and infrastructures to areas outside the risk range, wherever possible, during the post-disaster reconstruction process. The 2015 Guidance on Protecting People from Disasters and Environmental Change via Planned Relocation and the accompanying 2017 operational Toolbox seek to provide guidelines to states, communities, and other relevant stakeholders on the design and execution of natural disaster-induced relocation initiatives (UNHCR 2015). The International Law Association adopted the Sydney Declaration of Principles on the

Protection of Persons Displaced in the Context of Sea Level Rise in 2018, giving states guidance on avoiding, mitigating, and addressing displacement in the face of sea-level rise, including through evacuations and planned relocations (SFDRR 2015).

While the SFDRR highlights the significance of anticipatory relocation measures in reducing disaster risk, the Conference of Parties to the United Nations Framework Convention on Climate Change, meeting in Cancún in 2010, advocated relocation as a climate change adaptation measure (UNHCR 2015). An attempt to decipher relocation as a disaster risk reduction strategy calls for demarcation of relocation from evacuation measures. The term "evacuation" describes the quick physical transfer of people away from an immediate threat or impact of a hazard to a safer place. Evacuations are frequently defined by a short timeframe (ranging from hours to weeks) in which emergency procedures must be implemented to save lives and reduce exposure to danger (Freeman 2000). On the other hand, relocation refers to moving people from one location to another when it becomes too unsafe or impossible to stay in one. Relocation might be voluntary or forced, and it can be large-scale or small-scale. Even though moving away from a dangerous place may begin as an evacuation, relocations are meant to be permanent (UNHCR 2015).

Relocations that follow best practices also include a resettlement procedure to help relocated persons replace their housing, assets, livelihoods, land, and access to resources and services to sustain their communities and to improve, or at the very least restore their living conditions (IDMC 2017).

While relocation promises the reduction of communities' exposure and, ideally, vulnerability to the effects of extreme weather events, certain population groups tend to resist such relocation initiatives and opt for living with risk. These groups prefer residing in high-risk areas over complying with the requirements of relocation initiatives and shifting to safer areas (UNHCR 2015). It is against this background that this study seeks to explore the reasons behind the resistance of certain population groups to relocate.

7.1.1 Resistance to Relocation

Resistance to relocation may manifest in various forms. Resistance may range from more passive forms such as illegally remaining in their original settlements to the active form, for example, leveraging patron-client ties to influence government authorities, submitting letters of appeal to the relocation agent, or organizing non-violent demonstrations at the relocation project site (Ockey 1997; Oliver-Smith 2021). While often resistance is demonstrated at the early stages of relocation process, i.e., at the project planning and implementation stages, certain studies have illustrated how resistance was expressed post-relocation. For example, Fernando (2018), in a study conducted for 13 years following the 2004 Indian Ocean Tsunami relocation in Sri Lanka, has found that certain relocates have illegally returned to their original settlements situated within the buffer zone. This has been mainly owing to the inability of relocates to cope with the risks and stresses induced by relocation, particularly those which related to the loss of coastal livelihoods. A similar study has been conducted by Mitchell (1976) which elucidates how villagers relocated following the Gediz earthquake of 1970 had returned to their previous settlements to rebuild their homes.

Not all the people who are entitled to a relocation program resist relocation. Some people, of course, are attracted to the benefits of improved housing, better living conditions, high quality infrastructure, etc., and therefore accept the relocation process (Oliver-Smith 2021). Nevertheless, certain population groups do indeed resist the process and they are often at a "distinct disadvantage in terms of power and resources" (Oliver-Smith 2021, p. 148). Non-compliance of people with relocation programs challenges not only the effective implementation of relocation programs but also the extent to which relocation qualifies as a constructive disaster risk reduction strategy. For this reason, delving into the reasons why certain population groups resist relocation, what sets apart these groups from others who comply with relocation programs, and what can be done to address these sources of resistance is pivotal.

The limited number of studies carried out on the subject has put forward various reasons as to why certain people resist relocation. For instance, Rey-Valette et al. (2019) have proposed an index to measure resistance to relocation which encompasses three dimensions, namely (1) place attachment, (2) residential mobility, and (3) risk perception. While place attachment stems from the accumulation of emotional experiences as well as physical, social, cultural, and economic capital throughout a person's period of residence in a particular area (Luttik 2000; Faure 2006), residential mobility is determined by socio-demographic and geographical attributes of a population group such as age and income (Debrand and Taffin 2005; Cusin 2014). (VERB) Risk perception on the other hand entails people's subjective judgements of the likelihood of an event resulting in negative outcomes. Risk perception is influenced not only by the level of information that people have access to, but also by socio-cultural and psychological dynamics (Eiser et al. 2013; Lupton 1999). People further depend on their past experiences which in certain studies (e.g., Weiss et al. 2011) have been cited as "memory of past events," in deriving subjective interpretations of risk. When interpreting risk based on prior experiences, people often consider aspects such as "the severity of the threat, the magnitude of potential consequences, their ability to do something about the risk, uncertainties and ambiguities about the risk, and what they know about the hazards creating the risk in question" (Oliver-Smith 2021, p. 137). In a study carried out to examine the basis on which relocation has been resisted by coastal populations in the South of France, a higher degree of resistance to relocate has been observed among second-home owners and homeowners [as opposed to, e.g., tenants], persons who resided closer to the sea [as opposed to those who resided relatively far from the sea], retired people [in comparison with skilled self-employed persons and employees], and people who had greater awareness of weather alerts (Rey-Valette et al. 2019).

Oliver-Smith (2021) further draws attention to the potential risks that relocation itself carries and how particularly marginalized communities are often torn between mitigating risks associated with natural hazards via relocation and avoiding the potential negative consequences of relocation projects that are executed by the State which they barely trust. Relocation programs may result in a series of unfavorable consequences for relocates which often involve serious risks such as landlessness, marginalization, loss of livelihoods, poor health, dismantling of social networks, food insecurity, and loss of access to common resources (Cernea 2002; Scudder and Colson 1982). Pushing communities to further deprivation of post-relocation by depriving them of access to basic facilities, infrastructure, and psych-social support systems reflect a violation of basic human rights that people are eager to protect (Ploeg and Vanclay 2017). This largely relates to the quality of the relocation project. The extent to which the relocation project is designed to minimize the abovementioned risks coupled with the degree of control and understanding people have of the project impacts the tendency for people to resist the relocation process (Oliver-Smith 2021).

Apart from the reasons discussed above, studies have found that when new settlement is located further from the original place of residence (Raju 2013) and when the damage caused and experienced by a disaster is low (Mileti and Passerini 1996), people tend to exhibit higher resistance to relocation and vice versa.

7.1.2 Importance of the Study

The importance of this study can mainly be discussed on a global, regional, and local scale since resistance to relocation has been a universal problem in almost all the disaster-prone areas in the world.

Displacement is currently a significant issue associated with disasters worldwide (Burson et al. 2018). Mainly climatic disasters are more vulnerable and less predictive than any other disaster on the earth. Therefore, the relocation process is more important as a disaster risk reduction strategy. The reluctance of vulnerable communities to leave their original place of residence despite the high disaster risk has become a burning issue for most responsible authorities.

Nevertheless, most studies on natural disaster-induced relocation have discussed the effects of relocation while efforts to examine resistance to relocation are limited. The few studies that capture resistance to disaster-induced relocation have contributed to scholarship mainly by outlining possible causes or resistance, but attempts have not been made to address how such resistance can be overcome and the responsibility of relevant authorities in doing so. This study attempts to address this lacuna in research by (1) exploring the reasons underlying resistance to relocation with reference to a case of landslide-induced relocation in the Kegalle district of Sri Lanka and (2) proposing a balanced policy framework on resettlement to minimize such resistance. This kind of study will have an added advantage for the policy-makers in the planning of disaster risk reduction.

7.2 Study Site

In Sri Lanka, over 20% of the total land area of the country (65,610 km^2) has been recognized as landslide prone. Approximately 20,000 km^2 of Sri Lanka's total geographical area falls within the moderate-to-high danger category (MDM 2014). Landslides account for over 80% of natural disasters in the country, owing to poor land-use practices during fast expansion of building (Sugathapala and Vijekumara 2013). The districts of Badulla, Nuwara Eliya, Kandy, Matale, Ratnapura, Kegalle, Kalutara, Galle, Matara, and Hambantota are all prone to landslides (Bandara 2011). Kegalle District has been selected as the study site of this research mainly due to the reporting of several large-scale landslides in the recent past. A significant landslide occurred on the 17th of May 2016 in "Samsara," a mountain in Aranayaka, Kegalle District, Sri Lanka, due to excessive rainfall throughout May, rendering around 6000 families homeless and causing extensive damage to their properties. The responsible authorities and the government have taken necessary steps to relocate the victims of the landslides away from the landslide-prone areas to ensure their future security. However, a considerable number of families still reside in the high-risk regions of this district refusing the relocation process. Therefore, Kegalle has been selected as an appropriate study site for this study (Fig. 7.1).

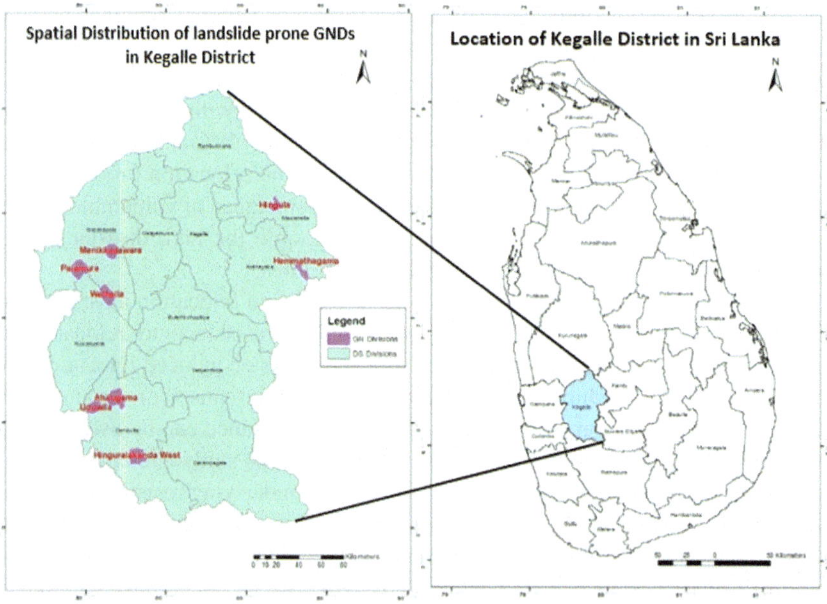

Fig. 7.1 Study area. *Source* Prepared by the authors, 2022

7.2.1 *Physical Setting of the Study Area*

Kegalle district is located in Sri Lanka's wet and intermediate boundary zone (7° 18′ N–80° 24′ E), which has a tropical climate. Kegalle experiences significant rains throughout the monsoonal (May–September) and inter-monsoonal (October–November) seasons and bright sunshine during the dry season (January–March). The district's average annual rainfall ranges from 2500 to 3000 mm. The average temperature is between 25 and 28 °C. The community of Kegalle has a total area of 1693 km^2.

Figure 7.2 shows the location map of the study area with general topography and landslide locations. Figure 7.2a depicts the position of Sri Lanka in the Indian Ocean and the location of Kegalle district in the intermediate zone of Sri Lanka.

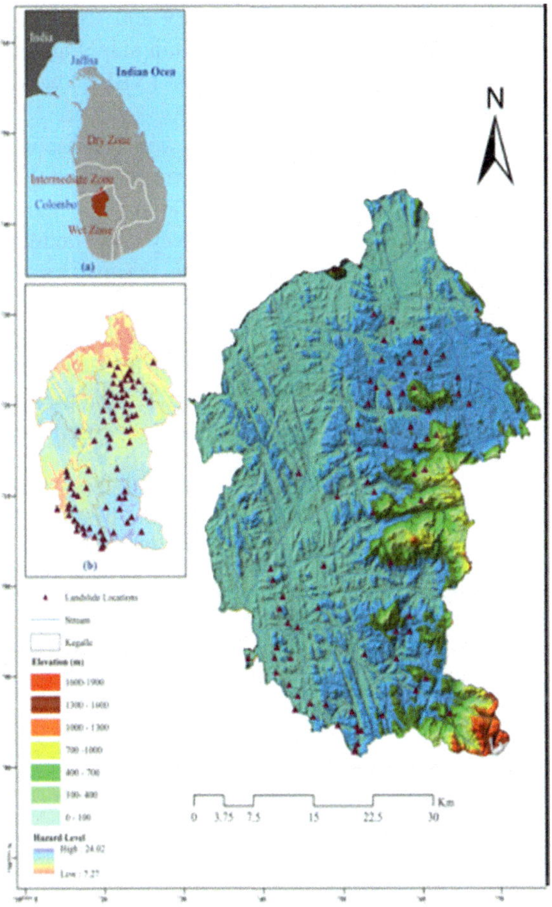

Fig. 7.2 Physical setting of the study area. *Source* Perera et al. (2019)

Figure 7.2b denotes the Hazard zonation map based on SMDM. The maps are based on 1:10,000 scaled maps (2016) of the Department of Survey, Sri Lanka. *Source* Perera et al. (2019).

7.2.2 Methodology

The methodology of this study mainly consists of data collection, data analysis, and information presentation. These sections can further be elaborated as follows.

7.2.2.1 Data Collection

The study depends on both primary and secondary data. The resource profile of Kegalle district and landslide zonation mapping manuals published by the National Building and Research Organization (NBRO) were used as the primary sources of secondary data.

Primary data were collected by deriving a purposive sample of 21 families in Kegalle district who refused to move away from the landslide-prone areas, resisting the relocation program launched following the Samasarakanda landslide which took place in 2016. A questionnaire survey was conducted in the study area among the selected households to identify the reasons for refusing the relocation program launched by the government following the landslides which occurred in the area in the recent past. Unstructured interviews were carried out selecting five members randomly from the area during the study to identify the socioeconomic causes of respondents' hesitancy toward the relocation program. The primary data collection methods used in this study have mainly focused on the following aspects.

Questionnaire Survey—To identify the main causes of refusing the relocation program launched by the government. This was done using a purposive sample of households who have totally refused the relocation.

Unstructured Interviews—To identify the successfulness of the planned relocation process and to get a general idea about the pros and cons of the program. This mainly targeted the old adults who have been living in the area for more than 50–60 years and the Grama Niladhari officer of the area.

Observations—Participatory observations were recorded at each sample point to witness the causes of refusing the relocation process, and it has helped the researchers to understand the real ground-level problems faced by the respondents.

7.2.2.2 Data Analysis

Data analysis was carried out using the Statistical Package for the Social Sciences (SPSS) software and Microsoft Excel (MS Excel). The data were subjected mainly to univariate analysis in descriptive statistics. Univariate analysis was conducted to trace trends and patterns in the primary data. Concerning univariate analysis, frequency distributions were run to explore the distribution of the sample of respondents. Priority matrix analysis was carried out to identify the priority drivers behind the reluctance of the informants to move away from their original place of residence. ArcGIS 10.5 software was used to prepare maps and illustrations showing the spatial aspects of the study. The absolute locations of the informants were recorded using a handheld GPS recorder for mapping purposes.

7.2.2.3 Information Presentation

Information received from the analysis has been presented using maps, charts, graphs, and diagrams.

7.3 Results and Discussion

7.3.1 Visual Illustration of the Communities Living with Landslide Risk

Kegalle district which belongs to the wet zone is identified as one of the most vulnerable districts for landslides in Sri Lanka. Agriculture is the primary source of income in the district. Export crops, such as tea, rubber, cocoa, pepper, clove, and nutmeg, contribute significantly to the local economy. Home gardens cover the most significant area of the district's territory (35%; 595 km^2). Rubber is the most common crop, and it covers the majority of the area (31%; 525 km^2). Paddy covers 6.3% of the total land area (107 km^2). Tea, coconut, and built-up area account for 2.5, 3.9, and 0.02% of the total (DCSSL 2016). The forest area has been exploited for agricultural and development activities due to lack of an adequate national land-use policy. Soil erosion was higher in agricultural regions than in natural vegetation areas (Dissanayake et al. 2018). As a result, the district is prone to significant soil erosion and slope instability (Jayasinghe 2016).

Since this area recorded no landslides in the recent past, people tended to use irregular slopes primarily for agricultural purposes with poor surface water management systems. As a result, people were less conscious of the potential for calamity. On the other hand, Paleo-landslides can be found all over the region. Paleo-landslides appear to have been active around 500–1000 years ago (Jayasinghe 2016); as a result, residents of the Kegalle district have had few landslide experiences in their lifetime.

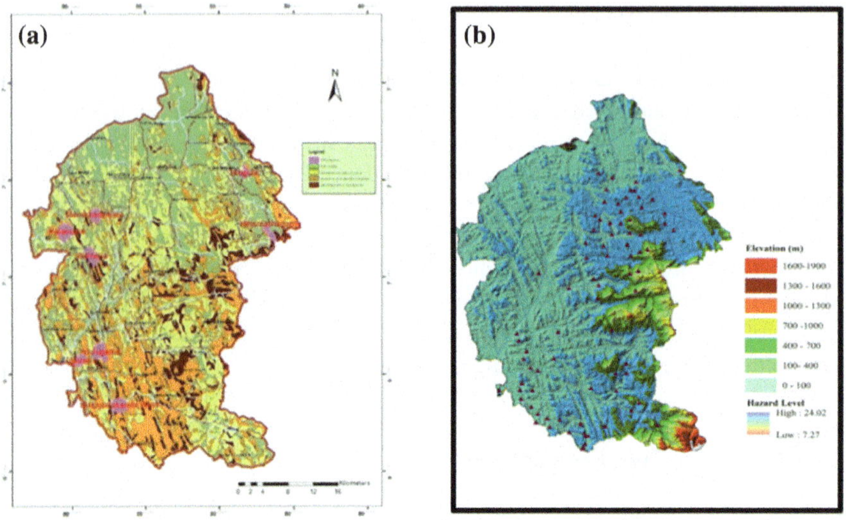

Source: Prepared by the authors, 2022 (Perera, et al., 2019)

Fig. 7.3 Location of the households in landslide-prone areas. (**a**) Location of GNDs in landslide vulnerablity map of Kegalle, (**b**) Elevation and hazard zonation map of Kegalle

Therefore, the people living in the areas prone to landslides were not aware of the landslide risk in their areas of residence. After the massive landslide that occurred in the Aranyaka area in May 2016, the hazardous experiences faced by the people compelled them to move away from the high-risk areas accepting the relocation programs launched by the Government of Sri Lanka with the objective of protecting people from future landslides. Unfortunately, the study could identify 21 families who have completely refused to move away from their original residences even though the life risk is immense. Figure 7.3 depicts the locations of these 21 households in their respective Grama Niladhari Divisions (GNDs).

The study revealed that these identified 21 households are currently located in 8 Grama Niladhari Divisions (GNDs), which are highly prone to landslides.

The identified highly vulnerable GNDs are Menikkadawara, Palamure, Welhella, Aluthgama, A Uduwila, Hinguralakanda West, Hingurala, and Hemmathagama. The reasons underlying the hesitation of these 21 families to relocate were assessed, and the findings will be discussed in detail in the preceding sections.

7.3.2 Identification of the Primary Causes Behind the Reluctance of Residents to Relocate

The study revealed that 33% of the dwellers have wholly refused to relocate, while 66% have a partial willingness to cooperate with the process in the future. However,

100% of the informants came up with their reasons for rejecting the relocation process carried out by the Government of Sri Lanka as a measure of disaster risk reduction following the landslide that took place in Aranayaka in 2016. The study revealed economic and socio-cultural reasons as the leading causes behind resistance to relocate among the respondents of the study. The reasons can further be elaborated as follows.

7.3.2.1 Economic Reasons

Economic reasons were identified as the most burning causes behind the rejection of the process of relocation by the vulnerable communities. 50% of the settlers who were hesitant to relocate were primarily dependent on land-based livelihoods. Therefore, their reluctance to relocate was mainly based on the unfavorable implications of relocation for continuation of their land-based livelihoods in their new settlements. Figure 7.5 depicts that nearly 50% of the victims are below the poverty line of Sri Lanka, and they mainly depend on the daily wage-based primary modes of income. Therefore, the relocation process has caused extra stress for these communities since they would be forced to find new agricultural lands for cultivation after relocation. The plot of land provided through the relocation was deemed inadequate to be used for land-based livelihood activities and the construction of a house to reside in both (Fig. 7.4).

According to Fig. 7.4, 60% of the informants have stated that they got relatively insufficient financial support from the government to construct houses through the relocation scheme. 10% of the respondents who depend on land-based economic activities such as agriculture have highlighted that the area of land that they were given in the relocation process was adequate only for the construction of a house. The respondents were therefore of the view that relocating to a new settlement would result in the loss of cultivation plots that they were using in their current settlement. Further, 30% of respondents claimed that they needed sufficient land for their cultivation activities since their main livelihood was agriculture and farming. Poverty

Fig. 7.4 Economic reasons to reject relocation

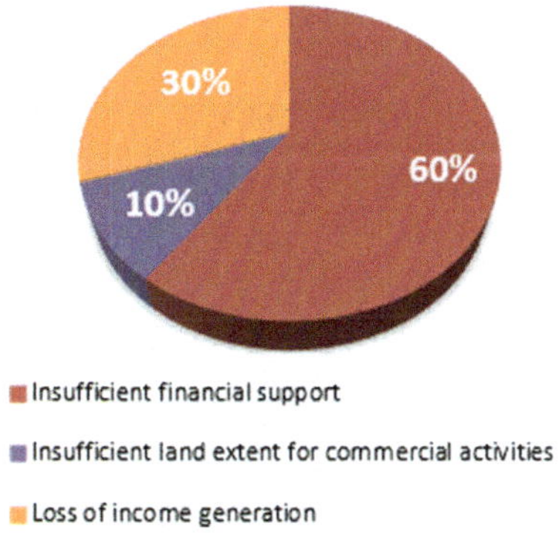

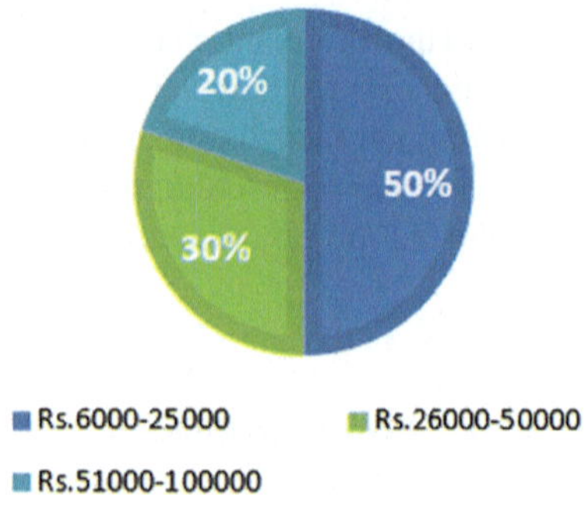

Fig. 7.5 Income from primary sources

is one of the salient threats they have had since birth. In the face of disasters like landslides, such households are pushed to further deprivation, thus trapping them in a vicious cycle of poverty.

7.3.2.2 Socio-Cultural Reasons

The study could identify some significant socio-cultural reasons behind the rejection of the proposed relocation program by the selected respondents. The majority of the respondents (35%) mentioned a lack of trust in the relevant authorities as one of the major reasons for their resistance to relocating. Those respondents reported that they lacked trust in the rules mainly due to the gap between the responsible authorities and the vulnerable communities. This gap further resulted in a considerable degree of unawareness among the respondents about the disaster risk reduction measures that should be followed in the area. The discussions with the selected communities further suggested that their unawareness of the scale of danger, risk reduction efforts, and post-disaster evacuations had largely led them to be victimized by landslides (Fig. 7.6).

20% of the respondents have dismissed the relocation program due to complexities in their relationships. Among the psycho-social issues, the majority were due to lack of adequate space for peaceful family life in the new living place. This concern was

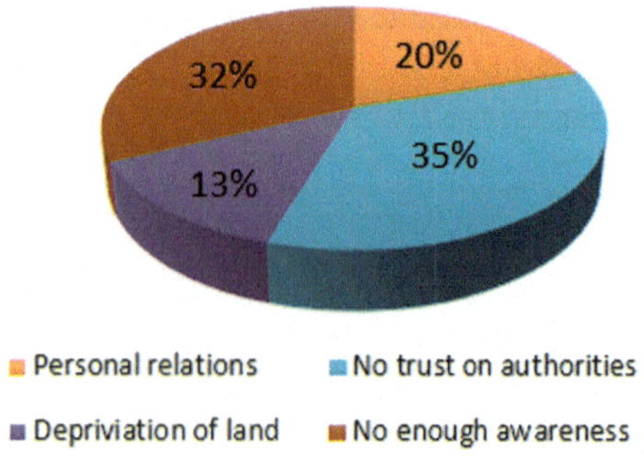

Fig. 7.6 Socio-cultural reasons to reject the relocation

mainly raised by households with extended families where two–three generations resided in the same household. Another major concern highlighted by the respondents was the distance between the new settlement and places of worship.

Further, the deprivation of land inherited by them was another problem raised by about 13% of the respondents. Most of their lands had been transferred to them as traditional assets from their previous generations. Therefore, they were the sole owners of the land since their origin, and thus, they were reluctant to abandon their original lands even though they were located in landslide-prone areas.

7.3.3 Investigation of the Prevailing Awareness of Dwellers on Disaster Risk Mitigation Measures

Lack of awareness among the residents about the disaster risk reduction processes was also identified as one of the major reasons behind the rejection of the relocation process by the vulnerable communities. 32% of the respondents stated that they were not aware of disaster mitigation measures. The study could identify lack of education as the main reason for the lack of awareness among the dwellers about disaster risk management.

According to Fig. 7.7, it is evident that the majority of the respondents (nearly 50%) had obtained an education up to or below the GCE ordinary level examination. The unstructured interviews had with them further emphasized their lack of knowledge regarding disaster mitigation measures and lack of attention paid to future predictions on disasters.

Further, the study could get an in-depth understanding about the education level of the respondents who have refused (33%) the process of relocation. Accordingly, around 70% of respondents had not completed their GCE Ordinary Level Examination (Fig. 7.8). Therefore, the education level of the people has affected the

Fig. 7.7 Overall education level of the respondents

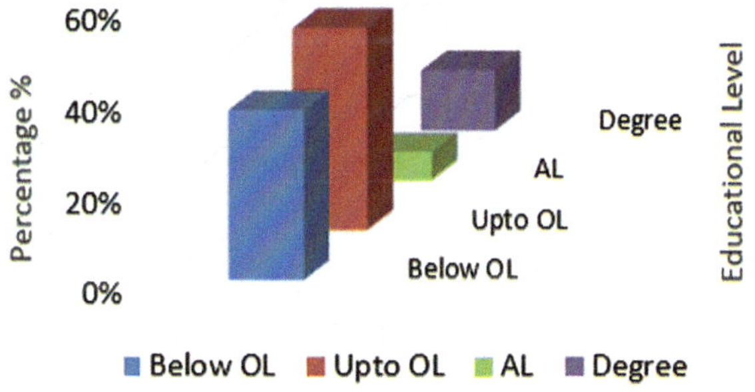

level of awareness on disaster risk reduction in the area considerably. The study further focused on evaluating the pre-preparation measures for disasters taken by the vulnerable communities who have refused to relocate. The findings revealed that 80% of respondents (Fig. 7.9) had no pre-preparation for disasters. Furthermore, the majority (75%) did not participate in the disaster risk awareness programs in the area (Fig. 7.10).

Accordingly, lack of awareness among the communities about the disaster mitigation measures and disaster risk reduction has resulted in the rejection of the relocation program by the majority of the communities living in the landslide-prone areas. Furthermore, their social status, educational level, and level of understanding about the authorities' measures for the benefit and welfare of the communities have been misinterpreted. Unawareness of the general public about the benefits that they are

Fig. 7.8 Education level of respondents who have refused relocation

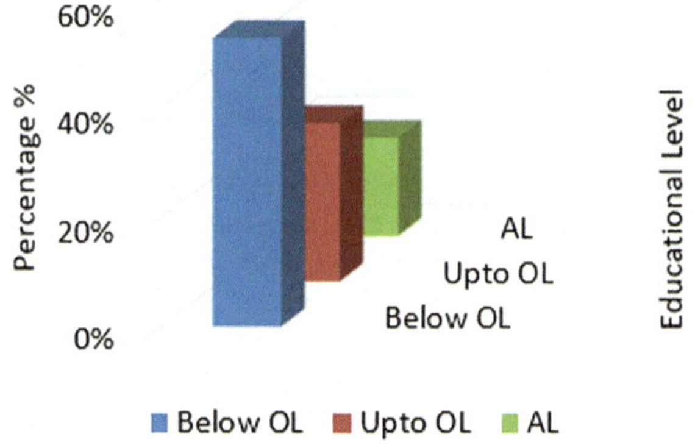

Fig. 7.9 Pre-preparation status of communities for disasters

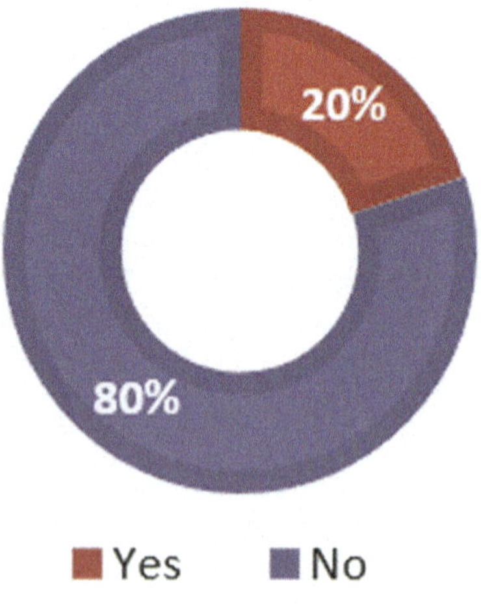

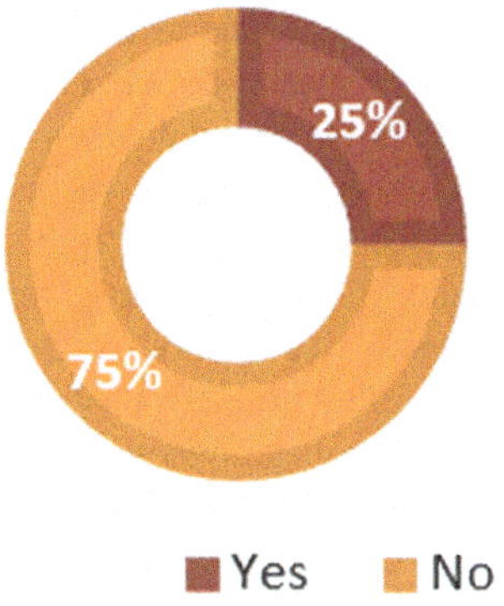

Fig. 7.10 Participation status of communities in disaster risk reduction awareness programs

entitled due to the adaptation of disaster risk reduction measures have made them the victims of disasters. Further, they have moved toward temporary means of disaster risk reduction rather than adopting feasible and permanent measures.

7.3.4 Evaluation of the Suggestions of Dwellers on the Process of Relocation

The study evaluated the recommendations of the dwellers to enhance the relocation process in a people-centered manner. Among the suggestions of the dwellers, the majority of the respondents commented on the nature and the selection criteria of the new lands that they were given in the process of relocation. Since many of those affected resided on the traditional land that they had received from their families, they preferred land of the same size and benefits in the new settlements (Fig. 7.11).

The unstructured interviews with the respondents further suggested that they did not have enough space for their home gardening activities in the new settlements.

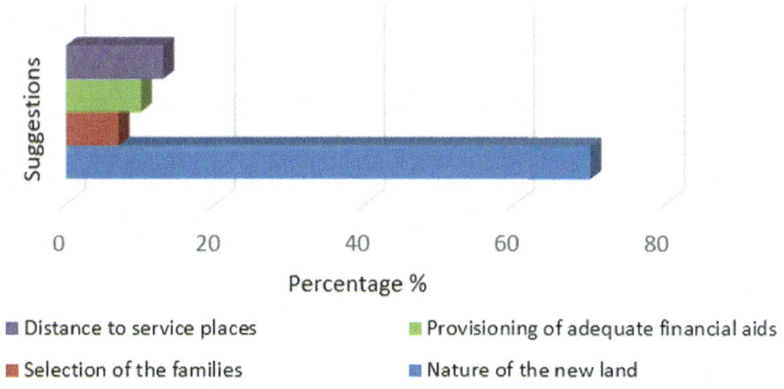

Fig. 7.11 Suggestions of the communities to enhance relocation program

Since 70% of the dwellers depended on home gardening and agro-based economies, the land's nature became significant in determining their place of settlement. Further, 15% of the respondents raised the concern that the new settlements were located far from services such as education and health which in turn would potentially make these services inaccessible from said settlements. On the other hand, 12% of the respondents claimed that the government's financial aid and installment basis payments in constructing the houses are not adequate to complete the house.

Furthermore, the unstructured interviews with vulnerable communities suggested that they have become the victims of the disasters as well as social, cultural, and economic hazards. Therefore, these vulnerable communities wanted only to live in their current places until the catastrophe destroyed them. The attitudinal perceptions of the communities on disaster risk reduction are identified as one of the major concerns that should be addressed in the country's relocation process.

7.3.5 Suggestions to Develop a Balanced Policy Framework of Rational Decision-Making on Resettlement Alternatives

This study which focused on identifying reasons for rejection of the relocation program launched by the Government of Sri Lanka following Samasarakanda landslide in Kegalle district could propose several significant suggestions to develop a balanced policy framework on resettlement. They can be elaborated as follows.

- Strengthening the integration between institutions in decision-making

Lack of integration within the institutional arena in disaster risk reduction has resulted in negative impacts in the process of resettlement. The delay in project implementation is due to the poor cooperation and accountability between the institutions responsible for disaster risk reduction. This system can be further strengthened by adequately monitoring and coordinating the institutional roles in disaster mitigation.

- Development of a multidisciplinary team of personnel in decision-making

The decision-making process is an essential task in disaster risk assessment and disaster risk reduction in Sri Lanka. The availability of a team of members representing multiple disciplines would be instrumental in investigating the various impacts of relocation such as economic, socio-cultural, emotional, environmental, and physical. The ideologies of a multidisciplinary team of members would help make the most realistic, appropriate, and holistic decisions in disaster risk reduction.

- Enhancing community participation in decision-making

Further, community participation should be encouraged in the process of decision-making. This will convince the community that the decisions taken by administrative authorities have addressed their problems. Encouraging community participation in the decision-making process would help understand the actual situation of the communities striving hard to adapt to the new environment while struggling for a living.

- Conducting robust Environmental Impact Assessments (EIA) and feasibility studies before the relocation process

EIA and feasibility studies should be done correctly to address the problems related to the nature of the lands given by the authorities to the vulnerable communities. The EIA assessments, which evaluate the environmental conditions of the new lands, should meet the requirements of the communities, especially those with respect to agricultural activities. Further, formal feasibility studies should be carried out on newly introduced lands before being distributed. The relocation process would be successful only if executed after a prior understanding of the land.

Accordingly, the above suggestions can be made to enhance the process of relocation carried out by the authorities of Sri Lanka as a strategy for disaster risk reduction. Institutional integration, a proper understanding of the requirements of the communities, and the ability to address the essential needs of the people in the relocation processes would encourage people to move toward safer places than living with the risk of hazards such as landslides.

7.4 Conclusion

Landslides have become a catastrophe that occasionally make thousands of people homeless in Sri Lanka. Most of the studies done in Sri Lanka on the relocation process have focused mainly on the socioeconomic impacts of relocation. This paper primarily focuses on the identification of the leading causes behind refusing of the relocation process by the most vulnerable group of people who are living with the risk. The resistance to relocation is identified as only one short-term consolation that they have adopted to address the prevailing relocation model. The risk is long-term, and their understanding of the risk is also apparent. Nevertheless, the study could reveal several significant drawbacks behind the failure of the relocation program. The study mainly points toward the importance of community participation in decision-making related to the relocation.

According to the findings, the area consisted of a community, most of whom depended on daily wage-based employment. More than 50% of the respondents relied on agro-based economies. Therefore, the land is one of the most important assets they had in their lives. The majority of the families were below the poverty line. So the communities identified as resistant communities were suffering from the risk of disasters and many socioeconomic complications. All the areas with potential for landslides should be mapped, detailed studies of the vulnerable areas as well as the proposed relocation areas should be made by a multidisciplinary committee consisting of NBRO, Geological Survey and Mines Bureau, Disaster Management Authority, Ministry of Lands, Department of Survey, Environment Authority, and

District/Divisional Secretariat, and any other relevant agencies is necessary. Awareness programs for those living in vulnerable areas should be conducted, and their attitudinal changes too have to be changed. Hence, a holistic approach that would address the necessities of these communities should be developed along with the policy frameworks related to the relocation of vulnerable communities following natural disasters.

References

Bandara R (2011) Landslides in Sri Lanka. NBRO, Sri Lanka

Brookings Institution, Georgetown University, United Nations High Commissioner for Refugees (UNHCR) (2015) Guidance on protecting people from disasters and environmental change through planned relocation, 7 Oct 2015

Burson B, Kälin W, McAdam J, Weerasinghe S (2018) The duty to move people out of harm's way in the context of climate change and disasters. Refug Surv Q 37:379–407

Cernea M (2002) Impoverishment risks and reconstruction: a model for population displacement and resettlement. In: Cernea M, McDowell C (eds) Risks and reconstruction experiences of resettlers and refugees. The World Bank, pp 11–55

CSSL (Department of Census & Statistics, Sri Lanka) (2016) Census of population, housing, and land use. Department of Census & Statistics Ministry of Policy Planning and Economic Affairs, Colombo, pp 187–280. http://www.statistics.gov.lk/PopHouSat. Accessed 25 Feb

Cusin F (2014) Les dynamiques urbaines au prisme de mobilités résidentielles longues. In: Fol S, Miot Y, Vignal C (eds) Mobilités résidentielles, territoires et politiques publiques [Residential mobility, territories and public policies]. Septentrion, Villeneuve d'Ascq, pp 65–95

Debrand T, Taffin C (2005) The structural and cyclical factors of residential mobility over the last 20 years. Econ Stat 381–382:125–146

Dissanayake D, Morimoto T, Ranagalage M (2018) Accessing the soil erosion rate based on RUSLE model for sustainable land use management: a case study of the Kotmale watershed, Sri Lanka. Model Earth Syst Environ 4:1–16. https://doi.org/10.1007/s40808-018-0534-x

Eiser J, Bostrom A, Burton I, Johnston D, McClure J, Paton D, Pligt J, White M (2013) Risk interpretation and action (RIA) report no. 1. Integrated Research on Disaster Risk, Beijing. http://www.irdrinternational.org/2013/01/25/ria-report-1

EM-DAT (2009) The OFDA/CRED international disaster database. Technical report. Universite Catholique de Louvain, Belgium

Faure S (2006) The social effects of housing demolitions: a study of public housing high-rises in Saint-Etienne, France. Espaces Soc 124–125(2–3):191–206

Fernando N (2018) Lessons learnt from long term impact of 2004 tsunami relocation: a case study of selected relocation settlements in Akmeemana Divisional Secretary Division in Galle District, Sri Lanka. Procedia Eng 212:47–54

Freeman PK (2000) Estimating chronic risk from natural disasters in developing countries: a case study on Honduras. Working paper

Global Compact for Safe (GCS) (2018) Orderly and regular migration, Dec 2018, para 18(j)

IDMC, GRID (2017) Global report on internal displacement, June 2017. IDMC, Geneva, p 11

Inter-Agency Standing Committee (IASC) (2011) IASC operational guidelines on the protection of persons in situations of natural disasters. Brookings-Bern project on internal displacement (IASC guidelines), Jan 2011. Available at: http://www.ohchr.org/Documents/Issues/IDPersons/OperationalGuidelines_IDP.pdf. Last visited 15 Aug 2018

Jayasinghe P (2016) Social geology and landslide disaster risk reduction in Sri Lanka. J Trop For Environ 6:1–13

Kahn ME (2005) The death toll from natural disasters: the role of income geography and institutions. Rev Econ Stat 87(2):271–284

Lupton D (1999) Risk. Routledge, London and New York

Luttik J (2000) The value of trees, water and open space as reflected by house prices in the Netherlands. Landsc Urban Plan 48(3–4):161–167

MDM (Ministry of Disaster Management, Sri Lanka) (2014) Sri Lanka comprehensive disaster management program. MDM, Colombo, pp 127–245

Mileti D, Passerini E (1996) A social explanation of urban relocation after earthquakes. Int J Mass Emerg Disasters 14(1):97–110

Mitchell W (1976) Reconstruction after disaster: the Gediz earthquake of 1970. Geogr Rev 66(3):296–313

Ockey J (1997) Weapons of the urban weak: democracy and resistance to eviction in Bangkok slum communities. Sojourn J Soc Issues Southeast Asia 12(1):1–25

Oliver-Smith A (2021) The choice of perils: understanding resistance to resettlement for urban disaster risk reduction and climate change adaptation. In: Johnson C, Jain G, Lavell A (eds) Rethinking urban risk and resettlement in the global south. UCL Press, London, pp 133–153

Perera E, Jayawardana D, Jayasinghe P, Ranagalage M (2019) Landslide vulnerability assessment based on entropy method: a case study from Kegalle district, Sri Lanka. Model Earth Syst Environ. https://doi.org/10.1007/s40808-019-00615-w

Ploeg L, Vanclay F (2017) A human rights-based approach to project induced displacement and resettlement. Impact Assess Proj Apprais 35(1):34–52

Raju E (2013) Housing reconstruction in disaster recovery: a study of fishing communities post-tsunami in Chennai, India. http://currents.plos.org/disasters/article/housingreconstruction-in-disaster-recovery-a-study-of-fishing-communities-post-tsunami-inchennai-india/

Rey-Valette H, Robert S, Rulleau B (2019) Resistance to relocation in flood-vulnerable coastal areas: a proposed composite index. Clim Policy 19(2):206–218

Scudder T, Colson E (1982) From welfare to development: a conceptual framework for the analysis of dislocated people. In: Hansen A, Oliver-Smith A (eds) Involuntary migration and resettlement: the problems and responses of dislocated people. Westview Press, Boulder, CO, pp 267–287

Sendai Framework for Disaster Risk Reduction 2015–2030, adopted by the Third United Nations World Conference on Disaster Risk Reduction and endorsed by the UN General Assembly in Res. 69/283, 3 June 2015, Annex II, para. 33(m) (Sendai Framework)

Sugathapala K, Vijekumara PA (2013) Importance of implementing landslide clearance process complementary to national physical plan 2030. In: 5th annual symposium: engineering in disaster resilience. National Building Research Organization, Ministry of Disaster Management, pp 78–86

Weiss K, Girandola F, Colbeau-Justin L (2011) Protection behaviors with regard to natural hazards: from resistance to commitment. Prat Psychol 17(3):251–262. https://doi.org/10.1016/j.prps.2010.02.002

Chapter 8
Internal Displacement in Nigeria: What Are the Preventive Measures?

Lilian Nwanyisonde Smart, Dilanthi Amaratunga, and Richard Haigh

Abstract Internal displacement resulting from conflicts and disasters has increasingly become one of the major global humanitarian and development challenges. The 2021 report on internal displacement shows that the conflicts and disasters of 2020 internally displaced about 40.5 million new persons across 149 countries and the number rose to 55 million by the end of 2020. In Nigeria alone, the 2020 statistics demonstrate that about 279,000 new persons were displaced by conflicts and 169,000 persons displaced by disasters. It was noted that about $20.5 billion was spent on the 2020 internal displacement providing the displaced persons with support for housing, education, health, security, and income. While these are massive costs of assistance and protective supports geared towards addressing the needs of the displaced, it appears not to be sustainable. Therefore, it can be argued that preventing internal displacement is very important and more sustainable than focusing more on assisting and protecting the displaced. It was based on the necessity to adopt a preventive approach to displacement that a High-Level Panel on Internal Displacement was established by the UN secretary-general in December 2020. This panel was saddled with the responsibility of identifying appropriate measures to ensure financing for long-term efforts to prevent and resolve internal displacement. Against this backdrop, this study seeks to identify the appropriate measures to prevent and address circumstances that might lead to internal displacement, particularly in Nigeria. The paper adopted a semi-structured interview technique with participants carefully selected from displaced communities and those at risk of displacement. The result of the study reveals that factors such as war, oil-induced armed conflicts triggered by the uneven distribution of resources, marginalisation of some ethnic groups, disasters, and development projects constitute the central causes of internal displacement in Nigeria. Accordingly, the paper made some recommendations which include that the governing authorities and other relevant institutions should as a matter of urgency address all the triggers of war, violent ethnic, and communal armed conflicts; and that the issue of marginalisation arising from political and economic factors should be addressed. It was equally suggested that the government should strategically relocate

L. N. Smart (✉) · D. Amaratunga · R. Haigh
Global Disaster Resilience Centre, School of Applied Sciences, University of Huddersfield, Huddersfield, UK
e-mail: lilian.smart@hud.ac.uk

© The Author(s), under exclusive license to Springer Nature Switzerland AG 2023
M. Hamza et al. (eds.), *Rebuilding Communities After Displacement*,
https://doi.org/10.1007/978-3-031-21414-1_8

residents in disaster-prone areas as well as take appropriate measures to reduce the impacts of disasters in the country that by so doing the risk of internal displacement will be drastically reduced in Nigeria.

Keywords Displacement · Internal displacement · Forced displacement · Preventive measures

8.1 Introduction

Presently, displacement has become one of the challenges that bedevil the world at large leading to an increasing number of forced migrants fleeing for their lives and safety. Statistics reveal that at the end of 2018, conflict and violence displaced about 41.3 million people within their countries coupled with about 20–25 million people being displaced by disasters yearly (OCHA 2019). The recent prediction of the Global Displacement Forecast report on 26 countries shows that in 2022 (this year), the total number of displaced persons will increase by 2.9 million, and in 2023, it will have an additional 3.9 million (OCHA 2022). The report notes that when this happens, the total number of displaced will have an additional increase of 6.8 million thereby doubling the number of displaced persons with an increase of more than 35 million people from 2014 to 2023. That is within a space of 9 years from now. What is perhaps the most worrisome aspect of this displacement was the one reported by the UNHCR (2021) that the number of people displaced due to persecution, conflict, violence, human rights violations, and other events rose to about 82.4 million by the end of 2020. This increase is both worrisome and disheartening mostly because it happened at the peak of the COVID-19 pandemic when the world was battling the outbreak of the deadly disease; when people were advised to stay in their homes to control the rapid spread of the disease. Yet such a massive population of people were displaced from their homes and communities to flee for their lives making them face more challenges as well as being more vulnerable to the dangers of the COVID-19 disease.

Some of these challenges as identified by the Intra-Agency Standing Committee (IASA 2010) include loss of lives, loss of homes, loss of property, and loss of lands which results in armed robbery attacks and abuse of the displaced thereby heightening their risks. Other challenges are inadequate provision for safety; family separation—including an increase in the number of children separated and unaccompanied by their parents resulting from sudden flee from safety; heightened increase in sexual abuse and gender-based violence which affects mostly women and under-aged children (IASA 2010). These displaced persons are classified as internally displaced persons, asylum seekers, stateless persons, and refugees (Porter and Haslam 2005). However, Becker (2022) classified the displaced into externally displaced persons and internally displaced persons. Becker (2022) explained that externally displaced persons are those who are forced to flee across national boundaries and internally displaced persons are those displaced persons that remained within their national borders.

Though both the external and the internally displaced persons are faced with significant challenges because there are no less vulnerable persons in displacement (Salama et al. 2001), this study focuses on internal displacement. This is mostly because the 2020 report on displacement shows that the 55 million people displaced by the end of 2020 were internally displaced persons (United Nations 2021). It was based on this rapidly increasing number of internally displaced persons that the United Nations report emphasises the fact that internal displacement is too important to neglect (United Nations 2021).

Nigeria is among the developing countries that are hugely affected by internal displacement with devastating consequences. Hence, this study focuses on internal displacement in Nigeria. Regarding internal displacement in Nigeria, the 2020 statistics demonstrate that about 279,000 new persons were displaced by conflicts and 169,000 persons displaced by disasters in Nigeria (UNHCR 2021). It was noted in the report that about $20.5 billion was spent on the 2020 internal displacement providing the displaced persons and their host communities with support for housing, education, health, security, and income (IDMC 2021). While this act of making these provisions is applaudable, the question begging for an answer is to what extent do these provisions adequately address the needs of the displaced. Also, can these measures be sustainable?

Indeed, the increasing vulnerability of the internally displaced persons, the various challenges they suffer as well as the deplorable living conditions they battle within the camps, and resettlement centres speaks volume of the fact that these provisions are not enough to address their needs adequately and reduce the challenges they face (Koch 2015; IDMC 2021). Moreover, while money can help to address some of the challenges of internal displacement, it is difficult for money to address most of these challenges, particularly the emotional and psychological challenges. Hence, doing everything possible to prevent circumstances that might lead to internal displacement in Nigeria can be considered a more appropriate and sustainable approach to internal displacement. This is important following the suggestion provided by Beida and Aroge (2019) on the need to adopt preventive measures through more effective legal and regulatory procedures to ensure that people are not displaced and forced to leave their homes in the first place.

Moreover, Zapater (2019) affirms that instead of focusing and waiting to rebuild after displacement, it is better to put in the work to prevent displacement as it is a better long-term response. Accordingly, the research problem is established as identifying the measures to be taken to prevent internal displacement in Nigeria. Consequently, this study aims to explore the measures that can be adopted to prevent the increasing rate of internal displacement in Nigeria. This will help to prompt a paradigm shift from focusing mostly on short-term assistance and protection of the displaced to taking actions towards achieving a sustainable approach which is the prevention of internal displacement-inducing situations. The chapter started with the research methodology, followed by the concept of displacement, then an understanding of internal displacement, its causes, and consequences. This is followed by the discussion of protracted displacement, internal displacement in the Nigerian context, the preventive approach to internal displacement, the discussion of findings, the summary of findings, and then the conclusion.

8.2 Research Methodology

Research methodology in research is as important as the research itself. More importantly, the identification of the right research methodology can determine the success and quality of the research (Thattamparambil 2020), and it strengthens the foundation of knowledge (Paul and Criado 2020). This research aims to explore the appropriate measures that can be adopted to prevent internal displacement in Nigeria. To achieve this, detailed literature reviews were conducted first to help to develop a good understanding and insight into the relevant previous and current research in the field of study and identify the research gap (Saunders et al. 2016; Kumar et al. 2020). This is mostly because a strong literature review is a foundation for sound empirical research (Eisenhardt and Graebner 2007). It is important to note that the type of literature review conducted in this research is the narrative or traditional literature review as it helped the researcher to explore the relevant existing body of knowledge critically to establish the research problem and identify the research gap. After the review, 15 different semi-structured qualitative interviews were conducted to generate non-numerical data grounded in the views of the interviewees (Saunders et al. 2016; Thattamparambil 2020). In research, knowledge may seem inexhaustible as a researcher can decide to spend significant time collecting data for a particular study. However, there is usually a point where new data does not bring additional insight to the research questions, this is a point where the researcher needs to stop data collection. Based on this, the researcher chose to conduct 15 interviews mostly because of the time constraint in writing the chapter and because there were no new insights coming from the additional data. Moreover, 15 research participants were all experts in the field of internal displacement as such the data collected from them were considered enough to write a chapter of this nature. The interviews helped to generate first-hand information to address the question: What are the preventive measures for internal displacement in Nigeria? Importantly, the participants of this study were carefully selected from displaced communities and those communities that are at risk of displacement. The researcher adopted the snowball sampling techniques in this study as it made it possible for the interviewees who already participated in the research to connect the researcher to other people who can contribute to the study. The data collected were analysed with the qualitative data analysis technique using a thematic approach which helps the researcher to categorise the data into themes (Saunders et al. 2016); these themes were, in turn, discussed to address the research question posed.

8.3 The Concept of Displacement

The rate at which people are being forced to flee their homes and communities is increasingly alarming. Morina et al. (2018) who referenced the 2017 report of the UNHCR on displacement state that about 65.6 million individuals were displaced worldwide because of armed conflicts alone. The report from the UNHCR (2020) demonstrates that 11.2 million additional people were displaced in 2020 including people displaced for the first time and people displaced repeatedly compared to 11.0 million additions in 2019. It is bad enough to be displaced for the first time, displacement happening for the second time and perhaps several times after can be seen as the worst-case scenario. This situation is even incredibly worst in cases where the displaced suffer long-term consequences, particularly when there is no adequate measure for proper resettlement of the displaced as well as the sustainable provision of their various needs, especially basic goods, and services. When this happens, it can be as bad as some of the displaced roaming about in the streets looking for where to lay their heads and have access to food and water.

On the 18 June 2021, the UNHCR released a worrisome pictorial representation of a long queue of mass displaced Congolese asylum seekers who were awaiting medical screening in Zombo near the border between Uganda and the Democratic Republic of Congo in 2020 (UNHCR 2021). This picture which was boldly represented on their website in 2021 was not only heartbreaking but also horrific as children, adults, and especially elderly and vulnerable ones were seen looking very exhausted and helpless yet standing in a long queue awaiting medical attention.

A critical look at the picture on their website will reveal how tired these displaced persons are, yet they are subjected to what can be termed a more demeaning situation under the sun just to receive medical screening. This can be seen as preposterous as the displaced whether internally displaced or asylum seekers deserve better treatment. Also, there need to be provisions to provide them with these medical screenings in their various camps and in a more comfortable environment. The lack of this provision reveals why the measures set in place to provide for those affected by forced displacement seem inadequate. This uncovers how much people suffer following displacement, and no wonder the UNHCR put the said picture on their website with the caption, "World leaders must act to reverse the trends of soaring displacement". Moreover, the picture reveals that people can be displaced in their masses. Becker (2022) confirms that most episodes of forced displacement considerably affect people on varying levels and scales. According to him, conflicts or natural disasters can affect small groups of individuals or can lead to mass expulsions of millions of individuals. Now, what is displacement?

Onifade and Osinowo (2019) define displacement as being forced to flee from one's home or place of habitual residence, particularly because of armed conflicts or to avoid its effects, situations of generalised violence, events that violate human rights, natural or human-made disasters which may occur within or across national borders. They also note that displacement should not be understood in relation to what happens to the displaced people during and after the flight alone, but the understanding should

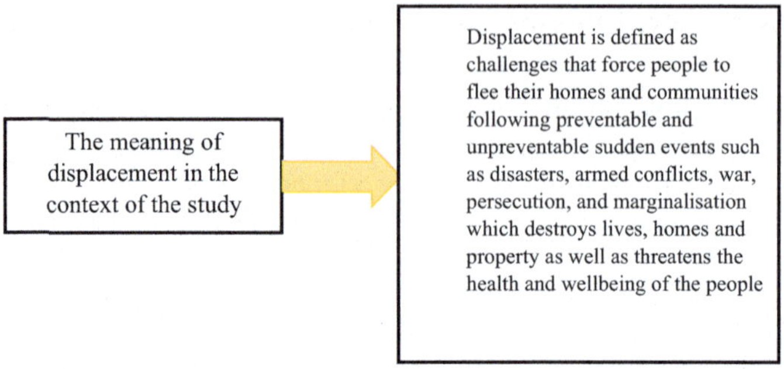

Fig. 8.1 Definition of displacement in the context of this study. *Source* Prepared by the author

extend to the movement of the displaced people from one place to another. This line of reasoning is interesting as it shows that in defining the displaced persons, care should be taken not to neglect first; their pre-flight situation (their condition before they move or the cause of their fight). Second is the flight situation (their condition when they are on the run). Third is the post-flight situation (their resettlement phases and conditions in their places of resettlement) (Siriwardhana et al. 2014).

Another definition of displacement as presented by Verme and Schuettler (2021) is that displacement is a situation where people flee due to violence, persecution, or high levels of insecurity which are largely unpredictable. They further state that forced displacement is less of a choice and less voluntary, and it is a decision taken quickly without careful preparation and plans because of the sudden nature of the shock. From this second definition, displacement can be seen as an unpredictable sudden occurrence that makes people flee their homes and communities unprepared and unplanned with little and no choice to remain. These are elaborate definitions of displacement covering details of its meaning. Accordingly, Fig. 8.1 depicts the working definition of displacement in the context of this study.

As stated earlier, the displaced can be refugees or internally displaced persons. According to Salama et al. (2001), refugees who are the displaced who flee to other countries for safety have special status under international law, while the internally displaced persons (IDPs) who flee their homes for the same reasons as refugees but did not cross internationally recognised borders do not have such status. The report from the GRID (2021) reveals that the number of IDPs has continued to increase globally for more than a decade, that as of the 31 December 2021, it increased more than twice the number of refugees. Unfortunately, these people did not just flee from their homes, they were forced from their homes, displaced from their livelihoods, and separated from their families and support systems (United Nations 2021). This makes it more pressing to investigate the causes of displacement.

8.3.1 Causes of Displacement

Several factors drive displacement, with most of these factors and their degree of intensity differing from country to country (United Nations 2021). It is important to note that the causes of displacement both internal (IDPs) and external (refugees) can be the same factors, though on varying scales. Below are the arguments on the causes of displacement.

War, political violence, and related threats: Porter and Haslam (2005) note that displacement is caused by such factors as war, political violence, or related threats.

Civil conflicts and natural disasters: While Becker (2022) acknowledged that displacement is a consequence of war, he also adds that civil conflicts and natural disasters contribute to displacement. Agreeing with this, the report from the United Nations (2021) explained that most of the conflicts and violence that ravage most countries of the world leave the people with no choice but to flee to protect themselves and their families.

Other causal factors of displacement as highlighted by the report are the **adverse effects of climate change challenges; disasters—technological disasters; criminal and gang violence; persecution and human rights violations and the effects of development mega-projects** (United Nations 2021). These causes of displacement are demonstrated in Fig. 8.2.

Figure 8.2 has shown that the central cause of displacement is such factors as war, civil conflicts, armed conflicts, political violence, climate change challenges, disasters—natural and man-made, criminal, and gang-induced violence, persecution, the violation of human rights, and development-induced displacement. It is important to state that the extent to which these factors drive displacement in countries varies, meaning that different countries have their own peculiar factors that cause mass displacement. For instance, while some countries are usually displaced by war and violent conflicts, other peaceful countries where there is hardly any war may continue to experience mass displacement because of disaster occurrences. Figure 8.3 shows the total number of people displaced by just conflict and violence by year.

8.3.2 The Consequences of Displacement

There have been increasing devastating consequences of internal displacement on mostly the displaced. The report from the United Nations (2021) notes that every year, people around the world face the devastating consequences of displacement. This according to the report is mostly because displacement in many contexts is sudden, unplanned, and arbitrary. Siriwardhana et al. (2014) note that most displaced persons are usually at a greater risk of developing mental disorders following exposure to traumatic events, daily stress, and impoverishment. Becker (2022) presented

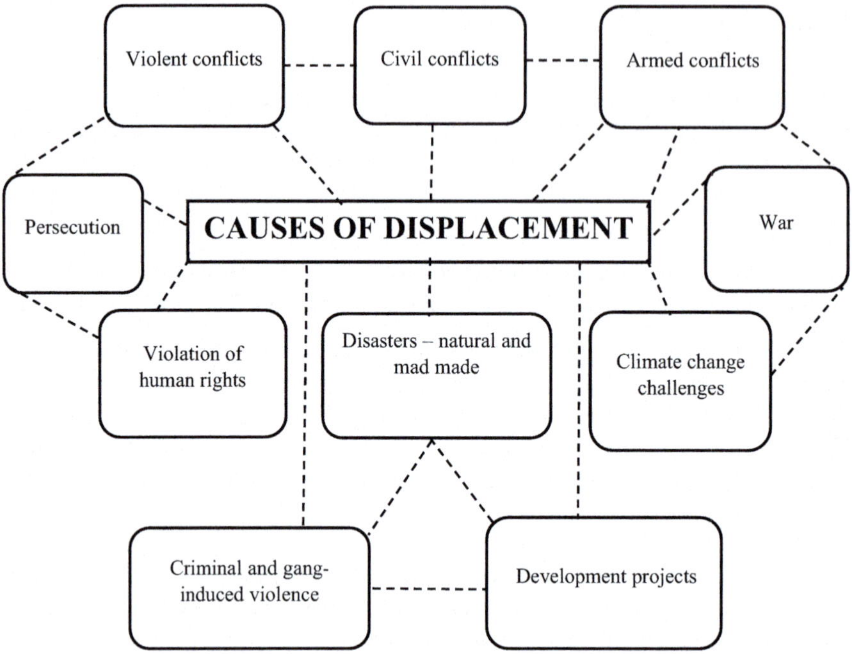

Fig. 8.2 Summary of the causes of displacement as identified from the literature reviews in Sect. 8.3.1. *Source* Prepared by the author with concepts derived from literature reviews

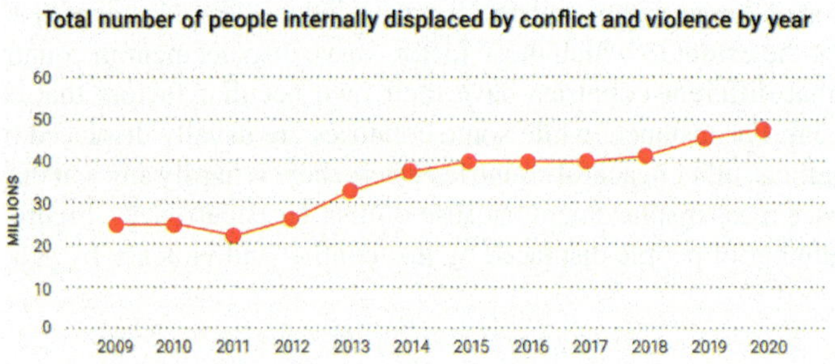

Fig. 8.3 Total number of people internally displaced by conflicts and violence by year. *Source* United Nations (2021)

a similar finding in a more recent study that the displaced suffers severe psychological distress, depression, and mental health challenges, adding that it increases mortality rates. Siriwardhana et al. (2014) clarified that this challenge is intensified by issues associated with pre-existing vulnerabilities, the stress resulting from forced migration, post-migration environments, trauma exposure, depression, and anxiety. It can be argued that the rate of displacement-induced increased mortality as put forward by Becker (2022) is hugely dependent on the cause of the displacement and the severity of its impacts.

Further to this, the 2021 report of the Special Rapporteur on the Human Rights of IDPs in the context of internal displacement states that the loss of land and loss of housing are among the consequences of internal displacement irrespective of its cause (United Nations 2021). According to the report, there is a lack of secure resettlement housing for the displaced. A situation that makes most of the displaced live in dangerous conditions where they face significant risks and violations of their rights, resulting in the women and girls being exposed to sexual abuse and gender-based violence as well as early and forced marriages.

The report further affirms that the IDPs usually lack sources of livelihood and are among the most food-insecure parts of the population with most of them living outside of camps and in urban areas where they eventually become among the poorest of the poor (United Nations 2021). It was clearly stated that older people and people with disabilities struggle to access effective assistance and support during the episodes of displacement (United Nations 2021). Another consequence of displacement is that it devastates those who are displaced, put a strain on the communities hosting them as well as has a negative impact on the development, stability, and societies at large. Regarding the impacts of displacement on the host communities, Verme and Schuettler (2019) affirmed that displacement typically results in population shock and expenditure shock to the host communities. According to them, the population shock is a sudden increase in the population of the host communities because of the inflow of the displaced population to these communities. Then the expenditure shock results from the increased financial flows that the displacement crisis may attract, including aid from international donors and increased expenditure from the government of the host communities.

These are significant challenges that displacement presents. It is worthy of note that while some of these challenges like shelter, food, security, and education can be addressed with money though not adequately, it is difficult to address others like the loss of the lives of loved ones, emotional, and psychological challenges with money. Hence, an urgent need to adopt preventive measures for internal displacement. This is urgent because most of these enumerated challenges can be worst in protracted displacement where the solution is not in view. With this, the need to discuss protracted displacement has emerged.

8.4 Protracted Displacement

Internal displacement can be temporary or protracted. Becker (2022) explained that displacement is temporary when the displaced find short transitory shelter in a safe place while waiting to return to their homes. Displacement is permanent or protracted when the displaced stay in host countries, communities, and families without any hope of going back to their country and homes (Becker 2022). According to the United Nations (2021), millions of IDPs are entrapped in protracted displacement and have spent several years and some decades, yet unable to find an end to their displacement or a way to go back to their homes. This is disheartening mostly because a recent report has shown that among the various people who are internally displaced, only a few people find lasting solutions for their situation, others are trapped in protracted displacement (OCHA 2019). It was confirmed that the reasons why the IDPs numbers remain high are because while there are new displacements, millions of people are trapped in protracted displacement and unable to re-establish a safe and lasting home (United Nations 2021). This results in newly displaced persons adding up to the number of already displaced persons trapped in protracted displacement.

Report demonstrates that protracted displacement can hugely depend on the cause of the displacement (United Nations 2021). According to this report, conflicts or disasters can result in protracted displacement depending on the cause of the conflicts and how long it takes to resolve the conflicts. The report explained that conflict-induced displacement can be protracted when the conflicts that displaced the people persist with no end in sight, and in some cases, it causes second and third generations of children of the displaced to be born into displacement and grow up with no idea of their family's original home. The report, however, demonstrates that displacement caused by disasters can be less protracted as it is often viewed as a short-term displacement. In all, both conflict-induced and disaster-induced displacement can cause protracted displacement depending on the cause of the conflicts and disasters, the level of its destruction, and how fast the rebuilding takes place.

The central thinking here is that if temporary displacement whether conflict-induced or disaster-induced can have devastating consequences, protracted displacement can have even more consequences. Therefore, to avoid the consequences of displacement and the dangers of protracted displacement, there is a need to adopt the principle of prevention rather than the principle of cure (Zapater 2010). This further necessitates the need for this study which centres on exploring and ascertaining ways to prevent the occurrences of internal displacement in Nigeria. The next section discusses internal displacement in the Nigerian context.

8.5 Internal Displacement in Nigeria and Its Consequences

The rate of internal displacement in Nigeria in recent times has been very unprecedented. It is noted that Africa has the largest number of IDPs in the world with Nigeria ranking third in Africa (Niyi-Gafar and Adelakun 2022). Unfortunately, women and girls are said to make up over half of the population of internally displaced persons, over 30.5 million are children and youth, about 2.5 million are disabled persons, and about 2.6 million are elderly persons (United Nations 2021). These are the most vulnerable groups in any country. Unfortunately, during and after episodes of displacement in Nigeria, children suffer more vulnerability especially because of the difficulty in accessing health care, education, water, food, and nutrition (Niyi-Gafar and Adelakun 2022). These scholars clarified that this situation makes the children even more vulnerable as they are usually quick to contact viral diseases such as water-borne diseases and respiratory tract infections. This level of vulnerability is something that calls for concern to protect the vulnerable groups among us. According to Jan Egeland, the secretary-general of the Norwegian Refugee Council, "We are failing to protect the world's most vulnerable people from conflict and disasters" (IDMC 2021). This arguably means that not taking adequate care of the most vulnerable displaced persons or not taking action to prevent displacements-inducing events and situations that expose these vulnerable groups to more vulnerability could indeed be seen as failing the most vulnerable group in our societies as stated by Jan Egeland.

Rapid rate of displacement in Nigeria leads to an overwhelming vulnerability of the already vulnerable groups in the country. According to Beida and Aroge (2019), Nigeria has been experiencing frequent mass internal displacement with different waves on different scales both small and big in every part of the country. Niyi-Gafar and Adelakun (2022) affirm that the threat of internal displacement resulting from increased armed conflicts, natural hazards, and development-induced displacement remains a major policy concern for the Nigerian government and the humanitarian sectors for the last decades. Notwithstanding the fact that the internal displacement of Nigeria has become a major policy concern as explained, Beida and Aroge (2019) found out that the IDPs in Nigeria are being abandoned by the country without the much-needed assistance and protection as far as International Law is concerned. That they also suffer a gross violation of their rights which may be a result of the fact that the country lacks the right capacity to manage the affairs of the displaced.

These situations enumerated herein show that the internally displaced persons in Nigeria particularly the vulnerable groups are made more vulnerable than they were already. This may be due to negligence and the lack of adequate provision of at least the basic goods and services such as shelter, food, security, and healthcare services. This still points to the fact that preventing internal displacement in Nigeria remains the last and most effective resort. In fact, the Nigerian government and other authorities should think of adopting preventive measures as a more sustainable approach to addressing internal displacement. This measure will help to avoid new displacements and the displacement of the already displaced for the second and third time. This paper emphasises the importance of resorting to adopting preventive measures rather than relying exclusively on reactive action (Zapater 2010).

8.6 Preventive Measures for Internal Displacement

The concept of the preventive approach to displacement is not new in the academic sphere. Zapater (2010) states that this approach is recognised by the Guiding Principles on Internal Displacement (GP20), the United Nations High Commissioner for Refugee (UNHCR), the United Nations Office for the Coordination of Humanitarian Affair (OCHA), and the Special Rapporteur on the Human Rights of IDPs. According to Zapater (2010), the launch of a three-year multistakeholder Plan of

Action which started from 2018 to 2020 and aimed at advancing prevention, protection, and solutions for internal displacement was fronted by these organisations. This concept was introduced in the early 1980s as a strategy to avoid new flows of displaced persons (Zapater 2010). To further explain the evolution of the preventive approach to internal displacement, Zapater states that the concept emanated from the suggestion by the Federal Republic of Germany on the need to create displacement preventive measures. Following this suggestion, a group of experts were appointed in 1981 by the secretary-general of the United Nations to explore ways to prevent new flows of refugees (Zapater 2010).

Though it was noted that at first, the UNHCR responded to the approach indifferently, but with the subsequent rate of displacement resulting in an increased flow of refugees, the preventive approach becomes the last resort. Resorting to preventive measures to displacement rather than relying solely on reactive action is exactly what is needed in Nigeria. This is mostly because one can say that it is less costly to prevent internal forced displacement and more costly not to do so. Moreover, considering the impacts of forced displacements on especially displaced communities, preventive measures can be a more sustainable approach to addressing the challenges of internal displacement (Zapater 2019). It was noted that the right step towards this prevention is to identify the local communities that are at risk of forced displacement, find out the possible causes, and strategically take action to address and prevent these risky threats with appropriate measures (Zapater 2010, 2019). These key preventive strategies as found in the literature are represented in Fig. 8.4.

The figure represents the literature perspective on the steps to adopt to prevent displacement in countries. Therefore, this study follows these recommended steps by first identifying the causes of displacement in Nigeria and second, identifying the appropriate measures to prevent these causes. Accordingly, the next section centres on the discussion of the research findings.

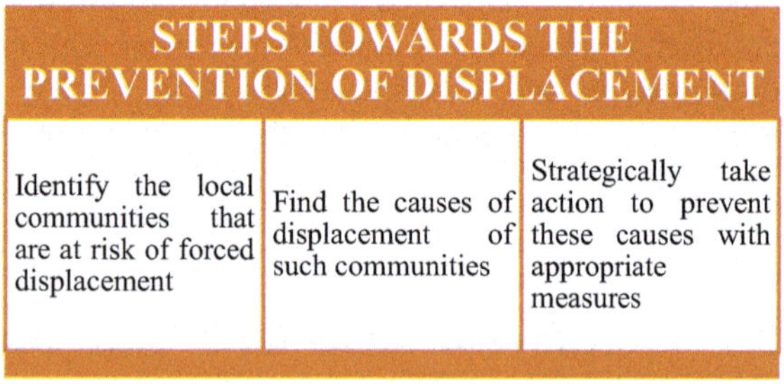

Fig. 8.4 Steps to prevent displacement. *Source* Prepared by the author with concepts from the literature reviews

8.7 Discussion of Findings—What Are the Causes of Displacement and Its Preventive Measures?

This section centres on discussing the findings of the study on what constitutes the preventive measures for internal displacement in Nigeria. For the sake of clarity, this section will first discuss the causes of displacement as identified from the literature reviews and then in the context of Nigeria as identified from the data. This will be followed by the preventive measures from the perspective of the existing literature and in the context of Nigeria as generated via interviews.

8.7.1 The Causes of Displacement

Discussing the findings on the causes of displacement is an important part of this study because knowing the causes of displacement helps to identify the ways to prevent it. According to all the interviewees to be able to propose ways of preventing internal displacement, it is important to understand what causes displacement. Consequently, literature reviews in Sect. 8.3.1 reveal the various causes of displacement in countries, and these findings were also represented in Fig. 8.1. The literature reviews presented about ten (10) causes of displacement in countries. These causes of displacement as enshrined in the literature are civil conflicts, armed conflicts, political violence, climate change challenges, disasters—natural and man-made, criminal and gang-induced violence, persecution, the violation of human rights, and development-induced displacement (see Porter and Haslam 2005; United Nations 2021; Becker 2022).

While these are acknowledged in this study as important factors that can trigger displacement in countries, it is important to note that different countries are faced with diverse challenges that cause displacement which may include all of the already enumerated factors and even more. Based on this, data were collected to ascertain the causes of displacement in Nigeria and the interviewees identified fourteen (14) causes of internal displacement in Nigeria. These identified causes are war, ethnic and communal conflicts, oil-induced violent conflicts, persecution of some ethnic groups, political factors that lead to the marginalisation of ethnic groups, mass kidnapping and assassination of people, the destructive activities of the Boko Haram, the Fulani Herdsmen and farmers crises, religious crises, development-induced displacements, disasters, lack of durable resettlement plans, neglected protracted displacement, and abject poverty—unemployment and hunger. These identified causes of displacement in Nigeria are represented in Fig. 8.5.

A look at the causes of displacement in the existing literature as represented in Fig. 8.1 and the causes of internal displacement identified by the interviewees which are also represented in Fig. 8.5 will reveal some similarities as well as differences among these findings. Table 8.1 represents the findings from literature reviews and

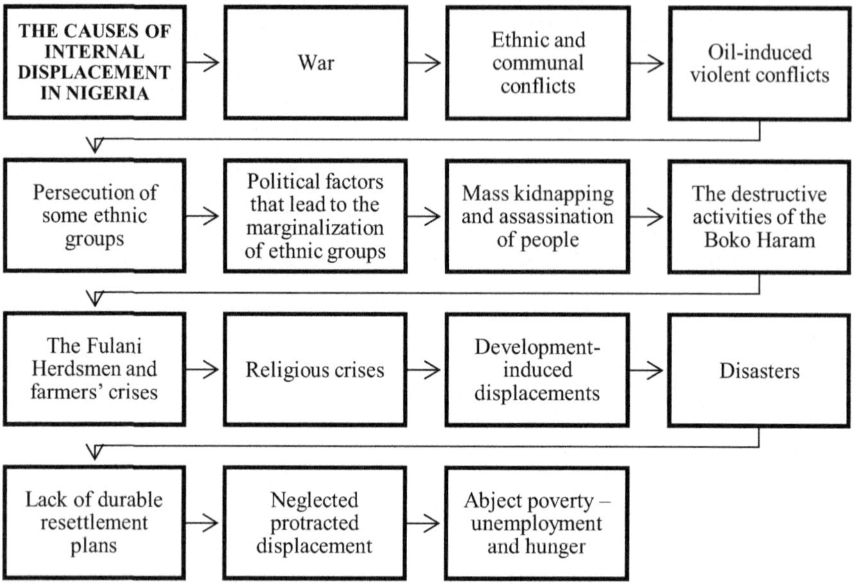

Fig. 8.5 Identified causes of displacement in Nigeria. *Source* Prepared by the author with concepts from the literature reviews

the findings on the causes of displacement in Nigeria as identified via interviews. It is important to note that similar findings were highlighted in yellow colour.

Table 8.1 shows that while the causes of displacement as found in the literature are ten (10) in number, the findings as generated from the data presented fourteen (14) causes of displacement in Nigeria. Interestingly, six (6) of the causes of displacement from the study findings were similar to the findings from the existing empirical literature (see the highlights in bold). These similar findings are war, civil conflicts, armed conflicts, disasters, persecution, and development-induced projects (United Nations 2021; Becker 2022). The rest of the factors found as the causes of displacement were peculiar to Nigeria and not found in the existing literature. Accordingly, these differences can be established as additional causes of displacement in Nigeria. These additional different factors reveal the fact that no two countries have the same causes of displacement. Nigeria is one of such countries that has diverse factors that contribute to its increasing rate of displacement. Having noted this, it is important to discuss the measures to prevent displacement, and this discussion will be carried out in the next section.

Table 8.1 Findings on the causes of displacement—literature reviews versus empirical findings from the data

S. No	Findings: literature reviews	Findings: empirical findings from data
1	**War**	**War**
2	**Civil conflicts**	**Ethnic and communal conflicts**
3	**Armed conflicts**	**Oil-induced conflicts**
4	Political violence	**Persecution**
5	Climate change challenges	Political factors that lead to the marginalisation of some ethnic groups
6	**Disaster**	Mass kidnapping and assassination of people
7	Criminal and gang-induced violence	The destructive activities of the Boko Haram
8	**Persecution**	Religious crises
9	The violation of human rights	**Development-induced displacement**
10	Development-induced displacement	**Disaster**
11	–	Lack of durable resettlement plans
12	–	Protracted displacement
13	–	Abject poverty—unemployment and hunger
14	–	The Fulani Herdsmen and farmers' clashes in Nigeria

Source Prepared by the author and compiled from both literature reviews and the themes generated from the collected data

8.7.2 What Are the Preventive Measures for Displacement?

In Sect. 8.6 of this study and Fig. 8.4, the steps to adopt to prevent displacement were enumerated in the literature which includes identifying the communities at risk of displacement, finding out the possible causes of their displacement, and taking actions to prevent such displacement. These steps are well followed in this study as first, the researcher identified and discussed the causes of internal displacement. This is essential as the interviewees emphasised that to adequately prevent displacement in Nigeria, the factors that cause it should be identified. Furthermore, this section is focused on discussing ways to prevent or address the factors that cause displacement in Nigeria. This is also critical because the interviewees affirmed that addressing the factors that cause displacement in Nigeria will facilitate the prevention of internal displacement in the country. However, it is important to note that while some of these factors may not be completely prevented, most of them are preventable. It becomes important and a great step to put in measures to prevent the preventable and work towards reducing the impacts of the unpreventable. This will help to stop new displacement from occurring thereby meeting the target of reducing new internal displacement by 2030 (United Nations 2021).

The discussion on the preventive measures for displacement is done in line with the steps proposed in the existing literature, particularly the step on taking action to prevent the causes of displacement with appropriate measures. Accordingly, Fig. 8.6 demonstrates the identified preventive measures for internal displacement in Nigeria.

The above preventive measures are discussed below.

a. Prevent war with the use of early and prompt negotiation
All the interviewees agreed that war is among the major causes of internal displacement in Nigeria. Regarding war, the interviewees stated that war is characterised by mass killings, sicknesses, diseases, disabilities, destructions, huge economic losses, and sexual violence with lasting emotional trauma. Some of these consequences of war were confirmed by the existing literature (see Rathi n.d.). Considering these factors, the interviewees emphatically stated that war is one of the major causes of internal displacement in Nigeria and needs to be prevented at all costs. The existing literature also confirmed that war is among the causes of forced displacement (Porter and Haslam 2005; United Nations 2021; Becker 2022). According to the interviewees, there is hardly any war without an underlying cause. Based on this, the interviewees stated that to reduce or prevent the war that induces displacement in Nigeria, the governing authorities should identify the central causes of these wars and address them promptly using early strategic negotiations. Accordingly, the prevention of war and its causes has been added as one of the ways to prevent and drastically reduce the rate of internal displacement in Nigeria.

b. Address the causes of ethnic and communal conflicts
Ethnic and communal conflicts are identified as among the factors that cause internal displacement in Nigeria. The interviewees noted that most of the communities in most parts of Nigeria are bedevilled with long-term communal violent conflicts which are caused by disputes on the ownership of lands, houses,

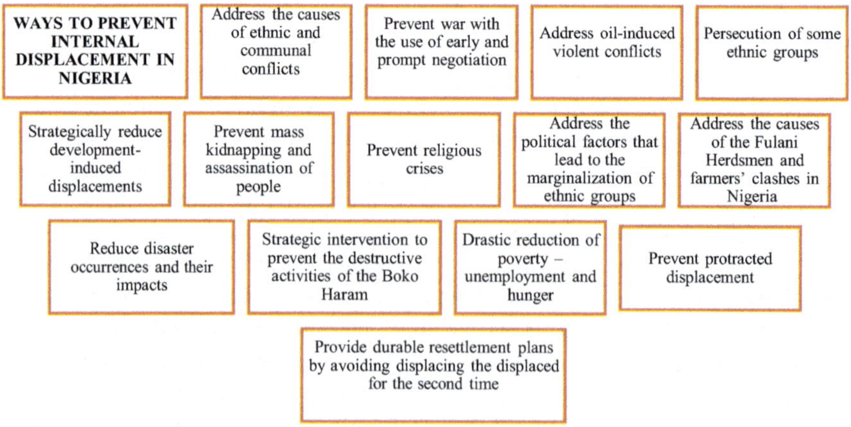

Fig. 8.6 Identified preventive measures for internal displacement in Nigeria. *Source* Prepared by the author using the themes generated from qualitative data

and properties. According to them, some of the conflicts lead to massive killings and destruction of houses and people's properties causing internal displacement. A similar finding was recorded in the report of the United Nations (2021) that tensions over land, housing, and property which cause internal displacement are on the rise and arise at all stages of displacement and that they drive conflict and displacement and pose obstacles to durable solutions. Based on this, the interviewees affirmed that to prevent ethnic and communal conflicts resulting from housing, land and property, the governing authorities should step up their strategies to address these issues permanently that this can be done with the use of negotiation or force where necessary to cut short these elongated conflicts and the consequent displacement. The report from the United Nations (2021) also confirms that addressing housing, land, and property issues in situations of displacement is therefore essential to preventing displacement, mitigating its impact, and achieving durable solutions. Indeed, to prevent ethnic and communal conflicts, the rudimentary cause of such conflicts needs to be traced and addressed promptly, and in addressing this, the need for a durable end to the conflicts should be considered.

c. Address oil-induced violent conflicts

It was noted by the interviewees that most internal displacement in Nigeria results from oil-induced violent conflicts. Explaining this further, the interviewees noted that the oil discovered in the Niger Delta region of Nigeria is what made Nigeria one of the oil-producing countries of the world today, yet the region has continued to suffer from abject poverty, lack, and negligence. That while the oil wealth generated from the oil exploration in the region is invested in other parts of Nigeria, the main oil-producing region has continued to be abandoned and backward and suffers much from oil-induced environmental degradations. Decried of the uneven distribution of resources, the interviewees stated that these factors have continued to fuel violent conflicts in the oil region leading to killings and destructions of houses and other essential infrastructural facilities thereby escalating the rate of internal displacement. This is in line with the findings of Opukri and Ibaba (2008) that oil exploration is the central cause of most of the internal displacements suffered in Nigeria. Based on this, the interviewees suggest that it is only fair that the oil resources be distributed evenly in the country to avoid subsequent conflict-inducing internal displacement. It can be said that the oil-induced conflicts can be easily addressed once the governing authorities start to engage in even distribution of resources as well as developing the oil-rich region of the country rather than neglecting them and their sufferings. This is critical to preventing further internal displacement in the country.

d. Persecution of some ethnic groups

There was a consensus among the interviewees that the increased persecution of some ethnic groups in Nigeria has fuelled conflicts that induces internal displacement. According to them, some displacement happening in places like A and B (NAMES WITHHELD) is mostly a result of the elongated hostility and ill-treatment that these communities have suffered in Nigeria. Most of the interviewees made it clear that the governing authorities are aware that certain ethnic groups in a country with over 250 ethnicities are treated like enemies and foreigners, because of mostly political or religious beliefs. Yet the complaints of these communities have continued to be swept under the carpet, hence, the outburst of riots and conflicts leading to burning down houses thereby causing internal displacement. It can be said that the level of displacement fuelled by persecution is high and that is why the UN Refugee Agency has urged world leaders to step up their efforts to foster peace, stability, and cooperation in order to stop and start to reverse the surging trend of displacement driven by persecution (UNHCR 2021).

e. Address the political factors that lead to the marginalisation of ethnic groups
All the interviewees agreed that the marginalisation of some ethnic groups induced mostly by political factors is one of the factors that cause displacement in Nigeria. According to them, when you visit the IDP camps in this country, you will be amazed how people especially vulnerable children are abandoned to suffer from diverse health issues that are not suffered by the refugees. This revelation is confirmed in the existing literature that the health status of IDPs calls for concern they have had some of the highest crude mortality rates (CMRs) in humanitarian emergencies for the past decades and they do not have access to essential services (Salama et al. 2001). It was noted by the interviewees that something needs to be done urgently to avert circumstances that put people in such demeaning and horrible situations. The interviewees noted that most of the issues that cause this horrific displacement in Nigeria like marginalisation can be addressed and prevented. That the governing authorities have the capacity to address the political factors that causes marginalisation of some of these ethnic groups; that once this is done, internal displacement resulting from marginalisation will be prevented. This is an acceptable position because the governing authorities have the capacity to find out about the marginalised ethnic groups, negotiate with them, and address their concerns.

f. Prevent mass kidnapping and assassination of people
All the interviewees confirmed that there has been a series of mass kidnappings and assassinations happening in Nigeria for a long time now. According to them, the kidnapping and assassination started with the kidnapping of individuals and the assassination of the kidnapped to the kidnapping of students in their masses, hijacking trains and kidnapping everyone inside the train, and so on. The interviewees stated that one of the causes of the kidnapping is the lack of security in the country and the lackadaisical attitude of the governing authorities to address the situation adequately. The interviewees strongly recommend that the government should provide adequate security to address these kidnapping and assassinating challenges, and that in doing so, internal displacement will be reduced. Zapater (2019) confirmed that the events that cause forced displacement lead to kidnappings and assassinations. Based on this, the interviewees strongly recommend that the governing authorities should take action to address the issue of kidnapping and assassinations sternly to avert internal displacement. This is important because without prompt intervention by the governing authorities, the situation can even get worst by the day.

g. Strategic intervention to prevent the destructive activities of the Boko Haram
The interviewees decried the destructive activities of Boko Haram and the subsequent increase in the rate of internal displacement in Nigeria. According to them, Boko Haram holds communities under siege, kills thousands of people, caused many casualties, and abducts some on daily basis leading to an increase in the internal displacement of the communities concerned. A similar finding on the impacts of Boko Haram in Nigeria was recorded in the work of Prieto Curiel et al. (2020) that Boko Haram has caused nearly 40,000 casualties in Nigeria, Niger, Cameroon, and Chad, making it one of the deadliest Jihadist organisations. Based on this, the interviewees strongly recommend that the governing authorities should strategically intervene to prevent the destructive activities of Boko Haram as well as to prevent further havoc and internal displacement. When asked if the governing authorities have not done anything in the past to prevent the activities of Boko Haram, the interviewees who acknowledged that the government has done something in the past stated categorically that what they have done is not enough. According to them, the governing authorities need to set politics aside and tackle the activities of Boko Haram accordingly. That unless this is done, their destructive activities will continue to cause havoc and displace people in their masses. Accordingly, it can be noted that the strategic intervention of the governing authorities in addressing and preventing the activities of Boko Haram will help to facilitate the prevention of internal displacement.
h. Address the causes of the Fulani Herdsmen and farmers' clashes in Nigeria
The interviewees decried the increasing destructions and losses incurred on daily basis following what they termed incessant clashes of Fulani herdsmen and farmers in various communities in Nigeria. They explained that though the Fulanis breed the goats, sheep, cattle, and other animals which are the source of meat in Nigeria, their activities and engendered conflicts result in massive killings of the Nigerians who are to consume the meat they produce. The interviewees pointed out that communities that are ravaged by the crises have complained that these Fulani herdsmen operate with guns, especially AK-47 and use them to terrorise and destroy lives and properties. A similar finding was recorded by Ajibefun (2018) that though the Fulani provides an affordable source of animal proteins in Nigeria, their attacks on communities with AK-47s and incessant killings are weird and unacceptable. The interviewees stated that the governing authorities know exactly what to do to prevent these crises and that they are being unnecessarily sluggish in taking the right actions. The interviewees, therefore, suggested that the governing authorities can bring them to round-table discussions to sort this out, and if this persists, they can sanction them and their business to the point that no Nigeria will be allowed to buy their meat. When this happens, they will be called to order, and the unnecessary crises will cease. This is important because some scholars affirmed that the Fulani herdsmen and farmers' clash are largely not necessary (Okereke 2012; Bello 2013). Accordingly, preventing the destructive uncalled for activities of the Fulani herdsmen has been identified as one of the ways to prevent internal displacement in Nigeria.

i. Prevent religious crises
Religious crises particularly crises between the Christians and the Muslims were identified as one of the factors that cause internal displacement in Nigeria and among the preventable factors. According to them, it is well known that Nigeria is a multi-religious country, and it is mostly the inability of every religion to accommodate and respect other religions irrespective of the diversity in beliefs that contributes to this crisis. The interviewees affirmed that the governing authorities know what to do especially with the use of the constitution and law to stop the escalating religious crises in the country. The interviewees, however, expressed concern about the fact that instead of the government to do the needful, they seem to "sit on the fence" on these matters allowing too much bloodshed and mass displacement in the name of religion. According to them, this peril has been ongoing for an elongated period. Beida and Aroge (2019) confirmed that religious clashes and their consequent killings and displacement have become unprecedented in Nigeria since 2012. Considering the consequences of the elongated religious crises and the displacement it causes, the interviewees advised that the government should take appropriate measures to end these crises to prevent displacement in the country. Accordingly, the prevention of religious crises has been identified as one of the ways to prevent internal displacement in Nigeria.

j. Strategically reduce development-induced displacements
Development projects were identified as one of the causes of internal displacement in Nigeria. According to the interviewees, such development projects as road construction, the building of industries, the construction of dams, and the exploitation of natural resources can cause internal displacement. It is worthy of note that Stanley (2004) recorded similar findings that development projects can displace people. Stanley explained that the construction of urban infrastructure and transportation are among the development projects that displace people. All the interviewees stated that though development projects are necessary for any country, there are ones that may not be overly necessary and those that do not need to be sudden and unplanned. They stated that the rate of development-induced displacement can be reduced with prior planning and preparation to ensure that those displaced by development do not need to suffer physical and emotional challenges as those displaced suddenly by disasters and conflicts. Some of the interviewees lamented how sudden the governing authorities destroy people's property and pull down houses built with millions of naira in the name of development with little or no prior notice. The interviewees emphasised that development-induced displacement should not be sudden, that if it must happen, there is a need for authorities to take their time to plan it properly. This plan includes informing the masses who are likely to be displaced by this development and making every necessary arrangement to resettle them before the development projects commence. According to them, this will help to reduce the trauma and shocks of being displaced suddenly without necessary plans. Becker and Ferrara (2019) confirmed that people displaced suddenly and forcibly may suffer lasting depression due to the physical or psychological trauma they

experienced. Accordingly, not carrying out sudden and unplanned development projects will help to facilitate the prevention of internal displacement in Nigeria.

k. Reduce disaster occurrences and their impacts
Disaster occurrences were identified as among the factors that cause internal displacement in Nigeria, especially flood disasters. The interviewees agreed that though disaster occurrences in Nigeria like every other country cannot be completely prevented, their occurrences and impacts can be reduced. Mentioning the fact that disasters can be natural or man-made, the interviewees affirmed that most disaster occurrences are the consequences of man's actions on their environment which damaged it and put it at the risk of disaster. According to the interviewees, there is hardly any disaster occurring in Nigeria without being induced by one human activity or the other on the environment. This argument corresponds with the one made by Kelman (2020) that there is nothing like natural disasters as disaster is a consequence of human and social creation. Hence, the interviewees advised that all efforts should be made by the governing authorities to prevent unnecessary environmental degradation that induces displacement. Also, the governing authorities particularly the urban and development planners should endeavour to prevent people from building, inhabiting, and investing in disaster-prone areas as it increases the occurrences of disasters and their impacts. While the argument that most of the disasters occurring in Nigeria are man-made is acknowledged, it is important to note that there are still natural disasters that were not induced by man's activities on the environment. Having noted this, it can be said that in any case, reducing disaster occurrences and their impacts in Nigeria is an important strategy to prevent internal displacement.

l. Provide durable resettlement plans by avoiding displacing the displaced for the second time
One of the measures identified by the interviewees to prevent second and subsequent internal displacements is the resettlement of the displaced in a safe environment that is not prone to disasters or any other calamity. This according to them is to prevent displacing the already displaced persons for the second and probably the third time. According to the interviewees, this will also help to provide a durable solution to the IDPs and can be sustainable. Similar findings were identified in the report of the United Nations where it was noted that IDPs need help to find a durable home (United Nations 2021). This report further states that scaling up efforts to help the IDPs to achieve a durable solution to their displacement is a crucial necessity; solutions that can help them return to their homes; integrate locally into their resettled areas or settle in another part of the country. The interviewees affirmed that the resettlement phase of displacement is an important phase of displacement and should be taken seriously to reduce further internal displacement. This according to them can go a long way to deciding the fate of the displaced. Accordingly, it can be said that the resettlement location of the displaced is very important in preventing subsequent displacement of the displaced.

m. Prevent protracted displacement
The prevention of protracted displacement was identified as one of the measures to address the increasing rate of displacement in Nigeria. According to them, the total number of the displaced in any country increases when the newly displaced add up to the number of those in protracted displacement. The interviewees lamented over the trauma and pains that the internally displaced people face and how that multiplies for those in a protracted displacement. According to the interviewees, one of the factors that cause rapid increase in the population of the displaced is when newly displaced persons add to the already increased number of people in protracted displacement. Regarding the increased number of people in protracted displacement, it was found in the existing literature that millions of people are trapped in protracted displacement and are incapable of having a safe and durable home (United Nations 2021). Based on this, the interviewees stated that one of the preventive measures of internal displacement is to ensure that protracted displacement is drastically reduced. This corresponds with the call made in 2016 by the UN secretary-general that efforts should be made to reduce protracted internal displacement by at least 50% by 2030 (United Nations 2021). Indeed, reducing protracted displacement is crucial and well noted in this study as one of the ways to prevent internal displacement in Nigeria, especially because the capacity of the communities who are willing to help and absorb the IDPs has continued to be strained by the protracted nature of most displacement, the lack of space and resources (Norwegian Refugee Council (NRC) n.d.).

n. Drastic reduction of poverty—unemployment and hunger
One of the factors identified as among the preventive measures to internal displacement is the creation of employment opportunities especially youth employment and the drastic reduction of abject poverty. According to the interviewees, some of the conflicts in Nigeria are triggered by high rate of unemployment and disheartening hunger resulting from food insecurity. A similar finding particularly on the food note that at the end of 2019, 135 million people across 55 countries and territories experienced acute food insecurity (GRID 2021). The report states that these results in 75 million children have stunted growth and 17 million suffering from malnutrition. The interviewees affirmed that most parents will hardly stand to see their children starve in a country where the money meant for the public good is continuously embezzled by the political leaders. The interviewees further noted that this situation of abject poverty worsens the predicament of the country at large as many atrocities and criminality continue to take place making the country more vulnerable to displacement. They said that these criminal acts and atrocities lead to massive killings and violent riots. GRID (2021) confirms that the interplay between hunger and poverty creates increasingly complex emergencies. Accordingly, it has been noted that internal displacement can be prevented with a drastic reduction of poverty and hunger and the creation of jobs.

8.8 The Summary of Findings

The study shows that internal displacement has been unprecedented in Nigeria with the displaced bedevilled with diverse challenges. Also, the study reveals that though a huge amount of money is being spent to provide for and protect the displaced, it is neither enough nor sustainable considering the heightened sufferings of the displaced. Accordingly, the study argues that preventing internal displacement is more sustainable than spending a huge amount of money in addressing the challenges that displacement presents. The interviewees identified 14 causes of internal displacement in Nigeria and proposed 14 preventive measures that need to be adopted in Nigeria. According to the interviewees, addressing the causes of internal displacement in Nigeria will help to reduce as well as prevent new internal displacements. In all, internal displacement can be best addressed by preventing it rather than spending resources that are hardly enough to address the challenges that emanate from displacement (Zapater 2010, 2019; UNHCR 2021; Becker et al. 2021).

8.9 Conclusion

The increasing rate of internal displacement in Nigeria calls for concern, and this study explored the measures that can be used to prevent it. This is mostly because of the resultant enormous challenges that ravages the displaced such as the massive loss of lives, destruction of houses, and destruction of valuable infrastructural facilities, including emotional and psychological impacts. Moreover, on a broader level, the loss and destructions incurred from internal displacement work as a setback to the achievement of the 2030 sustainable development goals agenda. No wonder, Zapater (2019) notes that a detailed understanding that internal displacement poses human, environmental, institutional, social, and economic challenges will hasten actions towards devising a more sustainable approach to address it. Indeed, there is a detailed understanding of the impacts of internal displacement on the displaced and the overwhelmingly increased vulnerability of the displaced, making the adoption of preventive measures mostly needed at this time to prevent further unprecedented displacement and its subsequent consequences. Importantly, while some of the causes of internal displacement in Nigeria can be preventable, others may not be entirely preventable. Therefore, it becomes important to prevent the preventable and put in the necessary efforts to reduce the impacts of the unpreventable. In any case, this paper argues that the prevention of these causes of internal displacement is a more important, adequate, and sustainable measure than spending hugely to address the challenges of internal displacement. It is based on this that Zapater (2010) states that internal displacement requires proactive measures rather than a remedial method. Accordingly, this study is a wakeup call to the governing authorities in Nigeria to respond to the strong recommendation from the UN Refugee Agency. The recommendation is that world leaders should step up their efforts to foster peace, stability, and cooperation to stop and start to reverse the surging trends of displacement driven by different factors (IDMC 2021).

References

Ajibefun MB (2018) Social and economic effects of the menace of Fulani herdsmen crises in Nigeria. J Educ Soc Res 8(2):133

Becker SO (2022) Forced displacement in history: some recent research. Aust Econ Hist Rev 62(1):2–25

Becker SO, Ferrara A (2019) Consequences of forced migration: a survey of recent findings. Labour Econ 59:1–16

Becker SO, Mukand S, Yotzov I (2021) Persecution, pogroms and genocide and lessons from the Jewishexperience in the 20th Century, Mimeograph

Beida OJ, Aroge PA (2019) Legal and institutional framework for the operations and management of internally displaced persons (IDPs) in Nigeria

Bello AS (2013) Herdsmen and farmers conflicts in north-eastern Nigeria: causes, repercussions and resolutions. Acad J Interdiscip Stud 2(5):29–139

Eisenhardt K, Graebner ME (2007) Theory building from cases: opportunities and challenges. Acad Manag J 50(1):25–32

Internal Displacement Monitoring Centre (IDMC) (2021) Global report on internal displacement 2021 (GRID 2021). Available online at: https://www.internal-displacement.org/global-report/grid2021/downloads/IDMC_GRID21_Final_HQ.pdf?v=2. Accessed 7/11/2021

Intra-Agency Standing Committee (IASA) (2010) Mental health and psychosocial support in humanitarian emergencies what should protection programme managers know? Available online at: https://www.refworld.org/pdfid/4e97d2d12.pdf. Accessed 06/05/2022

Kelman I (2020) Disaster by choice: How our actions turn natural hazards into catastrophes: Oxford University Press

Koch EJH (2015) Protracted displacement in Georgia: structural vulnerability and Existing not Living. 74(2):135–143

Kumar A, Paul J, Unnithan AB (2020) 'Masstige' marketing: a review, synthesis and research agenda. J Bus Res 113(C):384–398

Morina N, Akhtar A, Barth J, Schnyder U (2018) Psychiatric disorders in refugees and internally displaced persons after forced displacement: a systematic review. Front Psych 9:433

Niyi-Gafar OL, Adelakun OS (2022) The impact of viral diseases on the rights of vulnerable population: Covid-19 and the Nigerian internally displaced child. Soc Sci Humanit Open 100268

Norwegian Refugee Council (NRC) (n.d.) Living conditions of displaced persons and host communities in urban Goma, DRC. Available online at: https://www.nrc.no/globalassets/pdf/reports/living-conditions-of-displaced-persons-and-host-communities-in-urban-goma-drc.pdf. Accessed 11/06/2021

OCHA (2019) Reducing protracted displacement: a snapshot of successful humanitarian initiatives. Available online at: https://www.unocha.org/sites/unocha/files/Reducing%20Protracted%20Internal%20Displacement.pdf. Accessed 05/05/2022

Okereke D (2012) The remote immediate cause of crimes, insecurity, terrorism and instability in Nigeria and solution. http://www.nairand.com/1027585/remote-immediatecauses-crimes. Accessed 19 Jan 2018

Onifade V, Osinowo R (2019) Living conditions of internally displaced persons (IDPs) in northern Nigeria

Opukri CO, Ibaba IS (2008) Oil induced environmental degradation and internal population displacement in the Nigeria's Niger Delta. J Sustain Dev Afr 10(1):173–193

Paul J, Criado AR (2020) The art of writing literature review: what do we know and what do we need to know? Int Bus Rev 29(4):101717
Porter M, Haslam N (2005) Predisplacement and postdisplacement factors associated with mental health of refugees and internally displaced persons: a meta-analysis. JAMA 294(5):602–612
Prieto Curiel R, Walther O, O'Clery N (2020) Uncovering the internal structure of Boko Haram through its mobility patterns. Appl Netw Sci 5(1):1–23
Rathi A (n.d.) Psychological impact of victims of war and conflict. Available online at: https://www.apa.org/international/united-nations/un-matters/rathi-war.pdf. Accessed 14/05/2022
Salama P, Spiegel P, Brennan R (2001) No less vulnerable: the internally displaced in humanitarian emergencies. Lancet 357(9266):1430–1431
Saunders MNK, Lewis P, Thornhill A (2016) Research methods for business students, 7th edn. Pearson Education, Harlow
Siriwardhana C, Ali SS, Roberts B, Stewart R (2014) A systematic review of resilience and mental health outcomes of conflict-driven adult forced migrants. Confl Health 8(1):1–14
Stanley J (2004) Development-induced displacement and resettlement. Forced Migration Online
Thattamparambil N (2020) How to choose the research methodology best suited for your study. Available online at: https://www.editage.com/insights/how-to-choose-the-research-methodology-best-suited-for-your-study. Accessed 22/01/2022
UNHCR (2020) Global trends forced displacement in 2020. Available online at: https://www.unhcr.org/flagship-reports/globaltrends/. Accessed 03/05/2022
UNHCR (2021) UNHCR: world leaders must act to reverse the trend of soaring displacement. Available online at: https://www.unhcr.org/am/en/14639-unhcr-world-leaders-must-act-to-reverse-the-trend-of-soaring-displacement.html. Accessed 08/11/2021
United Nations (2021) Shining a light on internal displacement: a vision for the future report of the UN Secretary-General's high-level panel on internal displacement. Available online at: https://internaldisplacement-panel.org/wp-content/uploads/2021/09/HLP-report-WEB.pdf
United Nations Office for the Coordination of Humanitarian Affairs (OCHA) (2022) Global displacement forecast 2022. Available online at: https://reliefweb.int/report/world/global-displacement-forecast-2022#:~:text=The%20forecast%2C%20which%20covers%2026,total%20increase%20of%206.8%20million. Accessed 04/05/2022
Verme P, Schuettler K (2019) The impact of forced displacement on host communities: a review of the empirical literature in economics. World Bank policy research working paper 8727
Verme P, Schuettler K (2021) The impact of forced displacement on host communities: a review of the empirical literature in economics. J Dev Econ 150:102606
Zapater J (2010) Prevention of forced displacement: the inconsistencies of a concept. New issues in refugee research. Research paper no. 186. Available online at: https://www.unhcr.org/4bbb2a199.pdf. Accessed 12/04/2021
Zapater J (2019) Prevention of forced displacement: the inconsistencies of a concept. New issues in refugee research. Research paper no. 186. Senior Protection Officer UNHCR, Geneva. Available online at: https://www.unhcr.org/4bbb2a199.pdf. Accessed 7/11/2021

Part II
The Built Environment in Resettlement Planning

Chapter 9
An Architectural Analysis of Tsunami Re-settlement Villages of South of Sri Lanka

K. B. Seenapatabendige and Achini M. De Silva

Abstract Tsunami, displacement, fully and partially damaged houses, community infrastructure, and rehabilitation programs created a new design and build environment in the coastal areas of Sri Lanka. Nearly two decades later the life of re-settlements, engagement in fisheries and other livelihood options showed multiple shades. Present study aimed to perform an architectural analysis on the resettlements to investigate the perspectives of the displaced communities. Disaster induced forced migration of fishing communities and resettlement villages were investigated specially to identify the architectural design of foreign funded tsunami resettlement villages, to find out the influence of housing on individuals and families, and to propose sustainable architectural solutions for resettlements. Primary data collection locations were nine tsunami resettlement villages of Galle district. In-depth interviews of residents, and representatives of both government and non-government agencies involved in planning and implementing the resettlement programmers, on-site observations, and questionnaire survey (180 respondents) were the main data collection tools. Building plan, living space, build and social environment, housing allocation mode, housing expenditure and quality, community infrastructure, hygiene, participation, and socio-economic status were compared among resettlements. Single-story individual houses with back and front yard model were observed as best design fit and residents were able to manage quality of living. In contrast, both single- and two-story twin houses were showed design mismatch along with varied living conditions and social conflicts. Housing complex was the worst fit model. Poorly manage common spaces indicated the low level of social capital. Sustainable design solution included space needed by the family for private use, areas of intimate contact, and neighborhood spaces. Findings provided the future directions towards the establishment of re-settlements.

K. B. Seenapatabendige (✉)
Department of Architecture, Faculty of Built Environment and Spatial Science, General Sir John Kotelawala Defence University, Southern Campus, Kiribbanwewa, Sooriyawewa, Sri Lanka
e-mail: kbseenapatabendige@gmail.com

A. M. De Silva
Department Agribusiness Management of Faculty of Agricultural Sciences, Sabaragamuwa University, P.O. BOX 2, Belihuloya, Sri Lanka

© The Author(s), under exclusive license to Springer Nature Switzerland AG 2023
M. Hamza et al. (eds.), *Rebuilding Communities After Displacement*,
https://doi.org/10.1007/978-3-031-21414-1_9

Keywords Architectural design · Concept · Resettlements · Sri Lanka

9.1 Introduction

Traditional coastal housing and vernacular settlements are those formed by the people living and working in them by employing the wisdom, knowledge, and practices handed down from generation to generation. They are often well adapted to the geographical terrain, the society inhabiting the land, and their environmental contexts.

Along the 73 km lengthy coast of Galle district, there are 273 fishing villages (Correa 1989). Galle, third largest fish producer of Sri Lanka, accounts for 53,690 Mt fish production in 2020 (Correa 1989). There is one person (1.2 average fisherman per household) most of the fishing households perform as fisherman, and there are 11,390 active fishermen reside in Galle district. One individual fisherman is the only income earner to the family for whom fish is an essential produce to generate income and for food. Fishing vast Indian Ocean in an anticipation of a bountiful catch, surrogating with rough sea is a life treating experience of one's own life.

The coastal fishermen of Sri Lanka are generally considered as poor, and they are operating on extremely hard environment. The settlers along the coastal belt of Sri Lanka, many of which, are isolated from cities and urban centers while some villages are in the center of globally rewound tourist destinations. The state investment infrastructure such as water, electricity, sanitation, health, education, transportation, and communication facilities as well as infrastructure for fish landing and marketing remains quite low (Amarasinghe 2005).

Life of coastal village in Sri Lanka, especially the South coast, is different from its inland village setting. People used to live small- to medium-sized houses located in home garden and surrounded by coconut palms. Most of the houses are open to beach front, and other houses are scattered in open space where boundaries of land slots were not separated by fences or walls. Beach front and huge open spaces were used for various activities, such as net mending, fish drying, playground for both kids and youth, coir rope making, store fishing gear, and other ancillary items. Prior to the tsunami, people enjoyed the freedom of space and this open space allows them to perform various community activities without borders. They were engaged with various extra income generating activities, tourism support services, fish drying, coir rope making, livestock raring, etc. Fishing communities used to share part of their harvest among neighbors, and kids and elderly caring were shared responsibility which strengthens the community ties. On the one hand, alternative income generating activities ensure secure living for fishing families especially in off fishing season. On the other hand, female members were occupied in productive work, strengthen their personal capacities, and recognized as economic actor to the family. Moreover, domestic fish processing activities, drying, salting, and Maldive fish making, allow families to work together and share the know-how among young generation.

Tsunami displacement and relocation in re-settlement villages have created new culture. Government policy on 100 m buffer zone pushed the re-settlement villages away from coast. Most of the relocation sites were developed in government lands, and only available sites were away from the coastal belt. Re-settlement villages were established with the financial support of various funding bodies and foreign governments where plans and designs were adopted from foreign experiences. Houses were built in individual land slots, and limited public spaces were made available for common purposes, such as playground, community center, and temple.

Family activities were restricted into limited land slot and fences, and boundary walls made families physically away from each other. Lifestyles have to change to fit into new situation, and relocation led to a considerable reduction in earning opportunities. They were moved from shore-based activities, dry fish making, coir rope making, etc., and search for new livelihood options available in new locations. In general, most of the female members were lost their income generating opportunities. Elderly and kids care became the individual family responsibility, and kids' education was badly affected due to re-location. Further, underdeveloped community infrastructure, especially market access, search for food, and extra cost for transportation, were added pressure on household economy of the re-settled people. In the meantime, government policy on buffer zone released and most of the re-settlers were returned to their original lands on coast and setup temporary houses first and begun their usual life. Most of them abandoned their houses in re-settlement villages (Barenstein and Wickramagamage 2009), or sold out and re-built the houses on original land spaces.

Marine fisheries were the most affected by the great Indian Ocean Tsunami compared to other industries. Damage on fishing communities, fishing vessels and implements, landslides and other coastal infrastructure, as well as livelihood of the people was unaccountable (De Silva and Yamao 2010). Moreover traditional fish processing long the southern coastal affected so badly. Beyond the lost life of fisherman and their families, the tsunami wrecked much of the community infrastructure available to facilitate those directly affected (De Silva and Yamao 2010).

Widespread dissatisfaction is common among the residents of resettlements on housing designs, quality of material, living spaces, access to fishery, etc. Housing of fishing villages before tsunami was established in closer vicinity to beach and the location was convenient, and the housing plan and structure facilitate their livelihood options. But re-settlements have created new environment and individuals and families have to adjust into new living and neighborhood environment. Our aim was to perform an architectural analysis to explore the present situation of the tsunami re-settlement villages of South coast of Sri Lanka and propose participatory as well as sustainable resettlement design approach.

Return to original coastal locations, abandoned and/or sold the houses of resettlement villages, re-built houses in former locations, conflicts among neighbors and community dwellers, abandoned common community infrastructure, ignore participatory maintenance, and management of common resources at re-settlement villages were common observations from most of re-settlements along the coastal belt of Sri Lanka. Table 9.1 presents the summary of the published research on disaster-induced displacement, rehabilitation efforts, and re-settlements in various destinations and extracts of the publications; especially objectives and conclusions were analyzed to understand the nature of the problem, mitigation measures, and good practices towards to establish sustainable re-settlements. In general, published papers highlighted that disasters and the issue of internally displaced people need to be handled with special care and attention. Nearly two decades ago, most of the coastal communities of South and East lost their all belongings within few hours. Along with the pain of loss of family members, neighbors, as well as belongings, they have to move into temporary rehabilitation camps. Policy, administrative setup, as well as local preparedness was poor for such a massive disaster. Unplanned rehabilitation work added extra pressure on internally displaced people. Moreover, re-settlements were built with the aim of provide shelter physically for the displaced people. Unfortunately, most of the re-settlements were established without participation, involvement, or even consent of the affected people. At first instance, people moved into re-settlement villages, tried to fit into new location, house, neighbors, etc. Incompatibilities of the location, shelter, neighbors, community, infrastructure, livelihood options, etc., were made people away from resentments. Our focus is more on an architectural analysis of Fernando et al. (2019), Vijekumara (2015), Dias et al. (2016), Silva (2011), Chen (2011) and Ruwanpura (2009). Housing designs of re-settlement villages find out the problems of housing design and its impact on lifestyle of displaced communities and understand the ground realities of what displaced communities actually need? Is it simply a house or a shelter? Or a complete package of houses, community infrastructure, surrounding, small group of people, same way of living, livelihood option, etc. Table 9.1 discusses the results of quantitative analysis of published research on displacement, disasters, relocation, and resettlements.

Results of the content analysis of the objectives and conclusions of the published research on displacement, re-settlement, housing projects, problems, and issues, etc., explain in Figs. 9.1 and 9.2. In Fig. 9.1, the word cloud highlighted the key concerns of the objectives of the published research. Results revealed the key considerations of the analysis of the objectives. Re-settlement was the ranked first as most frequently appeared theme followed by disaster, relocation, and housing. Further, in Fig. 9.2, word cloud explains the results of content analysis of the key conclusions of the published research.

Content analysis of conclusions revealed the most commonly appeared conclusions of the published research on displacement, re-settlements, and living. Housing-related concluding remarks were identified as most frequently appeared concept followed by the re-settlements. Disaster, communities, environment, living conditions, socio-economic issues, place, and process were appeared next. Great Indian Ocean Tsunami, devastated coastal communities, displacement, temporary settlements, and re-settlement housing schemes were the chain of actions occurred last 16 years. Continuing research work on displaced coastal communities, open new research avenue based on the experience of displaced communities. Key questions were what kind of architectural designs are available in tsunami resettlement villages? How housing designs influence individuals and families? And what kind of housing design will fit with the individuals, families, and local culture? Key focuses of the

Table 9.1 Content analysis of the published research on re-settlements

Author/s	Published year	Country	Objectives	Conclusions
Correa et al.	2011	USA	Facilitate policy and decision-making of governments on disaster risk reduction and management Application of preventive resettlement programs as disaster risk reduction measures	Geological and hydrometeorological hazards shows a clear upward trend the frequency of those of geological origin remained fairly constant floods and storms account for a high percentage of the total and had a more rapid rate of increase than other hazards People are affected by natural hazards related to climatic events such as floods and storms Relocating a population, its economic activities, and its social networks and relations, as well as its natural physical and built environment (buildings, infrastructure, and facilities) is a complex process with significant impacts. Preventive resettlement as a corrective mitigation measure through intervention in an existing risk condition corresponds to the sphere of public policy, that is, it is based on recognition of the rights and responsibilities of public, private, and civil society stakeholders; is guided by principles of effectiveness, equity, and precedence of public wellbeing; and takes place in a specific economic
Fernando et al.	2019	Sri Lanka	To investigate the impact of the disaster-induced relocation process on the displaced communities; sub-objectives were, to explore various needs of victims in different stages of relocation, and to examine how relocation have altered the social, cultural, and livelihood dimension of the said needs of the victims and to understand the role of the built Investigate the environment in the relocation process	An action plan with regard to the given disaster induced displacement setting has been formulated in a rapid nature to cater quick relief for the displaced communities. Officials accepted the unavailability of action plan. There is a clear resilient housing policy in place. The major draw backs in operationalization of policy frameworks were given relocation process is the lack of consultation of the displaced communities, most of the policy decisions are restricted to the planning stage of the housing setting and has not Long-term impacts of the relocation (economic, social, and cultural needs). Resettlement programmed has only given importance to the physical housing structure that is also neglecting the subjective interpretations of the residents with regard to their housing satisfaction. Relocation programs are closely related to the mismatch between the needs of the displaced community. Major negative impact on the project as the needs of the residents tend to change along with time

(continued)

Table 9.1 (continued)

Author/s	Published year	Country	Objectives	Conclusions
Ess	2019	Germany	Focused on the resettlements as a result of the operations of the opencast mine of Welzow-Süd (Lusatia, southern Brandenburg), which witnessed the development of the resettlement practices from the GDR (German Democratic Republic) period from the late 1960s and after reunification up to today. Briefly described the villages, implications of the devastation of these villages, resettlement of their inhabitants on the rural landscape and settlement structure of the region discussed	Industrialization, lignite mining, the devastation of villages, and the resettlement of their inhabitants had, and continue to have, massive implications on the rural landscape and the settlement structure of the region. In both East and West Germany, authorities and planners considered resettlements as an opportunity to improve living conditions and infrastructure as well as to reorganize the settlement and territorial structure. GDR was common to resettle displaced inhabitants via single and group relocations into new Plattenbau housing on the outskirts of cities. Starting in the mid-1980s, regulations that facilitated the resettlement of displaced persons into detached prefabricated houses were put into place. Resettlement practices in the Rhineland acted as a prototype for those in Lusatia and both in West and East Germany collective, "socially responsible" relocations are being conducted
Li and Feng	2021	China	Measure the subjective well-being (SWB) of disaster-induced migrants and examine how housing resettlement affects individual subjective well-being	The original housing of disaster victims is mostly self-made and commonly lacks repairs, which poses a certain safety risk. In addition, their original village infrastructure is underdeveloped, causing inconvenience to their daily lives. After they move centralized resettlement communities, their housing conditions and community environment tend to improve to some extent, and their SWB slightly increases. Overall, the SWB of disaster-induced migrants was good, and most self-evaluated SWB values were above average. Housing resettlement has a significant influence on the SWB of disaster-induced migrants. Housing plays a significant role in the SWB of people Regarding the resettlement housing conditions, SWB was significantly influenced by the housing allocation mode, housing expenditure, housing quality, and living space. The SWB was negatively associated with housing expenditure and positively Associated with the housing allocation mode, housing quality, and living space. That is, the findings show that disaster migrants are likely to feel happier if they are satisfied with the housing allocation mode, housing quality and the living space, and the more housing expenditure related to the resettlement is, the less happy they tend to be. Study enriches the perceived association between housing and SWB. Dominant studies show that the per capital living space tends to positively affect SWB, finding further points out that the housing allocation mode can have a strong impact on SWB Among the resettlement community environment, SWB was not significantly associated with community facilities or community participation, but it was significantly positively associated with community hygiene and the cadre–mass relationship. That the community environment and public facilities have a positive relationship with SWB

(continued)

Table 9.1 (continued)

Author/s	Published year	Country	Objectives	Conclusions
Vijekumara	2015	Sri Lanka	To identify prevailing practices and reasons for effectiveness in terms of landslide disaster Resettlement, to identify the problems of resettlement projects in Sri Lanka, to examine and analysis prevailing landslide resettlement planning approaches used in local context, to propose suitable resettlement framework for successful landslide disaster resettlement practice	Finding out draw backs of existing resettlement process. Introducing a new framework for landslide resettlement in Sri Lanka that could serve as a guideline. These implications are of direct benefit the implementing institutions related to the landslide resettlement
Dias et al.	2016	Sri Lanka	Purpose to establish the indicators affecting long-term Satisfaction of communities in the context of post disaster permanent housing	The housing resettlement strategy should use a "systems approach" which integrates Socio-economic and environmental functions with the housing development Identifying housing resettlement as a master planning exercise would ensure that these different aspects are integrated, rather than merely providing permanent houses to people and then deciding about the provision of other facilities
United Nations High Commissioner for Refugees (UNHCR)	2014	Italy	Focused on planned relocation within national borders, not cross-border movements. Discussions centered on planned relocation made necessary by climate change-related disasters and other impacts of climate change	More people live in locations highly vulnerable to disasters and other climate change impacts, planned relocation will be one response. Well planned relocation can be both a form of disaster risk reduction and a form of climate change adaptation states may take to protect affected communities. The sustainability of planned relocation should be assured through adequate attention to site selection, livelihoods, integration (identity and culture), and host communities, among other factors
Silva	2011	Sri Lanka	Discuss key objectives of resettlement locations and investigate critical attribute of vernacular building process in order to derive a set of principles that can be followed in re housing effects	Understanding of the attribute of vernacular process of housing in resettlement planning is essential for the creation of successful resettlement location. Knowledge disaster privation and response is not transpired to design professionals who undertake resettlement housing projects. Resettlements should be built upon the values aspiration of the communities rather than any aesthetic references s of designers or visual evidence of vernacular form
Chen	2011	Sri Lanka	To investigate the role of Religious Non-Government Organization (RNGO) disaster recovery context and defined characteristic of RNGO housing projects	Resettlement housing is intended as an instrument for transforming people materially, socially, and spiritually. Buddhist to Buddhist charity model perceived as less inversive. RNGOs can address a pressing the filling and existing gap of human and spiritual connections, in international humanitarian assistance
Ruwanpura	2009	Sri Lanka	Interrogates of social and political geographies of resettlements and reconstruction of temporary and permeant shelters	Relief needs to recognize that the process of putting house in place should be embedded with in local social relations

(continued)

Table 9.1 (continued)

Author/s	Published year	Country	Objectives	Conclusions
Weerasinghe and Shigemura	2008	Sri Lanka	Focuses on the totally and partially damaged housing that necessitated relocation or rebuilding along the western coastline of Sri Lanka	One of the benefits associated with the recovery programs was that the inhabitants of shanty areas would receive guidance for building permanent housing; possibly the first such experience for many of these people. If successful, this instruction will help to mitigate the damage caused by future disasters and help alleviate problems associated with informal settlements along the coastline of the country. The reestablishment of temporary structures inside the buffer zone is deferring coastal housing and living environment problems into the future. The decrease in floor area and household occupancy could result in future housing quality problems. Due to the buffer zone revival policy, the number of temporary structures has increased. Furthermore, the distance from the coastline affects the level of damage. These findings show the necessity for clear regulations pertaining to coastlines. While the relocation and resettlement programs are effective, both need to be undertaken in a sensitive manner as community satisfaction is important. Relocation efforts will be more effective when they consider the livelihood, original community order, and the previous neighborhood environment. Housing measures and resettlement arrangements differ markedly between urban and suburban areas. Our observations revealed a wide gap between the quality and quantity of permanent housing and the progress of restoration efforts in urban areas compared to suburban areas. This implies that redevelopment plans should consider the differences between areas and community needs

Fig. 9.1 Content analysis of objectives of the publish research on resettlement housing

Fig. 9.2 Content analysis of conclusions of the publish research on displacement, resettlement housing, and issues

research were to identify the architectural design of foreign funded tsunami resettlement villages in Galle district of Sri Lanka, to find out the influence of housing on individuals and families and to propose sustainable architectural solutions for resettlements. Fish bone diagram (Fig. 9.3) is used to discuss the chain of causes and effects of tsunami and post-tsunami period. Timeline of the events created with tsunami, causes, and its effects highlighted the changes occurred during last 16 years. Baseline information on causes and effects lead to build the research concept.

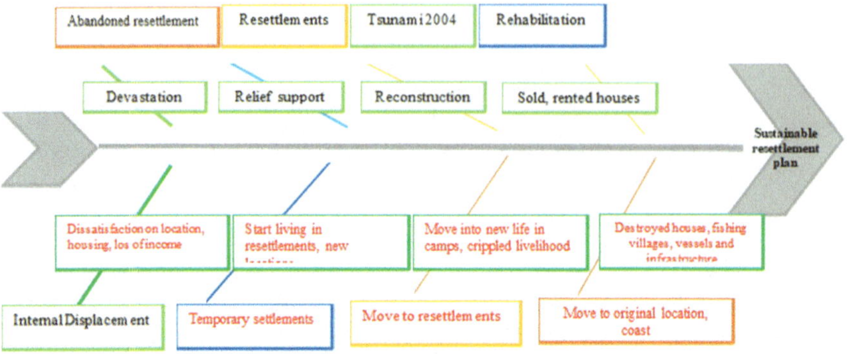

Fig. 9.3 Chain of causes and effects of timeline of disaster management: disaster, internal displacement, rehabilitation, re-settlement, and abandoned resettlement

Fish bone diagram brought the timeline of events, causes, and effects of Indian Ocean tsunami on coastal fishing communities of South of Sri Lanka. Conceptualization of the study begun from baseline of events, the major cause for internal displacement of fishing communities, and this was the first experience of tsunami for the coastal communities, administrators, government, and even the nature. Massive internal displacement establishes rehabilitation camps and temporary settlements for the people badly suffering from trauma and pain of loss of loved ones as well as all belongings. Construction and establishment of resettlement villages welcomed people into new housing schemes situated in new location. Field observation derived the research aim, and most of the resettled people first move to temporary houses in their original location, the coast. Conceptualization model explains the link between events, causes, and effects.

9.2 Methodology

Study locations were nine foreign-funded tsunami resettlement villages established in Galle district of Sri Lanka. Resettlement villages were selected based on diversity of housing designs and structure of villages. There was homeowner driven, housing reconstruction villages, and doner built housing construction and relocation. Study locations were located at Balapitiya and Ambalangoda DS division of Galle district. Research locations were undergone identical changes to their living and neighborhood environment, domestic spatial environment in 2004, after the tsunami, on set of relocation and resettlement programmers and long the timeline (Fernando et al. 2019).

Study was heavily based on primary data. Qualitative data collection began with the informal discussions with house owners, community leaders, and administrative officers of the area. Data collection tools were filed observations, in-depth interviews with key stakeholders of the community, focus group discussions with community members, and field survey along with structured questionnaires. Further, photographs of past, initial stage and recent, were used investigate living, neighborhood, and domestic spatial arrangements. Household survey conducted in nine villages where 30 households from each village were selected for the data collection. The first we have observed living and neighborhood environment and find out the design, housing structure, and available infrastructure facilities.

Secondly, in-depth interviews and focused group discussions (one focus group for each village) were conducted to investigate design, structure of the different resettlement, influence of housing on individuals and families, and problem and issues and, finally, conducted detail study of the resettlement with different housing design. Key variables were building plan, living space built and social environment, housing allocation mode, housing expenditure and quality, community infrastructure, community hygiene, community participation, and socioeconomic status in order to find out the compatibility of architectural design with local living. Respondents' satisfaction was tested by 5-point rating scale (5—highly satisfied; 1—highly not satisfied). Research locations (Fig. 9.4) were nine Tsunami resettlement villages established in Galle district, Sri Lanka. They were Erora village, UNSCO village of Kosgoda, Waturagama village, Induruwa village, Balapitiya village, Thalgasgoda village, Ambalangoda village, Okanda village, and Urawaththa village.

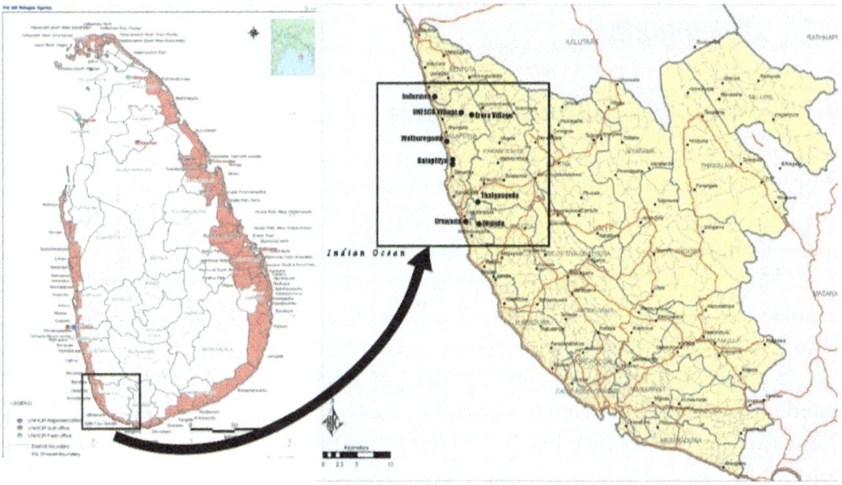

Fig. 9.4 Study locations, tsunami re-settlement villages in Galle district, Sri Lanka. *Source* Disaster Management Center

9.3 Results and Discussion

With the reference to government policy on 100 M, coastal buffer zone-affected communities have to move away from the land which they use to live generations. Unavailability of the public lands near by most the resettlements were established in far away from the beach. Coastal communities were scattered among different resettlements, and their living environment and neighborhood were changed very much. In general, they used to live in a small house with a large open space where they share open space for many activities. Resettlements were totally different environment where the limited space allocation for house construction and living. New neighborhood was hardly well come the coastal communities with their various socio-cultural issues between resettles and their neighborhood. Table 9.2 brings the pre-disaster living environment of the displaced people. Housing type, structure, construction materials, living environment, spatial environment, and neighborhood environment were key concerns. Information was sourced from elderly members of the resettled communities and community leaders. Temporary, semi-permanent, and permanent housing types with various vernacular architectural concepts were studied with the help of participatory discussions.

9.3.1 Pre-disaster Living Environment of Coastal Communities

9.3.2 Single Story Housing Designs

UNESCO and Erora villages, two resettlements, were identified for the category of single-story housing design. Housing design composed of two bedrooms, living space, kitchen, and toilet. With reference to in-depth interview with village management initial design from doners came with one bedroom, kitchen, living space, and veranda. In general, local families living with extended family grandparents are always living with the sons and daughters. Therefore, they have converted the design into two bedroom houses to cater the need of local families. Land space were allocated for back and front yards. UNESCO village of Kosgoda, Sri Lanka, established 5 km away from the beach but the village was established with all necessary infrastructure facilities. Therefore, houses of resettlement were not abandoned and village management enabled to provide necessary guidance and facilities even sixteen years after established. Figure 9.5 explains the housing plan and the front view of the single-story housing designs.

Table 9.2 Structure and nature of the housing designs: pre-tsunami

Housing type	Structural and material	Living environment	Domestic spatial	Neighborhood environment
Temporary houses	In expensive construction materials (metal sheet, waste wood) Unstable structure No proper foundation Floor—sand Roof—asbestos, metal sheet	Temporary fence Good connection with the community More open space for net mending, storage facility and fishing processing	Kitchen and living space	People of same religious, ethnicity, cast, and similar employment
Semi permeant house	Proper foundation (cement) Wall—wood, cement, brick No plaster applied Roof—timber frame covered by metal sheet, asbestos			
Permeant house	Good quality construction materials Proper foundation Permeant windows and doors Roof timber frame covered by clay tiles and asbestos	Permeant fence More space for private uses	Bedroom (1–2) Kitchen and dining space	

Source Filed survey, 2021

9.3.3 Duplex Housing Designs: Single-Story and Two-Story Houses

Wathuregama and Balapitiya (Cyril De Zoysa tsunami village) were two Tsunami re-settlements considered for the study. Two resettlements were identified with single-story duplex housing design and two-story duplex housing story duplex housing designs. Single-story duplex housing design composed of one bedroom, living space, kitchen, porch, and toilet. Figure 9.6 brings the house plan and the front view of the duplex houses.

Two-story duplex housing design composed of two bedrooms and small veranda in second floor and living space, kitchen, and toilet in first floor. The duplex housing

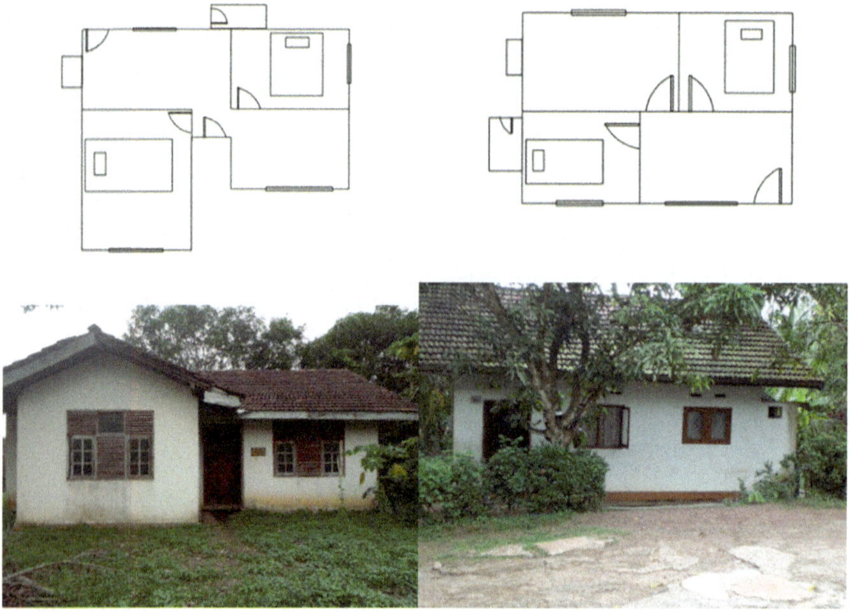

Fig. 9.5 Architectural designs of the single-story houses of the tsunami re-settlements: **a** plan of the house, **b** front view of the house

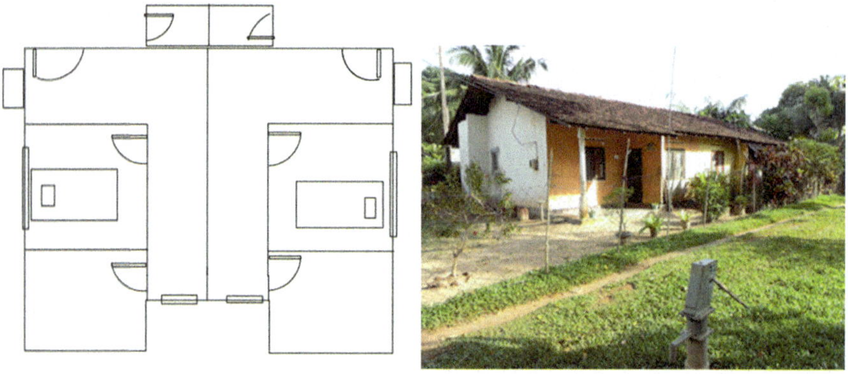

Fig. 9.6 Single-story duplex house of the tsunami re-settlement Balapitiya, Sri Lanka: plan and the front view of the house

design is not match with local culture, and incompatible neighborhood has created the various socio-cultural issues to the inhabitants. Figure 9.7 brings the plan of the house and front view of the duplex two story houses of the foreign-funded Tsunami re-settlement. Social issues, unrest, and incompatible neighborhood environment were common issues of most of the housing units. After 16 years of the establishment, consequences were able to observe not only among the residents as unrest but also

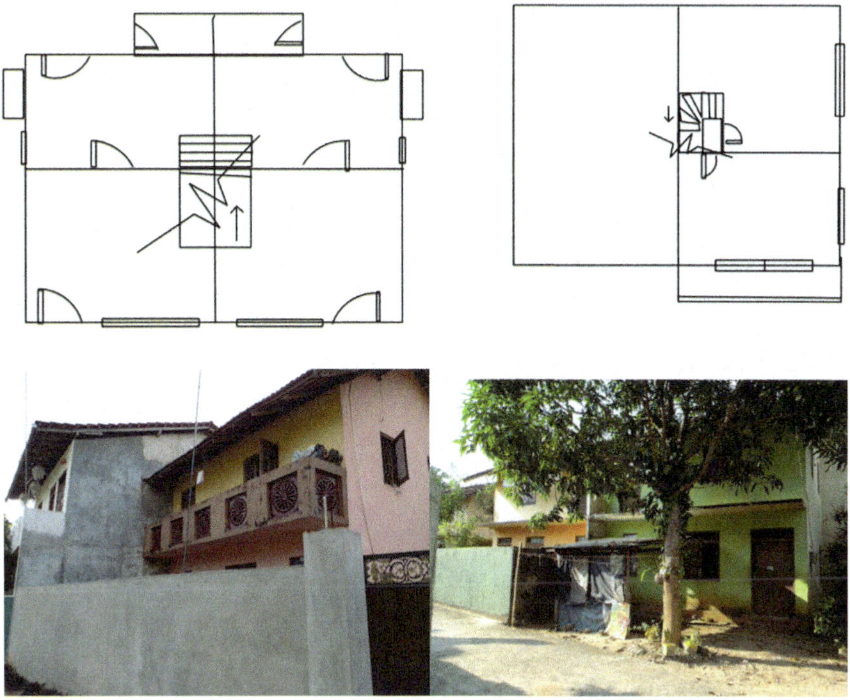

Fig. 9.7 Two-story duplex house of the tsunami re-settlement Wathuregama, Ahungalla, Sri Lanka: plan and the front view of the house

in physical structures. Of the study sample, some of the households were able to construct structure to separate their private living from the shared living spaces.

9.3.4 Housing Complex: Flat

Four story housing complexes, a Flat is not a common experience for many of the internally displaced rural coastal communities. Housing complex, a Flat was the only available solution to setup a re-settlement due to the limited public land. Most of the displaced people who used to live close to the Ambalangoda fishing harbor and their willingness to settle within the city limits directed towards to this decision. Social unrest, incompatibilities, and poor participation in common activities and political interventions disturb much on living environment. On the other hand, life within the city limits provided access to many essential logistic facilities including schools, transport, markets, and hospitals. Figure 9.8 brings the front view of the housing complex.

Planning and establishing resettlement for the communities those who displaced by natural disaster priority must give to the community participation discussion

Fig. 9.8 Four story housing complex, flat of the tsunami re-settlement Urawatte, Ambalangoda, Sri Lanka

making and finding the location similar to the land of origin (De Silva and Yamao 2010).

Some of the resettlement established far away from the coast were unsuccessful due to several reasons. Filed observation highlighted that most of the re-settlers moved to their original land and set up temporary housing there and form they livelihood activities. This is very common among resettled fishing communities. Moreover, most of the re-settlers rented out or sold their houses in resettlement villages and move to permanent and semi-permanent house built at the original land.

Figure 9.9 brings the social capital status of the resettled communities, and it was measured using rating scale. Bonding, bridging, and linking social capital levels were measured separately for different types of resettlements. Community government linkage is a power-sharing arrangement in modern democracies. Inter-communal relationship is a relationship which exists between two or more communities. One of the main objectives of any good governance is to bring government close to the citizens and provide a forum where local people can defend and develop their local interests. Intra-community relationship is a relationship within a single community. Community government linkages or link units of the duplex house made their own structure in social capital status; the resettlements were similar, expect flats. In flats, they maintain similar relationship with local and central government since they shared close neighborhood. Inter-community relationship or bridging social capital status of the SHB was recorded as the highest and lowest were SH, SDB, and TDB. Inter-community relationship or bonding social capital status of the SHB was the highest, and lowest recorded was TDB. Both SDB and TDB were identified as incompatible housing design with local culture. Of the sample responded were highlighted that duplex housing design was unsuccessful because families were different in attitude, values, behavior, and further structure such as walls and fence to separate from neighborhood.

Figure 9.10 brings that the satisfaction of housing design was measured using rating scale. Space for private use, quality of building materials: walls, floor, roof, method of waste disposal, arrangement of living space, status of sanitary facilities,

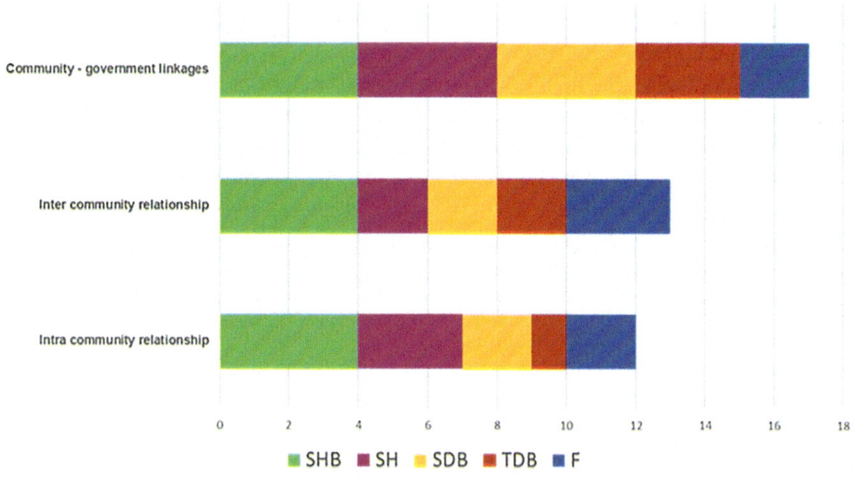

Fig. 9.9 Social capital status of the various resettlements. SHB—single story house with backyard, SH—single story house, SDB—single story duplex house with backyard, TDB—two story duplex house with backyard and F—flats

and arrangement of doors and windows: ventilation and natural light were used to measure the resident satisfaction on housing design. Space for private use and having front and back yards were the highly satisfied features of SHB. Highly unsatisfied housing design was F.

Figure 9.11 highlighted the average satisfaction of residents on built environment (housing design, spatial arrangements, etc.) and social or neighborhood environment of the resettlements, and the satisfaction levels were measured using rating scale.

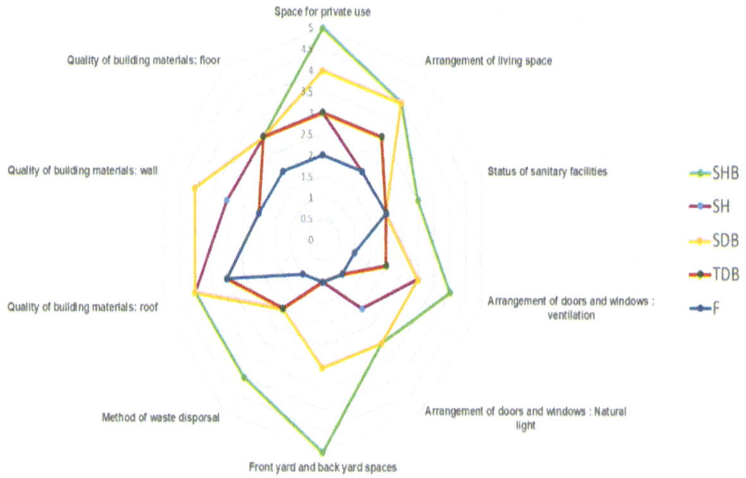

Fig. 9.10 Level of satisfaction of the residents on housing design

Variables were availability of natural environment, anti-social behaviors of the community members, such as gambling, condition of women, violence behaviors of community members, alcoholism and narcotic drug abuse, community events, employment opportunities, local transport, cost of living, regular maintenance and development, recreational facilities, shopping facilities, composition of resettlement population, acceptance by local people, public institution, public infrastructure, mutual trust, calmness, public safety, access to kindergarten, school, hospital, nearest city, and beach and past natural environment were used to measure the resident satisfaction on built and social environment of the resettlement. Further, local transport and access to the beach were the highly satisfied features of F. Gambling, condition of women, violence behavior of community members, shopping facilities composition of resettlement population, and public safety were the highly unsatisfied features of F. Four-story housing complex was totally new experience for fishing families and limited space, especially unavailability of space for private use; limited common space has created various social and cultural issues. Varying degrees of mismatch of neighborhoods and dissatisfaction over their behaviors made the housing type unpopular among all other designs.

With reference to Correa design (2011), we were able to design living environment and neighborhood environment. Design approach essentially needed to understand unique cultural tradition of living as well as requirements of the specific community. Coastal communities refer much large space neighborhood environment while maintaining space for privet living. Nature of their living essentially needed to consider during the resettlement planning process which was badly ignored. Correa identifies this hierarchical system as consisting of four major elements: space needed by the

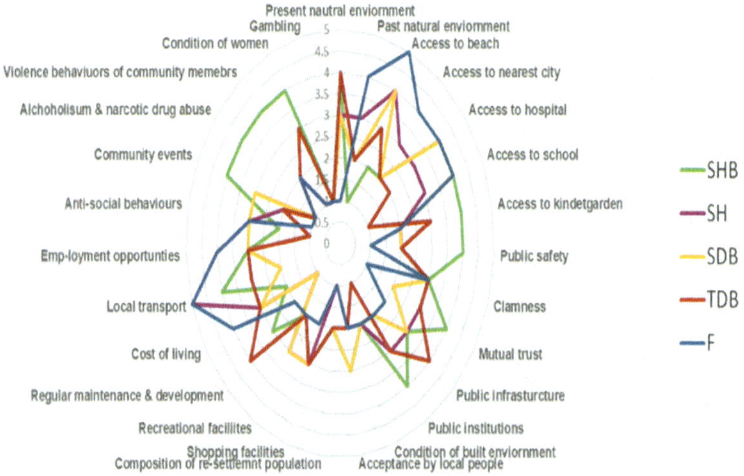

Fig. 9.11 Average satisfaction on built and social environment of the resettlements

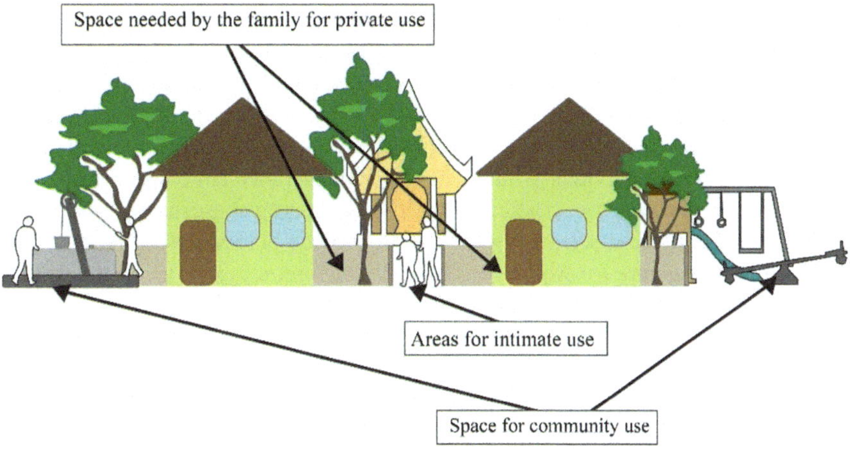

Fig. 9.12 Proposed spatial arrangements for the resettlement village complex. *Source* Adopted from Correa (1989)

family for private use, areas of intimate contact (e.g., the front doorstep where children play), neighborhood spaces, like a water tap, and urban area open space used by the whole city. Figure 9.12 explains the components of the living and neighborhood environment design for resettlement adopted from the Correa (2011).

With filed observations and in-depth interviews with stakeholders and community members, we were able to design suitable housing designs. There were three different mechanisms to facilitate housing infrastructure: foreign-founded design and built, foreign-founded design and built by the local people, and foreign-funded locally designed and built. Foreign design and built resettlement's housing design proposed by foreign donors and quality raw materials were used, but community participation was poor in decision-making. Most of the designs were planned for limited land space and maximum utilization of resources. Single-story duplex house and two-story duplex house and flat were ideal solution for resource limited resettlement design. Unfortunately, these designs were identified as the unsatisfied housing design for resettling coastal communities of south of Sri Lanka. In general, coastal communities of south of Sri Lanka used to live in multi-functional coastal zone where there living spaces design and built to fit with it. Traditional coastal settlements are those formed by people living and working in them by utilizing knowledge, skills, and traditions down from generation to generation. The coastal settlements of South as its own unique often well adapted to weather patterns (south-west monsoon) climate (heavy rains, cyclone, etc.) culture, tradition, and social contest. In general, houses are small in size but quite large back yard and comparably small front yard. Most of the houses use locally available raw materials, technology, and family labor which invariably contribute to sustainable architecture (Vijekumara 2015; Dayaratne 2018). Doors and windows were small in size but it enough to get ventilation and natural light which is required to household. Walls contain air holes and doors, and windows contain louvers to facilitate proper ventilation and managing heat.

Unfortunately, most of the houses in resettlements established in limited land space where the residence has to accept houses without front yard or back yard, shared back yard and front yard or small space for both front and back yard. On set of resettlements, space for private use was not considered because they were badly affected by the disaster but with the time, they were face various difficulties in their day-to-day activities without having considerably large space for private living. Most of the issues associated with alcoholism, drug abuse, as well as political problems. Resettlement schemes were able to provide better housing facilities and essential infrastructure to displaced community. Moreover, architectural design of new houses was ignoring the concept of vernacular architecture.

Therefore, we were able to develop housing design with the participation of displace communities to fit with their desires. Adaptability and flexibility provide open space for business and common gathering, and design ecology was key concept followed during the design process (Defiana and Mahendra 2017). Moreover, cultural, political, and economic identities of the affected communities are essential components in rebuilding process (Dombek 2016).

With reference to participatory design proses, three areas were considered most, living environment, domestic arrangement, and neighborhood. Living space were created for multiple uses based on the vernacular architecture. Living space were divided into two bedrooms, bathroom, dining, and living area. Two bedrooms were essential to cater the needs of extend family. Further, multifunctional roof design to fit with the livelihood options as well as climate variabilities. The roof proposed was concrete slab flat roof (normally made from concert structure), the space of the roof facilitate to store fishing gear, net mending, and fish drying activities. Wide door and windows were designed to facilitate cross and stack ventilation. Design includes space for exterior living, front and back yard and space for common living. Figures 9.13 and 9.14 show the single-story housing design and two-story hosing design that we proposed. Single-story housing design for space with front and back yard and two-story housing design for urban areas were proposed.

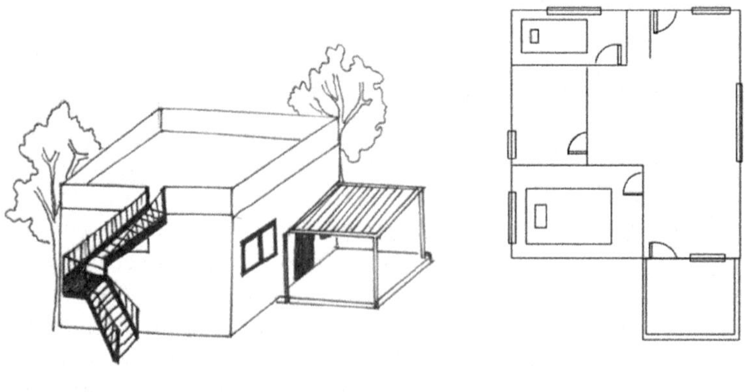

Fig. 9.13 Single-story housing design

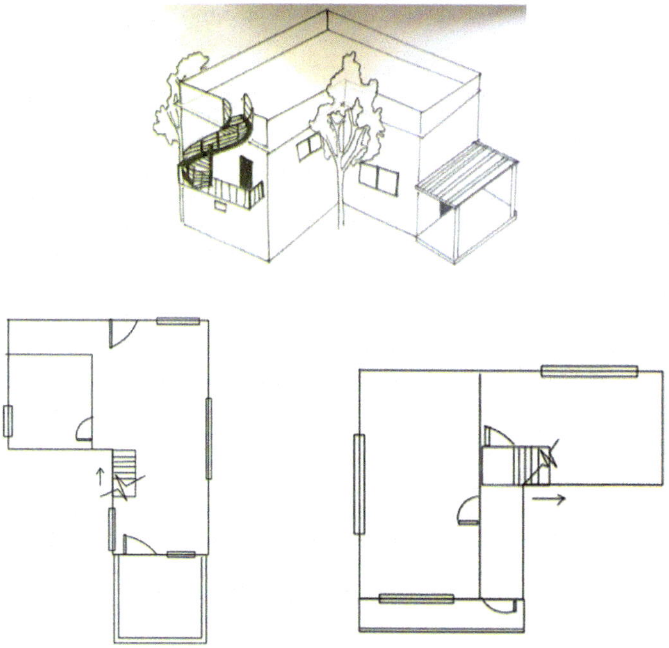

Fig. 9.14 Two-story housing design

Housing designs of resettlements are single-story house with backyard, single-story house without backyard, single-story duplex house, two-story duplex house with backyard, and flats. Resettlers were highly satisfied with single-story houses established close to the beach. Highly unsatisfied with single-story duplex and two-story houses. Social capital states of the resettlements highlighted that inter-community relationships (bonding social capital) was highest in SHB and lowest in TDB community relationships (bridging social capital): highest in SHB and lowest in TDB. Community–government linkages (linking social capital) was highest in SHB, SH, and SDB and lowest in F. Satisfaction on housing design space for private use and having front and back yards were the highly satisfied features of SHB. Highly unsatisfied housing design was F. Satisfaction on built and social environment of the resettlements were poor compared to previous setup. Local transport and access to the beach were the highly satisfied features of the F. Gambling, condition of women, violence behavior of community members, shopping facilities composition of resettlement population, and public safety were the highly unsatisfied features of the F.

9.4 Conclusions

Our approach was to explore the nature of housing units of the rural coastal communities those who internally displaced by the great Indian ocean Tsunami. Prior to the Tsunami, housing and community infrastructure were underdeveloped and most of the fish landing sites and markets were on poor and un-hygienic condition. Natural disaster leads to displaced coastal communities internally, and they were first settled in temporary settlements. Temporary, semi-permanent, and permanent housing units were the three main housing types available prior to the Tsunami, and the vernacular architectural concepts were utilized to match with the local conditions. Re-settlement schemes were established with the participation of various foreign donors, and study explores the architectural designs of various foreign-funded Tsunami re-settlements and the impact of housing designs on the residents. In general, most of the housing designs were brought from various foreign re-settlements and displaced people were ignored from the design phase.

In general, houses of fishing villages were located in huge open and common space where village structure is away from individualism. These shared open spaces help to build strong bonds among neighborhood that share and care the daily life of each other. Especially, fishers are away for fishing, fisher women, elderly, and children were cared and protected by neighborhoods. Moreover, fisher women engaged with various shore-based cottage industries such as fish drying, making coir ropes, and dress making. Housing designs were in line with the livelihood. Verandah was a common space which families use for engaged with cottage industries, small business, family gathering, placed of study for the kids, rest and relaxes place, and net mending. Most of the resettlement housing designs were ignored the verandah in the housing designs.

Dissatisfaction over the resettlement housing was mainly due to cultural mismatch and cultural misunderstanding, between donor and recipient. Therefore, a new concept of design for resettlement must emerge in which the problem of rebuilding after a disaster is approached as a partnership between these two entities. Satisfaction of the residents on resettlement housing depends on their personality traits as well as fit between the lifestyle and the features of housing, house design, functionality, accessibility, and neighborhood environment. Design mismatch was common in most of the resettlement housing and functionality issues such inconvenient housing structures disturbed very much on simple living style of the fishing communities. Moreover, access to common facilities was initially disturbed with the displacement and further complicated with new locations, especially schools, hospitals, and markets. In addition to that, incompatible environment and neighborhoods add extra pressure on their living in resettlements. Therefore, in-depth analysis of life prior to disaster and participatory design and built resettlement housing will mitigate the socio-economic consequences.

One of the most controversial issues with regard to rehabilitation policy was the introduction of a 100 m buffer zone which was intended to discourage people from rebuilding within a certain distance of the shore line. This considerably limited the amount of land which was available to build on, in most urban areas where it was already a challenge to find space for rebuilding. Architectural designs found among re-settlements were single-story houses with and without backyard, both single- and two-story duplex houses and housing complex, flat. Except single-story housing design with backyard, all other designs were ignored the local culture and living traditions. Duplex houses and flat were among the most dissatisfied housing designs. The present study proposes that a careful investigation of local culture, traditions, behaviors, and attitudes of various ethnic and religious groups along with the livelihood style is an essential component in designing resettlement villages and housing units. Community-based participially planning and design is foundation of sustainable resettlement.

References

Amarasinghe O (2005) Social welfare and social security in Sri Lankan fisheries. SAMUDRA monograph file: ///C:/Users/e-wis%E2%84%A2/Downloads/oscar-scoialwelfare.pdf. Accessed 15 Aug 2021

Barenstein JD, Wickramagamage P (2009) Coastal livelihoods and post-tsunami resettlement in Sri Lanka. Swiss Agency for Development and Cooperation

Chen TYS (2011) Religious NGO housing: Buddhist Tzu Chi Foundation's post-tsunami resettlement in Sri Lanka. South Asia J Cult 5–6:97–114

Correa C (1989) The new landscape: urbanization in the third world. Butterworth-Heinemann. ISBN 0408500719 (ISBN13: 9780408500715). file:///F:/Resettlement/The_New_Landscape_By_Charles_Correa.pdf. Accessed 15 Aug 2021

Correa E (2011) Populations at risk of disaster: a resettlement guide. In: Ramírez F, Sanahuja H. GFDRR. The World Bank, Washington, DC, 142 p: xii

Dayaratne R (2018) Toward sustainable development: lessons from vernacular settlements of Sri Lanka. Front Archit Res 7(3):334–3346. https://doi.org/10.1016/j.foar.2018.04.002

De Silva DAM, Yamao M (2010) Effects of the tsunami on fisheries and coastal livelihood: a case study of tsunami-ravaged southern Sri Lanka. Disasters 31(4):386–404. https://doi.org/10.1111/j.1467-7717.2007.01015.x

Defiana I, Mahendra AS (2017) Open building concept for fisherman housing on the north coastal Surabaya. J Archit Built Environ 44(1). https://doi.org/10.9744/dimensi.44.1.15-20. ISSN 0126-219X (print)/ISSN 2338-7858 (online)

Dias NT, Keraminiyage K, De Silva KK (2016) Long-term satisfaction of post disaster resettled communities: the case of post tsunami—Sri Lanka. Disaster Prev Manag Int J 25(5):581–594

Dombek KE (2016) Rebuilding after a natural disaster: housing strategies for minority communities in post-tsunami Sri Lanka. Architecture undergraduate honors theses. Retrieved from https://scholarworks.uark.edu/archuht/16

Ess J (2019) Re-location: urban and architectural analysis of resettlement practices in the brown coal mining area of Welzow-Süd in East Germany. SHS Web Conf 63:13002. In: Modernism, modernization and the rural landscape: proceedings of the MODSCAPES conference 2018 & Baltic landscape forum. MODSCAPES 2018. https://doi.org/10.1051/shsconf/20196313002

Fernando N, Senanayake A, Amaratunga D, Haigh R, Malalgoda C, Jayakody RRJC (2019) Disaster, displacement and relocation: an analysis of the needs and policy implications on a displaced community in Sri Lanka. In: 10th international conference on structural engineering and construction management (ICSECM). ICSECM2019-83, 12–14 Dec 2019. Earl's Regency Hotel, Kandy, Sri Lanka. ISBN 978-955-589-273-5

Li Y, Feng X (2021) Influence of housing resettlement on the subjective well-being of disaster-forced migrants: an empirical study in Yancheng City. Sustainability 13:8171. https://doi.org/10.3390/su131581

Ruwanpura KN (2009) Putting houses in place: rebuilding communities in post-tsunami Sri Lanka. Disasters 33(3):436–456. https://doi.org/10.1111/j.1467-7717.2008.01082.x. Epub 2008 Oct 31. PMID: 19178548

Silva KD (2011) Resettlement housing design: moving beyond the vernacular imagery. South Asia J Cult 5–6:117–135. https://kuscholarworks.ku.edu/bitstream/handle/1808/15270/Silva_2011_Resettlement.pdf?sequence=1&isAllowed=y

Vijekumara PA (2015) A study on the resettlement planning process applied in post-landslide disaster resettlement projects in Sri Lanka. In: NBRO symposium—innovations for resilient environment, resettlement planning

Weerasinghe WK, Shigemura T (2008) A study on transformation of living environment and domestic spatial arrangements: focused on a western coastal housing settlement of Sri Lanka after Sumatra earthquake and tsunami 2004. https://doi.org/10.3130/jaabe.7.28. https://kuscholarworks.ku.edu/bitstream/handle/1808/15270/Silva_2011_Resettlement.pdf?sequence=1&isAllowed=y

Chapter 10
Forced Displacement Following Reconstruction Approaches After 2005 Zarand Earthquake, Iran

Azadeh Shojaei and Alireza Fallahi

Abstract On February 22, 2005, just one month after the Bam catastrophe's first anniversary, another earthquake measuring 6.4 on the Richter scale hit Zarand, the north of Kerman province. Reconstruction approaches resulted in the displacement of many people and forced migration to the selected sites. This paper aims to study three villages reconstructed with taken approaches after sixteen years of the disaster. The method used is descriptive-exploratory; its approach is qualitative. The essential tools for this research include library documents, field studies, photographing, sketching, and identifying the vernacular architecture features, four in-depth interviews with authorities, and thirty-eight individual/group interviews with the affected community. The conclusions showed that the accommodation of IDPs regardless of family ties, the proximity of neighborhood units in their homeland, and the combination of local people with the immigrants without reasonable interpretation have led to the isolation of immigrant families, disruption of the rehabilitation process, and ultimately user's dissatisfaction. It seems that simulating a pre-disaster built environment may help the displaced people to get through this situation. However, implementing this idea requires a comprehensive knowledge of the affected community, which has been ignored in Zarand.

Keywords Forced displacement · Reconstruction approaches · Post-disaster rehabilitation · Population integration · Zarand earthquake

10.1 Introduction

House is usually the most valuable asset for people in developing countries and particularly, and the impact of disasters on them is much higher than in developed countries (Ahmed 2011). In developing countries, the forced displacement of communities is

A. Shojaei (✉)
Researcher at Post-Disaster Reconstruction Group, SBU, Tehran, Iran
e-mail: azadeh.shojaei1985@gmail.com

A. Fallahi
Chairman of Post-Disaster Reconstruction Group, SBU, Tehran, Iran

© The Author(s), under exclusive license to Springer Nature Switzerland AG 2023
M. Hamza et al. (eds.), *Rebuilding Communities After Displacement*,
https://doi.org/10.1007/978-3-031-21414-1_10

a severe issue for decision-makers in post-disaster activities. The idea of relocating a residential complex and the people living in it to solve environmental problems and vulnerability reduction is the philosophical foundation of displacement. However, the existing experiences prove that this idea and its implementation are not successful in the long term from public satisfaction (Fallahi 2006). Because the authorities' focus is mainly on the speed and cost-effectiveness of post-disaster reconstruction and hazard mitigation (Jigyasu and Upadhyay 2016). The researchers highlight that displacement deals with the issue of "moving away from people's lives," which needs careful planning as it builds new communities and built environments (Fernando and Punchihewa 2013). Although providing the favorable primary conditions for people to adapt to the new built environment lies at the core of the sustainability of the resettlement, identifying the relevant positive conditions at the right time is often challenging for authorities (Jigyasu and Upadhyay 2016). However, many of these new built environments were unsuccessful projects, resulting from implementing agencies or the government's lack of understanding of local needs. Most reconstruction projects have inadequate house sizes, indoor and outdoor spaces, building materials, and infrastructures (Ahmed 2011).

Disaster literature review shows that the general meaning of housing for displaced people due to the disaster is beyond just "living place" and includes broad concepts such as livelihood, peace of mind, and place attachment, because the people become "homeless" after the disaster, not just "building less" (Fallahi 2008). Relocating the affected community to a new site has a different nature from in situ reconstruction, and many sociocultural, economic, physical, and environmental factors are involved that require a meaningful and dynamic response (Lizarralde et al. 2009). Due to the identity of the affected area depends on the particular location, displacement changes its most important shaping factor, which is the "location." Such displacements cause changes in many aspects, and its inhabitants face many problems in unfavorable condition (Mortazavi 2006). Although the level of assistance and the type of accommodation play an important role in creating favorable initial conditions, forced relocation and post-disaster resettlement rarely can make an adoption between the built environment and the affected community and are often criticized for this inability (Sridarran et al. 2018a, b).

Resettlement is introducing an entirely new built environment to the people attributing resilience measures for rehabilitation, not only building houses (Barenstein 2015). Also, post-disaster resettlements embrace considerable differences from conventional methods in the layout, house design, building materials, and construction processes. For the homeless community, movement to the areas with the slightest resemblance to their previous place and their needs means that multi-hazard risk commonly "travels with" them by conflict. They exchange one form of risk for another, sometimes knowingly, sometimes unknowingly, but often with little natural choice (Few et al. 2021). The built environment and the affected people's needs are expected to adjust to each other. However, this adjustment cannot be expected in the short term without ensuring certain primary conditions, as it is an outcome of a long-term social process. In addition, the built environment modifies by the users to satisfy their needs (Barenstein 2015).

Although the different characteristics of permanent housing can have various effects on affected people's livelihood, forced relocation following a disaster often involves changes in the jobs and livelihood options of the displaced community. The impact of these changes is more on the government's policies and the host community's support than on their own efforts (Sina et al. 2019). The disaster, reconstruction, displacement, and all its consequences are not temporary events and should not be considered merely in terms of physical structure. Instead, it means a long-term social situation for the affected community and other factors should be considered (Cutter et al. 2008). However, the scope of reconstruction and relocation are often so high, and the time is so short that other factors are ignored. So, reconstruction is limited only to the physical dimensions of the affected area. As a result of these proceedings, increased vulnerability, decreased resilience, and lack of place attachment, which are the worrisome consequences of displacement, can be observed among these communities (Davis and Izadkhah 2006).

Place attachment is an influential factor that can be observed in post-disaster displaced communities or even immobile communities that do not tend to relocate. This is an important factor that many of the displaced communities demand can be understood through it. Place attachment clarifies why the population might not migrate even in the face of environmental or socio-economic risks or have the desire to return to their homeland. This multifaceted concept can potentially increase resilience and adaptive capacity in post-disaster situations only if used in the right shape and condition (Blondin 2021). The place attachment could have a conceptual model which involves three terms. Personal context such as place dependence and rootedness, community context such as neighborhood attachment and familiarity, and natural environment context such as environmental identity and affinity to nature are the three main parts of this model (Raymond et al. 2010). Although there are some fundamental differences in the level of place attachment between displaced people during gradual and forced displacement (Adams et al. 2015), neglecting this critical factor could create a new threat to society and often lead to physical or social crises in the long term (Dikmen 2006). Because dependence on ancestral lands, neighborhood patterns, and group life of relatives in the former place for the affected community is more critical than physical structure. In addition, the destroyed land's cultural, historical, and symbolic values are not easily transferred to the new land for its inhabitants (Fallahi 2006).

Participation in decision-making and the right to choose is another important factor in post-disaster activities (Fallahi 2018). "*The right to choose one's residence, to determine where to live, and to freedom of movement*" is illustrated as the critical aspect of the right to adequate housing in UNHABITAT (2009). Resettlement approaches need to consider more elements than just hazard exposure. In this process, people's much-needed participation is in planning, population integration, house designing, site selection, agricultural land access, or other alternative livelihood programs (Rieger 2021). Since most recipients have had a permanent home before and have a certain expectation level, the recipient's needs should be identified and addressed from the outset. Non-participation of displaced people and neglecting their opinions during the planning, design, and construction stages of resettlement projects means that authorities may not correctly understand users' needs. This subject can lead to evictions by recipients, higher development costs, and lower satisfaction levels with permanent housing and resettlement (Wijegunarathna et al. 2018). The lack of governmental collaboration with affected communities is a significant factor in why relocation programs become disordered and increase people's vulnerabilities (Rieger 2021). Therefore, the active participation of the community in this process from the beginning is a crucial need for future housing projects (Fallahi 2018).

Although the relocation is one of the long-term strategies of any disaster management plan, reaching a satisfactory relocation program would be a significant challenge in any circumstances (Sangasumana 2018). The quality of constructed houses and infrastructure and relocation process adaption can affect the success or failure of a relocation program and will influence vulnerability to the next disaster. Without considering the necessary indicators, relocation procedure cannot be succeeded (Hidayat and Egbu 2010). Indicators such as the scale and location of the disaster, vulnerability and physical, demographic, climatic, economic, cultural, political, environmental, as well as the coordination between local or national organizations, are effective in the type of planning, management, and adoption of reconstruction approaches. It is impossible to provide a single version for all reconstruction and rehabilitation programs. Approaches that do not pay attention to all the mentioned aspects sometimes lead to unnecessary forced displacement, population integration without knowing the guest and host community, and the rupture of their life and livelihood even for the next generation (Fallahi 2018). There is a comprehensive definition of post-disaster reconstruction from Lizarralde et al. (2009) which is stated that "*the post-disaster reconstruction is defined as the process of improvement of pre-disaster conditions, targeted to achieving long-term local development and disaster risk reduction through the pairing of local and external resources.*" In addition, if post-disaster reconstruction is considered an opportunity to improve the condition and development of the affected community, access to various innovative solutions will be possible. These innovations can reduce vulnerability and increase resilience in affected communities (Wijegunarathna et al. 2018). Built environments are constantly evolving to adjust to the changing needs of the displacement people (Sartipipoor and Asadi 2018). It seems that the post-disaster forced displacement will be successful if it can restore the various aspects of the affected community to the pre-disaster conditions or near or even better with the support of appropriate and principled policies (Sartipipoor et al. 2015).

10.2 Methodology

10.2.1 Background

Iran, located in the Middle East, is a disaster prone country, and among the forty different types of natural disasters observable in the world, thirty-one types have been identified in Iran such as earthquakes, floods, droughts, landslides, desertification, and deforestation (Bakhtiari 2014). The 2003 Bam earthquake (Kerman province), the 2018 Sarpolzahab earthquake (Kermanshah province), the 2020 Aqqala flood (Golestan province), and many other disasters happened in Iran and left a large number of internally displaced people with their problems.

Kerman is the largest province of Iran, and due to its location in the seismic zone, it has always witnessed mild and severe earthquakes that have sometimes turned into tragic human catastrophes. Eighteen significant faults have covered 98% of this province, and the Koohbanan is the most active fault which is located in Zarand city in the northwest of Kerman province (Fig. 10.1).

On March 22, 2005, just a few days after the first anniversary of the Bam earthquake in the east of the province, while the government was still reconstructing Bam, another earthquake measured 6.4 on the Richter scale shook Zarand, north of the province. More than 900 people were killed, 1400 people were injured, and thousands were left homeless. After this disaster, the reconstruction and resettlement of homeless people in the affected villages were handed over to the Housing Foundation of the Islamic Revolution, which is a government-affiliated institution. Based on the location and proximity to the fault, this organization adopted different types of

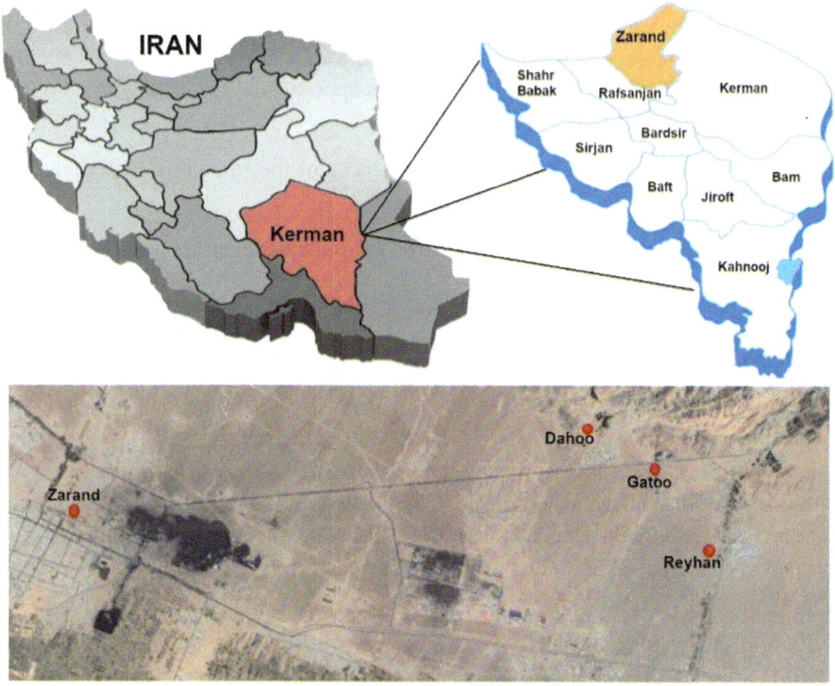

Fig. 10.1 Study area in Iran, Kerman province, and case studies in Zarand

approaches to reconstruct the affected areas. In this paper, three different approaches in the reconstruction of Zarand villages are studied:

1. Village relocation, forced displacement of residents to a new place, and integration of internally displaced people (IDPs) with post-disaster planning
2. Accommodation of the displaced people in the predicted area and integration of IDPs with pre-disaster planning
3. In situ reconstruction of some nearby villages, integrating them with new small towns, accommodation of displaced people in them, and turning this complex into a city with post-disaster planning.

As case studies, two villages and a city were selected which their locations are shown in Fig. 10.1:

- Gatkooyeh village (Gatoo) was planned with approach No. 1
- Dahooyeh village (Dahoo) was planned with approach No. 2
- Reyhanshahr city (Reyhan) was planned with approach No. 3.

Now, 17 years after the earthquake, we have gone between the affected community of those days to talk about how they integrated into the host community, how they interacted with the new built environment, the problems of their early days, and today. This research seeks to study the adaptation of displaced people to the new built environment and the long-term consequences of forced displacement after the 2005 Zarand earthquake in Iran and assess the various aspects of the affected people's lives in the form of post-disaster reconstruction approaches. The following two main questions guide us through this study:

Question 1. What are the long-term consequences of the forced displacement after nearly 17 years?

Question 2. What are the pros and cons of approaches taken to resettle displaced people?

10.2.2 Methods

The method adopted in this study was a descriptive-exploratory method with a qualitative approach to answer the questions. This method is flexible and allows understanding a wide range of expressions, observations, experiences, feelings, opinions, and perceptions that affect the community. It was performed in four main steps:

- Initial observations of case studies through six separate visits
- Field studies and informative data collection during eleven months
- Filling in the questionnaires and conducting individual, group and in-depth interviews with the people and officials
- Classification and analysis of data.

Years after the forced displacement, some villagers of this region, as soon as their mental and financial conditions improved, migrated from their forced residence for reasons such as lack of place attachment and use their houses as a seasonal home. Also, some moved to their former homes due to their occupation or air pollution caused by tar refinery factory. For these reasons, six separate visits were made in all seasons from June 2019 to April 2020 to complete field studies and interview more villagers. Our research tools for data collection included digital and library documents, articles, observation and field studies, interviews (individual, group) with filling the questionnaire, in-depth interviews with authorities, study of vernacular architectural features, photography of houses, and sketching. In addition, we used Excel for making pie charts and identifying the percentage of frequencies and Autocad and Sketchup software to create 3D views with separation of primary core and additional parts. This process helped us display the house expansion based on physical needs. The methodology flowchart is shown in Fig. 10.2.

The target community was selected from villagers who remembered the pre and post-disaster conditions and were forced into displacement at the time. We tried to select them from various layers of society and inhabitants of different parts in case studies to get fair results. Thirty individual interviews and eight group interviews with expanded families, neighbors, or colleagues were conducted, and the questionnaires were filled during the conversations (totally 51 people participated in group interviews). Due to being native and familiar with the residents 'dialect, people could feel close and intimate with the environment and the interviewer. So, they readily shared their experiences in this situation without worrying about local accents. It is worth mentioning that the interview questions were slightly different based on their occupation, gender, age, pre-earthquake location, the reason of forced displacement, and the resettlement approach. However, the primary format of the questions was similar to reach an acceptable and equivalent conclusion. In addition, four in-depth interviews were conducted with the authorities and decision-makers to access classified organizational information such as government decisions, displacement approaches, and their mistakes. The interviewee details in Table 10.1 and the obtained codes or themes in Tables 10.2, 10.3 and 10.4 are summarized based on case studies.

10.2.3 Theoretical Framework

Forced relocation after the 2005 Zarand earthquake and the changes in the villagers' life have many factors to consider. In this paper, some critical factors in post-disaster displacement were extracted by studying books, articles, and digital and library documents. Then, because the author is a native of this region and witnessed the disaster, reconstruction, rehabilitation process, and the implemented changes in the affected community's life in the last years, some of the more related factors to this community were selected. Finally, initial observations and field studies help us choose three main factors of the forced displaced community. These factors, which related

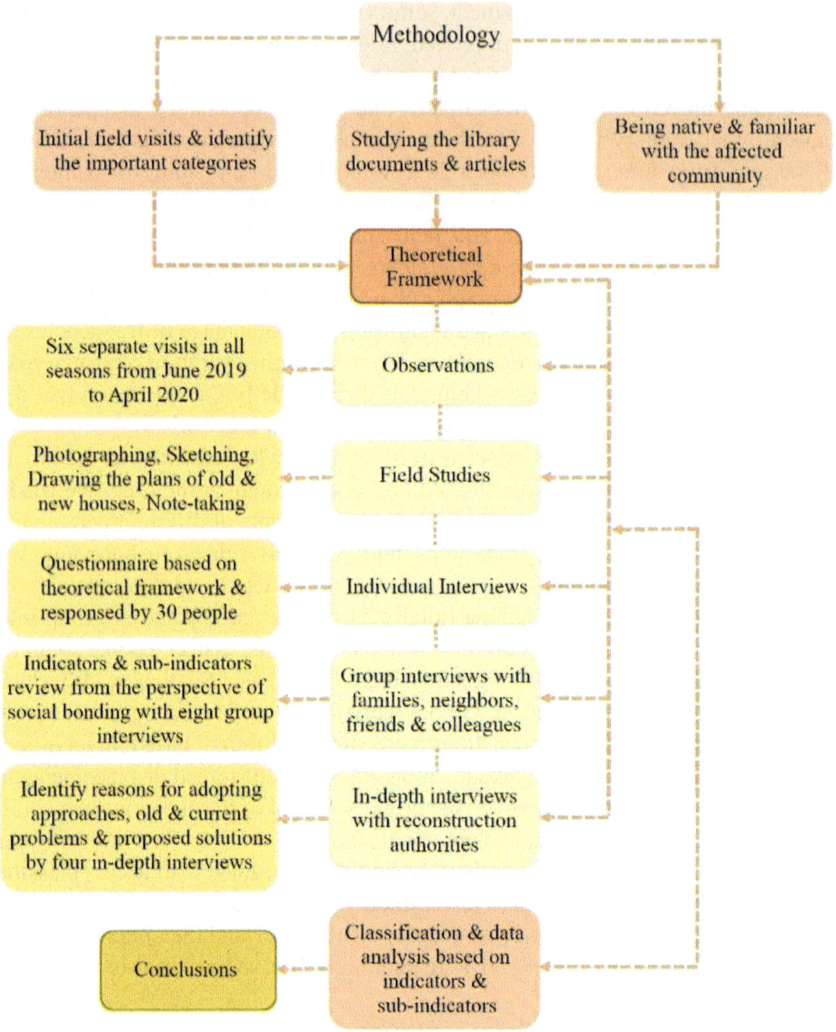

Fig. 10.2 Methodology flowchart

indicators and sub-indicators have been categorized, are shown as the theoretical framework in Fig. 10.3.

It should be noted that other factors such as environmental or psychological issues have also played an essential role in the displaced villagers' life. However, due to the wide range of factors, they have not been considered in this research. Building houses for IDPs, especially in developing countries, without recognizing and considering the various dimensions of society leads to negative consequences that its effects on society are visible in the short and long term. Since post-disaster relocation and its

Table 10.1 Interviewee details

Age			Gender		Career				
30–45	45–60	60–75	Male	Female	Housewife	Farmer	Rancher	Worker	Other
34%	44%	22%	32%	68%	37%	25%	16%	13%	9%

Case study	Individual interviews		Group interviews		In-depth interviews with authorities
	No.	Location	No.	Location	Position
Gatkooyeh	3	Old village	8 people	Farmland	The mayor of Gatkooyeh village
	7	Their homes in new village	4 people	Among driving them to the old village	
			4 people	A beauty salon	
Dahooyeh	10	Public places or their homes in new village	4 people	Husbandry	The mayor of Dahooyeh village
			6 people	Cemetery	
Reyhanshahr	3	Babtangal (see Fig. 10.6)	10 people	Women meeting in a park	The mayor of Reyhanshahr city
	3	Islamabad (see Fig. 10.6)	8 people	A mosque in Islamabad	
	4	New towns (see Fig. 10.6)	7 people	An under construction building in new town	Head of research center in the Ministry of Housing and Urbanism

subsequent exposure to various hazards are an essential topic in the disaster literature, studying the economic, sociocultural, and physical aspects of IDP's life after the 2005 Zarand earthquake can help the government respond appropriately to future disasters.

10.3 Case Studies

10.3.1 Gatoo Village

The first case study is the Gatkooyeh village, called Gatoo for short. It was located 15 km east of Zarand at the hillside of Koohbanan Mountain with an organic and tortuous form. The occupation of the villagers is agriculture, animal husbandry and working in the coke factory or tar refinery near the village. After the 2005 earthquake, 80% of the village was destroyed and uninhabitable. So, the Reconstruction Headquarters of the Housing Foundation decided to displace the village due to the fault location and its high vulnerability. In 2007, the new village 5 km southwest of the

Table 10.2 Summary of interviews and survey results in Gatoo

Interview	Respondent	Derived themes and codes
Individual interview	Natives	The small size of land and building, lack of privacy, adding more parts, ignoring livestock in design, long distance from old place, missing old social ties, desire to return to the homeland, God's help, none-participation in decisions, having past jobs
	Immigrants	Adding few parts to the house, lack of privacy, feeling loneliness not as much as the beginning, less place attachment, loosing job, believed to the God's will, none-participation in decisions
Group interview	Natives and immigrants	Imposition of additional costs, limited facilities, low loans, financial problems, allocating quotas in near factories, spending time in the homeland
	Natives	Historical conflicts with some immigrants, renovating old houses, penalty for renovation, dissimilarity between new house and rural form, feeling humiliated
In-depth interview	The mayor of Gatoo	Inappropriate decision in site selection, conflict over water sources, official request to return to the former village, secret result of the request, unsuccessful population integration, elder's depression

Table 10.3 Summary of interviews and survey results in Dahoo

Interview	Respondent	Derived themes and codes
Individual interview	Natives	Adding some parts before the disaster, less privacy, average-participation in decisions, short distance from old place, moderate place attachment, holy shrine as an critical element, having past jobs
	Immigrants	Changing function, adding required parts during the years, none-participation in decisions, getting used to the new place, working in gypsum mine, being a past tenant and today's stakeholder, role of religion
Group interview	Natives and immigrants	Importance of cemetery, ignoring vernacular architecture, choosing close plots by families and neighbors, making handicrafts by women, using immigrants as workers in farmlands and husbandries
In-depth interview	The mayor of Dahoo	Building small houses to reduce the reconstruction costs, changing people's lifestyle, conflicts over stealing livestock, site selection based on holy shrine and limitation, using pre-disaster, for other villages

Table 10.4 Summary of interviews and survey results in Reyhan

Interview	Respondent	Derived themes and codes
Individual interview	Natives in Babtangal and Islamabad	In situ reconstructing houses better than before, acceptable privacy, having place for livestock, adding few parts, high place attachment, having past social ties, Mosque's role, entrepreneurship
	Immigrants in new towns	The small size of land and building, adding few parts to the house, lack of privacy, desire to return to the homeland, less place attachment, far from the city center, surrender to the God's will
Group interview	Natives and immigrants	More entertainment and facilities, dissimilarity between new houses and vernacular architecture, using common resources, business development
	Immigrants in new towns	Refusing relocation for 2 years, unsecure environment, raising petty crimes, alien neighbors, feeling abandonment, desert landscape
In-depth interview	The mayor of Reyhan	Natives' satisfaction, filling the gaps between the three sections to have a uniform city, the wrong decision for forcing non-natives to relocate, creating more jobs to reduce petty crimes
In-depth interview (general)	Head of the Research Center in the Ministry of Urbanism	Site selection just based on fault location, getting involved people in reconstruction to improve the rehabilitation process, regardless of other aspects, limitations of time and budget, carrying out comprehensive studies to use in future disasters

old Gatoo was built in the middle of a desert landscape with a modular form divided into sections of 300 m^2. Authorities did not pay attention to the natives' opinion, which was to stay in the old place and build durable houses. The foundation decided to gather some IDPs from abandoned villages that were in a similar situation and integrated them with the 22 remained families of Gatoo. Therefore, the population would increase, and they could use the benefits of rural houses and facilities. During field studies and interviews, it was revealed that the villagers had decided to return to

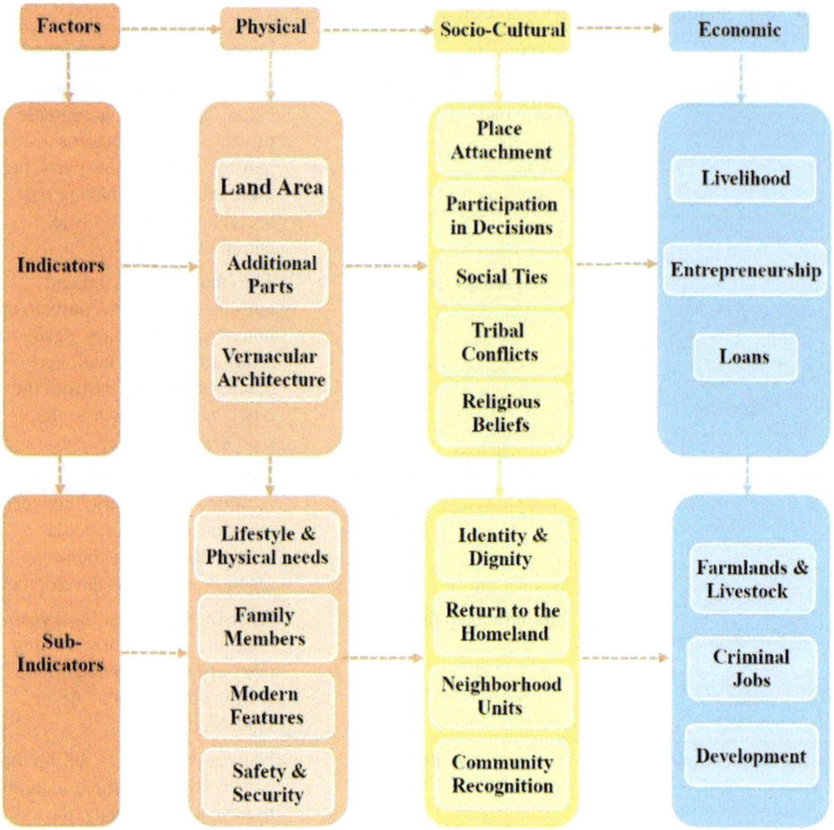

Fig. 10.3 Theoretical framework

their original village. They even submitted their request to the Housing Foundation through the village mayor, but the authorities refused it. The approach adopted for the reconstruction and displacement of Gatoo was approach No. 1: Relocation of the village, forced displacement of all residents to a new place, and integration of IDPs with post-disaster planning, and its old and new location is shown in Fig. 10.4. It should be noted that the distance between villages has been calculated from the start point of houses.

10.3.2 Dahoo Village

The second case study is Dahooyeh village, called Dahoo for short. It was located 17 km east of Zarand and surrounded by the mountains, except in the southwest. The occupation of the villagers is agriculture, animal husbandry, and work in a gypsum

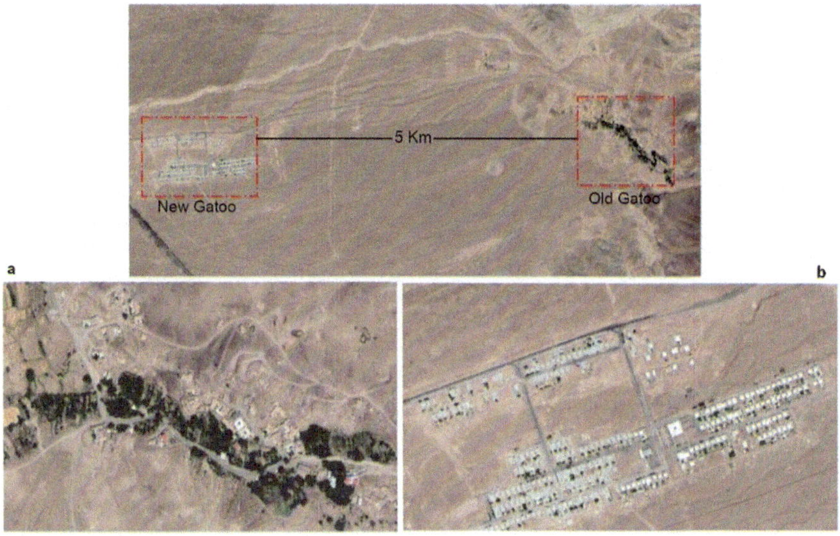

Fig. 10.4 **a** Old Gatoo, **b** new Gatoo

mine near the village. This village is one of the most vulnerable residential areas on the Koohbanan fault and has been the epicenter of two severe earthquakes in 1989 and 2005. According to the documents following the 1989 earthquake and the interviews, the authorities chose the southwest land of Dahoo to resettle the people and prevent further damage in future disasters. They divided the land and handed it over to the villagers to relocate and rebuild housing with the aim of pre-disaster planning. However, minor destruction to buildings, proximity to agricultural lands, and place attachment caused them to refuse displacement. In addition, valuable elements such as holy shrine, cemetery, aqueducts, mineral springs, organic form, and green landscape in the face of adverse conditions of the new site with modular segmentation patterns made them reluctant to move. Over time, due to the surrounding mountains, population growth, land needs, and the short distance between the two sites, construction on the new site was begun by the young Dahoo community. After the disaster, village faced more severe damages due to its location on the main fault and the earthquake's epicenter. Using unformed stones and layers of clay and mud to insulate the walls caused excessive weight gain and 100% destruction of the village body. On the other hand, debris, high removal costs, and a suitable location for displacement, determined for the same purpose in the previous earthquake, caused the evacuation and ultimately relocated the village to a new site entirely. In addition, rural homeless who were tenants in their villages were added to the natives and resettled in New Dahoo. So, the approach adopted for the reconstruction and resettlement was approach No. 2: Accommodation of the displaced people in the predicted area and integration of IDPs with pre-disaster planning. The Dahoo's location is shown in

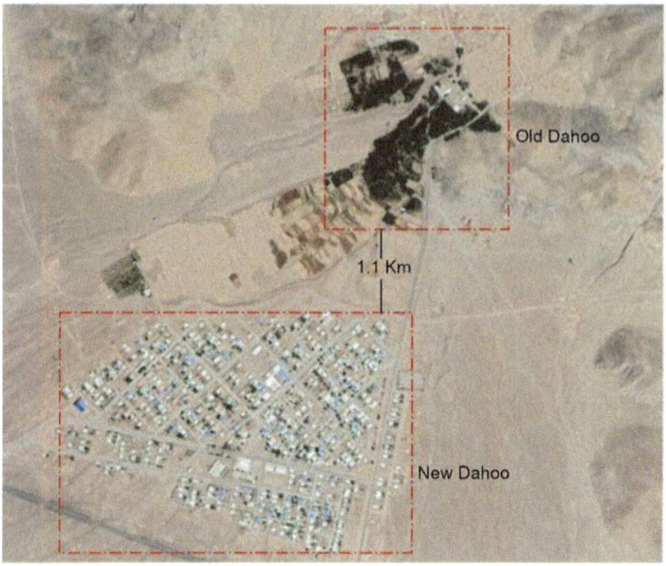

Fig. 10.5 Old and new Dahoo

Fig. 10.5. It should be noted that the distance between villages has been calculated from the start point of houses.

10.3.3 Reyhan City

The last case study is Reyhanshahr, which is called Reyhan for short. One year after the disaster, this new city was formed by integrating Babtangal and Islamabad villages as the primary core and two new small towns near them. This complex has known as a city in the political geography of the province in the name of Reyhan. Initially, the residents of the destroyed houses in nearby villages refused forced displacement and did not leave their ancestral lands. So, they lived in tents and small metal rooms in the ruins of their homes for about two years, and new towns were uninhabited until the remains of their houses collapsed following another earthquake in 2007. Despite residents' resistance, all of these areas have been declared uninhabitable due to their proximity to faults and high vulnerability. Finally, the homeless people have been forced to move to the new towns of Reyhan. As a result, the approach adopted for the reconstruction of Reyhan is approach No. 3: *In situ reconstruction of some nearby villages, integrating them with new small towns, accommodation of displaced people in them, and turning this complex into a city with post-disaster planning*. The Reyhan location is shown in Fig. 10.6. It should be noted that the distances have been calculated from the center to the center of the parts.

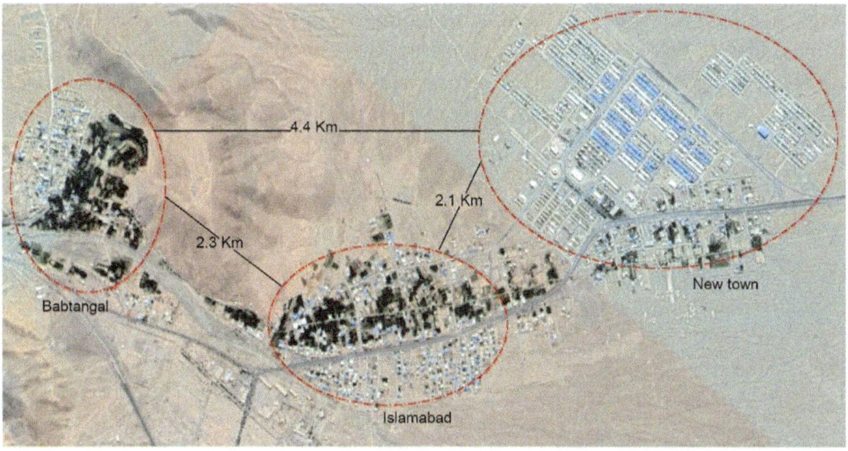

Fig. 10.6 Constituent parts of Reyhan

10.4 Results

Based on the theoretical framework, field studies, and interviews, three critical factors and their indicators have been discussed in the following parts to answer the main questions of the paper. Therefore, post-disaster forced displacement following reconstruction approaches in the physical, sociocultural, and economic aspects have been studied from the perspective of the affected community after 17 years.

10.4.1 Physical Factors

10.4.1.1 Land Area

The first indicator studied in physical factors is the land area. The pattern of land division into equal sizes of 300 m^2 was typical in all three case studies except the in situ reconstructed parts in Reyhan. From the justice point of view, it seems fair that all IDPs, whether local people or immigrants, benefit from the land equally. However, villagers believed the government should give them the plots as much as they had in the old village, which could have been possible due to the considerable desert land of new sites. The affected community also got used to living together in large houses with their children and grandchildren. Nevertheless, coordination between the land area, family members, and livelihood style was neglected. Sharing rooms, lack of privacy, especially in Muslim families, and changing functions such as living room to bedroom and bedroom to storage were other consequences of the tiny houses. In addition, most villagers' livelihood in this area depended on livestock, and they kept a few goats, sheep, chickens, roosters, and pigeons in their former house. However, no particular place was provided for this purpose in the new houses. When asked if the land area was sufficient for the needs, 78% answered no. 22% who answered positively were older people living alone, newly married couples without children or people who had added enough parts to meet their needs.

10.4.1.2 Additional Parts

The additional parts are now better understood by explaining the land area. The displaced people wanted to have their former lifestyle and faced the area limitation. Therefore, they started building additional parts on their land to solve this problem. They added bedroom, living room, warehouse, or a separated place for livestock in the yard. This indicator was studied with slight differences in all three case studies.

- In Dahoo, where the new village was prepared previously, some natives added required parts to their homes during the pre-disaster time. However, immigrants were among the last families who take their homes and completed them during the post-disaster time by adding a few parts.
- In Gatoo, the additional parts differed slightly. Some local people and immigrant's families have added required parts to their houses and completed them over the years. Even during our field studies, about 16 years after the earthquake, some residents were constructing and adding some sections to their houses. However, some inhabitants did not want to complete their homes due to the lack of place attachment, desire to return to their homeland, financial problems, or loneliness.
- In Reyhan, the condition is divided into two sections. In the first section, the stakeholders of Babtangal and Islamabad built their houses on their old lands and added the required parts in the first months after the earthquake. In the second section, where the new towns are located, immigrants refused to relocate there for two years. However, they were forced to displacement and complete their houses about four years after the disaster. Furthermore, some of them did not make any changes to the preliminary plan and still have a desire to return to their homeland. Figure 10.7 shows a permanent house plan in Dahoo. The primary core is presented in pink, and the additional parts are in yellow.

10.4.1.3 Vernacular Architecture

One of the forced displacement consequences, which was common in all three case studies, was the slight resemblance of the new houses to the vernacular architecture. This incoordination alienated the users from the newly built environment and disrupted the rehabilitation process. Based on the remains of destroyed houses and villagers' descriptions, a rural house is in this form: Rooms were located around a central yard, and the end of the building was a place for the toilet or keeping livestock. Each room had its function and belonged to a family member with a spouse and children. The largest room, closed to the kitchen, was like a living room called Orsi in the local dialect and belonged to the guests. The family also had another living

Fig. 10.7 Permanent house in Dahoo

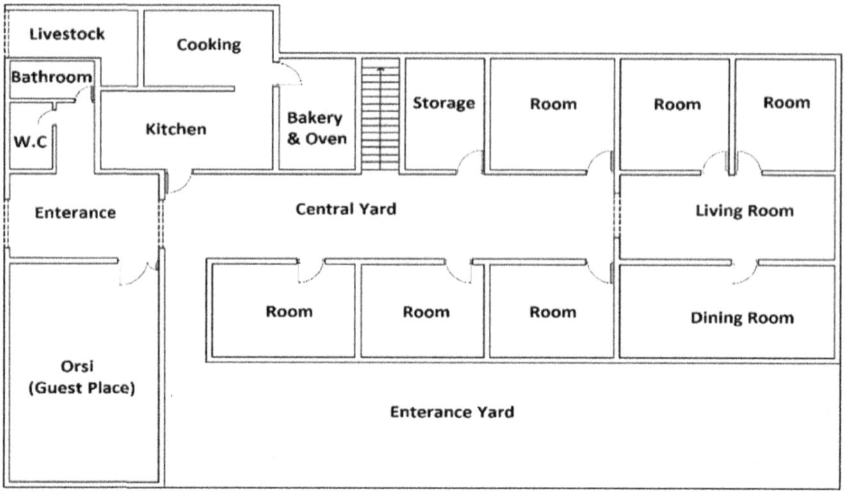

Fig. 10.8 Pre-disaster rural house in Reyhan

room used as a gathering place and dining room (Fig. 10.8). However, designers and authorities have entirely ignored this vernacular architecture and have not respected Islamic privacy. For example, if a stranger entered a pre-disaster house, he would enter the Orsi directly and could not see other house parts, but in new homes, the opposite is true. The next topic is the new house's kitchen, which was built in a modern and open form without a separate oven. Women of this area are used to baking the bread, so they have made a clay oven in the yard to meet their needs and make new houses more similar to their old rural homes.

10.4.2 Sociocultural Factors

10.4.2.1 Place Attachment

In all three case studies, place attachment was one of the most impressive items seen during the interviews. However, the authorities did not consider this indicator, its side effects on the rehabilitation process, and its consequences on the forced relocated community. According to the interviews in Babtangal and Islamabad parts of Reyhan, the rehabilitation process passed gradually. The affected community returned to routine life just a few months after the reconstruction was completed. Fatima, one of the interviewees from Islamabad, stated: "*We tolerated living in temporary shelters or small metal rooms as transitional accommodation just for being in our lands. We were sure that anyone did not force us to displace.*"

Nevertheless, in Gatoo, Dahoo, and the new towns of Reyhan, the natives and guest communities were homesick for their homeland and traveled there under any pretext. Reasons include visiting relatives and friends, agricultural land, keeping livestock there, holy shrines and mosques, family graves in the cemetery, weekend picnics, fresh air, or just walking and remembering memories. Despite the Gatoo's abandonment, some local people renovated their former houses. They are using them as a second house to keep livestock, agricultural products, a fodder warehouse, or a place to rest. Although the government removed infrastructures from old villages, some have even piped water from the aqueduct to their former houses even though they were fined.

During the interview with the Gatoo's villagers, it was revealed that they asked the mayor and the relevant authorities officially to return to their former village. They cited the wrong approach to their displacement, attachment to ancestral lands, and the difficulty of accessing farms or livestock due to the long distance as the main reasons for their request. The Engineering Organization of Kerman province commissioned a research group to review their request. Their report showed that this relocation is possible if the building codes, safety points, and building reinforcement against earthquakes are provided. However, the authorities kept the report secret to prevent chaos. Also, if the authorities accepted this request, it was not unlikely that other displaced villages would make such a request, which would have imposed additional costs for the government.

10.4.2.2 Participation in Decision-Making

The participation of IDPs in decisions for displacement and reconstruction is one of the most critical indicators of long-term satisfaction and accelerating the rehabilitation process. 94% of respondents in all three case studies stated that they had no involvement in site selection, design, or implementation. The remaining 6% are divided into several categories:

- In Gatoo, some immigrants joined the local people at the end of the resettlement process and were among the last units in the reconstruction program. They had to design and build their houses because the contractors had left the village at the time.
- In Dahoo, some natives relocated and built their houses on the new site before the earthquake. They designed and constructed them according to their needs, regardless of the proposed plans in some cases.
- In the new towns of Reyhan, during two years of resisting the homeless against relocation, many houses were left abandoned in the structure phase. Eventually, villagers were forced to relocate, and they built houses according to their needs and structure limitations.
- In Babtangal and Islamabad parts of Reyhan, local people could design their houses based on their needs on former land.

These residents did not participate directly but have made every activity to improve their situation. They also believed that they would be more satisfied with their houses and pay less money for the development, just if the site was selected by voting and their needs were considered.

10.4.2.3 Social Ties

Social ties are an influential indicator ignored during forced displacement and resettlement. Families live as a group in the Zarand local communities, and neighbors are so close such as relatives. Also, they know each other's life details and have common issues after the disaster. The women gather every evening and have many conversations that could relieve them (Fig. 10.9). These close ties would be helpful in post-disaster rehabilitation and satisfaction of displacement if only the government had knowledge about and paid enough attention to it.

- Gatoo's forced displacement separated small social communities and neighborhood units. Without consulting with IDPs about their characters, kinship, and neighborhood relations, the permanent houses were dedicated to them by lottery. Large families whose parents, children, and grandchildren lived in the same house were forced to separate and live in detached homes due to the limited land area. In addition, the unfamiliarity of the natives with some immigrants, new village enlargement, and the distance between the houses led to the limitation of long-standing social ties.
- In Dahoo, the situation was a little better. Before the earthquake, most families and neighbors chose close plots when the new village lands were divided. Some of them even agreed to live in the village suburb so they could take many plots to live near their relatives. However, immigrants in Dahoo met the same fate as immigrants living in Gatoo and were not even close to their compatriots in the same village.
- In Reyhan, the situation is divided into two parts: In the first part, which involves Babtangal and Islamabad, residents stayed in their old lands just with new houses

Fig. 10.9 Women meeting in a park, during the group interview, Babtangal

Fig. 10.10 New towns of Reyhan and the desert landscape

due to the in situ reconstruction approach without any relocation. So, they could keep their social ties and relations like before the disaster.

The second part involves new towns far from the city center and small local communities. The thugs' gangs, drug dealers, a desert landscape, and low lightness at night have reduced the security of the new towns. Also, there are some semi-built houses that no one is willing to move in and have become a gathering place for addicts. These conditions led to the fact that even the alleys were not secure for women and did not have any social place for local gatherings. So, the houses and residents gradually have turned to the members of an isolated environment and community (Fig. 10.10).

During these years, new communities and relationships have formed between neighbors, natives, and immigrants through marriages or joint problems, but this is not enough. If the planning and reconstruction approach had considered the social ties indicator, the forced displaced community would have shown more resilience and could have passed the rehabilitation process sooner.

10.4.2.4 Tribal Conflicts

Less attention to tribal conflicts in population integration was another negative consequence following the Gatoo and Dahoo reconstruction approaches. Two of this area's main disputes were over water resources for agriculture and livestock security. These conflicts have long roots, and many families in Gatoo, Dahoo, and several other nearby villages have been involved for years.

Water resources Most of the water required for agricultural lands in this area is provided by aqueducts, and each aqueduct is a common source between several villages with short distances. Due to the drought of recent years and the reduction of surface water, groundwater resources have been rationed. Based on agricultural products and land area, every farmer can use the aqueduct water at particular hours of the day and night. The farmers of these villages often quarreled over water resources and blocked the water line to each other under the pretext of high consumption of products in the neighboring village. This case has a long history between the Gatoo and Abpango, which have short distances. The presence of Abpango villagers among the displaced people was one of the mistakes in immigrants' resettlement in Gatoo's reconstruction approach. Especially in the post-earthquake conditions and the pressures and shortages caused by the disaster, this conflict turned bold. The natives considered the migrants as a reason for the deficiencies, the migrants accused the natives of discrimination, and old conflicts in the post-disaster conditions fueled the discrepancies.

Livestock One of the methods of keeping livestock in this region is the use of Bidkan, which is a small cave in the mountain. Some villagers, who are ranchers and have livestock without a place to keep it, excavate Bidkan. Before the earthquake, if livestock were lost or stolen, nearby villagers were accused due to previous disputes or financial problems. So, they had many conflicts over the livestock's safety during the years. The integration of those who were previously accused with the claimant people in Dahoo without knowing their conflicts' history caused additional problems for both groups, which continue to this day.

10.4.2.5 Religious Beliefs

According to the interviews, one of the critical topics in the rehabilitation process was their religious beliefs. The positive point of the reconstruction approaches in all three case studies was considering this indicator and reviving the religious gathering centers among the first places for reconstruction. This indicator is so considerable that the proximity of the people to the holy shrine and the convenient holding of religious ceremonies were among the reasons for site selection within a short distance in Dahoo. Religious beliefs made the homeless community calm down after the disaster. The Muslim affected community increased their resilience with the belief in God's existence, the perception that this disaster was God's will, they had no role in its occurrence, and that God gave them the power to continue.

Zarand earthquake caused less financial and human losses than the Bam catastrophe, which had more than 50,000 killed. Zahra is a 30-year-old woman who lost her father and sister in the Zarand disaster. She was 14 at the time of the earthquake and was forced to migrate to Reyhan with her mother. She stated: "*I put myself in the shoes of a 14-years-old girl who lost all her family members in Bam and was taken to welfare centers. It was God's will. It is true that I am far from my childhood homeland and memories or even far from my father and sister's graves. At least I have my mother, a small house in a village that was a stranger once a time and is familiar today, and more importantly, I have God. I go to the mosque on weekends and pray for them.*" The Gatoo Grand Mosque, Imamzadeh Abraham holy shrine of Dahoo, and the Martyrs of Reyhan Mosque are the most important centers which form gatherings to introduce people to the social community through religious ceremonies such as Muharram or Ramadan in the past and today.

10.4.3 *Economic Factors*

10.4.3.1 Livelihood

The post-disaster livelihood problems will be affected more critically if the reconstruction approaches are accompanied by forced displacement. In all three case studies, most men's occupations were agriculture, animal husbandry, and working near mines and factories. The agricultural lands were close to the village, and the livestock were kept either at home or in Bidkans. However, forced displacement in each case study and subsequent conditions changed the affected community's livelihood over time.

- In Gatoo, farmers and ranchers were forced to buy a car to reach their Bidkans and agricultural lands, which were located near the old village. Natives who could not afford this renovated their former homes to stay there for days and control watering orchards or grazing livestock. These items required additional costs for the villagers, which directly affected the family's livelihood and rehabilitation over time. Those unable to travel frequently or renovate their houses suffered severe damage from the livestock's death, theft, or the loss of crops. Some farmers were forced to build a shelter on their farmland to rest there, and the Agriculture Organization sentenced them to pay fines and destroy the shelters. However, people working in mines or factories with a stable monthly income suffered fewer livelihood problems. Immigrants, who had completely lost their jobs, worked as laborers or construction workers to alleviate financial problems. Many of them are still working as simple workers after 17 years.
- In Dahoo, access to agricultural land and Bidkans was not as easy as before but more comfortable than in Gatoo. Farmers, ranchers, and workers of the gypsum mine are still doing their previous jobs. However, after the displacement, male immigrants lost their occupations, and female immigrants decided to help and support their husbands. So, they started adding a room to the yard or changing the bedroom's function to a carpet weaving or embroidery workshop and improved the family's livelihood by making traditional handicrafts.
- In Reyhan, the livelihood indicator was slightly different. Immigrants living in the new towns stated that some of the neighbors, who are also immigrants, are drug dealers or sellers of alcoholic drinks, which is a crime under Islamic laws. Unfortunately, earning money from these petty crimes is easier than working on a farm or as a construction worker. The isolated environment, distance from the city center, desert landscape, and alienation between neighbors have made it easier for minor criminals without any afraid of being arrested. Nevertheless, ranchers and

farmers in the central part continued to do their pre-disaster jobs without many particular livelihood problems. In addition, some natives and immigrants engaged in new activities and took advantage of the new jobs created by turning villages into the city.

10.4.3.2 Entrepreneurship

The authorities tried to do their best during the disaster and reconstruction process. Although these efforts were not comprehensive and only limited to the lowest level of the Maslow pyramid, some helped improve the livelihood rehabilitation of the affected community. According to interviews with residents and authorities, one of the positive points of the adopted approaches was entrepreneurship for immigrants and natives, which is expressed separately based on the case studies.

- The Iranian Coal and Tar Refining Factory is located 6 km from Gatoo and has been in operation since 2007. The government has tried to help the affected people's livelihood by allocating quotas to the Gatoo's residents. However, the plant's emissions from the factory and the village's location in the prevailing wind direction have negatively affected the environmental comfort, respiration, and even the agricultural products. Although residents have repeatedly mentioned this in interviews, they have also benefited from the factory. Their employment and house rent to the managers and factory workers are examples of this win-win deal which made them tolerate unfavorable environmental conditions and air pollution.
- Several groups and NGOs were formed in Zarand to guide villagers in recognizing their abilities and using them to improve the community's economic stage. Identifying women who were stakeholders and encouraging them to do handicrafts hold exhibitions and find major or minor buyers for their products were the samples of those NGOs' responsibilities. These activities have motivated Dahoo immigrant and native women to set up their carpet workshops and earn an independent income over the years. Interview results show that they feel helpful to the family and the community, and their self-confidence increases. It should be noted that these groups have also been active in Gatoo and Reyhan. Also, they have trained many men and women and prepared them to enter the labor market by holding technical-vocational training workshops.
- Businesses have been developed in Reyhan following the transformation of villages into a city. The 3-km pathway from the main road to the city's entrance is replete with various small shops, including florists, supermarkets, bakeries, and car repairments. The ecotourism house, paragliding site, mineral water springs, language institute, restaurants, and many other city facilities have led to entrepreneurship, environmental and cultural development, and economic improvement. It seems that the authorities tried to use the disaster as an opportunity to improve the economic situation. So, the "Build Back Better" idea in Reyhan has been successful in terms of entrepreneurship indicator in economic factors.

10.4.3.3 The Loans

The reconstruction and resettlement approaches were taken in the first months after the earthquake when the affected people were still living in temporary shelters. Two loans were given to all affected communities involving forced displaced people, voluntary migrants, or natives. The loan amounts were equal for them. According to the interviews in all three case studies, the loans were insufficient to complete the house's primary core, and people could hardly build their tiny homes with them. However, a US $5500 reconstruction loan and a $1700 livelihood loan were given to villagers, which were insufficient to reconstruct houses and revitalize lost livelihoods. Also, the affected community could not pay small installments due to rising economic inflation. Many interviewees stated that they spent livelihood loans to complete the house and used public donations, charities, or borrowing to revive their livelihoods. Considering low loan amounts and the inability to return to pre-disaster conditions, some villagers waived their loans and chose to migrate to the cities. Despite the relatively stable situation today, interviewees stated that they have not yet forgotten the psychological and financial conditions of post-disaster days. They considered those small sums as a tool for humiliating them. They rightly believe that the affected community has lost only normal living conditions, and their human dignity has not been lost.

10.5 Conclusion

After studying three factors and their indicators of forced displacement after the 2005 Zarand earthquake, we return to the two main research questions and answer them.

Question 1. What are the long-term consequences of the forced displacement after nearly 17 years?

Answer Incoordination between the land area, family members, and their needs and neglecting their livelihood style led to the additional parts and costs imposition on families for a long time. Also, the change of priorities and the slight adaptation of permanent houses to the vernacular architecture caused alienation and increased dissatisfaction with the new built environment. The affected community would be more satisfied with their homes and spend less money on development if the users get involved in officials' decisions, such as participation in choosing sites or designing houses. The consequences of neglecting tribal conflicts, social ties, and neighborhood units in the old village have led to the disrupted rehabilitation process in the short term and the affected community's isolation in the long term. Also, the place attachment and spiritual value of the ancestral lands have not stopped people from moving to their former place under any pretext, even after 17 years. The forced displacement, regardless of the primary income source, led to being far from agricultural lands and Bidkans. So, the imposition of additional costs for transportation and security, which continues until now, has resulted in general dissatisfaction. The loans were spent on reconstruction instead of reviving their livelihoods. Therefore, many villagers were forced to change jobs, resulting in long-term economic dissatisfaction.

Question 2. What are the pros and cons of approaches taken to resettle displaced people?

Answer The answer to this question is provided as a table of pros and cons which is given in Table 10.5 based on each case study.

From all the long-term consequences, the pros and cons of the three different reconstruction approaches, and the lessons learned from forced displacement, it seems that considering the following points is effective for the successful displacement and population integration:

Table 10.5 Pros and cons of the reconstruction approaches

Case study	Approach	Pros	Cons
Gatoo	Relocation of the village, forced displacement of all residents to a new place, and integration of non-native IDPs	Proximity to the main road, attention to pilgrimage places, safe houses against earthquakes, proximity to factories and mines, allocating jobs to the villagers	Neglecting vernacular architecture, small land area, ignoring privacy in permanent houses, less attention to place attachment and social ties, ignoring tribal conflicts during population integration, long distance from the old village, decision-making without user's participation
Dahoo	Accommodation of the displaced people in the predicted area and integration of non-native IDPs	Proximity to the former village, preparedness for disaster, relocation before the disaster, average-participation in pre-disaster construction, average place attachment, assistance to women's self-employment	Neglecting vernacular architecture, population integration without knowledge of the tribal conflicts, small land area, ignoring privacy in permanent houses, less chance for immigrants to choose their favorite house location
Reyhan	In situ reconstruction of some nearby villages, integrating them with new small towns, accommodation of displaced people there, turning this complex into a city	The combination of in situ reconstruction and relocation, entrepreneurship and business development, evacuation of vulnerable areas, use of natural resources like mineral water springs to attract tourists, high place attachment and having social ties in city center	In new towns: isolation of residents, low security, less attention to social and family ties, decision-making without user's participation, small land area In general: neglecting vernacular architecture, raising petty crimes

- Identifying areas with high vulnerability and planning for them before the disaster
- Comprehensive knowledge of the critical indicators and the sociocultural, physical, environmental, and economic needs of the people
- Participation of people in all phases of post-disaster reconstruction
- Considering religious beliefs, neighborhood units, social ties, and place attachments of villagers
- Simulating the pre-disaster built environment to help accelerate the rehabilitation process and increase the resilience of the affected community.

References

Adams H, Alaniz R, Broner R et al (2015) Maintaining and building place through managed and forced community relocations: lessons for a climate changed world. UNU-EHS working paper, no. 16

Ahmed I (2011) An overview of post-disaster permanent housing reconstruction in developing countries. Int J Disaster Resil Built Environ 2:148–164

Bakhtiari A (2014) Country report of the Islamic Republic of Iran on disaster risk management. Iranian National Disaster Management Organization, Kobe, Japan

Barenstein J (2015) Continuity and change in housing and settlement patterns in post-earthquake Gujarat, India. Int J Disaster Resil Built Environ 6:140–155

Blondin S (2021) Staying despite disaster risks: place attachment, voluntary immobility and adaptation in Tajikistan's Pamir Mountains. Geoforum 126:290–301

Cutter S, Barnes L, Berry M et al (2008) A place-based model for understanding community resilience to natural disasters. Glob Environ Change 18(4):598–606

Davis I, Izadkhah Y (2006) Building resilient urban communities. Open House Int J 31:11–21

Dikmen N (2006) Relocation or rebuilding in the same area: an important factor for decision making for post-disaster housing projects. http://www.grif.umontreal.ca

Fallahi A (2006) Rehabilitation and reconstruction. SBU Publisher, Tehran, Iran

Fallahi A (2008) Temporary human settlements architecture after disasters. SBU Publisher, Tehran, Iran

Fallahi A (2018) Reflections on approaches and models of post-disaster reconstruction. Crisis Prevention and Management Organization, Tehran, Iran

Fernando N, Punchihewa A (2013) Relocating the displaced strategies for sustainable relocation. Friedrich Ebert Stiftung Publisher

Few R, Ramirez V, Armijos M et al (2021) Moving with risk: forced displacement and vulnerability to hazards in Colombia. World Dev 144:105482

Hidayat B, Egbu C (2010) A literature review of the role of project management in post-disaster reconstruction. Paper presented at the proceedings of 26th annual ARCOM conference, University of Salford, Manchester, 6–8 Sept 2010

Jigyasu R, Upadhyay N (2016) Continuity, adaptation, and change following the 1993 earthquake in Marathwada, India. In: Daly P, Feener M (eds) Rebuilding Asia following natural disasters: approaches to reconstruction in the Asia-Pacific region. Cambridge University Press, Cambridge, pp 81–107

Lizarralde G, Johnson C, Davidson C (2009) Rebuilding after disasters: from emergency to sustainability. Routledge, London

Mortazavi H (2006) Relocation forever. J Soffeh 42:123

Raymond C, Brown G, Weber D (2010) The measurement of place attachment: personal, community and environmental connections. J Environ Psychol 30:422–434

Rieger K (2021) Multi-hazards, displaced people's vulnerability and resettlement: post-earthquake experiences from Rasuwa district in Nepal and their connections to policy loopholes and reconstruction practices. Prog Disaster Sci 11:100187

Sangasumana P (2018) Post disaster relocation issues: a case study of Samasarakanda landslide in Sri Lanka. Eur Sci J ESJ 14(32)

Sartipipoor M, Asadi S (2018) The role of place attachment on social resilience due to relocation, case study: Dahooyeh village. J Hous Rural Environ 161:3–16

Sartipipoor M, Hashemian A, Hoseini M (2015) Explain the requirements of post-disaster reconstruction on relocation policy in rural settlements, case study: Estalkh village. J Hous Rural Environ 151:3–18

Sina D, Yan A, Wilkinson S et al (2019) What does the future hold for relocated communities post-disaster? Factors affecting livelihood resilience. Int J Disaster Risk Reduct 34:173–183

Sridarran P, Keraminiyage K, Amaratunga D (2018a) Enablers and barriers of adapting post-disaster resettlements. Procedia Eng 212:125–132

Sridarran P, Keraminiyage K, Fernando N (2018b) Acceptance to be the host of a resettlement programme: a literature review. Procedia Eng 212:962–969

UNHABITAT (2009) The right to adequate housing. Fact sheet no. 21/Rev.1, Nov 2009 edn. UN Office of the High Commissioner for Human Rights (OHCHR), Geneva

Wijegunarathna E, Wedawatta G, Prasanna L et al (2018) Long-term satisfaction of resettled communities: an assessment of physical performance of post-disaster housing. Procedia Eng 212:1147–1154

Chapter 11
Disaster-Induced Relocation of Vulnerable Households: Evidence from Planned Relocation in Sri Lanka

Nishara Fernando and Naduni Jayasinghe

Abstract Given the increasing incidence of natural, particularly hydrometeorological hazards and the extensive damages caused by them, planned relocation has come to be frequently utilized as a Disaster Risk Reduction (DRR) strategy in Sri Lanka. In the last couple of decades, many communities resident in hazard-prone areas have been shifted from their original settlements and resettled in risk-free areas. However, in an attempt to reduce the exposure of communities to the risk of hazards, most planned relocation programs have overlooked the disproportionate effects posed by natural disasters on vulnerable groups in turn resulting in relocation failures. This paper presents a set of evidence-based policy guidelines aimed at facilitating the successful relocation of vulnerable groups. The paper makes reference to five cases of planned relocation in Sri Lanka: one case of relocation following the Indian Ocean Tsunami in 2004; two cases of landslide-induced relocation; and two cases where relocation has been used as a flood risk reduction strategy. Based on evidence drawn from said cases, this paper proposes the following policy guidelines to be followed when relocating vulnerable groups: (a) place emphasis on livelihood recovery and restoration, (b) administer planned relocation as a long-term process, (c) account for the pre-existing vulnerability of beneficiaries in designing relocation programs, (d) incorporate understanding about the social structures and cultural characteristics into relocation decision-making, and (e) advocating participation and consultation. The paper shows that the failure of planned relocation efforts to accommodate an overarching view of the vulnerability of relocatees has rendered planned relocation (as a DRR strategy) counter-effective. The authors argue that planned relocation should not only seek to shift vulnerable groups from their hazard-prone physical locations but also attempt to eliminate the social, political, economic, and institutional roots of their vulnerability.

N. Fernando
Department of Sociology, University of Colombo, Colombo, Sri Lanka

N. Jayasinghe (✉)
Social Policy Analysis and Research Centre, University of Colombo, Colombo, Sri Lanka
e-mail: nadunikj@gmail.com

© The Author(s), under exclusive license to Springer Nature Switzerland AG 2023

M. Hamza et al. (eds.), *Rebuilding Communities After Displacement*,
https://doi.org/10.1007/978-3-031-21414-1_11

Keywords Planned relocation · Vulnerability · Sri Lanka · Hydro-meteorological hazards · Vulnerable groups · Policy

11.1 Introduction

11.1.1 Background

The United Nations office for Disaster Risk Reduction's (UNDRR) report: 'Human Cost of Disasters: An Overview of the last 20 Years' has demonstrated that evidence of a staggering rise is climatic disaster over the past twenty years. Climate-related disasters have almost doubled during the period 2000–2019 recording a total of 7348 reported disasters compared to only 4212 reported events in the preceding couple of decades (Centre for Research on the Epidemiology of Disasters and UNDRRa 2020). Human and economic losses caused by such events have also surged in the past 20 years compared to the period 1980–1999 (Centre for Research on the Epidemiology of Disasters and UNDRRa 2020) (see Fig. 11.1). The incidence of extreme weather events has not ceased to increase in the years that have followed. UNDRR report '2020 The Non-COVID Year in Disasters: Global Trends and Perspectives' reveals that in 2020, the number of reported climatic disasters amounted to 389 which is 5.7% higher than the 2000–2019 annual average (Centre for Research on the Epidemiology of Disasters and UNDRRb 2020). While the highest incidence of natural disasters in 2020 has been recorded in Asia (see Fig. 11.2), Sri Lanka ranked the 2nd among countries that are most affected by extreme weather events in the Global Climate Risk Index 2019 (World Bank 2019).

Sri Lanka is a relatively small island located in the Indian Ocean covering a total land area of 65,610 km^2 with a population of approximately 21.4 million (Asia Disaster Preparedness Centre and United Nations office for Disaster Risk Reduction 2019). The country's geographical location, its topographical features,

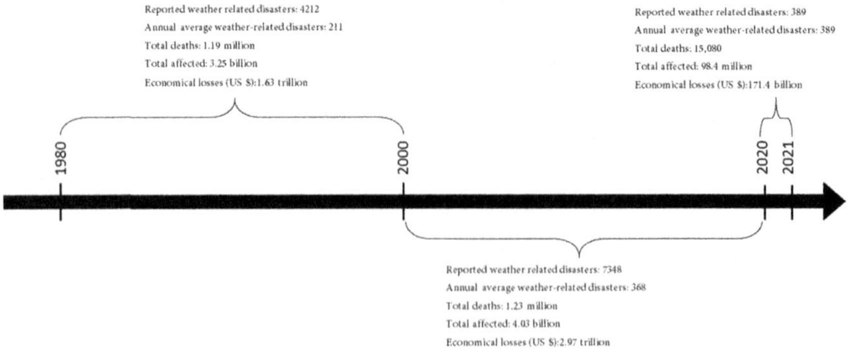

Fig. 11.1 A timeline depicting trends in weather-related disasters from 1980 to 2021 (*Source* Adapted from Centre for Research on the Epidemiology of Disasters and UNDRRa 2020)

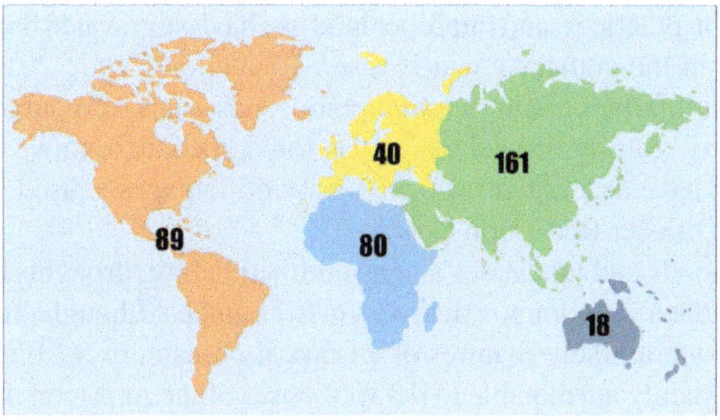

Fig. 11.2 Number of weather-related disasters by continent in 2020 (*Source* Adapted from Centre for Research on the Epidemiology of Disasters and UNDRRb 2020)

development activities, and recent socio-economic and demographic trends have increased its exposure and vulnerability to recurrent hazards, predominantly hydro-meteorological hazards with floods (56%), high winds (10%), droughts (18%), and landslides (16%) recording the highest frequency of occurrence (Disaster Management Centre 2016). Sri Lanka is situated to the Southwest of the Bay of Bengal which features the north-eastern part of the Indian Ocean. Eastern, Northern, and North-Central Provinces of Sri Lanka have been repeatedly threatened by cyclone activity in the Bay of Bengal which has in turn triggered floods and landslides causing human, economic, and social losses (Asia Disaster Reduction Centre 2019; Centre for Excellence in Disaster Management and Humanitarian Assistance 2021). For example, in 2017, Cyclone Mora which developed from an area of low pressure over the Bay of Bengal caused flooding and major landslides in fifteen districts in the South of Sri Lanka resulting in damages amounting to approximately US $415 million (Centre for Excellence in Disaster Management and Humanitarian Assistance 2021).

Further, given that Sri Lanka is located close to the equator, the country is characterized by a tropical climate featuring two main monsoon seasons: the 'Maha' monsoon season, also known as the North-east monsoon which runs from October through January or February affecting the northern and eastern regions of the country, and the 'Yala' monsoon season, also known as the Southwest monsoon which runs from May through August causing rainfall in the south, west, and south-central regions of the country (Department of Meteorology 2016). Monsoonal rains coupled with certain topographical features; particularly, the prevalence of 103 perennial rivers and around 34,000 tanks intensify the risk of floods and landslides in Sri Lanka (Gunawardena and Dayawansa 2020). Illustrating this, in May 2017, South-west monsoonal rains inundated, particularly the Central, Southern, and Western provinces of the island with flash floods and landslides causing more than 210 deaths and displacing approximately 73,600 people. This has been recorded as the worst disaster that struck the island since the 2004 Indian Ocean Tsunami (Ministry of Disaster Management 2017). Apart from the mentioned reasons, unsustainable consequences of development activities such as deforestation, absence of scientific soil conservation practices, and improper land use have aggravated the threat of floods and landslides in the country (Asia Disaster Reduction Centre 2019; Dissanayake et al. 2018). Similarly, certain demographic trends such as a surge in population and the resulting scarcity of land which has forced people to move to and reside in high-risk areas have accounted for the severity of damages caused by floods in the country (Asia Disaster Reduction Centre 2019).

Similar to floods and landslides and as outlined before, droughts have been identified as one of the most common disasters in Sri Lanka. Although prolonged periods of dry weather are unlikely features of a tropical climate, in Sri Lanka, occurrence of droughts is mainly attributable to delayed onset of the monsoon and the temporal variability of rainfall (Manesha et al. 2015). Drought liability tends to be higher in areas demarcated as the 'dry zone' encompassing the north, northwest, north-east, east, and southeast of Sri Lanka in comparison with areas in the Southwest of Sri Lanka, known as the wet zone (Manesha et al. 2015; Nilanthi 2011).

Based on observed meteorological data in the recent years and future projections of climate models, the National Adaptation Plan for Climate Change in Sri Lanka discerns a recorded and projected increase in the frequency and severity of extreme weather events in the country. This has been and is to be accompanied by an increase in ambient air temperature and changes in the distribution pattern of rainfall instigated by climate change (Ministry of Mahaweli Development and Environment 2016). The increasing frequency and severity of climatic events have resulted in massive losses and damages in the past couple of decades. For example, during the period between 1990 and 2018, the total damages caused by disasters have amounted to nearly US $7 billion and said amount excludes the costs that have arisen from unrecorded events such as regional floods (Asia Disaster Preparedness Centre and United Nations office for Disaster Risk Reduction 2019). The incidence and the impact of floods and droughts in Sri Lanka have particularly been on the rise in the recent years (Chandrasekara et al. 2017; Consortium of Humanitarian Agencies 2016). Between 2010 and 2018, close to 14 million people were affected by floods while nearly 12 million people were victimized by droughts (World Bank 2019). In terms of the years to come, the World Bank (2019) anticipates a 1.2% annual GDP loss by 2050 due to climate change-induced extreme weather events.

With the aim of mitigating the human, economic, and social losses caused by climatic shocks, countries including Sri Lanka have resorted to planned relocation as a Disaster Risk Reduction (DRR) strategy apart from other mitigation measures such as hazard mapping, carrying out public awareness and training programs on disaster preparedness and response, enforcement of land use planning guidelines, and improving mechanisms for early warning dissemination (United Nations office for Disaster Risk Reduction 2019; Ministry of Disaster Management 2014). Accordingly, several initiatives have been taken in the recent past to permanently move communities residing in hazard-prone areas to safer areas (Fernando 2016; Fernando 2018a, b, Norris et al. 2019; Piggot-McKellar et al. 2020; United Nations High Commissioner for Refugees 2014). In Sri Lanka, similar to many other countries in Asia and the Pacific, disaster-prone areas are predominantly occupied by vulnerable groups such as the poor. Elaborating on this, a recent post-disaster needs assessment carried out following the Southwest monsoon in 2017 has found that those affected by landslides in said year were substantially poor with an average poverty rate of 11.6% (Ministry of National Policies and Economic Affairs and Ministry of Disaster Management 2017). Further, the Sri Lanka National Report on Disaster Risk, Poverty and Human Development Relationship highlights that the urban poor in Sri Lanka are at a greater risk to floods, as they mostly reside in slums that are located in areas with poor drainage or adjoining solid waste dumping sites (Disaster Management Centre and UNDP 2009). Adding to this, such population groups lack the coping capacities that are required to bounce back from displacement by disasters as they have limited assets (including financial, physical social, natural, and human capital); fragile livelihoods; and minimal/no social protection (Ministry of National Policies and Economic Affairs and Ministry of Disaster Management 2017; Fernando and Punchihewa 2013).

Thus, there is a need for a set of evidence-based policy guidelines, particularly on relocating vulnerable groups to risk-free zones. Studies on relocation have largely focused on outlining relocation outcomes of specific relocation programs. Nevertheless, existing literature lacks an extensive discussion on how relocation outcomes are moderated by the vulnerability of target beneficiaries. Moreover, relocation scholarship has not been adequately oriented toward generating a policy informing discussion on specific measures to be taken when relocating groups with limited coping capacities. This paper attempts to address this knowledge gap by proposing a set of policy guidelines to be followed in designing and executing planned relocation programs for vulnerable groups, drawing evidence from five cases of planned relocation in Sri Lanka.

11.1.2 Planned Relocation as a DRR Strategy

The increasing incidence of climate-related hazards has urged a shift in disaster policy from disaster response to Disaster Risk Reduction (DRR). The Sendai Framework for Disaster Risk Reduction 2015–2030, the main global policy framework that governs disaster management at a global level, strongly advocates a DRR approach toward managing hazards (United Nations Office for Disaster Risk Reduction 2015). A DRR approach calls for the analysis and reduction of causal factors of disaster risk (e.g., exposure to hazards, vulnerability of people, and level of preparedness) with the aim of reducing disaster risk and losses of lives, livelihoods, and other aspects of human well-being (United Nations office for Disaster Risk Reduction 2020). It is against this backdrop that planned relocation has come to be both regarded and deployed as a DRR strategy (Piggot-McKellar et al. 2020; United Nations High Commissioner for Refugees 2014).

The term ‘planned relocation’ has been used specifically to indicate relocation following natural disasters (in contrast to development and conflict-induced relocation) and has been defined as ‘a solutions-oriented measure involving the State, in which a community (as distinct from an individual/household) is physically moved to another location and resettled there' (United Nations High Commissioner for Refugees 2014, p. 10). The United Nations High Commissioner for Refugees (UNHCR) (2014) draws a clear distinction between planned relocation and evacuation stating that evacuation occurs within a short time frame where the sole objective is to rapidly move people away from an immediate threat to safer locations as part of emergency procedures undertaken instantly following a disaster. On the contrary, planned relocation is a long-term process that requires significant investment in pre-planning and continuous monitoring (United Nations High Commissioner for Refugees 2014). The long-time frame associated with planned relocation could be further rationalized on the basis that the process should not only aim at assisting the movement of the displaced away from their original habitats and settling them in a safer location but also focus on the provision of conditions that are necessary to rebuild their lives in the new location (International Organization for Migration 2019). Warner et al. (2013) outline four instances, where planned relocation is executed:

- People inhabit areas which are prone to sudden onset natural disasters
- People's livelihoods are threatened by slow-onset impacts from climate change
- People reside on lands that are required for mitigation and adaptation projects
- People reside in a country or a part of country which may become inhabitable due to the adverse effects of climate change.

Although planned relocation has come to be utilized as a DRR strategy and therefore seeks to reduce the damages caused by increased exposure to natural disasters, it has often been used as a last resort strategy due to its tendency to pose detrimental outcomes for relocated communities. Past experiences of planned relocation have proven to be more harmful than beneficial for relocatees with the exception of a few stories of success (Arnall 2019; Correa 2011; de Sherbinin et al. 2011; Bergmann et al. 2021). Severe impacts of relocation have ranged from disruptions to community structures and social networks, increase in financial expenses and pressure to reconstruct livelihoods, psychological stress caused by emotional attachment to the previous place of residence to abandonment of former lifestyles, and cultural roots (Hino et al. 2017; Binder et al. 2019; McMichael and Powell 2021). Cernea (2002) in his Impoverishment Risks and Reconstruction Model for Resettling Displaced Populations (IRR) identifies eight impoverishment risks that may result from relocation, namely (1) Landlessness; loss of land in the original habitat of the relocated community; (2) Joblessness; loss of wage employment; (3) Homelessness; loss of housing and shelter; (4) Marginalization; demotion of socio-economic status due to loss of economic power; (5) Increased morbidity and mortality; decline in health conditions; (6) Food insecurity; hunger caused by inadequate intake of calorie-proteins; (7) Loss of access to common property resources; inaccessibility of shared assets such as like forests, pump water wells, and playgrounds; and (8) Social disarticulation; disintegration of community structures, social networks, and ties. Similarly, adopting a human rights-based approach, Gramilova (2014) states that planned relocation, although widely used both as a DRR and a climate change adaptation strategy, entails enormous human rights costs depriving people of certain rights such as right to life; right to adequate food; right to health; right to adequate housing; right not to be forcibly evicted; right to work; right to education; right to development; and the right to take part in cultural life.

Drawing evidence from past successes and failures of planned relocation in various regions, significant attempts have been made to propound factors that may enhance or hinder success of planned relocation efforts. Citing evidence from two cases of planned relocation in Peru, Bergmann (2021) observes how prevalence of poverty and lack of resources among community members to support themselves have deteriorated the well-being of relocated communities following relocation. Similarly, in a study of resettlement following 2008 Wenchuan earthquake in China, Wang et al. (2015) found that poor, unemployed households with a lower level of education were more disadvantaged in the post-disaster recovery process than affected households of a relatively higher socio-economic status. Several other studies have confirmed inverse associations between heightened states of vulnerability such as poverty and outcomes of relocation and/or migration following displacement from extreme weather events (Chambers 1989; Birkmann 2013; Ministry of National Policies and Economic Affairs and Ministry of Disaster Management 2017; Fernando and Punchihewa 2013; Krishnamurthy 2012; Black et al. 2013; Warner and Afifi 2014).

Further, drawing on evidence from Peru, Bergmann (2021) argues that while weak governance structures and institutional constraints (e.g., lack of institutional leadership; lack of technical and organizational capacities; and lack of inter-ministerial coordination) have hampered success of relocation efforts, livelihood restoration following relocation and people's agency in the relocation process have served to improve outcomes of relocation. Nevertheless, Bergmann (2021) posits that certain structural constraints such as poverty tend to decline people's agency in advancing the relocation process in turn hindering the possibility of positive relocation outcomes. McMichael et al. (2019) draws on evidence from cases of planned relocation in three villages in Fiji: Vunidogoloa, Vunisavisavi, and Narikoso to further comment on the importance of people's agency in achieving favorable relocation outcomes. The authors reveal that relocation in said villages has been supported by the relocated community as they have been actively involved in decision-making pertaining to timing of relocation, house design, location of new sites, and village layout. McMichael et al. (2019, p. 332) note that village-level consensus over such decisions had not been achieved overnight, rather, over a period of 'local level debate and discussion'.

With regard to livelihood restoration following relocation, authors have often raised concerns over relocated communities' ability to utilize immovable livelihood assets consequent to relocation (e.g., Artur and Hilhorst 2014; Zulauf 2013; Wilmsen 2018). Reasons for such concerns have been brought to light in a recent study on two rural farming communities in China relocated following the Wenchuan earthquake in 2008 (Chen et al. 2020). The study reveals that in one community farmland was not reallocated to farmers post-relocation while farmers of the other community were provided with both farmland and forestland. Chen et al. (2020) observe that availability of farmland following relocation significantly improved the livelihood outcomes and perceptions of well-being of the relocated communities. Further, based on an evaluation of planned relocation in the Austrian Danube catchment over five decades, Thaler et al. (2020, p. 8) assert the importance of implementing relocation as a long-term process of 'continuous adaptation, refinement and learning' allowing adequate time for pre-discussions, discussions and negotiations between affected households, policy-makers, and other stakeholders (e.g., private sector organizations; project authorities) involved. The authors also contend that flexibility in adapting to individual needs of relocatees over an extended period as opposed to favoring rigid rules and deadlines is key to harnessing relocatees' acceptance of relocation which could thus result in more positive relocation outcomes (Thaler et al. 2020).

Among the many factors that determine the success of relocation outcomes, it is evident that heightened states of vulnerability such as poverty in a community can significantly wear off the outcomes of relocation (Bergmann 2021). This coupled with the fact that disaster-prone areas are predominantly inhabited by vulnerable groups may almost render the successful implementation of planned relocation infeasible, unless of course an attempt is made to sensitize relocation efforts to the needs, capacities, and constraints of such groups. It is against this background that this paper scrutinizes planned relocation practices targeted at vulnerable groups in Sri Lanka and attempts to present a set of policy guidelines on relocating such groups.

11.1.3 Conceptualizing Vulnerability

A discussion on planned relocation of vulnerable households prompts the questions: 'who is vulnerable' and 'what constitutes vulnerability'. The concept of vulnerability stands significant in developing an overarching understanding of disaster risk, i.e., an orientation to perceive disaster risk in its entirety, not only in its physical form but also including its socio-economic and political dimensions. UNDRR (2022) defines 'vulnerability' as the conditions that stem from physical, social, economic, and environmental factors or processes, which aggravate the susceptibility of a community to the effects of hazards.

Chambers (1989), one of the pioneering scholars in the field of development and poverty research, defines vulnerability as the state of being exposed to contingencies and stress, and being faced with difficulties in coping with them, extensively building on studies pertaining to drought-induced crises and disasters. Chambers (1989) contends that vulnerability is of a dual nature. That is, vulnerability has an external side which is representative of risks, shocks, and stress to which an individual or household is exposed. Vulnerability also has an internal aspect which constitutes an individual's defenselessness meaning a lack of means to cope in a situation of disaster without damaging loss (Chambers 1989).

Informed by Chamber's (1989) conceptions regarding the dual nature of vulnerability, Bohle's (2001) 'double structure of vulnerability' framework resembles that vulnerability constitutes both an external and an internal aspect (see Fig. 11.3). In this framework, the external side of vulnerability represents an individual's or a community's exposure to risks of hazards and shocks in his/their environment. The conceptualization of the external side of vulnerability has been influenced by political economy and human ecology approaches and the entitlement theory. Hence, the external side of vulnerability extends beyond mere spatial exposure to hazards to include social, political, and institutional conditions and processes that increase a community's defenselessness against hazards (Bohle 2001). The external side of vulnerability represents the structural dimensions of vulnerability (Bohle 2001).

On the other hand, the internal side of vulnerability calls for the capacity of an individual, household, or a community to cope with and recover from the impacts of a hazard (Bohle 2001). This dimension of vulnerability focuses on the extent to which an individual, household, or a community can adopt coping mechanisms or take action to overcome or mitigate the impact of a hazard. In the framework, this aspect of vulnerability has been conceptualized based on models of access to assets, conflict, and crisis theory and action theory approaches. Hence, in this framework, vulnerability is expedited as the result of the interaction between an individual's coping capacities and the structural dynamics within his/her environment.

Similarly, Birkmann (2013) conceptualizes vulnerability as constituent of core factors such as exposure, susceptibility, and coping capacities that systematize the capability of a society to deal with adverse environmental conditions. Here, Birkmann (2013) draws a significant distinction between exposure and susceptibility stating that exposure refers to the extent to which a particular system (whether a community, city, or a building) falls within geographical parameters with a high risk of a hazard event. On the other hand, susceptibility is manifested through people's defenselessness (owing to poverty or lack of awareness about disaster risks) in the face of a hazard

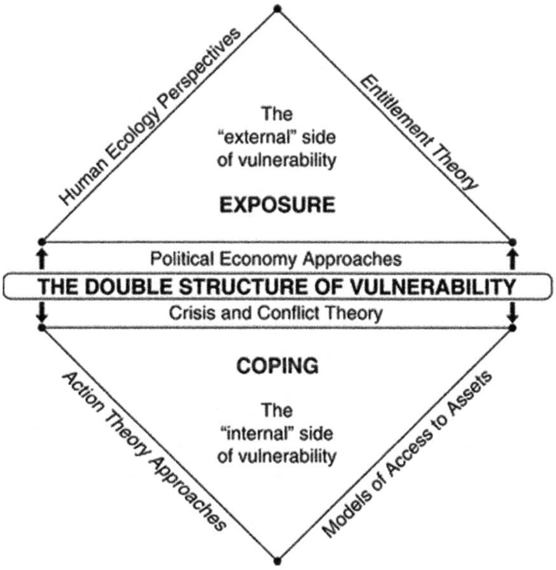

Fig. 11.3 The double structure of vulnerability (*Source* Bohle 2001, p. 4)

event. It features the likelihood of a community to experience damage, suffering, and loss when it is exposed to a hazard event. According to Birkmann (2013), susceptibility has several dimensions namely physical, social, environmental, cultural, and institutional. Birkmann (2013) also points out that exposure to a hazard event does not necessarily make a community susceptible. Similarly, not all people who are exposed to a hazard event are equally susceptible. On the contrary, coping capacities entail the strengths and resources that are available within a community to deal with the adverse consequences of hazards (Birkmann 2013; Intergovernmental Panel on Climate Change 2012; Turner et al. 2003).

The above discussion sheds light on the fact that vulnerability is a phenomenon that extends beyond a community's mere exposure to hazards; rather it encompasses the structural and personal conditions that moderate the extent of losses, suffering, and damages caused by a hazard event to a community. These structural and personal conditions stem from various economic, social, political, institutional, and environmental bases. Thus, in light of the above discussion, the term 'vulnerable groups' has been used in this paper to refer to those population groups who lack capacities and resources that are required to effectively cope with the unfavorable consequences of hazards which in turn leaves them more susceptible not only to the adverse effects of hazards but also to the risks of planned relocation that may follow.

11.2 Methodology

This paper makes reference to five cases of relocation: one case of relocation following the Indian Ocean Tsunami in 2004; two cases of landslide-induced relocation; and two cases where relocation has been used as a flood risk reduction strategy. The cases have been studied based on primary data collected by the lead author at different points in time from various stakeholders of relocation including the relocated community, host community, and key informants, particularly to capture the long-term outcomes of relocation. The case of relocation following the Indian Ocean Tsunami has been examined on the basis of in-depth interviews carried out with selected households (twenty relocated, ten host community, and twenty who have rented/ sold their donated house to move back into the buffer zone) in four relocation settlements namely, China Charity Village, Salzburg, Green Village, and Tea Garden, situated in the Akmeemana Divisional Secretariat Division in the Galle district located in the Southern part of Sri Lanka. The case has also drawn on data collected through in-depth interviews carried out with purposely selected government and non-government officials in the Akmeemana Divisional Secretariat Division. Data were collected for 13 years following relocation to capture how the status of the relocated community has been subject to change in the long run.

The two cases of landslide-induced relocation have sought to encapsulate the long-term impacts of relocation following the landslides: one that occurred in the Meeriyabedda tea estate situated in the Haldummulla Divisional Secretariat Division of the Badulla district of Sri Lanka and a relatively recent landslide that took place in the Aranayake region in the Kegalle district of

Sri Lanka. Primary data have been collected via a household survey administered across the occupants of the 75 housing units constructed by the Government of Sri Lanka (GoSL) for the purpose of relocating the landslide victims of the Meeriyabedda tea estate. On the other hand, the status quo of relocatees of the Aranayake region two years following relocation has been examined based on structured interviews conducted with households relocated into four resettlement villages: Duldeniya, PanapuraWatta, Wasanthagma, and China Friendship Village.

Relocation as a flood risk reduction strategy has been scrutinized with reference to two relocation programs: the Colombo City Flood Prevention and Human Environment Development Project (FPHEP) and the Lunawa Environment Improvement and Community Development Project (LEI and CDP). The long-term outcomes of the FPHEP have been evaluated based on interviews conducted with 284 relocated households in Badowita, the largest resettlement site established under the project. Similarly, for the purpose of delving into the relocation outcomes of LEI and CDP, 102 household heads relocated under the said project have been interviewed. In addition to this, key informant interviews have also been conducted with government and community-based development society officials affiliated with both flood-induced relocation programs mentioned. Table 11.1 summarizes the data collection techniques and the respondents used to explore the five cases of planned relocation cited in this paper (Fig. 11.4).

11.3 Results and Discussion

11.3.1 An Overview of the Relocation Programs

- **Relocation of Meeriyabedda Landslide Victims in Makaldeniya Estate, Sri Lanka**

On 29th October 2014, a massive landslide struck the Meeriyabedda tea estate located in the Haldummulla Divisional Secretariat Division of the Badulla district of Sri Lanka. The incident resulted in 37 deaths causing complete damages to 70 houses and leaving over 275 individuals homeless. The affected population constituted those occupying line rooms constructed on estate land and, thus, represented a group characterized by a high degree of socio-economic vulnerability. Following the landslide tragedy, steps were taken to relocate the landslide victims to government-constructed houses in the Makaldeniya Division of the Maskeliya planation situated 11 km away from the original settlement. The relocation program was executed under the supervision of the Ministry of Disaster Management in coordination with several agencies and authorities including the Ministry of Plantation Infrastructure Development; the National Building Research Organization (NBRO); the Urban Development Authority (UDA); Sri Lanka Army; Disaster Management Centre (DMC); and the District Secretariat, Badulla. The beneficiaries were selected based on interviews

Table 11.1 Data collection techniques and respondents used to explore the cases of relocation

Type of relocation	Relocation program/field of study	Data collection method	Respondents
The Indian Ocean Tsunami	Relocation settlements following the 2004 tsunami relocation settlements in the Akmeemana Divisional Secretary Division in Galle	Interviews	Relocated households, host community households and households who have rented/ sold their donated houses to move back into the buffer zone Government and non-government officials
Landslide induced	Relocation of Meeriyabedda landslide victims in Makaldeniya estate, Sri Lanka	Interviews	Relocated households
	Relocation following the 2016 Aranayake landslide	Interviews	Relocated households
Flood induced	The Colombo City Flood Prevention and Human Environment Development Project (FPHEP)	Interviews	Relocated households Government and community-based development society officials
	Lunawa Environment Improvement and Community Development Project (LEI and CDP)	Interviews	Relocated households

held with the landslide victims and a point system which sourced from the 'voters list' and the reports provided by the estate. Owing to delays in the preparation of land and construction of houses, the selected victims were provided with temporary shelter at the Makanda Tea factory. The victims were relocated to the Makaldeniya estate after a period of two and a half years.

- **Relocation of Underserved Settlers in the City of Colombo as a Flood Risk Reduction Strategy**

In the '2018 Revision of the World Urbanization Prospects', the urban population in Sri Lanka was estimated 18.58% of the total population in 2019 (The World Bank 2021). The Colombo city situated in the Western Province of Sri Lanka has the highest urbanization rate in the country (The World Bank 2012). One of the main issues facing this city is the exposure of the poor to the risk of floods. A significant proportion of the urban poor resides in underserved settlements such as slums and shanties which are located near canal banks or on low lying land in turn exposing these populations to floods. Relocation has been adopted by the Government of Sri Lanka as a flood risk reduction strategy to relocate such underserved settlers into safer

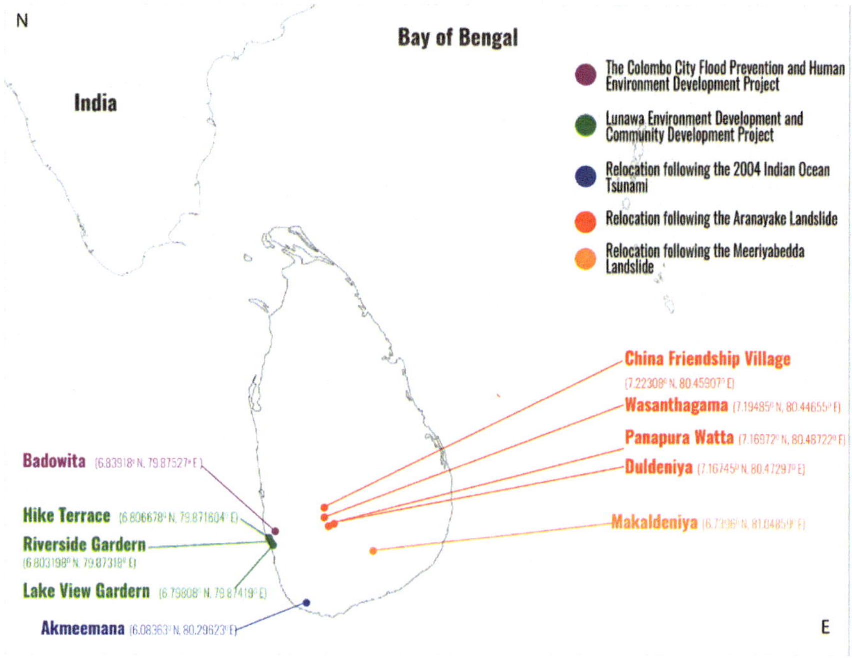

Fig. 11.4 Map of Sri Lanka indicating the relocation sites/areas where data collection was carried out (*Source* Developed by authors)

areas. A number of relocation programs have been executed in this regard. In this paper, attention has been directed at highlighting the long-term outcomes of two such relocation programs: (1) The Colombo City Flood Prevention and Human Environment Development Project (FPHEP) and (2) Lunawa Environmental Improvement and Community Development Project (LEI and CDP).

The Colombo City Flood Prevention and Human Environment Development Project (FPHEP) was implemented by the Government of Sri Lanka (GoSL) with the dual purpose of relocating underserved populations residing in flood-prone areas in Colombo and repairing the canals in the city to control floods. The project began in the 1990s with funding from the Japanese government. Under said project, shanty dwellers residing on canal embankments within the Colombo Municipal Council (CMC) Area were relocated to seven areas in the immediate suburbs of the CMC. The relocates were given 1–2 perches of land and LKR 20,000 as an interest-free loan to construct houses. The findings presented in this paper have been derived from data collected in 'Badowita', the largest relocated settlement under the FPHEP project.

The Lunawa Environment Improvement and Community Development Project (LEI and CDP) was executed by the GoSL with the financial support of the Japan Bank for International Cooperation (JBIC) to relocate urban underserved settlers who were encroaching the Lunawa lake catchment area situated in the South of

Colombo, Sri Lanka. These underserved settlements lacked proper sewage facilities as a result of which drains were constantly clogged during heavy rains and the community was exposed to frequent floods. The underserved settlements resembled a socially and economically vulnerable group whose livelihoods were mainly dependent on informal sector employment or illegal income-generating activities such as the trade of drugs and prostitution. The said project was implemented in 2000 with the objective of relocating identified underserved communities (which included 870 households) to upgrade their living conditions and mitigate the damage caused by floods through the improvement of urban drainage and canal systems cum transforming the settlements around the Lunawa lagoon into a sustainable and planned settlement. Under the LEI and CDP, identified households were given three relocation options: (1) settle in four relocation sites prepared by the project (namely Hike Terrace, Green View Garden, Riverside Garden, Lake View Garden); (2) settle in lands purchased by the identified households (self -relocation); and (3) settle in the original site consequent to regularizing the plots (onsite relocation). The facts presented in this paper have been sourced from data gathered from 102 household heads of families residing in three relocation sites namely, Lake View Garden, Hike Terrace, and River Side Garden.

- **Relocation Following the 2016 Aranayake Landslide**

In May 2016, a landslide of large magnitude occurred in Aranayake, a mountainous region located 80 km north-east of the Colombo district of Sri Lanka. The landslide displaced 513 families, completely destroying 72 houses while exposing the remaining houses to a significant risk of subsequent landslides. The region was home to a rural population that was largely dependent on cultivation in their home gardens for food and consistent income. Income and employment levels in this region were generally low. Subsequent to the landslide, families affected by the catastrophe and those families at risk of being exposed to potential landslides were relocated to safer areas. This paper has drawn on data gathered from four resettlement villages: Duldeniya, PanapuraWatta, Wasanthagma, and China Friendship Village. While Duldeniya and PanapuraWatta contained owner-built houses, where families had been provided with funding to construct their own houses, Wasanthagma and China Friendship Village accommodated donor-built houses where the houses had been built by donors (e.g., the GoSL, Chinese Government, and private donors).

- **Relocation Settlements Following the 2004 Tsunami Relocation Settlements in the Akmeemana Divisional Secretary Division in Galle**

Masses of people, mostly those residing in the coastal belts of Sri Lanka, were victimized by the Indian Ocean Tsunami which occurred on 26th December 2004. The disaster took over 35,000 lives in the country, leaving thousands homeless. Following the buffer zone regulations imposed by the Government of Sri Lanka in 2005, exceeding 70,000 tsunami-affected families were relocated into new areas via various relocation programs funded by local and foreign donors. In this paper, attention has been paid to four relocated settlements: China Charity Village, Salzburg,

Green Village, and Tea Garden located within the Akmeemana Divisional Secretariat in the Galle district of Sri Lanka, focusing on relation outcomes 13 years after relocation.

11.3.2 Policy Guidelines on Relocating the Vulnerable

All cases of planned relocation presented in this paper not only exemplify the execution of relocation as a Disaster Risk Reduction (DRR) strategy but also demonstrate instances where vulnerable groups who lack coping capacities have been displaced by natural disasters and hence, relocated to safer areas. In this section of the paper, a set of policy guidelines for relocating vulnerable groups have been outlined. These policy guidelines have been derived through the analysis of case data pertaining to the five cases of planned relocation cited in this paper. In analyzing case data, attention was paid to the identification of best practices and lessons learned from each case of planned relocation. These were subsequently subject to scrutiny and evaluation to propose policy guidelines for the effective execution of planned relocation programs targeted at vulnerable households. Accordingly, the need for these guidelines has been justified in this chapter with reference to the said cases.

(a) **Place Emphasis on Livelihood Recovery and Restoration**

One of the most common risks of relocating vulnerable groups such as poverty-stricken communities is that following relocation, such groups are most often subjected to heightened vulnerability leaving them trapped in the cycle of poverty, sometimes across generations. Most relocation programs executed in Sri Lanka have often considered relocation as the sole act of uprooting population groups from disaster-prone areas and relocating them in areas free from the risk of natural hazards. In this regard, considerable attention has also been paid to the construction of disaster-resilient housing. However, the possible damage caused to the livelihoods of the vulnerable in the course of relocation has been overlooked to a great extent. Illustrating this, evidence from the relocation of Meeriyabedda landslide victims shows that the majority of relocatees were satisfied with moving into larger, new housing units with improved access to basic services such as electricity and water from relatively smaller line rooms in their previous settlement. Similarly, most relocatees demonstrated satisfaction with housing conditions such as the floor area (57%); ventilation of the house (81%); materials used in building the housing unit (79%); orientation (80%); and privacy of the housing unit (77%). Apart from this, the perceived social status of the resettlers significantly improved after relocation mainly because they were able to enjoy legal title to houses.

However, complying with Cernea's (2002) conceptions on marginalization, the majority of relocatees (85%) experienced economic difficulties which were attributed to a loss of harvest, livestock, and other assets as a result of both the landslide and the relocation. The new settlement has created a sizable distance of 10–20 km to the workplaces of relocatees which in turn has increased their expenditure on fuel and transport. Distance to workplaces has in certain cases resulted in a loss of employment. For example, prior to relocation, 15 respondents had served as laborers at the Meeriyabedda estate but, following relocation only 9 respondents have continued to do so due to the distance to the estate and resulting rise in transportation expenses. Further, 85% of the respondents expressed dissatisfaction with the space available in their new settlement to continue the income-generating activities that had been carried out in their previous settlement. For instance, most respondents revealed that they lacked garden space for animal husbandry and that the adjoining lands could not be used for such activities as said lands were owned by the estate. This implies that the respondents had been more susceptible to the risk of landlessness and subsequent marginalization (Cernea 2002). The majority of the respondents (85%) claimed that their income had decreased consequent to relocation owing to reasons such as distance to the previous workplace; lack of employment opportunities close to the new settlement; and lack of financial assets to commence self-employment activities.

Similar findings have been reported in examining the long-term outcomes of relocation following the landslide tragedy in Aranayake. As a result of a loss of access to previous cultivation lands coupled with unavailability of land suitable for cultivation in the new settlement, the income level of relocatees has significantly declined subsequent to relocation. Most relocatees have lost access to a stable source of income while simultaneously, household expenditure has risen given that new expenses on food, water, electricity, and gas have emerged. Elaborating on this, food expenditure of households had been relatively low in their previous settlement as they owned large blocks of land which were partly utilized for subsistence agriculture. Further, most respondents had relied on their own springs to obtain water and used firewood for cooking in their previous settlement. However, loss of access to said natural assets following relocation has given rise to additional expenses on water and gas.

In contrast, new settlements established under the LEI and CEP were situated closer, i.e., within 5 km of their previous settlements. This was beneficial for the relocatees as their livelihood activities, which were centered on infrastructure, support systems, and assets in the previous settlement could be continued without disruption. Illustrating this, the majority of relocated respondents (78%) expressed their satisfaction with household income after relocation relative to the situation before relocation. Therefore, it can be deduced that placing emphasis on livelihood recovery and restoration is crucial to achieve positive relocation outcomes, particularly in the long term. Livelihood recovery and restoration may call for measures such as relocating vulnerable groups in settlements that are located in close proximity to their original settlements, providing training to increase the employability of youth, provision of self-employment opportunities backed by financial and advisory support to materialize self-employment ventures, etc.

(b) **Administering Planned Relocation as a Long-Term Process**

Most planned relocation programs implemented in the country have concentrated on the stages: before relocation and soon after relocation thus, deviating from UNHCR's

(2014) emphasis on the long-term orientation of planned relocation. Project authorities have largely failed to engage in effective evaluation and monitoring of the long-term outcomes of relocation. Inadequate investment of time and resources in monitoring, evaluating, and constructively addressing (if necessary) relocation outcomes in the long-term have often resulted in relocation failures.

Data collected from relocation settlements in the Akmeemana Divisional Secretary Division in Galle 13 years after relocation illustrates this. Following relocation, the relocatees had been removed from their familiar socio-economic and cultural conditions and introduced to a new socio-economic and cultural environment to which they found it difficult to adapt. The relocation settlements were established far from previous settlements of the relocatees, adhering to the buffer regulations imposed in 2005 following the tsunami. Livelihood activities (some of which included fishing, three-wheeler driving, selling lunch packets, driving tourists) of many of the relocatees were disrupted following relocation owing to the distance from the relocation settlements to the sea and the city. As a result, most resettlers had not only been pushed to chronic poverty but also been subject to economic cum social marginalization. This has compelled most relocatees to illegally move back to the buffer zone, abandoning the hazard-free relocation settlements. In such cases, the housing units in the new settlements were closed, sold, or rented. This has been illustrated in Table 11.2.

The case cited above shows that while planned relocation has been deployed to reduce the exposure of population groups to the risk of hazards, relocation itself has introduced a new set of risks as asserted by Cernea (2002). These risks are a greater detriment to vulnerable groups as they lack the coping capacities necessary to effectively mitigate the risks. Hence, it is crucial that project authorities exert adequate time, resources, and effort to constructively manage relocation-induced risks in the long term. Failure to do so may result in counter-effects as demonstrated in the above case where over time the resettlers had closed, rented, or sold the housing units in resettlement sites and moved back to the buffer zone in turn, re-exposing them to the risk of coastal hazards.

(c) **Account for the Pre-existing Vulnerability of Beneficiaries in Designing Relocation Programs**

Table 11.2 Number of housing units closed, rented, or sold by relocatees in relocation sites[a]

Relocation site	Number of housing units	Number of housing units closed, rented, or sold
Salzburg	79	27
Green village	90	51
China charity	135	22
Tea garden	92	18

Source Fernando (2018a)

[a] Pseudonyms have been used to indicate the relocation sites ensuring anonymity of the respondents

It is important that planned relocation programs are tailored to address the socio-economic characteristics (e.g., level of income, occupation, level of education, number of dependents in the family) of beneficiaries. Aspects of program design including the program structure, the amount of funds provided, the manner in which the funds are provided (e.g., whether they are provided in the form of a loan, a one-off payment or in installments), what is compensated, and what is not compensated should be ideally decided based on the socio-economic characteristics of the target beneficiaries. In the context of relocating vulnerable groups, an orientation as described above calls for adequately acknowledging the pre-existing vulnerability of such groups, or in other words, taking into account the fact that these groups lack coping capacities to recover and rebuild following displacement and relocation. Evidence shows that relocation programs that have emulated program designs disregarding the pre-existing vulnerability of beneficiaries have often resulted in relocation failures.

For example, in the case of FPHEP project, it was evident that for many of the relocatees, constructing a house in the plot of land provided by the government had been financially challenging. This is because the government's allocation of LKR 20,000 was paid in installments (i.e., LKR 6000; LKR 4000; LKR 4000; and LKR 6000) and the households neither had savings nor access to bank loans to fund the construction of houses. Owing to the financial difficulties of constructing houses, some families had used the allocated funds alternatively, such as for sending a family member abroad for foreign employment. Such financial difficulties had compelled some relocatees to sell the plot of land to outsiders and return to their original settlements, regardless of the fact that it was illegal to do so. At the time of data collection, the resettlement site, 'Badowita', was inhabited not only by former settlers (families who had been living in the area before the execution of the FPHEP project) and resettlers (families who had been resettled under the FHEP project) but also new settlers (families who had purchased land from the resettlers or have obtained them from politicians) and unauthorized settlers (new settlers who had bought land from second or third generation resettlers or others who did not have legal ownership to land).

Similarly, vulnerable groups relocated following the landslide which occurred in Aranayake were not given any financial assistance to purchase furniture and essential household items. Further, the same amount of funding was provided for all families who opted for owner-driven housing regardless of their family size. Thus, larger families were left with no alternative but to depend on their personal financial assets to construct a house that was able accommodate all family members. Complying with Cernea's (2002) arguments, these expenses drained household savings in turn pushing the households into further social and economic marginalization following relocation.

In contrast, the entitlement package that was offered under the LEI and CEP could be deemed as adequately scalable and flexible to address the pre-existing vulnerability of the underserved settlers who constituted the target beneficiaries of said project. The beneficiaries were given a bottom-line package of 50 sq. m parcel of serviced land in a resettlement site free of costs along with a minimum of LKR

400,000 (in four installments) to build a house with the size of 35 sq. m. However, if the total replacement cost of the house exceeded LKR 400,000, the additional cost was paid for by the project. 82% of the respondents relocated under LEI and CEP also stated that they had received additional livelihood support through the government mechanism.

These cases illustrate that the design of the relocation program should seek to capture the pre-existing vulnerability of target beneficiaries, particularly when relocating vulnerable groups to avoid relocation failures.

(d) **Incorporate Understanding About the Social Structures and Cultural Characteristics into Relocation Decision-Making**

The social structure and the cultural characteristics of both the relocated community and the original community are hardly taken into account in arriving at critical decisions related to relocation such as 'where to relocate to' and 'how to relocate (individually or as a community)'. However, these factors have a significant bearing on the extent to which the relocatees are able to establish new social networks and feel at home in the new settlement following relocation. Planned relocation experiences in Sri Lanka depict scenarios of prolonged friction between resettled and host communities that has been built upon certain invisible elements of social organization and culture. For instance, field experiences from the relocation settlements in the Akmeemana Divisional Secretariat Division in Galle show that the relocatees were labeled as 'tsunami villagers' or referred to based on their place of origin (e.g., Gintota community, Pettigalawatta community, or Magellan community) by the host community. The relocatees were often associated with deviant behavior (e.g., alcoholism, drug addiction, and theft) by the host community which ultimately led to the social marginalization of resettlers in the new community.

The friction between the relocatees and the host community has partly stemmed from urban/rural differences given that the relocatees have been relocated from an urban setting to a rural setting. It was also observable during data collection that complementing urban/rural differences, resettlers were marginalized based on the caste tied to their nature of employment as fishermen. These bases of marginalization are embedded in the social structure and culture of the host community, and as evident, these factors have the possibility of largely influencing the host community's behavior and attitudes toward relocatees.

Marginalization of relocatees as exemplified above implies that the potential for building social capital through the establishment of new support networks, which are critical for recovery and rebuilding following exposure to disasters (Bohle 2001; Cernea 2002; Chambers 1989), could be low among the resettled population. In cases of relocating vulnerable groups, marginalization may serve to aggravate their pre-existing vulnerability by restricting access to social capital while further abating their psycho-social well-being. Therefore, it is necessary that relocation-related decision-making is backed by a comprehensive understanding about the social structure and cultural characteristics of both relocated and host communities, especially when relocating vulnerable communities.

(e) **Advocating Participation and Consultation**

As mentioned, earlier, planned relocation itself carries various risks to which vulnerable groups who lack coping capacities are more susceptible. These risks can be minimized to a great extent through encouraging the participation of beneficiaries in the decision-making process. It is important that inputs from beneficiaries are taken and seriously considered in making decisions on key aspects such as selection of the relocation site, the design of the housing unit, the material, and labor for house construction. For example, in planned relocation executed under LEI and CDP, there was a substantial level of interaction between project authorities (including both project and government officials) and the beneficiaries throughout the relocation process. Elaborating on this, steps were taken by project authorities to provide adequate information about the relocation program to beneficiaries. For instance, the relocatees were advised and educated about the housing designs that can be constructed within funding allocated. This prevented the relocatees from having to expend personal funds on the construction of houses which would have otherwise pushed relocatees to further economic marginalization.

Apart from this, 99 and 97% of the relocated respondents stated that they had received comprehensive information about individual facilities in the new house and the common facilities in the new settlement, respectively. Most importantly, the relocatees of the LEI and CDP had a greater degree of agency and choice where they were able to actively participate in the relocation process. This is signified by the fact that nearly 70% of the respondents had decided to relocate by themselves in their preferred locations while only 28% of them were requested to relocate by project officials in given locations. Similarly, beneficiaries were not only provided with the opportunity to construct their houses to suit their needs and preferences but also vested with the authority to oversee and select the personnel and material required for construction. Data indicate that underserved settlers who were relocated under LEI and CDP successfully adapted to the new environment in less than five years following relocation.

As shown in the case of LEI and CEP described above, emulating a participatory approach to planned relocation is not only conducive for the successful adaptation of vulnerable groups to the resettled environment but also necessary given the limited coping capacities vested with such groups.

11.4 Conclusion

In this paper, an attempt has been made to present a set of policy guidelines that are geared toward the effective design and implementation of planned relocation targeting vulnerable groups. Planned relocation has often been utilized as a DRR strategy in Sri Lanka thus, marking its validity in national progress toward achieving certain Sustainable Development Goals (SDGs) such as goal 13: take urgent action to combat climate change and its impacts and goal 11: make cities and human settlements inclusive, safe, resilient, and sustainable. Nevertheless, most planned relocation efforts have aggravated the vulnerability of vulnerable groups as such efforts have predominantly focused on settling the communities in safer areas to reduce exposure to hazard events but, failed to consider and account for the limitations in the capacities of certain population groups to successfully counter the adverse impacts of hazard events and planned relocation that has followed. In some cases, failure to do so has produced counter-effects where people have eventually returned to their original settlements which in turn has re-exposed them to the risk of hazards. Evidently, failure to apply an overarching understanding of vulnerability, i.e., a view that acknowledges the social, political, economic, and institutional determinants of vulnerability, to the design and implementation of planned relocation programs has rendered planned relocation as a DRR strategy counter-effective.

Given the recent as well as projected increase in extreme weather events in Sri Lanka and elsewhere in the world, planned relocation would be an inescapable choice as far as communities at risk are concerned. However, this paper portrays that it is necessary for planned relocation efforts to not only seek to shift vulnerable groups from their hazard-prone physical locations but also attempt to eliminate the social, political, economic, and institutional roots of their vulnerability. The policy guidelines that have been proposed in this paper have sought to address this necessity. The integration of planned relocation programs with development projects such as poverty reduction programs, creation of an institutional framework that is conducive to reducing vulnerability (e.g., extending social security schemes for vulnerable groups that do not currently benefit from social protection), and mobilizing community-based organizations for continuous monitoring of relocation outcomes are certain practical measures that can be undertaken in line with the proposed policy guidelines.

This chapter emphasizes on the need to account for the socio-economic and political dimensions of disaster risk by presenting key policy guidelines to be followed in relocating vulnerable households as a DRR strategy. However, further research is needed to explore concerns to be taken into consideration when relocating population groups with specific vulnerabilities (e.g., women, children, the aged, etc.). Further, this chapter provides the impetus to delve into (1) how far vulnerability and long-term resilience-building aspects have been incorporated into DRR governance mechanisms in Sri Lanka; (2) synergies between DRR, sustainable development, and vulnerability reduction; and (3) the utilization of an asset-based/strength-based approach to vulnerability reduction.

References

Arnall A (2019) Resettlement as climate change adaptation: what can be learned from state-led relocation in rural Africa and Asia? Climate Dev 11(3):253–263. https://doi.org/10.1080/17565529.2018.1442799

Artur L, Hilhorst D (2014) Floods, resettlement and land access and use in the lower Zambezi, Mozambique. Land Use Plan 36:361–368

Asia Disaster Preparedness Centre, United Nations office for Disaster Risk Reduction (2019) Disaster risk reduction in Sri Lanka: status report 2019. https://www.unisdr.org/files/68230_10srilankadrmstatusreport.pdf. Accessed 20 Jan 2021

Asia Disaster Reduction Centre (2019) Visiting researchers program—FY 2019B: Safer communities and sustainable development in Sri Lanka. https://www.adrc.asia/countryreport/LKA/2019/Srilanka_CR2019B.pdf. Accessed 15 June 2021

Bergmann J (2021) Planned relocation in Peru: advancing from well-meant legislation to good practice. J Environ Stud Sci. https://doi.org/10.1007/s13412-021-00699-w

Binder SB, Barile JP, Baker CK, Kulp B (2019) Home buyouts and household recovery: neighbourhood differences three years after Hurricane Sandy. Environ Hazards 18(2):127–145. https://doi.org/10.1080/17477891.2018.1511404

Birkmann J (2013) Measuring vulnerability to promote disaster-resilient societies and to enhance adaptation: discussion of conceptual frameworks and definitions. In: Birkmann J (ed) Measuring vulnerability to natural hazards. United Nations University, Tokyo, pp 9–79

Black R, Arnell N, Adger N, Thomas D, Geddes A (2013) Migration, immobility and displacement outcomes following extreme events. Environ Sci Policy 27(1):S32–S43. https://doi.org/10.1016/j.envsci.2012.09.001

Bohle H (2001) Vulnerability and criticality: perspectives from social geography. IHDP update 2/2001, Newsletter of the International Human Dimensions Programme on Global Environmental Change, pp 1–7

Centre for Excellence in Disaster Management and Humanitarian Assistance (2021) Sri Lanka disaster management reference handbook. https://reliefweb.int/report/sri-lanka/disaster-management-reference-handbook-sri-lanka-march-2021. Accessed 17 June 2021

Centre for Research on the Epidemiology of Disasters and UNDRRa (2020) The human cost of disasters—an overview of the last 20 years 2000–2019. https://reliefweb.int/report/world/human-cost-disasters-overview-last-20-years-2000-2019. Accessed 16 June 2021

Centre for Research on the Epidemiology of Disasters and UNDRRb (2020) 2020: the non-COVID year in disasters—global trends and perspectives. https://reliefweb.int/report/world/2020-non-covid-year-disasters-global-trends-and-perspectives. Accessed 16 June 2021

Cernea M (2002) Impoverishment risks and reconstruction: a model for population displacement and resettlement. In: Cernea M, McDowell C (eds) Risks and reconstruction experiences of resettlers and refugees. The World Bank, pp 11–55

Chambers R (1989) Editorial introduction: vulnerability, coping and policy. IDS Bull 20(2):1–7

Chandrasekara S, Prasanna V, Kwon H (2017) Monitoring water resources over the kotmale reservoir in Sri Lanka using ENSO phases. Adv Meteorol 1–9. https://doi.org/10.1155/2017/4025964

Chen Y, He L, Zhou D (2020) Consequences of post-disaster policies and relocation approaches: two communities from rural China. Disas Preven Manage Int J 30(3):340–352

Ciurean R, Schröter D, Glade T (2013) Conceptual frameworks of vulnerability assessments for natural disasters reduction. In: Tiefenbacher J (ed) Approaches to disaster management: examining the implications of hazards, emergencies and disasters. Intech Open 2768

Consortium of Humanitarian Agencies (2016) Impacts of disasters in Sri Lanka, report of the consortium of Humanitarian Agencies (CHA), Colombo, Sri Lanka

Correa E (ed) (2011) Preventive resettlement of populations at risk of disaster: experiences from Latin America. World Bank and GFDRR, Washington, D.C., The United States of America, pp 26–115

Department of Meteorology (2016) Climate of Sri Lanka. http://www.meteo.gov.lk/index.php?option=com_content&view=article&id=94&Itemid=310&lang=en&lang=en. Accessed 15 June 2021

de Sherbinin A, Castro M, Cernea M, Fearnside P, Krieger G, Lahmani S, Oliver-Smith A, Pankhurst A (2011) Preparing for resettlement associated with climate change. Science 334(6055):456–457

Disaster Management Centre, UNDP (2009) Sri Lanka National Report on Disaster Risk, Poverty and Human Development Relationship. https://www.preventionweb.net/english/hyogo/gar/background-papers/documents/Chap3/Asia-overview/Sri-Lanka-DRAFT-march-09.pdf. Accessed 20 Jan 2021

Disaster Management Centre (2016) Impact of disasters in Sri Lanka. The Consortium of Humanitarian Agencies (CHA), Colombo, Sri Lanka, pp 6–27

Dissanayake P, Hettiarachchi S, Siriwardana C (2018) Increase in disaster risk due to inefficient environmental management, land use policies and relocation policies. Case studies from Sri Lanka. Procedia Eng 212:1326–1333

Fernando N, Punchihewa A (2013) Relocating the displaced: strategies for sustainable relocation. Friedrich Ebert Stiftung, Bonn, pp 24–72

Fernando N (2016) Bridging the gap: people centered policy guidelines to minimize relocation failures: case of Sri Lanka. Sri Lanka J Adv Soc Stud 5 & 6(1& 2):41–53

Fernando N (2018a) Lessons learnt from long term impact of 2004 tsunami relocation: a case study of selected relocation settlements in Akmeemana Divisional Secretary Division in Galle District, Sri Lanka. Procedia Eng 212:47–54

Fernando N (2018b) Voluntary or involuntary relocation of underserved settlers in the city of Colombo as a flood risk reduction strategy: a case study of three relocation projects. Procedia Eng 212:1026–1033

Gramilova M (2014) Revisiting planned relocation as a climate change adaptation strategy: the added value of a human rights-based approach. Utrecht Law Rev 10(1):76–95. https://doi.org/10.18352/ulr.258

Gunawardena DLK, Dayawansa NDK (2020) Assessment of water availability for agriculture and other uses—development of indices for DeduruOya River Basin in Sri Lanka. In: Haque A, Chowdhury A (eds) Water, flood management and water security under a changing climate. Springer, Cham, pp 187–200. https://doi.org/10.1007/978-3-030-47786-8_13

Hino M, Field CB, Mach KJ (2017) Managed retreat as a response to natural hazard risk. Nat Clim Change 7:364–370. https://doi.org/10.1038/nclimate3252

International Organization for Migration (2019) International migration law: glossary on migration. International Organization for Migration, Geneva, pp 155–160

Intergovernmental Panel on Climate Change (2012) Managing the risks of extreme events and disasters to advance climate change adaptation. A special report of working groups I and II of the intergovernmental panel on climate change. In: Field CB, Barros V, Stocker TF, Qin D, Dokken DJ, Ebi KL, Mastrandrea MD, Mach KJ, Plattner G-K, Allen SK, Tignor M, Midgley PM (eds). University Press, Cambridge, pp 41–70

Krishnamurthy P (2012) Diaster-induced migration: assessing the impact of extreme weather events on livelihoods. Environ Hazards 11(2):96–111

Manesha S, Vimukthini S, Premalal K (2015) Develop drought monitoring in Sri Lanka using standard precipitation index (SPI). Sri Lanka J Meteor 1:64–71

McMichael C, Powell T (2021) Planned relocation and health: a case study from Fiji. Int J Environ Res Public Health 18(8):4355. https://doi.org/10.3390/ijerph18084355

McMichael C, Katonivualiku M, Powell T (2019) Planned relocation and everyday agency in low-lying coastal villages in Fiji. Geogr J 185(3):325–337

Ministry of Disaster Management (2014) Sri Lanka comprehensive disaster management programme 2014–2018. http://www.disastermin.gov.lk/web/images/pdf/slcdmp%20english.pdf. Accessed 15 June 2021

Ministry of Mahaweli Development and Environment (2016) National adaptation plan for climate change impacts in Sri Lanka. Climate Change Secretariat, Ministry of Mahaweli Development and Environment, Battaramulla, Sri Lanka, pp 26–45

Ministry of National Policies and Economic Affairs, Ministry of Disaster Management (2017) Sri Lanka rapid post-disaster needs assessment: floods and landslides. Ministry of Disaster Management & Ministry of National Policy and Economic Affairs, Colombo, pp 20–70

Nianthi R (2011) Drought risk reduction in the dry zone of Sri Lanka. In: Shaw R, Nguyen H (eds) Droughts in Asian monsoon region (community, environment and disaster risk management, vol 8. Emerald Group Publishing Limited, Bingley, pp 97–120

Norris L, McNamara K, Fernando N, Westoby R, Abeywardhana D, Vijekumara A, Prasanna J (2019) Resettlement stories of loss and barriers to recovery following the 2016 Aranayaka landslide. In: Sri Lanka, proceedings of the 10th annual research symposium—2019 equitable resilience, Colombo, Sri Lanka, 17–18 December

Perry R (2018) Defining disaster: an evolving concept. In: Rodríguez H, Donner W, Trainor J (eds) Handbook of disaster research-handbooks of sociology and social research. Springer, Cham, pp 3–22

Piggott-McKellar A, Pearson J, McNamara K, Nunn P (2020) A livelihood analysis of resettlement outcomes: lessons for climate-induced relocations. R Swedish Acad Sci 49(1):1474–1489

Simons M, Minson S, Sladen A, Ortega F, Jiang J, Owen S, Meng L, Ampuero J, Wei S, Chu R, Helmberger D, Kanamori H, Hetland E, Moore A, Webb F (2011) The 2011 Magnitude 9.0 Tohoku-Oki Earthquake: Mosaicking the Megathrust from seconds to centuries. Science 332(6036):1421–1425

Thaler T, Seebauerb S, Schindeleggerc A (2020) Patience, persistence and pre-signals: policy dynamics of planned relocation in Austria. Glob Environ Chang 63:102122. https://doi.org/10.1016/j.gloenvcha.2020.102122

Turner B, Kasperson R, Matson P, McCarthy J, Corell R, Christensen L, Eckley N, Kasperson J, Luers A, Martello M, Polsky C, Pulsipher A, Schiler A (2003) A framework for vulnerability analysis in sustainability science. Proc Natl Acad Sci 100(14):8074–8079

United Nations High Commissioner for Refugees (2014) Planned relocation, disasters and climate change: consolidating good practices and preparing for the future. UNHCR, Sanremo, Italy, pp 8–17

United Nations Office for Disaster Risk Reduction (2004) Living with risk: a global review of disaster reduction initiatives. UN Publications, Geneva, pp 28–85

United Nations office for Disaster Risk Reduction (2015) Sendai framework for disaster risk reduction 2015–2030. United Nations Office for Disaster Risk Reduction, Geneva, pp 9–20

United Nations office for Disaster Risk Reduction (2019) Global assessment report on disaster risk reduction. United Nations office for Disaster Risk Reduction, Geneva, pp 220–292

United Nations Office for Disaster Risk Reduction (2020) Disaster risk reduction. https://www.undrr.org/terminology/disaster-risk-reduction. Accessed 20 Jan 2021

UNDRR (2022) Online glossary. https://www.undrr.org/terminology/vulnerability. Accessed: 20 Feb 2022

United States Agency for International Development (2012) Haiti—earthquake and cholera fact sheet #8, Fiscal Year (FY) 201. https://reliefweb.int/sites/reliefweb.int/files/resources/06.27.11%20-%20USAID-OFDA%20Haiti%20Earthquake%20and%20Cholera%20Fact%20Sheet%20%238.pdf. Accessed 20 Jan 2021

Wang Y, Zou Z, Li J (2015) Influencing factors of households disadvantaged in postearthquake life recovery: a case study of the Wenchuan Earthquake in China. Nat Hazards 75(2):1853–1869

Warner K, Afifi T (2014) Where the rain falls: evidence from 8 countries on how vulnerable households use migration to manage the risk of rainfall variability and food insecurity. Climate Dev 6(1):1–17

Wilmsen B (2018) Is land-based resettlement still appropriate for rural people in China? A longitudinal study of displacement at the Three Gorges Dam. Dev Chang 49(1):170–198

World Bank (2019) Sri Lanka strengthens its climate resilience. https://reliefweb.int/report/sri-lanka/sri-lanka-strengthens-its-climate-resilience. Accessed 20 Jan 2021

World Bank (2012) Turning Sri Lanka's urban vision into policy action. World Bank: Colombo

Worldometer (2021) Sri Lanka population. https://www.worldometers.info/world-population/sri-lanka-population/. Accessed 20 Jan 2021

Zulauf A (2013) The impact of landlessness on rural livelihoods after post-Mitch resettlement in Nicaragua. In: Barenstein JD, Leemann E (eds) Post-disaster reconstruction and change: communities' perspectives. CRC Press, Boca Raton, pp 259–280

Chapter 12
How are Tamil Villages Reconstructed? Ethnography of Place-Making in Post-war Reconstruction in Sri Lanka

Pathmanesan Sanmugeswaran

Abstract This paper illustrates how community rebuilding is occurring in a gravely damaged, post-conflict society. Specifically, people in two villages in Tamil, Hindu, Jaffna, Sri Lanka, are using their "sense of place" and "place-making practices" or what I call here their "village-temple consciousness" or "village consciousness," to maintain and rebuild their communities after war to make them, once again, places in which they feel a comfortable sense of belonging. Thus, Tamil Hindus use the concepts of "village" and "village-temple consciousness," as *models of* or *for*, to reconstruct post-war communities on the Jaffna Peninsula, using a phenomenological ethnographic approach. Further, this study found that the memory of or nostalgia for village consciousness as it existed in the pre-war past is being used in the reconstruction of communities. An ethnography of the role of "place-making" in community rebuilding in Jaffna illuminates how communities after war must reconstitute themselves not just as physical locations but as places in which people can once again truly feel at home. An existential analysis shows *ūr* (village) nostalgia engages with everyday life and the reconstruction process in post-war Jaffna villages must be seen through spatial, sensory, temporal, discursive, and moral dimensions.

Keywords Village-temple consciousness · Post-war reconstruction · Place-making · Phenomenological ethnography · Belonging · Nostalgia

12.1 Introduction

This paper discusses how Tamil Hindus use the concepts of "village" and "village-temple consciousness," as *models of* or *for*, to reconstruct post-war communities on the Jaffna Peninsula, using a phenomenological ethnographic approach. In other words, I argue that "village" or "village-temple consciousness" is enacted as forms of life and practice in post-war recovery in Sri Lanka and. In particular, this paper illustrates and compares how people in two Tamil Hindu villages in Jaffna, Sri Lanka,

---

P. Sanmugeswaran (✉)
The Open University of Sri Lanka, Colombo, Sri Lanka
e-mail: spath@ou.ac.lk

© The Author(s), under exclusive license to Springer Nature Switzerland AG 2023
M. Hamza et al. (eds.), *Rebuilding Communities After Displacement*,
https://doi.org/10.1007/978-3-031-21414-1_12

are using village consciousness or village-temple consciousness to deal with post-war circumstances. "Village-temple consciousness" consists of the culturally organized and placed-based memories and sets of beliefs, practices, and culturally constructed emotions regarding their places of origin. An ethnography of the role of "place-making" in community rebuilding in Jaffna illuminates that communities after war must reconstitute themselves not just as physical locations but as places in which people can once again truly feel at home.

The overarching argument/research question of this paper is that in post-war Jaffna Tamil Hindu communities how people reconstructed their villages, their *ūr* (village), in the aftermath of a prolonged civil war by using their nostalgia for, and memories of, pre-war *ūr*. In other words, how the memory of or nostalgia for *ūr* (village) consciousness as it existed in the recent pre-war past is being used in the reconstruction of communities affected or destroyed by the war," and then arguing that this is part of a larger theoretical claim that in any post-conflict society where communities have been damaged or completely destroyed, an important part of what must be reconstructed is the "sense of belonging and place" that existed before the crisis. In addition, this paper also concentrates on the sub-research questions of how do people relate their village-temple consciousness to temple-centered sacred landscapes, and how are such landscapes being used/reconstructed in postwar circumstances?

The post-war reconstruction of these empirical physical properties needs to be understood in light of more anthropological investigations about nostalgia and about peoples' anxiety about social control (*kaṭṭupāṭu*), and how diaspora and transnational networks have influenced temples, villages, and places in all these regards. Understanding the reconstructions of villages, temples, and places, temple disputes, animal sacrifice, and temple entry issues was a very important objective of this study, because some of the issues of *ākama1* and non-*ākama* divides, pollution and purity complexes, restructuring caste hierarchy and caste geography, the restructuration of the priestly hierarchy, vegetarianism versus non-vegetarianism, and the emergence of Tamil Saiva Hindu nationalism in Jaffna are all part of, and re/emerged in, the reconstruction of villages, temples, and places after the war. This is a comparative study because Inuvil and Naguleswaram were affected differently by the Sri Lankan civil war. That is, while Inuvil was physically damaged and socially disrupted by the conflict, which ended in 2009, it nonetheless remained intact throughout; Naguleswaram, on the other hand, was completely raised to the ground by government forces, and its scattered inhabitants are only now being allowed to return after twenty-six years to rebuild their homes and Hindu temples.

12.2 Methodology

This paper uses data drawn from my Ph.D. dissertation research project fieldwork that was an intensive multi-sited ethnographic study of the post-war Tamil villages of Inuvil and Naguleswaram in Jaffna Peninsula, Sri Lanka. When it comes to ethnographic research practice, ethnographic methods may include participant observation, life history, case study, interviewing, and other participatory research methods. Thus, this study conducted 110 in-depth interviews with peoples in those villages. These include interviews with temples officials of the four key temples in the two villages, as well as 50 in-depth interviews in Inuvil and 60 in-depth interviews in Naguleswaram. Of these 110 interviews, 88 interviews were recorded; the rest were written up in my fieldnotes as those informants did not wish to be recorded.

12.3 Research Field Sites

This ethnographic study was undertaken in two villages of Inuvil and Naguleswaram in Valikāmam, one of four named regions of the Jaffna Peninsula, Sri Lanka. Geographically, the Jaffna Peninsula (*Yāzhppānam*) is part of Sri Lanka's Northern Province. Jaffna's population is largely Tamil Hindu, although there are also a significant number of Tamil Christians and Protestants (Holmes 1997 [1980]). After the war, the current Jaffna population is 614,541 of whom 605,224 are Tamil, 373 are Sinhalese, and 8, 944 are Muslims, a number greatly reduced by the war during which most Muslims were forcibly exiled by the Liberation Tigers of Tamil Eelam (LTTE) in 1990 (Jaffna Secretariat 2014; Thiranagama 2011).

Both Inuvil and Naguleswaram are called "ancient" villages (*pazhaiya urkal)* because they appear in Jaffna Tamil origin myths and are sites of religious pilgrimage (Pushparatnam 2002; Rasanayagam 1999 [1926]). Inuvil is located in the southern Valikāmam region and is divided into four village divisions populated by 5486 persons (Jaffna District Secretariat 2015). Inuvil, though affected by the war in many ways, was never completely depopulated or destroyed as Nagulewswaram was. Inuvil is occupied by multiple castes, has ongoing temple disputes (Whitaker 1999; Appadurai and Breckenridge 1976), temples of both high and non-high-caste people, is a site of much self-conscious artistic and religious activity, and is exhibiting new religious practices, as well as active in absorbing interactions with the Tamil diaspora and increased cyber-connectedness (Boswatte 2012; Whitaker 2006, 2004). For instance, members of the Swiss Tamil diaspora recently built a new lord Shiva temple in Inuvil, and other diaspora people have funded a public library, a computer center, various temple renovations, and other village infrastructure developments after the war (Fieldwork 2016, 2017–2018). However, Inuvil's temples were not physically affected by the war, even as many temples in HSZ villages (such as Neguleswaram) were destroyed and had their temple idols stolen.

Naguleswaram is in the former HSZ in northern Valikamam, on the northern seacoast of the Jaffna Peninsula and is now being rebuilt. Naguleswaram includes four sub villages: Karukampanai, Keerimalai, and the New and Old Colonies. This village now has only 989 returnees (Jaffna District Secretariat 2015) but likely had a population in size much like that of Inuvil before the war. Both villages are inhabited by people of multiple castes and were dominated by Vellālar prior to the war. But in Naguleswaram Kōviyar caste people have become the numerical majority since the war. Contrariwise, Vellālar caste people are still demographically and politically dominant in Inuvil.

Gunasinghe (1985) has pointed out that market-oriented farming made many Jaffna villages prosperous before the war. This was true of both Inuvil and Naguleswaram, though both villages also participated in industry. Prior to the war, people in Inuvil cultivated vegetables, tobacco, and other crops and worked at white color jobs and other businesses; they continue to do so now. But Inuvil's locally famous cigar and tobacco industry was disrupted during the war, and many Inuvil people have moved to Colombo or migrated abroad. A similar pre-war economic prosperity was found in Naguleswaram. In that village, the growing of onions, beetle, and grapes, fishing, and a cement factory were sources of prosperity prior to the war. But Naguleswaram and other villages in the HSZ were destroyed by the war, and their people were exiled to other parts of Jaffna and Sri Lanka for 26 years.

During this exile, people from Nagulswaram were barred from entering the area to worship at their Naguleswaram temple (Sivathamby 2005). This temple, destroyed and occupied by the Sri Lanka army for 26 years, was recently rebuilt and now hosts public worship. Indeed, several temples in Naguleswaram have been rebuilt and now boast newly built pilgrim guest houses (Fieldwork 2016, 2017–2018). Further, as military restrictions were removed, people begun to return to rebuild their homes, and, hypothetically, reinvent their sacred landscape. But these efforts have raised new tensions about land tenure (Fonseka and Raheem 2011). That is, since physical markers of individual and communal boundaries that defined ownership in the old village are gone, securing control over previously owned land has sparked struggle between people who remember the pre-war village differently (Fonseka and Raheem 2011).

12.4 What are "Village and Village-Temple Consciousness?"

By "village-temple consciousness," then, I mean the kind of consciousness of themselves that Jaffna Tamils have through their attachments to places of origin as mediated through their enacted affiliations, memories, embodiments, and "placing" (Kingsolver 2011, p. 13), as well as their sense of rightful ownership (*urimai*—see Appadurai and Breckenridge 1976, 1988; Whitaker 1999) of local temples, new and old gods, (Whitaker and Sanmugeswaran 2015), local landscapes, and each other (Kingsolver 2011, pp. 13–16).

Above all, this paper focuses on the questions of what it means to have village or village-temple conscious for Tamil Hindus themselves in Jaffna, not only when they are worshiping their *kulateiyvam*2, participating in village-temple festivals, or involved in village-temple disputes, but even when they are in small spaces like their houses (*vīṭu*) re/constructing their daily lives, or when they are rebuilding their houses, dealing with consequences of war and internal displacement, and interacting with family members in diaspora. These questions cannot be answered by asking informants either in Inuvil or Naguleswaram "what does it mean to be "village" or village-temple consciousness"? As previously explained, there is no specific Tamil term for "village" or "village-temple consciousness" in Jaffna Peninsula. However, other insights and other expressions collected by interacting with people as they work, worship, and talk enabled me to learn how people re/create spaces for their being at home: temples, villages, states, and in diaspora. It implies that we learned abstract concepts like race and sex in daily life similar to the way Tamils learn "village" or "village-temple consciousness."

Another way of putting this would be to say that "village-temple consciousness" consists of the sets of beliefs, practices and culturally constructed emotions (Clark-Decès 2005; Lutz 1986; Lutz and White 1986; Ring 2006) and memories regarding their places of origin that Jaffna people have available in their everyday lives. However, the consciousness of *ūr* is not exactly the same as having an identity, for often the anthropological conception of identity is discussed in terms of ethnicity, class, caste, sex or gender (Eller 1999; Giddens 1993), and while village consciousness involves these identities too, it more involves how people ground themselves (and their identities) and their understandings of the world in a complexly imagined locale.

12.5 Problems and Observations Encountered in Post-war Reconstruction

In the wake of that reconnection, there were many studies of Jaffna as a post-war community looking at post-conflict reconciliation, social harmony, peace building, resilience, mental health, and suffering (Gunaratne 2013; Keerawella 2013; Somasundaram and Sivayokan 2013; Nakagawa 2012; Phillips and Goldberg 2011; Hogg 2011; Goodhand and Walton 2009; Cochrane et al. 2009; Mehta 2010). There has also been much written about Sri Lankan Tamils on trauma, dislocation, and reconstruction efforts (Somasundaram 2014, 1998; Hogg 2011; Mehta 2010). However, this work has focused on state-level, collective, reconciliation, social harmony, and peace building for the Tamil community as a whole, or on the resilience, mental health, and suffering of Tamil individuals.

Other literature on post-war Sri Lanka has looked at issues of infrastructure development or economic livelihood (Jayatilaka et al. 2015; Gerharz 2014). None of the above, however, has looked at the recovery of villages as such, or at how temples and related practices of place-making might be playing a role in their postwar recovery. For example, Gerharz (2009, 2014) has written about the role of remittances and the "Western" practices of diaspora Tamils in Jaffna's recovery. She notes that Jaffna people sometime accept or reject these diaspora practices and money, and that this sometime creates conflicts between generations which cut across caste, gender and class lines. She does not note, however, that this flowing is mostly directed by diaspora people to their respective *ūr*. Hence, this study has looked at how such *ūr* directed diaspora flows interact with in situ village notions of social order (*kaṭṭupāṭu*) in Inuvil and Naguleswaram.

Ultimately, the literature on post-conflict recovery in Sri Lanka has not addressed the issue of how people recover their sense of feeling "at home" in a place shattered by war, or how a remembered sense of belonging might be used by them in the process of community recovery. Recovery literature, both in Sri Lanka and generally, mostly pays attention to physical, economic, and psychological recovery or, more widely, to "reconciliation" between the warring parties at the level of the state (as in the "truth commissions" in South Africa and so forth) (Hucklesby and Travis 2002; Anguelovski 2014; Barakat 2010; Wassel 2009; Truth and Reconciliation Commission 1999). Clearly, all of this is important. Neglected in all this, however, is a sociology of how people in shattered communities remake them so that they might once again feel at "home" in them; and, further, how people might be using a remembered sense or practice of belonging as a model of and for community recovery in radically altered postwar circumstances—such as the newly globalized circumstances that have arisen in Jaffna since the war.

Moreover, most studies of reconstruction have not looked at how Jaffna Tamils (or people in other post-conflict societies) determinedly maintain and utilize their localized consciousness during and after war. Hence, more useful to this research than the literature on war and trauma are concepts of "place-making," and "sense of place" (Kingsolver 2011; Muehlebach 2001; Feld and Basso 1996) drawn from anthropologists working on the cultural construction of belonging. These notions relate directly to the imagined place of belonging that I call "village" or "village-temple consciousness." Finally, it is also clear that because earlier studies dealt with the Tamil community from a whole-Jaffna-Peninsula rather than from an *ūravan* (village person) perspective, the role of village consciousness in this historical process was neglected in favor of issues of Tamil nationalism and ethnic conflict. This postwar negligence of village study in the anthropology of Sri Lanka is another gap in the literature on South Asia and Sri Lanka this study has contributed to fill.

12.6 Place-Making Through "Village" and "Village-Temple Consciousness"

"Village" or "village-temple consciousness" is enacted as forms of life and practice in post-war recovery in Sri Lanka. As already said, many studies of post-conflict societies have concentrated on infrastructure development, the reconciliation process, trauma, peacebuilding, resilience, and many other attributes, which are all the focus of State, NGO, and INGO projects and plans for post-war reconstruction. But none of these plans, so far as I am aware, tend to address the need for village or community reconstruction from the point of view of locally cultivated cultural knowledge about what "home" or "being at home" means for the people who live, or lived, in such places.

By local cultural knowledge, I mean that the knowledge Jaffna Tamilians use to construct their *ūr;* that is, the specific cultural knowledge and practices. I found that Tamils in Jaffna in 2017–18 paid great attention to such specific forms of cultural knowledge and practice when they were reconstructing their villages and renovating their temples, houses, and other physical dwellings. I also observed in other pre-war inhabited villages recently released from the status of high security zone (HSZ), that returning village residents and *ūr* members in diaspora both first turned their attention to renovating and reconstructing their destroyed Hindu temples. These temples were very important to their post-war recovery process because they anchored in place, for these people, their free-floating memories and socio-spatial imaginings, and hence became starting points of village-temple consciousness for both people recently returned and those still displaced. In this way, the post-war recovery became an ongoing, local (albeit, given the diaspora, sometimes globally local) cultural process wherein community identity construction occupied a significant position.

Here, identity construction did not only involve the Tamil ethnic identity construction but also village or *ūr* identity construction, hence the focus of people on specific temples. For instance, temples provide villages identity (or, perhaps, identities). Temples are, of course, places for worship, but for Tamilians, they are also more than places of worship. There are also places where people and groups (castes, priests, Saivite nationalists, and so forth) may show their *kouvravam* (prestige/honor) their publicly acknowledged authority and power. This is hardly a new role for temples. Temples played a similar role from at least the eighth century AD onwards in South Indian Tamil history where Tamil kings tended to build larger temples than palaces. As many scholars have noted, this was because temples were how South Indian Tamil kings showed their God-sanctified power, sovereignty, and *kouvaravam* (Dirks 1993; Stein 1980).

Hence, in some Jaffna villages, people reconstructed their settlements or houses in parallel with renovating or reconstructing their temples. In other villages, people renovated their temples first, sometimes even before moving back to their original inhabitation. Yet next to the reconstruction of temples, the reconstructing or renovating of *vīṭu* (of house compounds as homes) was an equally important cultural project for post-war recovery. Here, however, *vīṭu* also occupies a significant position because people's post-war efforts to rebuild them reveals something important about the danger of ignoring local village consciousness when thinking about post-conflict rebuilding, let alone when constructing—as states and INGOs tend to do, concrete plans for how this should be done.

Now in Jaffna just as Hindu temples were rebuilt according to the prescriptions of a specific *ākama* scripture, so Tamil Hindus or Saivites there generally wanted their lost houses rebuilt using cultural and astrological knowledge derived from the *maṉaiyaṭi cāstiram* (*shastra*) or *vāstu castiram* (an ancient Hindu scientific scripture about architecture) (Kumar 2002; Arya 2000). Here, the Tamil term, *maṉai* and the Sanskrit term, *vāstu* mean house. The activity of choosing a site for building a house is locally known as *nilaiyam pārttl3* in Jaffna; and people usually consult an astrologer to find out where a house should be located on any particular parcel of land. But the point, here, is that for Jaffna Tamil people the project of rebuilding either temples or houses involved a complex mixing of memory and text-guided orthodox practice—a village-temple consciousness—that was, for the most part, completely ignored by officially sanctioned state, NGO or INGO projects of community recovery.

This was one reason why the INGO and state model house plans, which villagers accepting their aid were mandated to follow, could not compensate people, as they promised to do, for the real value of their former dwellings lost during the war. For those mandated model architectural plans ignored the *maṉaiyaṭi cāstiram,* the Hindu architectural text, which post-war Jaffna villagers, anxious to return to a remembered or imagined pre-war physical and moral order, needed to use as a guide to house design. Thus, while people in Naguleswaram still tried to have their pre-war houses rebuilt according to Hindu architectural rules, they nonetheless generally either had to accept model houses or, otherwise, rebuild within the compensation values and rules laid down by the state and the INGOs. Yet others used their personal savings along with the housing scheme grants to build their houses according to the Hindu architectural rules. And others who accepted model houses, soon renovated them according to advice from astrologers based on Hindu architectural rules. Some claimed that they had suffered numerous problems in terms of economy and health after moving into these new model houses.

For Tamil Hindus in Jaffna strongly believe in Hindu architectural rules and pay attention to them when building their houses, mindful not only for the betterment of just one generation, but for many generations to come. Some people, however, had to accept the model houses just as they were instead of making any changes to them as per their wishes because they did not have the money to do otherwise. Another important thing I found related to some families who had never owned a house before the war received one under this system. For them too, of course, their desire to have their houses built following Hindu architectural rules and building houses were ignored by state and INGO-driven housing projects and their specific rules for building houses. All in all, then, rebuilding a house—a *vīṭu*—in postwar Jaffna was never a simple matter of putting up a structure.

Furthermore, this paper emphasizes the importance of the Tamil term *iṭam* or place in relation to village-temple consciousness. Knowledge of one's own place (or *conta iṭam*) or place is an important part of identity in a Tamilians' life, and this place-based identity indicates that they belong to a particular physical locality or dwelling. This belongingness tied to a particular place implies the importance of land ownership and village membership. In this regard, temples and places are closely connected because temples become key sources of identity for *ūr,* dwelling places, and neighborhoods. There were many instances where when I asked people where they were from, they not only specified a dwelling place but also mentioned various nearby big and small temples as landmarks. Also, in Inuvil, temple youths (*kōvil iḷaiñarkaḷ*) or young temple devotees (*iḷam toṇṭarkaḷ*), including both married and unmarried men, when talking to each other, addressed their identity by invoking their different temple affiliations. They did this not just because they belonged to different temples but because, in this way, they could present a finely calibrated account of who they were. In a way, for them, you are your place; and places in this identity-defining sense always include, somewhere, identity and place anchoring temples.

Thus, one's lived memory of a place is connected to a plethora of collective elements and temples are one of them. In this aspect, Tamil peoples' belongingness to a place is also revealed through the temples and temple festivals of their places. However, this is not the only form of place-belongingness wherein people are embedded into their temple-related consciousness, for there are other forms of belongingness as well. That is, belongingness is also revealed when we look at Tamil peoples' lived places, and this is why this study used the second analytical category, "village consciousness," in concert rather than in equation with the term "village-temple consciousness," to explicate people's experiences as these are connected to different identity-anchoring elements other than temples. For example, people also experience their places through the stories they tell, or though their memories of their own experiences of war, violence, trees, food, schools, family, commodities, playgrounds, and so on. Some Marxists in Inuvil remember their village by focusing on their past activities as youth mobilizers and social service activists at village community centers rather than on the village-temple alone. Further, while people rarely say *ūr,* but they often mention the place they lived, because place is a more accurate indicator of one's own dwelling. Thus, place as a form of consciousness, as *iṭam* consciousness, is invested in any place where Tamils dwell.

Dwelling denotes a certain culturally and emotionally organized set of practices that is defined as "village consciousness." Heidegger's (Cighi 2008) notion of dwelling brings people closer than anything else where people make their place of mere existence into a home. Further, dwelling comprises of "a sense of continuity, community, and homeness" (Seamon 1984: 4) which, this paper argues, are precisely what the efforts of post-war Jaffna Tamils having been trying to achieve. That is the notion to examine how post-war Tamils in Sri Lanka portray their way of life to themselves in a phenomenological fashion. In phenomenological terms, then, this study has designed its theoretical argument to show people look and think in their own fashion and style rather than when being intentionally motivated to perceive their world of being. Thus, this paper has clearly shown that experiencing *ūr* in the post-war context of Jaffna cannot be analyzed in terms of a set of physical properties alone. Rather, physical properties like temples, villages, and places should be analyzed through the culturally organized memories and emotions of the Jaffna people. Moreover, these terms "village consciousness" and "village-temple consciousness" despite being separate consist of the sets of beliefs, cultural practices, culturally organized and place-based memories, and emotions regarding their places of origin that Jaffna people have available in their everyday lives.

Hence, Jaffna Tamils do not perceive *iṭam* just as a physical setting, but look beyond any soil, wood, and other commodities to their own primordial sentiments of attachment to the place where they have lived, remember experiences, created past memories, and told family stories (*kuṭumpa kataikaḷ*; what Suseela aunt shared with me), and to where new stories are being imported and circulating from the diaspora. So, I discovered how Jaffna Tamils construct the cultural meanings of places and people, and how local (and non-local) knowledge is interwoven with this place-making process in post-war Jaffna. Thus, were people strongly connected to their own places; connections which, I think, they later brought to a national level to fight for their lands and rights during the civil war in Sri Lanka. A person's natal/birthplace is very important in his/her lifetime, and they do not want to give that up even though they have been displaced within the region or have migrated to foreign countries.

12.7 "Village-Temple Consciousness" as a *Model of* or *Model for* Post-war Reconstruction

The argument for using "village" or "village-temple consciousness" as a *model of* or *model for* postwar reconstruction is part of a larger theoretical claim that in any post-conflict society, it is likely that destroyed or damaged communities undergoing reconstruction will be using suitably modified pre-war place-making practices (Kingsolver 2011; Feld and Basso 1996) as *models of* or *for* their efforts (Geertz 1973, p. 93). My findings about "village-temple consciousness" in two Jaffna villages, namely Inuvil and Naguleswaram, address a general anthropological problem: the reinvention of destroyed hometowns and associated senses of place in post-conflict societies. Furthermore, understanding the conjoined village-temple sense of place that underlies social life and its reconstruction in Jaffna requires recognizing the conjunction of the physical and the sacred implied by this compound term. The outcome of this paper has been to show that the postwar social and communal rebuilding ongoing in these two villages depends upon people using their village-temple consciousness as *models of* or *for* such endeavors. Also, this ethnography of the role of "place-making" in community rebuilding in Jaffna, Sri Lanka illuminates the more general issue of how communities anywhere after war reconstitute themselves not just as physical locations but as places in which people can once again truly feel at home.

In a larger context, the Tamil *ūr*-sacred landscape does not only denote a form of religious life or *bhakti* (devotional) style religious devotional practices, but also the Tamil *ūr* has within it ongoing forms of social life such as violence, war, family disputes, village disputes, temple festivals, domestic rituals (rite of passage), family stories, village folk tales, village aesthetic values, village culture (in the Tamil sense, i.e., *kalāccāram*), Tamil *kalāccāram*, Tamil nationalism, Tamil *Eelam* (*īḻam*) ideology, and Tamil politics. Indeed, Tamils constructed a form of consciousness in their lived locales or *ūrs* through their religious and social lives in the pre-war period and during the war in Jaffna. The Jaffna Tamil Hindus' embodied cultural practices were generated through their religious and social lives, which, when people were in exile, constituted nostalgia for their pre-war *ūr*. The unique importance of the knowledge and experience of Tamil Saivites, and their nostalgia for pre-war *ūr*, is vastly neglected in studies of post-war reconstruction in Tamil Sri Lanka. For people in Naguleswaram, *ūr* consciousness was driven from a mixture of collective memory and individual memory. The collective memory is composed of historical imagination represented in two famous religious tales or myths. There are terminological problems as to whether they are part histories or religious myths, or even whether those religious myths are part of creating the sacred landscape or constructing the history of the Jaffna kingdom and Tamilness through the framework of a larger religious awakening. However, caste-politics, ethnic-politics, and Eelam ideology were all fabricated within this larger historical narrative.

Likewise, people have embedded the above attributes in many stories. As a result, a place can be defined by many versions of stories; for instance, the Jaffna Peninsula was portrayed as a war zone for three decades. Further, in the past, the Jaffna Peninsula was symbolized as a "place," which was a part of Tamil Eelam and ethnically segregated as a "place" of Tamils' land. In addition, the Jaffna Peninsula was projected as a terrorist zone full of Tamil tigers, and then as a high security zone. Many villages in Jaffna have gone through a great social transformation brought on by the war, militarization, and the global market economy. In this aspect, the war's targeted destructions (through violence, militarization, and displacement) and the global economy's "creative destruction" (Zukin 1991) have transformed the *ūr* landscape enormously. When Escobar (2008) talked about *place* in his book, *Territories of Difference: place, movements, life, redes*, he looked at the Pacific region as a place. In a similar vein, I perceive the Jaffna Peninsula as part of the larger South Asian region and Indian Ocean, which has re/constructed geo-political and geo-religious spaces in Sri Lanka. In addition, along with the historical processes of political movements (the Sri Lanka government's militarization, the LTTE's freedom struggle, civil war, and postwar violence), historical processes of capital flows and accumulation have greatly transformed the post-war Jaffna Peninsula.

By *ūr* landscape, I did not only mean the usual imagination in the geographical sense of one's "physical surroundings," but I also meant an assemblage of material and cultural practices and their symbolic representations (Zukin 1991). However, in a broader sense, the study has covered caste, class and gender in my analysis, which included expressions of the vernacular of the powerless in the representing the place. Rather than *meta descriptions* of places, this paper has addressed *local representations* of places, temples, and villages. The great social transformation that has occurred since the war has socially and culturally reconstructed spaces in post-war Jaffna. I would like to place the "post-war Jaffna Peninsula" in a global context to rethink the transformation of the region. For place-based identity, history, memory, culture, and narratives have been too long left out of investigations of post-war reconstruction in the Jaffna Peninsula.

12.8 Analysis

All of the ethnographic cases reveal the social, economic, religious, cultural, and historical importance of *conta iṭam* (own place) of "our place," in post-war reconstruction. The stories of post-war returnees reveal that multiple sufferings and challenges were present throughout their war and displacement journeys. For example, many individual stories described pre-war remembrances and post-war resettlement challenges. In Naguleswaram, the daily survival of people was not only filled with livelihood challenges but also with other encounter in post-war Jaffna. Upgrading one's own, or one's groups, *kouvravam* (prestige/honor) through foreign remittances, for example, was a new phenomenon in post-war Jaffna. Likewise, many of the ethnographic case studies in this study described the multiple challenges facing peoples' lives, but also the strong belief that many people had that, as one women put it to me, her *kulateiyvam* (family or caste or clan deity) protected her children from the war and its violence, and that her *kulateiyvam*'s grace had supported their efforts to reconstruct life in their *conta iṭam.*

The reconstruction of *conta iṭam* also revealed other challenges. For instance, the reemergence of caste as a factor during the reconstruction of the village and scared landscapes of Naguleswaram. At some point, diaspora and transnational connections also changed the post-war reconstruction process and post-war Tamil geography. Hence, with regard to my interest in the role of village consciousness in the reconstruction of post-war Naguleswaram, the notion or practice of *conta iṭam* clearly occupies a significant position discoverable through the stories and remembrances people told me. In them, *conta iṭam* was expressed through emotional attachments to the land, houses, wells, temples, schools, farmland, festivals, kinship, neighborhoods and all else found or remembered there. Additionally, *conta iṭam* for people indicated a territory or place which important to their identity construction and sense of belongingness. Moreover, *conta iṭam* denotes a "localism" which is shaped within economic, environmental/ecological, political, cultural and moral dimensions. Furthermore, "our place" remained important to culture and identity construction and even though a "delocalization of social life" had taken place due to displacement, diaspora, migration, and movement. That is, as Escobar (2008) puts it, "embodiment and emplacement" still could not be denied. For example, embodiment and emplacement experiences of animal sacrifice, at least as remembered, remained interconnected for people with day-to-day life in post-war Naguleswaram. Moreover, the geography of power relations is also very much relevant in postwar reconstruction. I also want to bring it to the fore in this paper, which was to show how people are being to reset power relations in post-war reconstruction.

Indeed, high-caste power was deeply established in most of the villages during the pre-war time, whereas low-caste people were not much established in villages during the pre-war time. Post-war community reconstructions and flows of opportunity have reconfigured the power relations among women, low-caste people and the younger generation in post-war villages. I found an ethnographic case that Velan (an 88-year-old man), one of my low-caste informants, states "to us all towns are our own, everyone our kin." His statement explicates how he perceives different villages as his own because displacement has also taught him lessons by living in multiple houses and multiple villages throughout his displacement journey. At the same time, I had to look at how a particular *ūr* has been more important to a person in terms of being powerful in pre-war time. That lived memory is very important to post-war community reconstruction. The changes of *kaṭṭupāṭu* clearly demonstrate how people of the younger generation are dealing with life in post-war villages. The *kaṭṭupāṭu* (control/power) also shapes and reconstructs *ūr* consciousness in post-war villages. Due to the war and long-term displacement, there has been a substantial attitudinal change among both older and younger people, which has effectively influenced the practice of *kaṭṭupāṭu* to change.

As far as I noticed in Jaffna villages, high-caste Veḷḷāḷar's power has not much changed. The notion of *matta āṭkaḷ* (other persons) is a category which means "other people" was invented by the high-caste Veḷḷāḷar to label the low-caste people in a derogatory way. The ability to make such categories and to place groups in them displays high-caste power and domination over Jaffna subjects (Foucault 1979). I was very eager to investigate caste in present day Jaffna because the presence of caste is identified "as a silenced public discourse" (Rao 2006, p. 10). In the global, national, and democratic contexts, we need to look at how the post-war Jaffna constructs and shapes the subject positions of people through their caste identity.

What I found was a kind of reemergence of caste or a reinvention of caste in post-war Jaffna. How is caste representation produced in post-war Jaffna? There, a shared narrative is circulated among the people that "caste related grievances have been vanished after the war in Jaffna," but I would argue that since the end of the war, remaking caste identity and a reformulation of high-caste domination have been rediscovered in different forms. Although a prolonged civil war, internal displacement, encampment, migration, and the Liberation Tigers of Tamil Eelam's (LTTE) past administration and eventual defeat have altered social life in the Jaffna Peninsula in many ways, caste consciousness still plays a pivotal role in creating and maintaining religious spaces and religious power under the leadership of the "dominant" castes. In fact, the Veḷḷāḷar caste has remained in a powerful position in Jaffna in terms of land ownership, temple ownership, and ownership of ritual authority. In other words, they continue to be the dominant caste there in Srinivas's sense. This is nothing new of course. The Veḷḷāḷar were already long advantaged by the strengthening of caste inequalities and the solidification of the caste system occasioned through colonial power and benefits (Pfaffenberger 1982, 1990).

Low-caste people are blocked from accessing many opportunities, but many of these issues and their grievances are either not properly communicated or not properly reported. For instance, Saraswathi, from the low-caste background, claimed that some low-caste people maintained a good relationship with the high-caste people for their personal reasons, and those who maintained such mutual relationships with high-caste people control their own community's issues without revealing their high-caste ties to the public. This reveals that different forms of power have emerged in present day Jaffna, and so we need to further explore the dynamics of caste-based relations and the silent form of domination that occurs within the same caste. Furthermore, as high-caste domination still exists in Jaffna, low-caste people do not have the freedom to express their opinion and are not allowed to participate in community activities. For instance, Kanthan, my informant, pointed out that in Veḷḷāḷar caste dominated temples, low-caste people are not allowed to carry the *vākanam* (vehicle of the deities) and cannot obtain the festival rights in Veḷḷāḷar temples. Even though there is a kind of mutual exchange services among different castes, this mutual contract is also a form of domination that leads to powerlessness. This contract-induced domination establishes caste supremacy, like Veḷḷāḷar supremacy in Jaffna, which enacts inequality and discrimination for the oppressed castes.

Anthropologists pointed out an ambiguity about power in human activities (Schultz and Lavenda 2014). Domination and hegemony can create ambiguity in terms of defining the power. Gramsci (2013) restructures the society by splitting it into two levels corresponding to two different functions: hegemony and direct domination. He suggests that everybody is an intellectual but not everybody plays this role. Gramsci explains domination by coercive rule (Schultz and Lavenda 2014); once Veḷḷāḷar's domination played out as a force or in a coercive manner to force low-caste people to accept Veḷḷāḷar domination as legitimate practices in Jaffna in early days (1950s and 1960s).

Though Veḷḷāḷar no longer project their domination in a coercive form, now they use schools, cultural institutions like temples, community centers, youth organizations, and other institutions in villages to disseminate Veḷḷāḷar ideology among low-caste people in Jaffna. Hence, Veḷḷāḷar's power has worked by establishing a preeminent position in Jaffna society while censoring the consciousness of low-caste people. In a Gramscian sense (2013), the persuasion of low-caste people to accept Veḷḷāḷar's ideology while preserving Vellalar's privileged position is called *hegemony*. According to him, knowledge is hegemonic because it is imposed, and one has to accept it. In this, the Veḷḷāḷar domination-possessing bundle of ideas imposes *hegemony* that limits people's ability to think further. Hence, while low-caste people of course can think, they cannot operate their ideas or make them live because the ideology and *mantra* of the bourgeois Veḷḷāḷar have become *hegemonic* in Jaffna society.

The sense of *ūr*, however, was not central to this kind of Tamil nationalistic discourse. That is, while, in the wake of the war, such national forms of identity have become, temporarily anyway, difficult to support, the sense of *ūr* remains. It could be argued, then that the forms of village and temple-focused belonging described in this study now provides a refuge, and a "place" to be Tamil in, and thus forms of belonging alternative to ethnicity per se., at a point in time when other forms of Tamilness are temporarily too dangerous. It is possible that research on other post-conflict populations, elsewhere in the world, may also show them focusing for a time on changing their own practices of local belonging in similar ways, to handle the challenges posed by the national and the global forces confronting them—particularly in times of political weakness and danger.

However, I found such shifts in focus from the national to the local among post-conflict people, whether in Jaffna or elsewhere, are temporary. For I want to briefly point out something else about my anthropological discussion of "village consciousness," in relation to notions of a Tamil nation state and Tamil sovereignty in post-war Sri Lanka. The Tamil word *conta iṭam* (own/natal place) refers to a form of consciousness associated with Tamil *ūr;* but at the same time, *iṭam* refers to land (*nilam*) as well. The notion of a Tamil nation state has a long history in South Asia beginning in the *Sangam* period (300 BCE–200 CE) where Tamil sovereignty had its own philosophical foundations. For instance, during the *Sangam* period, a literary convention was developed in which the *Sangam* landscape was conventionally divided into five different geographical territories: *kuṟiñci* (mountain regions), *mullai* (forest region), *marutam* (agricultural or farming land), *neiytal* (seashore region), and *pālai* (desert areas) (Balambal 1998). These five kinds of land had, in turn, five different types of sovereign deities and kings associated with them. The notion that Tamil land refers to a separate territory, which further developed to be notion of a Tamil nation state with its own philosophy, is thus of long standing.

Ultimately, once resident Jaffna Tamils rallied to Dravidian identity and Tamil consciousness. But for many Tamils, in the wake of the war, these forms of belonging felt too overtly political and nationalist, and too focused on a national homeland, or Tamil Eelam, for pursuit in post-war public circumstances. Hence, the shift to a focus on *ūrness* (one's sense of village), for *ūrness* and the national state are two separate entities, and a sense of *ūr* was not central to Tamil separatist nationalism, and thus avoids posing a reminiscent challenge to the Sri Lankan state. Moreover, this shift provided practical ways for people to respond to the tremendous changes that have accompanied postwar life. But none of this suggests that Tamil nationalist feeling have gone away.

Furthermore, reconstruction and recovery have created so many other problems, like, the naming and renaming of places, shifting actual boundaries between villages, expanding and shrinking former villages, writing and rewriting the history of places, demolishing cultural monuments in some places, and many other activities controlled by the state, local politics, and also by the region's high-caste, Veḷḷāḷar elite. This kind of place transformation occurred in Jaffna history through the respective colonial regimes and the state control (after the independence from 1948). For example, in Jaffna as among the Apache, less powerful people have little voice in publicly defining habitation, place-making and naming (Basso 1996). In Jaffna, all of the place-making/cultural construction of the home place, and official displays of cultural symbols, are dominated by Veḷḷāḷar caste discourse.

However, as Kingsolver (2011) has pointed out, the literatures on consciousness and place-making have paid less attention to interpretive and linguistic analysis (Douglas 2002 [1966]; Geertz 1973; Wittgenstein 1974) than they should. Yet language and meaning are central to the interpretive methods required to make sense of them. For example, one could usefully say village consciousness consists of related "forms of life" (Wittgenstein 1974); in other words, various enacted practices of cultural belonging that show a certain "family resemblance" to one another. But, as Basso has said (1996), such similar practices will still differ from territory to territory because groups of men and women will have invested themselves in a place through their thoughts, values, and "collective sensibilities." In this way, people make up differences between "our territory" and "their territory." These divides are what make local consciousnesses distinct from one another for the people who construct them as living places. These distinctions, in turn, suggest a reason why village consciousness might provide some people with a kind of refuge from other, sometimes more dangerous, forms of allegiance and identity based on ethnicity and nationality even while, of course, potentially having their own dangers.

12.9 Conclusion

Post-war community reconstruction is not a new phenomenon around the world and is mostly implemented through development, government and non-government agencies; but these agencies' reconstruction projects are completely different from how Tamil people attempt of reconstructions based purely on their own imaginations of their Tamil *ūr*. People are returning back to their places not only for redefining village history and identity to show their involvement in *ūr* but also to reinstate Veḷḷāḷar domination, *kouvravam* (honor), and heroism in the village community reconstruction.

The rebuilding now going on in Jaffna Sri Lanka, especially in the village in the high security zone, offers an important opportunity to investigate how people there are using their "senses of community" and practices of "place-making" (in this case, their village-temple consciousness) in the task of community reconstruction. I argue that their struggles to do this using "village-temple consciousness" offer obvious theoretical insights into the more general problem of community reconstruction in post-conflict societies. Furthermore, I have found it useful to deploy the Jaffna practice of "belonging"—village-temple consciousness—to the consideration of post-conflict readjustment (in the case of Inuvil) and reconstruction (in the case of the former high security zone village, Naguleswaram). But reconstructing a community is a problem in all post-conflict societies that had wars as violent as Sri Lanka's was, so this research is now addressing a general theoretical problem rather than a specifically Sri Lankan or South Asian one.

In particular, villages like Naguleswaram, which are in the process of reconstruction, have a major task at hand. They must concern themselves with new ways of life in post-war circumstances, demographic and structural disruptions (in caste demography). Nostalgia itself is impossible because Naguleswaram had undergone dramatic changes since lived dwellings (houses) and places were destroyed. My informants tried to remember their places and life, but they could not remember because they have even lost their old photos of their places. People have strategically rebuilt their *ūr* through discovering *ūr* nostalgia and domesticating new flows of opportunities. Thus, Naguleswaram constructs community continuity through the discontinuity of forms of life and the invention of forms of life through new connections. That is, people, who were uprooted from Naguleswaram, were disconnected from their places and pre-war life. They had to adopt a new form of life through new connections and opportunities when they displaced to new villages.

By comparing the efforts of people in these two villages—one (Inuvil village) where people are using an ongoing but mutating sense of place to adjust to changed circumstances, and another other (Naguleswaram) where people, of necessity, are using memory to reconstruct a place once lost—this research is attempting to understand the role Jaffna Tamil people's "village-temple consciousness" is playing in the process of local social healing and community reconstruction that is ongoing in this rapidly changing post-war society. It does so by comparing how people in these two villages are using their senses of place and practices of place-making—their "village-temple consciousness"—to do so. In the case of Tamil Sri Lanka, the relevant practices of place and belonging happens to be contained within the notion of "village-temple consciousness" and "village consciousness" since, there, it happens to be the case that when people experience belongingness and place, they do so in terms of villages as home places anchored by temples as ritual and social centers. As a Tamil from Jaffna, I have experienced and found that the recent removal of restrictions on people's movements and access to communication technologies, and the reopening of the A9 highway link between Jaffna and the rest of Sri Lanka, has been increasingly shaping and is reshaping the sense of *ūr* that infuses everyday life in Jaffna. My findings determine, thus, exactly how people in both villages are altering their practice and experience of village consciousness in the face of war and a post-war surge of globalization.

I think we must understand that place-making is always a socially and culturally significant activity. That is, while it varies from culture to culture, "place-making is a universal tool of the historical imagination" (Basso 1996, p. 5). Thus, "place-making involves multiple acts of remembering and imagining which inform each other in complex ways" (Ibid). For instance, in Jaffna Tamil culture, the belongingness or owning of a place is expressed through the term *conta iṭam* (our place). Owning land or a place does not only show one's identity or ownership (*urimai*) or belongingness, but also displays one's *kouvravam,* one's publicly acknowledged status. That is why, Jaffna Tamils always demand a house or a piece of land when they talk about dowry in marriage proposals. Also, I found many ethnographic cases during my fieldwork, and even before fieldwork, in which Sri Lankan Tamil people in diaspora were investing, and are still investing, their money into buying houses or lands in Jaffna. Though it is mostly the Jaffna Veḷḷāḷar mostly who do this, my informants of other castes confirmed that non-Veḷḷāḷar other Jaffna people in diaspora do this as well.

However, people have different perceptions of place; therefore, may have more than one version of experience toward a place. Further, place-making is not just the construction of a place, but also the construction of its history and the reinvention of a place. For example, in Naguleswaram and Keerimalai, I observed young children learning about the historical background and mythological narratives associated with Keerimalai and Naguleswaram and saw them transfer this knowledge to others who visited their villages. In Inuvil, when I was a child, I did the same. But the current generation of both villages were also aware of those stories even though their families were uprooted from their original places for more than two decades. Hence, to grasp the meaning of a place, it is also important to see how local knowledge is interwoven with the place-making process.

References

Anguelovski I (2014) Neighborhood as refuge: community reconstruction, place remaking, and environmental justice in the city. The MIT Press, Cambridge
Appadurai A, Breckenrdge CA (1976) The South Indian Temple: authority, honor and redistribution. Contrib Ind Sociol 10(2):187–211
Appadurai A (1988) Putting hierarchy in its place. Cultural Anthropology: J Soc Cult Anthropol 3(1):36–49.
Arya R (2000) Vaastu: the Indian art of placement: design and decorate homes to reflect eternal spiritual principles. Destiny Books, Rochester
Balambal V (1998) Studies in the history of the Sangam age. Kalinga Publications, Delhi
Barakat S (2010) After the conflict: reconstruction and development in the aftermath of war. I.B. Tauris, London
Basso KH (1996) Wisdom sits in places: landscape and language among the Western Apache. University of New Mexico Press, Albuquerque
Boswatte A (2012) Effects of Migration on Jaffna Tamil families: a case study assessing the effects of migration on a sample of tamil families living in the Jaffna Peninsula in Sri Lanka. UCL Migration Research Unit Working Papers, No. 2012/2, London: Migration
Cighi CI (2008) Senses of place. Master Thesis, University of Massachusetts, Amherst
Clark-Decès I (2005) No one cries for the dead: Tamil dirges, rowdy songs, and graveyard petitions. University of California Press, Berkeley
Cochrane F, Baser B, Swain A (2009) Home thoughts from abroad: Diasporas and peace-building in Northern Ireland and Sri Lanka. Studies in Conflict and Terrorism 32(8):681–704. Research Unit, University College of London.
Dirks NB (1993) The hollow crown: ethnohistory of an Indian Kingdom. University of Michigan Press, Ann Arbor [Mich]
Douglas M (2002) Purity and danger: an analysis of concepts of pollution and taboo. Routledge, London
Eller JD (1999) From culture to ethnicity to conflict: an anthropological perspective on international ethnic conflict. University of Michigan Press, Ann Arbor
Escobar A (2008) Territories of difference: place, movements, life, redes. Duke University Press, Durham
Feld S, Basso KH (1996) Senses of place. School of American Research Press, New Mexico
Fonseka B, Raheem M (2011) Land in the Northern Province: post-war politics. Policy and practices. Center for Policy Analysis, Colombo
Foucault M (1979) Discipline and punish: the birth of the prison. Vintage Books, New York
Geertz C (1973) Interpretation of cultures. Basic Books, New York
Giddens A (1993) Modernity and self-identity: self and society in the late modern age. Polity Press, Cambridge
Gerharz E (2009) Re-constructing Jaffna: transnational tamil activism at local interfaces. In: Gellner DN (ed) Ethnic activism and civil society in South Asia. SAGE Publications, Los Angeles
Gerharz E (2014) The politics of reconstruction and development in Sri Lanka: transnational commitments to social change. Routledge, New York
Goodhand J, Walton O (2009) The limits of liberal peacebuilding: international engagement in the Sri Lankan peace process. J Interven State Build 3(3):p303-323
Gramsci (2013) Intellectuals and hegemony. In: Charles L (ed) Social theory: the multicultural and classic readings. Westview Press, Philadelphia, pp 202–203
Gunasinghe N (1985) Peasant agrarian systems and structure transformations in Sri Lanka. In: Abeysekere C (ed) Capital and peasant production studies in the continuity and discontinuity of agrarian structure in Sri Lanka. Social Scientists Association, Colombo
Hogg CL, Royal Institute of International Affairs (2011) Sri Lanka: prospects for reform and reconciliation. Chatham House, London
Holmes WR (1997 [1980]) Jaffna, Sri Lanka, 1980. Atkins Duane International, Inc., Costa Mesa

Hucklesby S, Travis J (2002) Reconstructing communities in the context of large refugee populations: impact assessment. Policy Dev Res Needs Transform 19(2):97–107
Jaffna District Secretariat (2014) Population report. Jaffna District Secretariat, Jaffna
Jaffna District Secretariat (2015) Jaffna district handbook. Jaffna District Secretariat, Jaffna
Jayatilaka D, Amirthalingam K, Gunasekara S (2015) Conflict, displacement, and post-war recovery: a community profile of Passaiyoor East in Jaffna, Sri Lanka, working paper no-07. International Center for Ethnic Studies, Colombo
Keerawella G (2013) Post-war Sri Lanka: is peace a hostage of the military victory? Dilemmas of reconciliation, ethnic cohesion, and peace-building. International Centre for Ethnic Studies, Colombo
Kingsolver AE (2011) Tobacco town futures: global encounters in Rural Kentucky. Waveland Press Inc., Long Grove
Kumar V (2002) Vastushastra. R.R. Sheth, Mumbai
Lutz C (1986) Emotion, thought, and estrangement: emotion as a cultural category. Cult Anthropol 1(3):287–309
Lutz C, White GM (1986) The anthropology of emotions. Annu Rev Anthropol 15(1):405–436
Mehta A (2010) Sri Lanka's ethnic conflict: how Eelam War IV was won. Centre for Land Warfare Studies, New Delhi
Muehlebach A (2001) Making place at the United Nations: indigenous cultural politics at the U.N. Working Group on indigenous populations. Cult Anthropol 16(3):415–448
Nakagawa T (2012) The post-conflict reconciliation process. Truth Commissions. CUNY Academic Works [online]. Available at https://www.academicworks.cuny.edu/cc_etds_theses/98
Gunaratne PR (2013) Post-war reconstruction and peace building in Sri Lanka: a case study of Jaffna. Institute for Peace Science, Hiroshima University
Phillips LD, Goldberg BD (2011) Curriculum fostering social harmony among university students in Sri Lanka. The Ministry of Higher Education the Democratic Socialist Republic of Sri Lanka
Pfaffenberger B (1982) Caste in Tamil culture: the religious foundations of Sudra domination in Tamil Sri Lanka. Vikas Publishing House, New Delhi
Pfaffenberger B (1990) The political construction of defensive nationalism: the 1968 temple-entry crisis in Northern Sri Lanka. J Asian Stud 49(1):78–96
Pusparaṭṇam P (2002) Ancient coins of Sri Lankan tamil rulers. Bavani Patippakam, Jaffna
Rasanayagam MC (1999) Ancient Jaffna: being a research into the history of Jaffna from very early times to the Portuguese period. Asian Educational Services, New Delhi
Rao A (2006) Gender and caste. Women Unlimited, New Delhi
Ring LA (2006) Zenana: everyday peace in a Karachi apartment building. Indiana University Press, Bloomington
Seamon D (1984) Heidegger's notion of Dwelling and one concrete interpretation as indicated by Hassan Fathy's architecture for the poor. Geosci Man 24:43–53
Schultz EA, Lavenda RH (2014) Cultural anthropology: a perspective on the human condition. Oxford University Press, New York
Sivathamby K (2005) Being a Tamil and Sri Lankan. Aivakam, Colombo
Sivayokan S, Somasundaram D (2013) Rebuilding community resilience in a post-war context: developing insight and recommendations—a qualitative study in Northern Sri Lanka. Int J Ment Heal Syst 7(1):p1-25
Somasundaram D (2014) Scarred Communities: Psychosocial Impact of Man-made and Natural Disasters on Sri Lankan society.
Stein B (1980) Peasant state and society in Medieval South India. Oxford University Press, Delhi
Thiranagama S (2011) In my mother's house: civil war in Sri Lanka. University of Pennsylvania Press, Philadelphia
Truth and Reconciliation Commission (1999) Truth and Reconciliation Commission of South Africa Report. Macmillan, London
Wassel T (2009) Protecting housing rights for IDPs in Sri Lanka, forced migration review. Refugee Study Center, Oxford Department of International Development, University of Oxford

Whitaker PM (1999) Amiable incoherence: manipulating histories and modernities in a Batticaloa Hindu Temple. VU University Press, Amsterdam
Whitaker MP (2004) Tamilnet.com: some reflections on popular anthropology, nationalism, and the internet. Anthropol Q 77(3):469–498
Whitaker PM (2006) Internet counter counter-insurgency: TamilNet.com and ethnic conflict in Sri Lanka. In: Landzelius K (ed) Native on the Net: indigenous and diasporic peoples in the virtual age. Routledge, London, pp 255–271
Whitaker PM, Sanmugeswaran P (2015) Searching for Sakti: new gods, site and choices in postwar Tamil Sri Lanka. In: A research paper presented at the 44th annual conference on South Asia, The University of Wisconsin, Madison, 22–25 Oct 2015
Wittgenstein L (1974) Tractatus logico-Philosophicus. Routledge and Kegan Paul, London
Zukin S (1991) Landscapes of power: from Detroit to Disney World. University of California Press, Berkeley, California

Part III
Governance of Displacement

Chapter 13
Reproducing Vulnerabilities Through Forced Displacement: A Case Study of Flood Victims in Galle District, Sri Lanka

H. Unnathi S. Samaraweera

Abstract Resilience has become a buzzword in most disaster governance documents where its operationalization and evaluation raise questions in practical contexts (Cannon, in Why vulnerability still matters: the politics of disaster risk creation, pp 68–88, 2022). This chapter aims to explore whose, what and where "disaster resilience" has been achieved using a case study of a family who have been forcefully displaced through local government policy because their land is vulnerable to flooding. The chapter is developed from a larger fieldwork project which studied disaster resilience among two flood prone communities in western and southern Sri Lanka after the 2016 and 2017 floods. During fieldwork, the researcher only found one family which was forcefully relocated so their situation was further studied as a single-case study. In-depth interviews, direct observation and photographs were used as data collection methods and techniques in the study as a whole. Interview information was analysed using the thematic analysis method. Because of the extreme vulnerability to flooding of this specific family, they were provided with new land to rebuild on 16 km away in a nearby village, but they still live in a rented house in the same flood prone neighbourhood. They did not move because of a number of constraints that were not considered when the land was allocated. These include proximity to children's schools, lack of informal work opportunities, unavailability of funds to rebuild a new house on the land they were given, and loss of social support networks if they relocate. Further, if relocation is the best solution in the face of increasing flood risk, the question arises as to why the entire neighbourhood has not been offered assistance to relocate. Thus, this chapter argues that disaster governance mechanisms aiming for disaster resilience in this context have become a contributing factor to production and reproduction of a number of new vulnerabilities and failures. Solutions have overlooked whose disaster resilience was to be achieved, by what process and in which location.

H. U. S. Samaraweera (✉)
University of Canterbury, Christchurch, New Zealand
e-mail: unnathisamaraweera@yahoo.com

University of Colombo, Colombo, Sri Lanka

© The Author(s), under exclusive license to Springer Nature Switzerland AG 2023

M. Hamza et al. (eds.), *Rebuilding Communities After Displacement*,
https://doi.org/10.1007/978-3-031-21414-1_13

Keywords Forced displacement · Vulnerabilities · Disaster resilience · Disaster governance · Flooding in Sri Lanka

13.1 Introduction

Today, the term disaster resilience plays a crucial role in most disaster governance policies and strategic mechanisms globally, including in Sri Lanka. However, its operationalization as well as evaluation in the disaster context raises certain questions. Although the concept has been widely applied, the outcomes have not always been sustainable. Strategic relocation has been recognized as a long-term solution for post-disaster reconstruction and disaster mitigation under the buzz word of building "disaster resilience". Every so often, resettlement/relocation programmes have been carried out with the objective of developing disaster resilience among climate change-induced disaster affected communities, but these often neglect how these initiatives may lead to creating further disaster vulnerabilities. Due to reasons such as lack of planning and design, partly administered and/or inadequate funding, Oliver-Smith (2009) indicates that resettlement/relocation programmes have largely become development disasters.

One should not forget "…that discourses on disaster are inherently a Western invention as they are mediated by concepts, methods, scientific narratives, norms and values that are those of the West. They contribute to depict, capture and make tangible the very diverse experiences of millions of people across locations and cultures under the supposed universality of the nature/hazard–culture/binary" (Gaillard 2022, p. 22). Similarly, involuntary displacements and resettlement of disaster affected people have been mainly necessitated by development-oriented programmes, and they indicate how policy should guide solutions to the matter (Cernea 1997, 2004). After the involuntary displacement and resettlement process takes place, livelihoods must be reconstructed (Asthana 1996), which produces and reproduces a number of risks and vulnerabilities. Even though terms such as involuntary resettlement and relocation are commonly used, the current study prefers the term forced displacement to emphasize how such involuntary resettlement/relocation processes carried out by the Sri Lankan government lead participants to experience these as involving force or threat of force.

There is a large disaster scholarship worldwide which has explored the risks, vulnerabilities, livelihoods and wellbeing of relocated communities (Christensen and Harild 2009; De Jesus 2018; Few et al. 2021; Muggah 2008). Still, the association between forced displacement, disaster governance and disaster resilience has been given relatively less research attention including in Sri Lanka. Thus, this chapter explores how the process of forced displacement is being incorporated as "good" disaster governance under the notion of "building disaster resilience".

It is evident that the South Asian region including Sri Lanka experiences a high probability for displacement due to various kinds of disasters including war, land sliding, flooding and tsunami (Ginetti and Lavell 2015). Planned relocations as post-disaster reconstruction initiatives were mainly based on two approaches in Sri Lanka: donor-driven and owner-driven approaches (Senanayake et al. 2022). Apart from involuntary/voluntary resettlements as a result of the three-decade long war and 2004 tsunami, the vast majority of the literature has been based on land slide-induced involuntary/voluntary resettlement. According to the Sri Lankan National Policy Framework, precautionary mechanisms should be in place, not only to protect lives and properties but also to minimize economic losses due to disasters (National Building Research Organization (NBRO) 2022).

Recent data shows landslide has caused the most damage to the built environment, costing lives and economic losses; thus, resettlement programmes are mainly focused on high risk landslide disaster-induced resettlement (NBRO 2022). The NBRO (2022) has identified 12,126 families living in landslide high risk areas of Sri Lanka and recommended their resettlement to safer areas in order to save their lives. So far, only about 4,700 families have been resettled in safe places under the "Home Owner-Driven" housing construction approach (NBRO 2022). Using involuntary relocation as a Disaster Risk Reduction (DRR) strategy, NBRO also attempts to resettle other disaster risk associated communities to safer lands (NBRO 2022). However, the progress of the implementation programmes has been slow due to a number of reasons such as difficulty in identifying suitable lands in safer areas, difficulties faced by the communities in house reconstruction, increasing cost of house reconstruction including the material and labour prices and inadequate practical knowledge of project administrators about the implementation process (NBRO 2022). NBRO's goal is to resettle approximately 10,000 families via NBRO approved disaster-resilient housing plans in the coming years.

In addition, with the goal of assisting the people displaced due to the process of land acquisition for development projects, the Government of Sri Lanka adopted the National Involuntary Resettlement Policy (NIRP) in 2001 (Ministry of Land and Land Development with assistance from the Asian Development Bank 2013). This policy has also adopted Cernea's (1997, 2004) risk patterns in identifying impacts of the resettlement process. Since the policy mainly focused on communities affected by donor funded development projects, there was no uniform policy to address concerns of the displaced communities due to the disasters and conflicts (Fernando 2018).

It has been noted in a number of studies that most planned relocation projects have aggravated vulnerabilities of relocated families and communities in Sri Lanka (Fernando and Punchihewa 2013; Fernando et al. 2020). Within the limited number of studies available on flood-induced resettlement programmes in Sri Lanka, Fernando (2018) focuses on long-term impacts of involuntary or voluntary relocation of three unauthorized, underserved flood prone communities in the city of Colombo during the 1990s. Internationally, governments and relevant post-disaster relief agencies often overlooked, overgeneralized and masked social vulnerability during the relocation processes (Taiban et al. 2020). Thus, this process has resulted in forced displacement for disaster affected relocated communities. Such resettlement efforts have largely failed due to disaster governance mechanisms' inability to incorporate the disaster affected population's actual needs while addressing their grassroots level vulnerabilities. Thus, such resettling/relocating efforts have resulted in failures reflecting the gaps in the overall DRR strategy in Sri Lanka.

With this backdrop, this chapter primarily aims to explore whose, what and where "disaster resilience" has been achieved specifically using a case study of a forcefully displaced, flood affected family in Galle district, Sri Lanka.[1] The chapter argues how certain disaster risks and vulnerabilities have been reproduced in the post-flood disaster context as a result of the forced displacement process experienced by a flood affected household within the flood prone community. The chapter first introduces the methodology followed by theoretical and conceptual orientations based on the structure and agency debate and linking to relevant concepts such as disaster governance, disaster resilience, forced displacement and vulnerabilities. Then the chapter introduces the specific case study in detail. The next section of the chapter explores how risks and vulnerabilities have further been reproduced through government resettlement/relocation initiatives (which in the eyes of the particular households, however, are tantamount to forced displacement) in the particular flood prone community and returns to the limitations within the overall Sri Lankan disaster resilience and disaster governance mechanisms in the concluding remarks. The chapter suggests how disaster governance, while intending to build disaster resilience, has been neglecting the agency of disaster affected people.

13.2 Methodology and Limitations

The research problem mainly focuses on identifying how disaster-related socio-economic vulnerabilities have been reproduced through forced displacement following the 2017 flood disaster in a flood prone community in Galle district, Sri Lanka. Even though the chapter is a part of a larger research project, which is focused on exploring disaster resilience among two flood affected communities, this chapter is based on single-case study method. During fieldwork, the researcher explored the only family forcefully relocated due to their flood and economic vulnerability in this specific flood prone geographical locality. However, they still lived in a rented house in the same flood prone neighbourhood during fieldwork in 2020 because they had been given new land but no new house. Thus, the only available relocated household was studied in-depth as a single-case study in order to understand how disaster-induced forced relocation decisions by the authorities are unable to address grassroots level requirements of local disaster affected people. In-depth interview method, direct observation and photographs were used as data collection methods and techniques. These provide the essential context to the case examined. The interviews were analysed using thematic analysis. Further, both primary data and

[1] In order to maintain confidentiality of the research respondents, the research field is only introduced as Galle district without providing the specific geographical locality including the names of the community and the flood prone river.

secondary data have been used when analysing and presenting data and information in this chapter. [2]

Single-case study can be valuable depending on the context of the study (Simons 2015; Yin 2018). It often provides justifiable platform to explore an extreme or unusual case (Yin 2018), as in this case of an only forcefully relocated family in the entire flood prone and flood affected community. Further, it has the potential of examining a phenomenon from various perspectives where the focus lies on the process (rather than the results), the context (rather than the precise differences) and the discovery (rather than the evidence) (Merriam 2019).

It is significant to note that single-case study provides particularization (not generalization) emphasizing a uniqueness of the case (Palmer et al. 2014). Thus, it provides important aspects of the forcefully relocated family while bringing in-depth new insights of the grassroots level realities. Since it is significant to explain the context and boundaries of the single case (Palmer et al. 2014; Simons 2015), the chapter extensively describes the single-case study in a forthcoming section first before moving on to reflections and discussion.

Though some scholars indicate there is a possibility of generalization of findings even using a single-case study method (Simons 2015), this study brings it more as a limitation of the research study. Hence, it indicates since the chapter is being based on single-case study there might not be a possibility of generalization of research findings to the larger community. In order to overcome the limitation, this chapter aims at exploring related new insights with deeper theoretical and conceptual analysis avoiding generalizations.

13.3 Theoretical and Conceptual Considerations

This section introduces the sociological considerations used for developing the book chapter followed by the conceptual framework.

13.3.1 Theoretical Underpinnings: Structure and Agency

Since the structure and agency debate in social sciences is of long duration, different academic viewpoints of its key concepts have been developed. On the one hand, structure simply means frequent forms of arrangements in a macro-level setting which affect and/or limit existing choices and opportunities available. On the other hand, agency means capacities and capabilities of individuals to perform autonomously while making their own free choices (Elder-Vass 2010). When considering the way in which society is being influential towards human agents, it is significant to note that structure and culture shape the social context where individuals operate. Yet,

[2] In order to maintain confidentiality, pseudonyms are used in the entire book chapter.

this overlooks individual capacity which can be argued as responses towards social conditioning. Thus, this dualism becomes an important part of the structure and agency debate (Archer 2003).

Scholars central to the debate such as Margaret Archer claim that people's actions are influenced by the social structure, which is analytically distinct from agency, yet in the eyes of some, she has neglected the study of how their actions have been guided and controlled through their other social characteristics (Elder-Vass 2010; McKeown et al. 2022; Thomas 2018). Elder-Vass's (2007a) argument is centred on the idea of complex causality which is understood through the unpacking of emergence as a more effective approach to what has been quite a dualistic position by Archer (2003). Elder-Vass (2007a) explores how to establish more nuanced understandings of structure and agency while applying the concept of emergence. Bringing emergence into the debate, he suggests a practical alternative to the dualism that Archer's (2003) argument is based on.

Understanding of emergence should be focused on internal relations between a structure's essential parts that makes it greater as a whole (Thomas 2018). For instance, one cannot totally understand forced relocation process merely studying casual tendencies of specific relocation structures. Instead, it should be extended to study highly complex socio-economic-political and cultural interactions of individual households which experience forced displacement and the nexus between various structural interventions including policy and government authorities and singular agency of grassroots level local people.

There are strong influential factors intertwined with the social structure which impact on singular agency in terms of making free choices and playing their roles in the larger society (Elder-Vass 2007b, 2010; Thomas 2018). Thus, social structures can be understood as influencing people's behaviours and outcomes as well as such practices and knowledge also influencing change within the social structure (Elder-Vass 2010). The argument suggests that individual agency is heavily affected by the larger social forces where there is a strong connection between the two (Elder-Vass 2008; McKeown et al. 2022). However, this does not necessarily mean that structure and agency should be recognized as binary concepts (Elder-Vass 2008). When human activity is seen as an outcome of the dual influence of structure and agency, there are a number of influencing factors including stratified social relations and level of singular agency (Bouchard 2020). Consequently, structure and agency are mediated through socio-cultural forces (Bouchard 2020).

In the disaster context, in terms of forced displacement, the government institutions, DRR policies and frameworks act as a structure manipulating the entire involuntary resettlement process claiming it to be a DRR strategy that builds disaster resilience among disaster affected communities. However, ground level realities including the forcefully displaced communities' narratives reveal how their individual capabilities and capacities as well as autonomous decision-making have been overlooked in the process of forced displacement globally as posited by Taiban et al. (2020). This has been the similar situation in the Sri Lankan context. This suggests there is a disconnect between the structure and agency which has been the theoretical underpinning for this chapter.

Disaster affected then forcefully displaced people's narratives suggest how their agency has been neglected via the role of the DRR policies, frameworks and the government decision-making process which are key to the social structures that shape their experience. However, one should not forget that both structure and agency are problematic concepts (Elder-Vass 2007b, 2010). Since some argue that problems occur due to the causes inherent in the structure, some other thinkers argue the impact is due to human reflexivity where agency exercises the power of decision-making (Elder-Vass 2010). "Social structures have causal powers in their own right, which arise from the combination of individuals and relations that constitute them, but which are different from the causal powers that would be possessed by these same individuals if they were not organized into these social structures" (Elder-Vass 2007a, p. 40).

Thinkers see this process as results due to the structure claiming that there are sets of characteristics that are developed through the social context which become social phenomena in terms of social forces or individual ones in relation to the structure (Elder-Vass 2010). Among such debates, this chapter mainly focuses on Elder Vass's (2010) argument on structure and agency debate in order to theoretically explore the operation of the agencies on the disaster affected family that has been studied.

Following the sociological theoretical underpinning of the structure and agency debate within the DRR context, the chapter is mainly based on four key concepts: disaster resilience, disaster governance, forced displacement and vulnerabilities. Thus, the next section briefly outlines the conceptual framework of the chapter.

13.3.2 Conceptual Framework

13.3.2.1 Disaster Resilience

"Resilience—A term that has roots in ancient thinking and was developed in mathematics and engineering—is slowly coming of age in today's arena of environmental policy-making" (Bodin and Wiman 2004, p. 33). The term resilience has been unpacked in various disciplines such as psychology, engineering, business studies, environmental studies and social sciences with numerous interconnected meanings and criteria (Bahadur et al. 2013). Within such diversity, the term disaster resilience is being commonly defined as an ability to return to the previous state following a disaster within a minimum possible time duration (Plough et al. 2013), while using the available individual and communal capacities and showcasing adaptation (Manyena 2006). Disaster resilience is often focused on returning the community to its previous status, and historically policy and programmes sometimes have overlooked the historical context, put the responsibility and blame of disaster consequences on disaster affected people and communities, and neglected larger disaster root causes and vulnerabilities inherent in the process (MacKinnon and Derickson 2013). In order to avoid these problems, it is important to establish good disaster governance, which has been recognized in Sri Lanka, but there have been a number of failings in the implementation of policy.

Hence, disaster resilience is linked with the disaster governance within contemporary neoliberal Sri Lankan context where adaptive disaster governance contributes to the building disaster resilience and achieving DRR.

13.3.2.2 Disaster Governance

Disaster governance, adaptive disaster governance and disaster risk governance are all similar concepts adopted in the literature in order to describe how disaster risks have been managed, and systematically categorized so they can be dealt with collectively while addressing multi-layers of meaning and action by/of the various establishments in a collaborative manner (Lane and Hesselman 2017).

"A commonly accepted view of the concept of collaborative governance is that it refers to "the processes and structures of public policy decision-making and management that engage people constructively across the boundaries of public agencies, levels of government, and/or the public, private and civic spheres in order to carry out a public purpose that could not otherwise be accomplished" (Emerson et al. 2012, Cited in Tierney 2012, p. 344). Considering that disaster governance is an evolving concept in the disaster scholarship, one should not forget there is a close connection between risk governance, environmental governance and disaster governance. Disaster governance is shaped by a number of global and local forces including global DRR trends, social inequality and sociodemographic characteristics of locals.

"… Disaster governance consists of the interrelated sets of norms, organizational and institutional actors, and practices (spanning pre-disaster, trans-disaster, and post-disaster periods) that are designed to reduce the impacts and losses associated with disasters arising from natural and technological agents and from intentional acts of terrorism" (Tierney 2012, p. 344). Tierney's definition suggests that disaster governance covers the whole system without limiting it to the organizational level and including community leadership patterns of membership, involvement of religious institutions and the role of Non-Governmental Organizations (NGOs) working at the grassroots level.

Further, disaster governance contains diverse characteristics without limiting it to collaboration, inclusive stakeholder contribution, transparency, clear flows of information and administrative and financial flexibility. A vast range of the literature connects disaster resilience and effective disaster governance. Since disaster governance is not limited to managing existing land-use patterns, developing mechanisms to measure and control disaster risks and supporting government institutions to hold responsibility, it should address the requirements of the affected communities while engaging them in planning and implementation processes for disaster recovery plans. This indicates there is a direct and indirect connection between disaster governance strategies and disaster resilience. Thus, this chapter primarily links how forced displacement (involuntary relocation) has been used as a DRR and disaster resilience building strategy in the Sri Lankan disaster governance process in order to critically evaluate how this process is experienced by the flood affected communities.

13.3.2.3 Forced Displacement

While examining empirical data on disasters, Cernea (1997, 2004) developed a theoretical model of displacement and reconstruction. The importance of this model is that it focuses on socio-economic conditions of both forced displacement and reestablishment processes which was not the focus of other models in the disaster-related relocation scholarship. The model introduces seven key areas of risks and disadvantages involved in the processes in displacement. They are: (a) landlessness; (b) joblessness; (c) homelessness; (d) marginalization: (e) food insecurity; (f) loss of access to common property resources; (g) increased morbidity; (h) community disarticulation. The model proposes that restructuring and refining the livelihood of displaced people needs risk reversals through clear tactics supported by acceptable funding. The model can be used not only as a diagnostic tool and a predictive tool but also as a problem-resolution tool and a research-guidance tool for disaster mitigation practitioners and researchers (Cernea 1997, 2004).

Following the forced resettlement of Egyptian Nubians during the 1960s, Scudder (2016) developed a four-stage theoretical framework to understand the dam-induced involuntary resettlement and its impact on resettled communities. The four-stage process for achieving relatively successful resettlement is "(Stage 1) planning for resettlement prior to physical removal, (Stage 2) physical removal and coping with the initial drop in living standards frequently following removal, (Stage 3) initiation of socio-economic development and community formation activities that are necessary to improve living standards of first generation resettlers and (Stage 4) handing over a viable resettlement process to the second generation of resettlers and to non-project authority institutions" (Scudder 2016, p. 6).

This chapter attempts to explore Cernea's (1997, 2004) and Scudder's (2016) conceptual models through the available case study to show how application of policy has resulted in a failure while reproducing further risks and vulnerabilities. However, the small number of available studies on the topic in Sri Lanka emphasizes how and why essential conceptualizations of forced relocation remain challenged due to three different existing influences: development, conflict and natural disasters which problematize the role of the disaster governance further (Muggah 2008). This government rhetoric often neglects how certain negative labels were constructed and positioned against resettling communities while overlooking the impacts on these communities (Muggah 2008). Such relocation/resettlement often culminates with separation of the community from their social support networks without addressing their grassroots level realities including associated risks and vulnerabilities (Muggah 2008).

13.3.2.4 Vulnerabilities

In early disaster scholarship, the term vulnerability is defined as: "the characteristics of a person or group and their situation that influence their capacity to anticipate, cope with, resist, and recover from the impact of a natural hazard (an extreme natural event or process). It involves a combination of factors that determine the degree to which someone's life, livelihood, property and other assets are put at risk by a discrete and identifiable event (or series or "cascade" of such events) in nature and in society" (Wisner et al. 2004, p. 11). This early focus has been later criticized by other disaster scholars claiming that it solely focuses on negative outcomes of vulnerable experiences by highlighting individual insufficiencies where the responsibility is simply laid upon individual victims of disasters without claiming the larger impact within the vulnerability discourse. Bankoff (2001) claims this early focus ignores possibilities for transformative change itself.

When considering vulnerability as a concept, it supports the need for assessing disasters within the socio-economic, political and environmental contexts of locals which was neglected earlier in the disaster scholarship (Dziaek et al. 2019; Hilhorst and Bankoff 2022). In addition, the concept has been used in formulation of DRR policies and relief activities. "Yet, despite the undoubted conceptual and methodological advances it represents on previous thinking, its utility and practical application is still hampered by a one-dimensional construction of the processes that transform a hazard into a disaster" (Bankoff 2001, p. 30). Further, Bankoff (2001) claims that the term vulnerability is not able to adequately analyse the connection between a social vulnerability and cultural adaptation, and it overlooks local knowledge, coping strategies and historical root causes.

Subsequently, the majority of scholars focused on theorizing how vulnerabilities are produced, by whom and with what consequences (McKeown et al. 2022). Further, it is important to note that " … vulnerability is not a static entity—a state—but rather an evolving set of conditions driven by a variety of forces, including the dynamic of the world system…" (Tierney 2019, p. 121). Thus, further substantive scholarly work on structure being influential to create and reproduce vulnerabilities is required (Dickinson and Burton 2022).

Disaster scholarship during the 1970s and 1980s discussed the concepts of vulnerability and disaster resilience mainly in relation to risk in society (Tierney 2019). Since the vulnerability paradigm was introduced within such a discourse, it disregarded the societal root causes of disasters in the face of vulnerability. Thus, the negative outcomes of the direct and indirect disaster-related vulnerable experiences are discussed as individual insufficiencies where the responsibility is also placed at individual disaster "victims". The development of early stage neoliberal policies in the 1980s in Sri Lanka further exacerbated the situation. Therefore, this chapter emphasizes the importance of going beyond a reductive conceptualization of vulnerability as weakness and deficiency. It brings out the possibilities for agency and transformative changes inherent within disaster affected communities by examining the strategy of an individual in the face of structural inflexibility.

13.4 A Case Study of the Forcefully Displaced Family

Vinod is a 44-year-old Tamil Hindu male. He lives in a flood prone community within the Galle district, close to a medium sized river. The community often experiences small-scale flash floods (at least annually), and it is acknowledged as one of the worst flood affected communities within the Southern Province after the 2017 high-scale flood. Vinod and his wife Rani who is 38 years old live with their four children aged 16, 11, 7 and 4. The eldest child is a daughter: Meena and the rest are sons: Suraj, Thiru and Guru. The three elder children are at school. A child less than school age is at home with Rani. Vinod has studied up to grade 5 and Rani has studied up to grade 8. Since their education qualifications do not guarantee any formal work opportunity in the area, Vinod is working as an unskilled labourer. He often does whatever labouring work he can get on daily wages and will take many kinds of work: help in house construction sites, cutting trees, digging drainage works, digging holes in farms and for neighbours, spraying fertilizer for tea plantations and other agricultural lands. Rani mainly is a housewife.

Until experiencing the 2017 flood disaster, the family lived in a house very close to the river. This house was entirely flooded and partially damaged in the 2017 flood which was exceptionally high. As happened in the entire flooded neighbourhood, their house was covered by flood water which was 10–12 feet higher than the flood line. The normal small-scale flood causes minor damages in the area and is one foot above the line. During the recovery phase after 2017, Vinod was informed that they cannot live in this flooded house again. The *Grama Niladhari* (GN) officer of the area informed him that the government has started a flood relocation programme where highly flood-vulnerable communities will be relocated to new lands. Upon the government's instructions, the GN informed Vinod that his family has been selected for the relocation programme due to the severe flood risk and vulnerability associated with the land on which their house is located. Even though the entire neighbourhood is facing the same flood risks and vulnerabilities, given their proximity to the river and due to their lower-economic status, Vinod was told that being selected to the relocation programme should be treated as an achievement and was asked to relocate according to the government relocation plan.

According to the government's relocation plan, Vinod was given a 12 perch area of land by the government. The new land is located nearly 16 km away from Vinod's original property. The GN officer had informed Vinod that the government will grant a specific amount to build a new house in the new land too. Vinod and family were told that if they continue to live in the original flood affected house and if they experience a future flood and any flood-related damages, no compensation will be provided by the government as they have already been given land to relocate. Yet, when the researcher first met Vinod and his family during the fieldwork conducted in 2020, they were living in a rented house in their original flood prone neighbourhood without relocating to the government provided land as per the government's instructions.

When the researcher asked why Vinod and his family had not relocated to the new place, Vinod revealed a number of reasons for staying. Vinod said: "the government only gave us a piece of land. We do not have money to rebuild a new house there. Even though the government promised to give money to build the house we have not got any money yet for the house construction" (In-depth interview: 24/02/2020).

Apart from not being in a financial position to build a new house in the given land, he revealed further concerns such as lack of informal work opportunities for him in the new location as nobody knows him, distance to children's school and fear of interruption to the children's education, unavailability of social support networks in new location, distance to all familiar day to day social institutions such as banks, market and hospital. Further, only local NGOs were active in the area and they only provided short term help without any long-term DRR preparations. The narrative of Vinod's family raises a number of concerns which problematize the disaster relocation process not only as a DRR strategy but also as a disaster governing practice intended to build disaster resilience among disaster affected communities. The problematic nature of disaster governance and building disaster resilience is further explored in the next section of the chapter.

13.5 How Does Forced Displacement Lead to Reproduction of Further Risks and Socio-economic Vulnerabilities?

The research findings suggest how the relocated family which was unable to choose the area they relocated to is facing further risks and socio-economic vulnerabilities. This section attempts to explore several important questions including, why do they continue to live in the same flood prone neighbourhood? and how the forced relocation process has become a contributing factor towards creating development disasters as Oliver-Smith (2009) argues.

Before the loss of access to their original property, the family lived in a house they owned. Their land included a private well giving them access to good quality drinking well water. A small tea plantation plot was attached to the land, which was mainly maintained by the unpaid labour of Vinod's wife Rani, giving them a secondary income source. Now, they are currently paying rent and paying for household water while their income has reduced without their tea plantation plot. Even though the government's aim was to reduce disaster risk and vulnerabilities, currently the family lives in the same flood prone neighbourhood with reduced income, extra cost and expanded disaster risks and vulnerabilities as they can no longer claim flood compensation in future floods. Their narrative suggests they experienced forced displacement not assistance, and although their neighbours were not forced off their land, they also received inadequate compensation for their losses and nobody has the time or skills to advocate for Vinod to the government. Other flood affected people living in the same neighbourhood shared their dissatisfaction on how Vinod and his family have been treated in relation to the whole relocation process.

This case study illustrates a number of gaps and limitations in the relocation process. The main problem is the aforementioned process of building disaster-resilient communities conceptualized by Scudder (2016) and Cernea (1997, 2004) was completely neglected. In addition, there is a lack of attention paid to the requirements of the beneficiary households showing that the process has been based on an entirely top-down approach neglecting the grassroots level realities. There was no prior consultation with the direct beneficiaries. This scenario is not limited to the South Asian region or to the Sri Lankan context. Numerous research studies on flood recovery schemes throughout Europe also indicate that not only were those programmes ineffective in building resilience (Priest et al. 2016; Slavíková 2018) but also the processes have increased vulnerabilities among the disaster affected communities (Sandink et al. 2016). There is inadequate empirical evaluation of the present financial flood recovery schemes and recovery mitigation outcomes in many European contexts (Suykens et al. 2016). As yet, there is no agreed way of achieving an efficient and successful recovery process (Slavíková et al. 2020). This brings home the argument that neglecting flood affected people's requirements in various disaster phases is common worldwide, where the legislation, administration of programmes, provision of funding and the implementation process maintained by disaster governance continue to dominate decision-making while overlooking the grassroots realities as is the case of Vinod's family facing displacement today.

The existing Sri Lankan legal framework for relocation does not consider the consequences of the post-land acquisition context. Further, there is no procedural mechanism to handle complaints related to land acquisition compensations, other disaster-related compensations such as damages of the built environment, destroyed crops and economic losses (Senanayake 2019). These reasons further extend Vinod and his family's vulnerabilities as they do not have a specific legal framework or an institution to go and complain to.

Further, relocation is a process which includes various physical-infrastructural, psychological and social impacts (Cernea 1997, 2004). In the same vein, Vinod explained during one of his interviews, coming from an ethnic and religious minority how it would be difficult for them to relocate to the given land, as their relatives, friends and all close social support networks have been extremely supportive not only throughout the flood disaster phases but also in their everyday lives. His account highlights social trust and cohesion between host community and resettling community as well as social capital being important yet often neglected social dimensions throughout the disaster relocation programmes. Disaster governance mechanisms should not ignore that the process of relocating a population not only involves loss of land, housing and livelihood but it also impedes people's survival and social support networks including all social, economic, political and cultural needs. The problem for the rest of the community in Galle was that nothing was offered to them where they did not get agency acknowledged either in the relocation process. Further, this case study reveals how such a holistic approach is absent in the entire involuntary relocation process that has been experienced by the flood affected community as forced displacement. It simply overlooks the various socio-economic-cultural impacts behind the process. It further suggests that the government at the structural level made the decision neglecting the agency of the disaster affected people. In addition, it indicates how the government DRR strategy does not incorporate evaluation programmes of the post-disaster contexts including the subsequent implementations in the post-relocation context.

In addition, this study suggests a broader vulnerability assessment is necessary which goes beyond the safe geographical locality, some land, and a sum of money to rebuild the house, and it should focus on larger socio-economic-cultural-political conditions and factors such as ethnicity, gender, income variations and employability facilities. The available studies on the topic suggest that there is a void when it comes to evaluating the implications of the implementation process in relation to resettlement programmes in Sri Lanka (Senanayake 2019). In particular, families who have been asked to resettle as in the above case study may have a number of practical difficulties in resettling and adapting in the new locality. Practical realities of the resettled families should be properly addressed in such government relocation programmes without further reproducing vulnerabilities.

Implementation of relocation programmes needs to incorporate practical mechanisms for addressing risks and vulnerabilities associated with disaster affected peoples' lives with their input. It further indicates Sri Lankan disaster governance mechanisms lack a practical lens towards disaster relocation from the beginning and before implementations, thus not attempting to address disaster resettlement-induced risks and vulnerabilities at the grassroots level properly. Other socio-economic aspects such as supporting economic stability and livelihood restoration, employability opportunities and encouraging secondary income generation mechanisms are also significant areas to be addressed during the process. These areas have been clearly neglected further reproducing vulnerabilities. In addition, there is a serious concern on nationally standardized temporary and permanent housing constructions covering overall infrastructure facilities such as accessible roads, proper drainage systems, proper garbage disposal plans, sustainable and affordable electricity and water supply plans towards disaster affected people which were certainly missing in most relocation programmes in Sri Lanka.

The government's disaster governance plans should extensively cover the practical needs of disaster affected people without limiting it to a mere programme which attempts to shift the blame and responsibility towards disaster affected people as happened with the current case study. The implications of it were the government was doing very little and the indirect outcome was to shift blame for future losses onto the individual. The DRR policies should clearly address the needs of disaster affected people, while including their capacities and capabilities in the process which should not overlook the agency of the disaster affected people. Implementing and monitoring DRR strategies well in the practical context, addressing grievances of disaster affected people should also be a part of disaster resilience building.

13.6 Questions and Concerns Raised in Forced Displacement

Certainly, disaster affected people who have been asked to relocate by the government should be involved in the disaster-induced relocation process from the planning to the post-relocation context as this act entirely changes their lives (Cernea 1997, 2004; Oliver-Smith 2009; Scudder 2016). Yet, what is indicated in Sri Lankan disaster governance is that the government and donor perspectives have been dominant and the voices and experiences of the disaster affected people are overlooked. Further, the government—through both statements and actions being the structure—is undermining the agency of ordinary flood affected people by expecting flood affected people not to challenge the disaster governance process.

The disaster governance perspective of involuntary relocation as a DRR strategy to support "vulnerable" disaster prone communities often highlights disaster affected people as weak, helpless and always a group of people who need and are looking for support. Thus, this perspective disregards the capabilities, capacities and individual agency of the disaster affected people. The vulnerability paradigm today should go beyond such reductive understandings and treating disaster affected individuals as passive receivers and passive observers. Flood affected people are often portrayed as passive recipients of money and goods, but not as key stakeholders in DRR planning to prevent the next disaster event and mitigate the disaster risks and associated vulnerabilities (Boustan et al. 2012; Seifert et al. 2013). The agency of disaster affected people should be brought into the DRR planning and implementation process in order to achieve better disaster resilience of these communities without forgetting whose disaster resilience is being discussed, planned and attempted in the processes.

The vulnerability paradigm has been influential in maintaining a focus on oppression when it critiques structural injustices in society. However, some scholars challenge it claiming that it reduces its subjects to a homogenized group which is powerless, poor and oppressed (Brown 2004; Fekete et al. 2014; Thompson and Wildavsky 1986). Individuals and groups who have been identified as lacking, weak and in need of help are expected to simply change (as in move away from risky areas) or be changed (do personal preparation for the move). The harmful structures and government institutions who benefit from them remaining passive and/or relocating still remain intact and unchallenged as powerful authorities claiming to be developing disaster-resilient communities. This has been illustrated in the forced displacement case study analysed in this chapter, where the government's management structures have taken the decision and implemented it without properly addressing the incentives, planning and stakeholder requirements. This has led to displacement with further increased vulnerabilities for the flood affected family. On the other hand, the structures that led them to experience poverty, low-income status, poor education, health and employment, ethnicity and religion-based discrimination and finally gender inequality remain unchanged.

Current vulnerability scholarship emphasizes the importance of focusing on various impacts of social and economic conditions on different disaster affected groups. This perspective supports how disasters have been affecting communities in terms of ethnicity, religion, region, geographical locality, gender, age, physical abilities and disabilities and economic status. This further suggests why and how multiple and holistic perspectives about vulnerability are required to explore the socio-cultural-historical root causes, larger disaster effects, causes, long-term impacts and complexities to which less attention has been paid. These are important social conditions which contribute to our understanding of the vulnerability paradigm. A number of disaster-related studies worldwide indicate how post-disaster recovery has emerged differently in relation to the availability of monetary and collateral assets and social support networks (Slavíková et al. 2020). This further suggests that disaster affected people's agency plays a crucial role in the disaster context including disaster resilience building which should not be neglected within disaster governance in Sri Lanka.

There has been a significant shift in disaster scholarship and the vulnerability discourse particularly from measuring and quantifying disaster-related casualties and built environment losses to exploring post-disaster recovery of the affected communities and building disaster resilience (Paton and Johnston 2001; Wisner 2020). This shift enables research into flood affected people's agency which was often overlooked in earlier scholarship. Thus, disaster governance should focus on supporting the flood affected people's agency to deal with disasters without limiting and undermining their capacities and creating further risks and vulnerabilities. The next section further explores gaps and limitations related to relocation/resettlement in Sri Lankan disaster governance and disaster resilience building processes.

13.7 Disaster Resilience Oriented Disaster Governance? Drawing Conclusions

It is evident from the earlier sections that disaster governance approaches to achieve disaster resilience should focus on socio-economic conditions and structural inequalities which leave certain communities more vulnerable to disasters. There needs to be more effort for developing community capacities and capabilities at the grassroots levels (Oliver-Smith 2013; Singer et al. 2015). In addition, research suggests enhancing communities' planning, preparation, self- mobilization and adaptation while developing relevant capacities, capabilities and capitals in the disaster context (Norris et al. 2008). However, this notion of disaster resilience has been criticized in disaster scholarship due to the problematic outcomes of over emphasizing the significance of returning to a stable pre-disaster status. This constructs a specific notion of development discourse-related resilience which can lead to overlooking structural inequalities and other socio-economic-cultural conditions influencing disaster affected people's responses as well as their recovery processes (Barrios 2016). Disaster affected people's responses to the particular social conditions created by the disaster are dynamic, but disaster governance has not paid adequate attention to this in plans to increase disaster resilience among the affected communities.

Hence, disaster governance aims at applying the governance principles to reduce disaster risks and enhance disaster resilience where the focus is being based on interventions towards environmental-related incentives, knowledge, establishments, decision-making and population behaviours and responses (Tierney 2012). Disaster governance should pay attention to all the disaster phases which include pre, during and post. Attention should be paid to reducing disaster vulnerabilities, planning land-use patterns and regulations beforehand, also developing early warning systems, capacity building, awareness raising and relevant training. Lack of effective disaster governance often leads to the failure of the entire disaster resilience building process creating further uncertainties (Tierney 2012).

Further, if the best solution in a highly flood prone geographical location is considered relocation, the question in this case arises as to why it has targeted one particular household not the entire flood prone neighbourhood. If relocating is required, creating minimal disruption to economic, social and cultural support systems for those relocated should be a priority. Thus, this chapter raises a serious concern about developing DRR strategies more efficiently and taking a more critical approach to policies based on defining some groups of people as "vulnerable" in Sri Lanka. It has been widely accepted that disaster magnitude and frequency have been increasing worldwide where the disaster governance becomes important and this has also been identified as an issue for Galle. Increased attention to forward planning to increase both disaster resilience while aiming to reduce poverty and increase development is essential and governance must support this. Building and improving disaster resilience among disaster affected communities must be addressed at different levels from macro to micro. Enhancing DRR policies and frameworks, enhancing disaster affected individual, household and community capacities including coping mechanisms and post-disaster recovery mechanisms are only a few suggestions. In this process, adaptation is also playing a pivotal role as it encourages disaster affected people to be flexible while engaging in decision-making processes and the set of actions to deal with disaster-related alarming situations. In addition, the chapter raises the question following reflexivity argument in structure and agency debate to raise if individuals within the government agencies are using their human reflexivity in exercising their powers, would the local government officials who decided to place the family 16 km away have made a different decision? Or do they merely focus on ticking boxes as per orders receiving from the higher authorities?

Disaster resilience simply stands as an ability to foresee risk, limit its impact, and rapidly recover while adapting, evaluating, surviving and changing with a sustained stand point. A research study of post-flood responses in Guyana, South America, showed how traditional fishing villages revealed better resilience to flooding compared to contemporary cities developed through expert planners' suggestions (Pelling 1999, 2011). The traditional fishing community has established effective flood control systems while adapting their lifestyles accordingly (Pelling 1999, 2011). This example indicates that disaster affected people's knowledge, community-based land management systems as well as rituals and belief practices in relation to their environment support enhancing disaster resilience among themselves. There are other research studies which suggest that such indigenous ecological knowledge helps disaster prone communities to continuously adapt, develop coping mechanisms and disaster-resilient strategies (Boillat and Berkes 2013; Gómez-Baggethun et al. 2013). Further, some disaster studies suggest culture is the driving and influencing factor in disaster recovery processes (Jang and LaMendola 2006). Thus, there is an essential requirement to incorporate these approaches into the Sri Lankan disaster governance mechanisms including the relocation programmes.

To conclude, this chapter primarily showed how disaster governance mechanisms aiming at disaster resilience in the Sri Lankan disaster relocation context have become contributing factors to reproduction of a number of new vulnerabilities and failures during their process of relocation. Disaster governance has overlooked whose disaster resilience they were attempting to achieve, in what process and in which location. There is an unequal relationship as well as a disconnect between the government, donors including international and local NGOs, and disaster affected people. Hence, the question remains whose disaster resilience is decided by whom in which locale (Cutter 2016; Hayward 2013). Even though the concept itself and practices related to disaster resilience may have their own gaps and limitations, disaster governance should focus on bridging the gaps between the processes of building disaster resilience and post-disaster recovery processes demonstrating resilience. This will require operationalizing and implementing disaster affected people's agency during these processes.

Acknowledgements I would like to acknowledge my Ph.D. supervisors Dr. Alison Loveridge and Dr. Ed Challies from the University of Canterbury, New Zealand, organizers of the "International Symposium on Rebuilding Communities after Displacement" and anonymous reviewers for their valuable comments to improve the chapter. Further, the New Zealand Commonwealth Scholarship Programme under the Ministry of Foreign Affairs and Trade funds my Ph.D. studies and the University of Colombo, Sri Lanka, has granted study leave to pursue my Ph.D. studies.

References

Archer MS (2003) Structure, agency, and the internal conversation. Cambridge University Press, Cambridge

Asthana R (1996) Involuntary resettlement: survey of international experience. Econ Polit Week 31(24):1468–1475. https://www.jstor.org/stable/4404268

Bahadur AV, Ibrahim M, Tanner T (2013) Characterising resilience: unpacking the concept for tackling climate change and development. Climate Dev 5(1):55–65. https://doi.org/10.1080/17565529.2012.762334. Accessed 12 Mar 2022

Bankoff G (2001) Rendering the world unsafe: "vulnerability" as western discourse. Disasters 25(1):19–35. https://doi.org/10.1111/1467-7717.00159

Barrios RE (2016) Resilience: a commentary from the vantage point of anthropology. Ann Anthropol Pract 40(1):28–38. https://doi.org/10.1111/napa.12085

Bodin P, Wiman B (2004) Resilience and other stability concepts in ecology: notes on their origin, validity, and usefulness. ESS Bull 2(2):33–43. https://bit.ly/37r7B7q

Boillat S, Berkes F (2013) Perception and interpretation of climate change among Quechua farmers of Bolivia: indigenous knowledge as a resource for adaptive capacity. Ecol Soc 18(4):21. https://doi.org/10.5751/ES-05894-180421. Accessed 10 Mar 2022
Bouchard J (2020) The resilience of native-speakerism: a realist perspective. In: Houghton SA, Bouchard J (eds) Native-Speakerism: its resilience and undoing. Springer Singapore, Singapore, pp 17–45. https://doi.org/10.1007/978-981-15-5671-5_2. Accessed 27 June 2022
Boustan LP, Kahn ME, Rhode PW (2012) Moving to higher ground: migration response to natural disasters in the early twentieth century. Am Econ Rev 102(3):238–244. https://doi.org/10.1257/aer.102.3.238
Brown H (2004) Violence against vulnerable groups, Integrated project "responses to violence in everyday life in a democratic society". Council of Europe Publishing, Strasbourg. https://www.corteidh.or.cr/tablas/r25587.pdf. Accessed 10 Mar 2022
Cannon T (2022) What must be done to rescue the concept of vulnerability? In: Bankoff G, Hilhorst D (eds) Why vulnerability still matters: the politics of disaster risk creation. Routledge, New York, pp 68–88
Cernea MM (2004) Impoverishment risks, risk management, and reconstruction: a model of population displacement and resettlement. In: UN symposium on hydropower and sustainable development, vol 27. https://citeseerx.ist.psu.edu/viewdoc/download?doi=10.1.1.574.3457&rep=rep1&type=pdf. Accessed 08 Mar 2022
Cernea M (1997) The risks and reconstruction model for resettling displaced populations. World Dev 25(10):1569–1587. https://doi.org/10.1016/S0305-750X(97)00054-5. Accessed 10 Mar 2022
Christensen A, Harild N (2009) Forced displacement: the development challenge. The World Bank, Washington D.C. https://documents1.worldbank.org/curated/en/837561468337294162/pdf/638920WP0Force00Box0361531B0PUBLIC0.pdf. Accessed 18 June 2022
Cutter SL (2016) Resilience to what? resilience for whom? Geogr J 182(2):110–113. https://doi.org/10.1111/geoj.12174. Accessed 10 Mar 2022
DeJesus KM (2018) Forced migration and displacement in Africa: contexts, causes and consequences. Afr Geogr Rev 37(2):79–82. https://doi.org/10.1080/19376812.2018.1447332. Accessed 18 June 2022
Dickinson T, Burton I (2022) Disaster risk creation: the new vulnerability. In: Bankoff G, Hilhorst D (eds) Why vulnerability still matters: the politics of disaster risk creation. Routledge, New York, pp 192–205
Dziaek J, Biernacki W, Konieczny R, Fiedeń L, Franczak P, Grzeszna K, Listwan-Franczak K (2019) Understanding flood preparedness: flood memories, social vulnerability and risk communication in Southern Poland, 1st edn. Springer International, Cham
Elder-Vass D (2007a) For emergence: refining archer's account of social structure. J Theory Soc Behav 37(1):25–44. https://doi.org/10.1111/j.1468-5914.2007.00325.x. Accessed17 June 2022
Elder-Vass D (2007b) Social structure and social relations. J Theory Soc Behav 37(4):463–477. https://doi.org/10.1111/j.1468-5914.2007.00346.x. Accessed17 June 2022
Elder-Vass D (2008) Integrating institutional, relational and embodied structure: an emergentist perspective. Br J Soc 59(2):281–299. https://doi.org/10.1111/j.1468-4446.2008.00194.x. Accessed 17 June 2022
Elder-Vass D (2010) The causal power of social structures: emergence, structure and agency. Cambridge University Press, Cambridge
Fernando N (2018) Voluntary or involuntary relocation of underserved settlers in the city of Colombo as a flood risk reduction strategy: a case study of three relocation projects. Procedia Eng 212:1026–1033. https://doi.org/10.1016/j.proeng.2018.01.132. Accessed 10 Mar 2022
Fernando N, Punchihewa AG (2013) Relocating the displaced: strategies for sustainable relocation. Friedrich Ebert Stiftung, Colombo
Fernando N, Senanayake A, Amaratunga D, Haigh R, Malalgoda C, Jayakody R (2020) Impact of the disaster-induced relocation process on the displaced communities in Kegalle district, Sri Lanka. Eur Sci J 16(39):33. https://doi.org/10.19044/esj.2020.v16n39p33. Accessed 10 Mar 2022

Fekete A, Hufschmidt G, Kruse S (2014) Benefits and challenges of resilience and vulnerability for disaster risk management. Int J Disaster Risk Sci 5(1):3–20. https://doi.org/10.1007/s13753-014-0008-3. Accessed 10 Mar 2022

Few R, Ramírez V, Armijos MT, Hernández LAZ, Marsh H (2021) Moving with risk: forced displacement and vulnerability to hazards in Colombia. World Dev 144:105482. https://doi.org/10.1016/j.worlddev.2021.105482. Accessed 18 June 2022

Gaillard JC (2022) The invention of disaster: power and knowledge in discourses on hazard and vulnerability. Routledge, London

Ginetti J, Lavell C (2015) The risk of disaster induced displacement in South Asia. The Internal Displacement Monitoring Centre, Châtelaine. https://www.internal-displacement.org/sites/default/files/publications/documents/201504-ap-south-asia-disaster-induced-displacement-risk-en.pdf. Accessed 06 Mar 2022

Gómez-Baggethun E, Corbera E, Reyes-García V (2013) Traditional ecological knowledge and global environmental change: research findings and policy implications. Ecol Soc 18(4):72. https://doi.org/10.5751/ES-06288-180472. Accessed 6 Mar 2022

Hayward BM (2013) Rethinking resilience: reflections on the earthquakes in Christchurch, New Zealand, 2010 and 2011. Ecol Soc 18(4):37. https://doi.org/10.5751/ES-05947-180437. Accessed 12 Mar 2022

Hilhorst D, Bankoff G (2022) Introduction. In: Bankoff G, Hilhorst D (eds) Why vulnerability still matters: the politics of disaster risk creation. Routledge, New York, pp 1–12

Jang L-J, LaMendola W (2006) The Hakka spirit as a predictor of resilience. In: Paton D, Johnston D (eds) Disaster resilience: an integrated approach. Charles C. Thomas, Illinois, pp 174–189

Lane L, Hesselman M (2017) Governing disasters: embracing human rights in a multi-level, multi-duty bearer, disaster governance landscape. Polit Governan 5(2):93–104. https://doi.org/10.17645/PAG.V5I2.899. Accessed 6 Mar 2022

MacKinnon D, Derickson KD (2013) From resilience to resourcefulness: a critique of resilience policy and activism. Prog Hum Geogr 37(2):253–270. https://doi.org/10.1177/0309132512454775. Accessed 12 Mar 2022

Manyena SB (2006) The concept of resilience revisited. Disasters 30(4):434–450. https://doi.org/10.1111/j.0361-3666.2006.00331.x. Accessed 10 Mar 2022

McKeown A, Bui DH, Glenn J (2022) A social theory of resilience: the governance of vulnerability in crisis-era neoliberalism. Eur J Cultur Polit Sociol 9(1):112–132. https://doi.org/10.1080/23254823.2021.1997616. Accessed 29 June 2022

Merriam SB, Grenier RS (2019) Qualitative action research. In: Merriam SB, Grenier RS (eds) Qualitative research in practice: examples for discussion and analysis, 2nd edn. Jossey-Bass, San Francisco, pp 335–337

Ministry of Land and Land Development with Assistance from the Asian Development Bank (2013) Land acquisition and implementation of the national involuntary resettlement policy: a guide for public officials on good practices. Ministry of Land and Land Development, Battaramulla, Sri Lanka. https://countrysafeguardsystems.net/sites/default/files/Land%20Acquisition%20and%20Implementation%20of%20National%20IR_Eng.pdf. Accessed 10 Mar 2022

Muggah R (2008) Relocation failures: comparing internal displacement and resettlement regimes in Sri Lanka. Zed Books, London

National Building Research Organization (NBRO) (2022) State Ministry of National Security and Disaster Management, Colombo, Sri Lanka. https://www.nbro.gov.lk/index.php?option=com_content&view=article&layout=edit&id=317&lang=en. Accessed 10 Mar 2022

Norris FH, Stevens SP, Pfefferbaum B, Wyche KF, Pfefferbaum RL (2008) Community resilience as a Metaphor, theory, set of capacities, and strategy for disaster readiness. Am J Commun Psychol 41(1–2):127–150. https://doi.org/10.1007/s10464-007-9156-6. Accessed 12 Mar 2022

Oliver-Smith A (2013) Disaster risk reduction and climate change adaptation: the view from applied anthropology. Hum Organ 72(4):275–282. https://doi.org/10.17730/humo.72.4.j7u8054266386822. Accessed 10 Mar 2022

Oliver-Smith A (2009) Development & dispossession: the crisis of forced displacement and resettlement, 1st edn. School for Advanced Research Press, Santa Fe

Palmer L, Carlsson G, Brunt D, Nystrom M (2014) Existential vulnerability can be evoked by severe difficulties with initial breastfeeding: A lifeworld hermeneutical single case study for research on complex breastfeeding phenomena. Breastfeed Rev 22(3):21–32. https://doi.org/10.3316/ielapa.029108831521039. Accessed 27 June 2022

Paton D, Johnston D (2001) Disasters and communities: vulnerability, resilience and preparedness. Disaster Prev Manag 10(4):270–277. https://doi.org/10.1108/EUM0000000005930. Accessed 10 Mar 2022

Pelling M (1999) The political ecology of flood hazard in urban Guyana. Geoforum 30(3):249–261. https://doi.org/10.1016/S0016-7185(99)00015-9. Accessed 12 Mar 2022

Pelling M (2011) Adaptation to climate change: from resilience to transformation. Routledge, New York

Plough A, Fielding JE, Chandra A, Williams M, Eisenman D, Wells KB, Law GY, Fogleman S, Magana A (2013) Building community disaster resilience: perspectives from a large urban county Department of Public Health. Am J Public Health 103(7):1190–1197. https://doi.org/10.2105/AJPH.2013.301268. Accessed 12 Mar 2022

Priest SJ, Penning-Rowsell EC, Suykens C (2016) Promoting adaptive flood risk management: the role and potential of flood recovery mechanisms. Paper presented at the E3S Web of Conferences; Les Ulis, vol 7. https://doi.org/10.1051/e3sconf/20160717005. Accessed 10 Mar 2022

Sandink D, Kovacs P, Oulahen G, Shrubsole D (2016) Public relief and insurance for residential flood losses in Canada: current status and commentary. Can Water Resour J 41(1–2):220–237. https://doi.org/10.1080/07011784.2015.1040458(Accessed10/03/2022). Accessed 10 Mar 2022

Scudder T (2016) Aswan high dam resettlement of Egyptian Nubians. Springer Singapore, Singapore. https://doi.org/10.1007/978-981-10-1935-7. Accessed 10 Mar 2022

Seifert I, Botzen WJW, Kreibich H, Aerts JCJH (2013) Influence of flood risk characteristics on flood insurance demand: a comparison between Germany and The Netherlands. Nat Hazard 13(7):1691–1705. https://doi.org/10.5194/nhess-13-1691-2013. Accessed 12 Mar 2022

Senanayake AC (2019) A review of existing laws and policies applicable to disaster induced resettlement in Sri Lanka. In: Fernando N, Siriwardana C (eds) Capacity building for research and innovation in disaster resilience: selected papers. University of Moratuwa and University of Colombo, pp 3–21. shorturl.at/amyV8. Accessed 10 Mar 2022

Senanayake A, Fernando N, Wasana M, Amaratunga D, Haigh R, Malalgoda C, Jayakody C (2022) Landslide induced displacement and relocation options: a case study of owner driven settings in Sri Lanka. Sustainability 14:1906. https://doi.org/10.3390/su14031906. Accessed 8 Mar 2022

Simons H (2015) Interpret in context: generalizing from the single case in evaluation. Evaluation (London, England. 1995) 21(2):173–188. https://doi.org/10.1177/1356389015577512. Accessed 26 June 2022

Singer J, Hoang H, Ochiai C (2015) Post-displacement community resilience: considering the contribution of indigenous skills and cultural capital among ethnic minority Vietnamese. Asia Pac Viewp 56(2):208–222. https://doi.org/10.1111/apv.12057. Accessed 12 Mar 2022

Slavíková L (2018) Effects of government flood expenditures: The problem of crowding-out: effects of government flood expenditures. J Flood Risk Manage 11(1):95–104. https://doi.org/10.1111/jfr3.12265. Accessed 6 Mar 2022

Slavíková L, Hartmann T, Thaler T (2020) Financial schemes for resilient flood recovery. Environ Hazards 19(3):223–227. https://doi.org/10.1080/17477891.2019.1703624. Accessed 12 Mar 2022

Suykens C, Priest SJ, van Doorn-Hoekveld WJ, Thuillier T, van Rijswick M (2016) Dealing with flood damages: will prevention, mitigation, and ex post compensation provide for a resilient triangle? Ecol Soc 21(4). http://www.jstor.org/stable/26269992. Accessed 12 Mar 2022

Taiban S, Lin H-N, Ko C-C (2020) Disaster, relocation, and resilience: recovery and adaptation of Karamemedesane in lily tribal community after Typhoon Morakot, Taiwan. Environ Hazards 19(2):209–222. https://doi.org/10.1080/17477891.2019.1708234. Accessed 12 Mar 2022

Tierney K (2019) Disasters: a sociological approach. Polity Press, Cambridge
Tierney K (2012) Disaster governance: social, political, and economic dimensions. Annu Rev Environ Resour 37(1):341–363. https://doi.org/10.1146/annurev-environ-020911-095618. Accessed 12 Mar 2022
Thomas NJR (2018) Sustainability marketing. The need for a realistic whole systems approach. J Mark Manag 34(17–18):1530–1556. https://doi.org/10.1080/0267257X.2018.1547782. Accessed 26 June 2022
Thompson M, Wildavsky A (1986) A poverty of distinction: from economic homogeneity to cultural heterogeneity in the classification of poor people. Policy Sci 19(2):163–199. http://www.jstor.org/stable/4532076. Accessed 10 Mar 2022
Wisner B (2020) Five years beyond Sendai—can we get beyond frameworks? Int J Disas Risk Sci 11(2):239–249. https://doi.org/10.1007/s13753-020-00263-0. Accessed 12 Dec 2021
Wisner B, Blaikie P, Cannon T, Davis I (2004) At risk: natural hazards, people's vulnerability and disasters, 2nd edn. Routledge, London
Yin RK (2018) Case study research and applications: design and methods, 6th edn. SAGE, Oaks

Chapter 14
Verticalized Slums, Governmentality and Pandemic Governance: A Critical Hermeneutical Analysis of Governance Practices in a Selected Urban High-Rise in Colombo, Sri Lanka

Anjali Korala and Veronika Cummings

Abstract During the Colombo Regeneration Project since 2010, thousands of inhabitants of the underserved settlements in Sri Lanka's capital were forcibly displaced and involuntarily relocated into high-rises, planned and constructed at the government's discretion. These high-rises turned the previously horizontal slums into vertical ones, creating new socio-economic and political problems for the relocated communities. In March 2020, during the COVID-19 pandemic, the government decided to lock down some of these high-rises to reduce the risk of community spread due to the increasing number of infected inhabitants. Within such a context, this research investigates how the built environment of these high-rises aided the government to monitor/supervise and control the movement of inhabitants during the lockdown period in a particularly opaque way. Methodologically, the research follows post-positivist thought. Empirical data were collected through in-depth qualitative interviews with relocated dwellers to gain the insiders' perceptions about their experienced reality. Using Michel Foucault's concept of panopticism, this paper argues that the government has been able to govern the movement of the inhabitants through structural surveillance and a specific form of 'institutionalized network' that further strengthened the processes of observation and examination of individuals and social groups.

Keywords Materialized governance · Panopticism · Built-environment · Pandemic Injustice · Inhabitants of underserved settlements · Sri Lanka

A. Korala (✉)
Department of Political Science and Public Policy, University of Colombo, Colombo, Sri Lanka
e-mail: anjali@poli.cmb.ac.lk

V. Cummings
Chair of Human Geography, Institute of Geography, Johannes Gutenberg University of Mainz, Mainz, Germany
e-mail: cummings@geo.uni-mainz.de

© The Author(s), under exclusive license to Springer Nature Switzerland AG 2023
M. Hamza et al. (eds.), *Rebuilding Communities After Displacement*,
https://doi.org/10.1007/978-3-031-21414-1_14

14.1 Introduction: Pandemic Governance—Pandemic Injustice?

The outbreak of the COVID-19 virus in December 2019 from Wuhan, China altered the world we knew completely. Within five months the virus spread to 213 countries around the world turning it into a global pandemic (Alfano and Ercolano 2022). Its global spread challenged all states irrespective of whether they were developed or still developing, not sparing any aspect of the state. In most countries, the health sector was not prepared for such a sudden surge of patients with a highly transmissible disease (World Health Organization 2020b). Governments were forced to lock down highly dense cities or in some cases entire countries to control the transmission of the virus (Onyishi et al. 2021). Countries struggled as people had to adapt to the 'new normal' (Corpuz 2021). States were pushed to take drastic measures or adopt debatable policies to control the spread of the virus and minimize its effects on their nation. These governing strategies included non-pharmaceutical measures such as testing, contact-tracing, isolation, and treatment and containment measures such as hygiene practices, social/physical distancing, and travel restrictions (Onyishi et al. 2021). Collectively, these strategies were referred to as 'pandemic governance', and they were often questionable when it came to civil liberties, justice, equality, and equity.

It can be argued that the level of "success" of states around the world in controlling the virus instead of being controlled by it has been based entirely on their pandemic governance strategies and how people perceived the role of the state and its power during such an uncertain time. Certainly, in such times of vulnerability due to very poor knowledge about a new virus, we often seem to rely on new patterns of solidarity and—or instead—on the state to navigate and provide action plans on how to control the situation and prevent it from worsening. Such times also bring to the fore controversial debates on the limits and privileges of the government to use its newfound state power under extraordinary conditions to determine the freedom and mobility of the individual citizen or of social groups. Alfano and Ercolano (2022) point out how people perceive the quality of governance (such as the capacity of the healthcare system to handle a pandemic) can positively impact their level of trust in governance, which can eventually influence the effectiveness of policies, especially with regard to pandemic governance strategies, as in of a higher level of individual compliance to policies such as lockdown. This interconnection is clearly visible in Reich's comparison of the United States of America's (USA) and Japan's pandemic governance strategies. Using Takemi's 'Control-Tower'[1] metaphor

[1] Reich explains the 'Control-Tower' metaphor as follows: "Think about the control tower at an airport. From there, a team of people monitor and guide the airplanes in the air, on the ground, taking off, and landing. The control tower also tracks other salient factors over which it has little control: birds, weather, potential security threats, time of day, and demand for air travel at that location and time. In addition, the control tower must understand the many other activities going on at the airport, including passengers arriving and departing, airport workers going about their

on pandemic governance to compare how USA and Japan responded to the COVID-19 pandemic, Reich (2020) argues that governance was the critical element, which made Japan's response to pandemic a comparative success while making the USA's a striking failure (2020, p. 1). Similar kinds of success stories are reported from Latin America as well. Cyr et al. (2021) point out that countries that had a high-level of inter-governmental collaboration and social collaboration were able to face the COVID-19 pandemic more successfully compared to the countries which had a lower level of inter-governmental and social collaboration.

Various studies from the so-called Global Urban South (Nyashanu et al. 2020; Parnell 2020; Connolly et al. 2021) showcase the neuralgic points of urban that mainly result from a lack of infrastructure and participation in governance, even more in governance in pandemic times. Urban self-organization, especially amongst socio-economically less privileged groups or informal settlements, often offsets uneven power structures that are counter-productive for social justice and equity when it comes to public services and fair democratic governance of urban spaces (ref. Brazil, South Africa).

Studies also point out how integrated governance-coordination between different actors and sectors as well as coordination between multi-level governing bodies-is detrimental in controlling the spread of the virus from highly dense cities to the countrysides, as well as in ensuring social justice. As Sharifi and Khavarian-Garmsir (2020) point out, while countries like Vietnam were able to flatten the curve with the assistance of local governments in providing economic support for the poor, disadvantaged and vulnerable communities, countries like Bangladesh were struggling to address the pandemic successfully, as there was no proper coordination between multiple actors and sectors (2020, pp. 7–8). While some countries succeeded and some struggled in implementing these pandemic governing measures, disparities were visible in the way these measures affected certain groups, especially in urban areas. One of the main arguments around the execution of pandemic governing strategies was that it highlighted already existing inequalities in society (Enright and Ward 2021). For instance, in their study on how the Austrian government's COVID-19 regulations revealed inequalities in the organization of paid work, Dobusch and Kreissl (2020) point out how the governance of im/mobility by the Austrian government allowed certain professions to stay home while some professions were obligated to be present at the work to keep the system running (2020, p. 711). Onyishi et al. argues that the spread of COVID-19 in African countries and the governing measures such as lockdowns followed by the authorities revealed "the extent of systemic fragility,

tasks, food deliveries, air cargo, and security checks. The control tower a governance mechanism—exists in this complex environment to ensure that the risks of air transport can be reduced to an acceptable level… Let me suggest four points about the role of a control tower in a pandemic… First, every control tower operates within a particular air space, just as pandemics occur within different societies and cultures, that is, communities with sets of shared beliefs and behaviors… A second critical factor for governance is the social values context in each country… The third important aspect of control-tower governance is the intentions and actions of the national… A related fourth aspect of the control-tower metaphor is the use and clarity of information provided by authorities." (Reich 2020, pp. 1–3).

underdevelopment, and pre-existing inequalities in cities, such as access to water, sanitation and hygiene" and "the extent of neglect and exclusion of the urban poor, who depend on daily income to survive and lack access to health insurance, paid leave, and functional clinics" (2021, 1231f). Moreover, Keil (2020) points out that these issues regarding spatial organisations, e.g., the inaccessibility and inability to reach institutions, cannot be separated from the spatial *and* social question of density and crowding. He explains this based on the fact that different municipal and regional jurisdictions in Ontario, Canada were 're-opened' on different dates, the rural communities being opened first and the urbanized areas much later. The main reason was the difference in density between rural areas and urban areas. Therefore, it should not be assumed that these governing strategies have a similar effect on the whole population equally. The pandemic situation emphasized prevailing social cleavages based on race, class, locality, etc., and certain pandemic governing strategies could reproduce and perpetuate these inequalities.

In the case of Sri Lanka, the Government of Sri Lanka (GoSL) followed most of the World Health Organization (WHO) recommendations in battling the spread of the COVID-19 virus. On March 11, 2020, the Ministry of Health (MoH) announced the first local case of COVID-19. To control the spread of the virus, on March 20, 2020, the Government of Sri Lanka (GoSL) imposed a lockdown that lasted for eight weeks (World Health Organization 2020a). After this, the GoSL continued to take different measures from time to time to control and manage the spread of the virus.

The overall response of the government to the COVID-19 pandemic was based on the suggestions of four lines of operations by the World Health Organization (WHO): (1) Tri-Forces (Sri Lanka Army, Sri Lanka Navy, Sri Lanka Airforce), police, and intelligence, (2) medical and health care, (3) community engagement, and (4) economic and psychosocial well-being (World Health Organization 2020a).

According to the *Sri Lanka Preparedness and Response Plan—Covid-19* issued in April 2020, the GoSL identified Sri Lanka as a country in **stage 3**, a country experiencing case clusters in time, geographic location, and/or common exposure (clusters of cases) out of the four transmission scenarios[2] set out by the WHO (Ministry of Health and Indigenous Medical Services 2020). The government response plan was under nine interlinked priority areas.[3]

[2] Stage 1—Countries with no cases (no cases), Stage 2—Countries with 1 or more cases, imported or locally detected (sporadic cases), Stage 3—Countries experiencing cases clusters in time, geographic location and/or common exposure (cluster of cases) Stage 4—Countries experiencing larger outbreaks of local transmission (community transmission).

[3] (1) Country-level coordination, planning, and monitoring; (2) Risk communication and community engagement; (3) Surveillance, rapid response, and case investigation; (4) Points of entry; (5) National laboratories; (6) Case management; (7) Infection prevention and control (8) Operations support and logistics; (9) Maintaining essential and emergency health services during an outbreak (Ministry of Health and Indigenous Medical Services 2020).

Under the priority areas of case management and infection prevention and control, measures were taken to prevent Sri Lanka from reaching **stage 4**, community transmission[4]. The basic strategy of the GoSL in managing the cases can be pointed out under four actions: (i) treating and monitoring confirmed COVID-19 cases[5] at treatment facilities, (ii) home quarantine of close contacts[6] of confirmed cases, and, if necessary, sending them to quarantine centers established in various districts in the country, (iii) subjecting people traveling from overseas (both locals and foreigners) to a mandatory quarantine of 14 days at quarantine centers, and (iv) transferring any confirmed COVID-19 cases from the quarantine centers to treatment facilities. The quarantine centers around the country were managed by the tri-forces. Furthermore, in order to mitigate the spread of the virus, the GoSL advocated public health measures such as hand hygiene, respiratory etiquette, and practice of social distancing. (Ministry of Health and Indigenous Medical Services 2020). The first line of operation; *detection, isolation, and tracing*, was primarily undertaken by the military, police, and intelligence services. They were responsible for tracing back the close contacts of the confirmed COVID-19 case and overseeing the process of transferring them to the quarantine centers established by the government (World Health Organization 2020a).

However, when the number of confirmed cases was increasing, especially when the clusters of COVID-19 patients from the same geographical area were starting to appear rapidly in society, the strategies used by the GoSL to prevent and manage the spread of the virus changed. The GoSL began to impose island-wide lockdowns for longer periods, implement inter-district, and inter-provincial travel restrictions, and isolate entire administrative divisions such as *Grama Niladhari Wasam* (Grama Niladhari Division*)* or *Police Divisions* to prevent people from leaving and entering potentially high-risk areas. The Department of Government Information, in its press releases officially announced the areas to be isolated and areas where the isolation status was to be lifted. This measure of isolating entire administrative divisions became a frequent government measure in 2020 and 2021 especially during the first and second waves of COVID-19 on the island.

A close examination on how these pandemic governing strategies, such as isolation of certain administrative areas, were executed shows that the experience of residents was quite different depending on the locality. As mentioned cases like Ontario, Canada, the isolations were comparatively longer and stricter especially when it

[4] Outbreaks with the inability to relate confirmed cases through chains of transmission for a large number of cases, or by increasing positive tests through sentinel samples—Source: WHO/COVID19/Community Actions/2020.2 (Cited as in *Sri Lanka Preparedness & Response Plan—Coivd-19* issued in April 2020).

[5] According to the 'Updated Interim Case Definitions on Covid-19 and Advice on Initial Management of Patients (version dated 4 April 2020) a 'Confirmed Case' is 'A person with laboratory confirmation of COVID-19 infection, irrespective of clinical signs and symptoms.

[6] According to the 'Updated Interim Case Definitions on Covid-19 and Advice on Initial Management of Patients (version dated 4 April 2020) a 'Close Contact' is 'A person staying in an enclosed environment for $>$ 15 min' (e.g. same household/workplace/social gatherings/traveling in same vehicle). OR who had direct physical contact.

came to the highly dense urban areas like city of Colombo. Even within the city of Colombo, residents in some areas were more vulnerable under isolation compared to others due to lack of social and physical infrastructure to support them through the period of isolation. In other words, when the pandemic hit, "it mattered where in the city you lived (Keil 2020, p. 5). It mattered because, if the residents do not posses the necessary social and physical infrastructure to face such a prolonged isolation, their level of resilience would deteriorate over time. In the case of the city of Colombo, one of the vulnerable communities were the ones who lived in government-built high-rises. They were formerly inhabitants of underserved settlements within Colombo who were relocated into government-built high-rises under the Urban Regeneration Project (URP) implemented by the Urban Development Authority (UDA). During the COVID-19 pandemic, these communities were continuously placed under isolation for comparatively longer periods (Dias et al. 2021) and a total number of 13 high-rises were placed under isolation (AdaDerana 2020). Over time, the inhabitants of these high-rises started protesting over the prolonged isolation and the restrictions on their mobility. For instance, inhabitants of the 'Methsanda Sevana' high-rise in Henamulla, Modara launched a protest on 6th December 2020 against the isolation measures (Dias et al. 2021).

In such a context, this chapter inquires how the built environment of the high-rises aided the government perfidiously in governing the movement of the inhabitants during the lockdown period. Thus, it requires a closer look at the possible mechanisms used by hidden power structures to produce and justify a subaltern citizenship. The objectives of this chapter are to unravel how the inhabitants in these government-built high-rises were governed, how the structure of the building and the governing mechanisms related to it created new power relations between the government agents and the inhabitants, and especially how these power relations became more asymmetrical and obscure during the period of isolation the inhabitants experienced. This chapter is organized as follows; In the second section, *Background: The social costs for structural beautification and modernization*, the vulnerabilities of the inhabitants of the high-rise are briefly explained. In this section the relationship between how their relocation to the high-rises and the issues in the resettlement process are explained. The third section, *Conceptual framework: The high-rises revisited in the light of Foucault's concept of panopticism* sets out the main theoretical foundation of the chapter. This section lays out the main arguments of Michel Foucault's concept of panopticism and how it relates to the current discussion on the link between the built-environment of the high-rise and governance. Under the fourth section *Methodology,* the main methodological approach and the methods used in the research are discussed. In the fifth section the two main findings of the research are discussed under two sub-sections; *Institutionalized network of control: The medical clinic and police post at Mihijaya Sevana* and *The Corridor: A modern spatial arrangement of panopticism for internal surveillance.* In the final section, the *Conclusion* the arguments of the chapter are briefly summarized.

14.2 Background: The Social Costs for Structural Beautification and Modernization

The Sri Lankan narrative of development after the end of the civil war in 2009 was primarily centered around infrastructure and urban development (Abeyasekera et al. 2019), especially highlighting structural modernization and improvement. While one set of development projects were initiated in the Northern and Eastern Provinces rebuilding the war-torn areas, the urban development projects initiated by the government were concentrated in the Western Province (Jayawardhana 2010). The Urban Regeneration Project (URP) by GoSL and the Metro Colombo Urban Development Project (MCUDP) by the World Bank and GoSL, were the two major development projects initiated in Colombo, Western Province (Perera 2020b).

The Urban Regeneration Project (URP) was initiated in 2010 by the Urban Development Authority. The project aimed to eliminate shanties, slums, and other dilapidated housing from the city by relocating dwellers to modern houses (Urban Development Authority 2018). The main aim behind the project was to free approximately 900 acres of valuable prime land in underserved settlements, which were home to approximately 68,000 families. Through the Urban Regeneration Project, UDA planned to construct 68,000 housing units to relocate these families and release the land to urban development (Perera 2020b). It is important to note that providing 'modern' houses to the inhabitants of the underserved settlements can be defined as a state policy inspired by the 'modernization paradigm', a state perspective quite common for the Global South. However, *what* is to be modernized, *how* it will be modernized or *who* decides on the terms and conditions of modernization have not been questioned much in the Sri Lankan context, especially with regard to urban development projects concerning the urban poor. Moreover, it can also be argued that these modernization projects indicate how the state reproduces its power at the ground level with its "capacity to construct and reconstruct the categories of legitimacy and illegitimacy" (Roy 2007, p. 149), in this case defining legal and illegal housing in the urban context.

Although URP has managed to provide former underserved settlement inhabitants with better living conditions, there are many criticisms against it, mainly on the forced displacement, involuntary relocation, and the resettlement process associated with it. Basically, this uncovers the hidden practice of withdrawing any recognition of socially grown networks or social aspects of place attachment. These criticisms of the visible formal arrangement are formulated from a critical perspective of social justice and following ideas of post-modern urbanism and a current paradigm of the Urban (Brenner and Schmid 2015; Roy 2015), revisiting also our understandings of the Urban Global South (Parnell and Robinson 2013) and the ideas of plannability of informality (Roy 2007; Ghertner 2010). It is important to investigate how these issues reported especially on relocation and resettlement processes, would enlighten our discussion on high-rise inhabitants' vulnerability during the COVID-19 pandemic.

14.2.1 Criticisms on the Displacement and Relocation Process of URP

The main criticism against the URP was *forced* displacement and *involuntary* relocation. The inhabitants of underserved settlements were removed from their houses forcibly and relocated to new places involuntarily. While the Urban Regeneration Project was aimed at relocating the underserved settlement inhabitants to modern houses with better infrastructure facilities, the affected families were not properly consulted during the relocation process regarding their needs, desires, and socio-economic problems as well as their self-organization structures (community networks, neighborhoods, place identity, place attachment, etc.). Therefore, the lack of a relocation policy that complies with the national and international standards in the URP made involuntary relocation a pre-condition to access better housing and living conditions, from a structural perspective (Perera et al. 2014).

Although Sri Lanka had already adopted a National Involuntary Resettlement Policy (NIRP) in 2001, the high-rise projects which were carried out before 2018[7] under the URP were not in compliance with NIRP (Gunatilleke and Nathaniel 2014; Collyer et al. 2017). There is no evidence that these principles were used when conducting the relocation. The Mews Street case that took place in May 2010, where 20 houses which homed 33 families were demolished by UDA after providing legal notification just three days prior to demolition and without providing any alternative accommodation and compensation (Perera et al. 2014; Gunatilleke and Nathaniel 2014) is one of the prime examples of forced displacement that took place under the URP. Moreover, it can be argued that these affected people have not been viewed as equal citizens in a democratic state.

In the case of Java Lane, where even residents with title deeds that run back several decades could not secure themselves from being displaced by UDA; where they were not provided with any written documents which state their rights, the obligations of the UDA, or the estimated timelines; nor being informed about the design, size, and specifications of the apartments where they were supposed to be relocated (Perera et al. 2014, pp. 29–30) it becomes evident that the affected communities were not fully informed or included in the relocation process.

While there were many issues regarding the displacement of inhabitants, there were more problems associated with the relocation process. Perera summarizes the socio-economic issues related to the relocation process as follows:

1. The size of the apartments (450 square feet) was too small.
2. The one million rupee payment for each apartment was beyond the means of most households and no proper assessment of affordability was done.

---

[7] In 2018, the Ministry of Megapolis and Western Development introduced a *Resettlement Policy Framework for Colombo Urban Regeneration Project*, for the Colombo Urban Regeneration Project, the second phase of the Urban Regeneration program as a pre-condition set by the Asian Infrastructure Investment Bank (AIIB), which partially funded the Colombo Urban Regeneration Project.

3. The URP did not consider those who had bigger houses and multiple families living in one house as eligible for more than one apartment.
4. Receiving compensation for their land and house was not an option, the relocation to an apartment was the only option given.
5. Those with tenure were not compensated as per the Land Acquisition Act.
6. The 100,000 Sri Lankan Rupees (LKR) that families had to pay upfront in order to collect keys to their apartments was beyond the means of most households, which most often had to borrow that money within a short amount of time at high interest, or pawn assets.
7. Households had no documents that were given to them regarding their apartment and only had an allotment letter, which said they were not allowed to rent, mortgage, or sell the apartment until they had paid the one million rupees.
8. Communities were not given a copy of any of the documents, they were asked to sign. Furthermore, documents were only available in Sinhala and translations were not made available to Tamil communities even though they were forced to sign multiple documents.
9. Even though the UDA said each apartment was worth seven million rupees and given to the people at one million rupees, there was no information available about how the value of seven million rupees was derived and this figure has varied over the years.
10. Communities had to pick their apartments at random, which meant that existing family and community networks were broken up and distributed all throughout the vast complexes.
11. Design wise, the high-rise complexes were criticized as unmanageable for communities in the long run with multiple lifts servicing twelve floors breaking down frequently and water shortages due to limited water infrastructure available in some areas; the layout of the complexes made public areas insecure for women and children, in particular, no play areas for children, lack of space for shops.

(Perera 2020b, p. 4)

One of the striking facts which emerge from both displacement and relocation issues related to URP is the mode of compensation. The affected slum and shanty dwelling families were compensated mainly with a housing unit in the high-rises (Collyer et al. 2017). While they were not presented with the opportunity to choose where they want to be relocated, they were also deprived of the right to be relocated to a space similar to the one from which they were displaced. When they were previously living horizontally on land, they had physical space to create their own lived space. Most importantly, they had the freedom to make necessary alterations to the house and to have vegetation and flowers if they had a garden. Nevertheless, when these families were relocated to the high-rises, they were situated vertically. This made them lose many advantages they had when they were living on the land horizontally.

All these points toward severe problems emerging from the forced relocation carried out by the Urban Development Authority during the Urban Regeneration

Project pursuing the overall target to beautify the city—at very high social costs. The issues regarding relocation were exacerbated due to the problems the affected families had to face after relocation, especially when resettling themselves in the high-rises.

14.2.2 Criticisms on the Resettlement Process of URP

Relocation and resettlement are two processes that go hand in hand.[8] Although bureaucracy considers these to be separate processes that have no impact on each other, in real life, it is evident that when relocating affected communities, it is essential to pay attention to how they are going to be resettled socially and economically within the host community. In the case of the Urban Regeneration Project, the relocated families had several severe issues when it comes to resettling in the new environment. The nature of the built environment was one of the major reasons behind these issues.

As it was mentioned above, the inhabitants of the underserved settlements were previously residing on separate plots of land, and their settlements were spread horizontally. This horizontal spread of the settlement had created a particular lived space for them and once they were relocated into high-rises a clear structural change took place. As they were relocated vertically, the nature of the spread of their settlements altered the lived space to which they had become accustomed. These alterations in the lived space are evident primarily in their social and economic relations. Perera explains this process of relocation as turning the high-rises into "vertical slums" (2020a, p. 55).

According to the National Involuntary Resettlement Policy (NIRP), the relocation processes should also consider the resettlement of the affected communities, especially regarding their successful integration into the host communities, providing opportunities to develop their social and economic prospects within the new place and in improving the living standards of the vulnerable groups[9] (Sri Lanka National Involuntary Resettlement Policy 2001).

---

[8] Relocation and resettlement are two distinct but inter-connected concepts. While relocation means the process of moving people from one place to another resettlement means the comprehensive process which entails providing assistance to the relocated people to adapt to the new place. "Resettlement can be defined as the comprehensive process of planning for and implementing the relocation of people, households and communities from one place to another for some specific reason, together with all associated activities, including: (a) the provision of compensation for lost assets, resources and inconvenience; and (b) the provision of support for livelihood restoration and enhancement, re-establishment of social networks, and for restoring or improving the social functioning of the community, social activities and essential public services." (Vanclay 2017, pp. 5–6).

[9] A precise definition of those who are classified "vulnerable" does not appear in the text, but under "definitions of terms" as follows:

"Distinct groups of people who might suffer disproportionately from resettlement effects such as the old, the young, the handicapped, the poor, isolated groups and single parent households" (Sri Lanka National Involuntary Resettlement Policy 2001).

Nevertheless, as mentioned before, the Urban Regeneration Project did not have any policy framework on relocation and resettlement until 2018, and the relocations and resettlements that took place prior to that had not followed the prevailing NIRP. There is much evidence on how hard the resettlement processes were for the affected communities due to the manner in which relocation took place and how the built environment of the high-rises hindered a smooth process of resettlement and rehabilitation. For instance, during a survey conducted by the Center for Policy Alternatives (CPA) at three high-rises built under the URP (Mihindusenpura, Sirisara Uyana, and Methsara Uyana), the respondents were asked whether they were happy about the size of the apartment they had received. 46.5% of the respondents said that the size of the apartment was small (Perera 2016). This size of the apartment paves the way for many problems when it comes to the expansion of the apartment. Since they were placed vertically, the residents have no way of improving or expanding the apartment as they prefer (Perera 2020a). This has forced the inhabitants to create a lived space in a physical space that was given by the government that has no means of modifying it according to their needs. This had severe effects on the inhabitants' social and economic relationships.

Another harsh impact on social relationships was the disruption of the kin and social networks (Collyer et al. 2017). Both happened due to the way the apartments were distributed. According to UDA's policy on the compensation of the lost houses, they could be replaced with an apartment. It was not 'one apartment for one family', but one apartment for a house (Perera 2015). This meant that even though, previously, there were several families living in the same house, they were not given separate houses. As there was no space for them to expand the house by adding an additional room or another story when their families got bigger the residents had to seek alternative housing options as they could not expand their houses as they had done when they were living on the land horizontally (Perera 2020a). On the other hand, the apartments were distributed among the affected families using a lottery. This minimized the possibility of the same community or relatives being relocated to the same floor to continue their life-supporting social ties and bonds (Abeyasekera et al. 2019).

Another major impact on social relationships was the security concerns caused due to the nature of the built environment. In their previous community or *watta*, where their houses were situated horizontally, they were more confident about their security, especially the women and children. They had their own system of security and social surveillance. They could keep their front doors open and the children could play outdoors freely. It was not easy for an unknown person to walk into the *watta* without being noticed and questioned by someone in the community (Perera 2016). Nevertheless, in the new high-rises, the residents were reluctant to send children out to play and imposed severe restrictions on children's movement, like general curfews at 6.00 p.m. This was due to a lack of trusting relationships within their new community (Development Planning Unit 2018). This indicates that being relocated into a vertical living space with strangers, where the inhabitants cannot monitor the people who visit the premises or rely on socially grown ties of familiar neighboring settings, as was the case in a horizontal living space, has created security concerns, which would eventually disrupt proper resettlement in the built environment.

On the other hand, the relocated families had difficulties in resettling due to the economic constraints they faced in the new built environment. The residents who had previously run a small business from their homes had to improvise or find alternative income-generating mechanisms. Despite the restrictions set by the UDA on using the apartment for business purposes (Perera 2015), most such residents have turned their living room space into shops, either in its entirety or by partitioning half of it. Nevertheless, they are unable to achieve conditions comparable to those under which they lived previously. Their clients are now spread out and sometimes they must compete with other inhabitants who run the same business (usually grocery shops) (Perera 2015; Development Planning Unit 2018).

These types of issues on relocation and resettlement processes related to the URP strongly indicate the pertaining problems of the inhabitants of the high-rises in terms of resettling in the new environment. This resettling phase can be challenging due to various factors, and it can be argued that these challenges have been intensified due to the built environment of the high-rises as well. People who have been living on the land, horizontally, now have been resettled into high-rises, that are vertical, and this can have a clear effect on their day-to-day activities and social relationships. This built environment of the high-rises has made a great impact on the relationship between the citizens and government and especially on the governing methods used by the government. It must be admitted that place attachment and sense of belonging need time to evolve. However, this is hindered if the built environment does not offer any fertile ground for them to grow, and they are already nipped in the bud instead.

During the COVID-19 pandemic, the above-mentioned specific conditions in the built environment of the high-rises and the socio-economic challenges associated with it were crucial in deciding the inhabitants' ability to be resilient in the face of the pandemic. Therefore it can be argued that when the high-rises were placed under special pandemic governing strategies like isolation, the problems faced by the inhabitants of the high-rises were different and, to some extent, more severe compared to the other communities placed under the same measures.

14.3 Conceptual Framework: The High-Rises Revisited in the Light of Foucault's Concept of Panopticism

The main conceptual framework of this chapter is based on the ideas of 'Panopticism' by Michel Foucault. We believe that Foucault's thought on how the organization of space can be used effectively to control the behavior of the people is quite useful in this research as it provides us with a strong conceptual base to analyze how the built environment of the high-rises was instrumentalized by the government officials during the isolation period to control the mobility of the inhabitants of the high-rise.

The concept of 'panopticism' was introduced by Michel Foucault in his book *Discipline and Punish* published in 1975. This book centers around the transitions which took place between the late eighteenth century and early nineteenth century, especially on how brutal punishments were used in Europe before the eighteenth century. Foucault starts the chapter titled "Panopticism" in his above-mentioned oeuvre with a description of an order published by the French authorities at the end of the seventeenth century. This order contained the public safety measures to be taken by town authorities at a time of plague. While extensively explaining the public health measures to be taken by the authorities (which highly resemble the current modes of governance used by governments all around the world during the pandemic), Foucault points out the extent to which 'segmentation' as a governing measure was more advanced compared to 'separation' or 'exile' of lepers during the spread of leprosy and how, interestingly, nineteenth-century authorities managed to combine and use these two measures as disciplinary actions in some medical and correctional institutions (Foucault 1975, pp. 195–199). It is through the evolution of these disciplinary measures that he presents the concept of 'panopticism'.

According to this order, Foucault specifies how authorities were recommended to use multiple separations, individualizing distributions, an organization in the depth of surveillance and control, an intensification, and a ramification of power, to minimize the spread of the plague. Upon the recommended guidelines of the order, inhabitants of the city were prohibited to leave town and the town was divided into distinct quarters which were governed by an intendant. Every street was to be placed under the surveillance of a syndic and the town gates were to be used as observation posts. The suggested surveillance system was based on a system of permanent registration, in which reports included name, age, sex, condition, deaths, illnesses, complaints, and irregularities of the inhabitants, passed from syndics to intendants and from intendants to the magistrates or mayor. The magistrates were to have full personal control over who would enter and leave the city and who would visit the inhabitants for medical or religious purposes (1975, p. 196). These measures can also be described as some pre-forms of social profiling/stigmatization of people affected by the plague. During the plague, towns turned into ideal disciplined societies, which Foucault called "the utopia of the perfectly governed city" (1975, p. 198). This was different from the massive binary division practiced during the spread of leprosy. The leper was usually rejected from society and pushed towards exile, which marked the governing measure of 'separation' as something based on a binary division between one set of people and others, in this case, lepers and others (Foucault 1975, p. 198).

Foucault merges these two projects; the separation of lepers and segmentation during the plague and argues that, if lepers gave "rise to the rituals of exclusion", it was the plague that gave "rise to the disciplinary projects" (1975, p. 198). Furthermore, he points out that by the nineteenth century, the authorities had combined these two measures and used them as disciplinary powers in institutions like the psychiatric asylum, the penitentiary, the reformatory, the approved school, and to some extent even in hospitals (Foucault 1975, p. 198). Foucault explains this combination as follows:

> Generally speaking, all the authorities exercising individual control function according to a double mode; that of binary division and branding (mad/sane; dangerous/harmless; normal/abnormal); and that of coercive assignment, of differential distribution (who he is; where he must be; how he is to be characterized; how he is to be recognized; how a constant surveillance is to be exercised over him in an individual way, etc.). (Foucault 1975, p. 198)

It is with Jeremy Bentham's panopticon that Foucault explains how these disciplinary measures are presently used by the authorities. Panopticon was a model penitentiary proposed by Jeremy Bentham (1748–1832). This architectural design for a prison or a fabric allowed continuous surveillance of all inmates and Bentham named it "panopticon", the Greek roots of which means "seeing all". His assumption was that if the prisoner is under the impression that he is being watched all the time, it would be sufficient to reform his behavior (Strub 1989, p. 40). Foucault explains this composition as follows:

> Bentham's Panopticon is the architectural figure of this composition. We know the principle on which it was based: at the periphery, an annular building; at the center, a tower; this tower is pierced with wide windows that open onto the inner side of the ring; the peripheric building is divided into cells, each of which extends the whole width of the building; they have two windows, one on the inside, corresponding to the windows of the tower; the other, on the outside, allows the light to cross the cell from one end to the other. All that is needed, then, is to place a supervisor in a central tower and to shut up in each cell a madman, a patient, a condemned man, a worker or a schoolboy. By the effect of backlighting, one can observe from the tower, standing out precisely against the light, the small captive shadows in the cells of the periphery. They are like so many cages, so many small theaters, in which each actor is alone, perfectly individualized and constantly visible. The panoptic mechanism arranges spatial unities that make it possible to see constantly and to recognize immediately. In short, it reverses the principle of the dungeon; or rather of its three functions-to enclose, to deprive of light and to hide-it preserves only the first and eliminates the other two. Full lighting and the eye of a supervisor capture better than darkness, which is ultimately protected. Visibility is a trap. (Foucault 1975, p. 200)

The major effect of the panopticon is that it puts the inmate in a state of conscious and permanent visibility, which in turn guarantees an automatic functioning of power. The central tower hints to the inmates that they are constantly being watched by a guard, even if such surveillance were not continuously in action, making the inmates behave properly. This makes even the actual surveillance unnecessary as the architectural design or the built environment creates a power relation that does not depend on the person who exercises the power. Thus, the inmates are in the middle of a power situation in which they themselves are the bearers (1975, p. 201). This points out a major principle of Bentham's panopticon: "…power should be visible and unverifiable" (ibid).

In this way, the panopticon provides an ideal way to observe the inmates through the central tower or allows them to believe that they are being observed at all times. Their inability to verify whether they are being observed or not pushes them to regulate themselves or behave accordingly. In simple, to discipline themselves. Nevertheless, what Foucault argues is that panopticon should not be understood as an architectural design that contains specific functions of observation, which make the inhabitants disciplined. It contains a figure of political technology (1975, p. 205), which can be extended to the outer world, to the most particle in the social structure. The generalization of the disciplinary mechanisms used in the panopticon toward the entire society is what Foucault terms 'panopticism'. According to him, discipline can be two-fold. At the one end, we can see the **discipline-blockade**, where the enclosed institutions are established away from society that is turned inwards for negative functions, where the inhabitants are not allowed to communicate with each other. On the other end, we can see the **discipline-mechanism**, which comes with panopticism, where the functioning of the power is made light, rapid, and effective (1975, p. 209). Following a historical evolution, now these discipline mechanisms or procedures are spread as centers of observation disseminated throughout society.

Foucault points out that 'discipline' should not be identified with an institution or with an apparatus. According to him, it is a type of power, a modality for its exercise. It is comprised of a whole set of instruments, techniques, procedures, levels of application, targets. He terms 'discipline' as the 'physics' or the 'anatomy' of power, a technology (1975, p. 215). He describes five major avenues where disciplinarization is used in society or where panopticism is visible.

1. 'Specialized' institutions like penitentiaries or 'houses of correction' in the nineteenth century in central Europe,
2. Institutions that use discipline as an essential instrument to reach a particular end, such as schools or hospitals,
3. Pre-existing authorities may find it as means of reinforcing or reorganizing their internal mechanisms of power like intra-familial relations,
4. Apparatuses which have made discipline their principle of internal functioning like the disciplinarization of the administrative apparatus during the Napoleonic period,
5. State apparatuses in which the major function is to ensure that discipline reigns over society like the police (1975, pp. 215–216).

Thus, Foucault states that panopticism exists not because the disciplinary modality of power has replaced all the other mechanisms, but because these disciplinary modalities have managed to infiltrate the others, sometimes undermining the existing ones, but acting as an intermediary between them, linking them with each other, expanding them to bring the effects of power to every nook and corner of society (1975, p. 216).

This concept of panopticism can be used in understanding how the inhabitants of the high-rises were governed during the COVID-19 pandemic. Just as Foucault explained the governing measures city officials used during the seventeenth-century plague, and the use of a central tower by officers to observe and surveil inmates in prison, the government of Sri Lanka used many surveillance methods to ensure the disciplined behavior of the inhabitants of high-rises during the pandemic.

14.4 Methodology

The research was carried out using a post-positivist methodological approach. The life experiences of the inhabitants of high-rises during the pandemic were collected through interviews. A total of 25 inhabitants from Block C of the Mihijaya Sevana high-rise were interviewed from April 2021 to October 2021. During the interviews, the main focus was placed on identifying how governance strategies were exercised by government officials within the high-rise during the isolation period and how inhabitants experienced the built environment during the isolation period. Furthermore, secondary data from previous research carried out on the relocation and resettlement process of high-rise inhabitants were also used to provide more clarity to the arguments made and views expressed in this chapter. Michel Foucault's concept of panopticism was used to elaborate the arguments presented theoretically.

The selected study location was the Mihijaya Sevana high-rise building, situated in Henamulla, Modara. Inhabitants were relocated into the high-rises in different stages beginning in September 2019. Mihijaya Sevana consists of four main blocks, named A, B, C, and D. The entire high-rise consists of 1645 housing units, and by the time of research only 640 housing units were occupied.[10] (Silva 2020, personal communication, 25 October). Since the study population was too large only the residents of Block C were selected for convenience. Block C consists of 434 housing units of which only 232 units were occupied by the time of research.

According to the inhabitants of Mihijaya Sevana block C, initially, they were locked down for around 3 months, from March 22, 2020, to June 2020. Then during Iftar in July the lockdown was somewhat released. But then again from October 22, 2021, to January 04, 2021, they were under lockdown (Meena 2021, personal communication, 4 April). Nevertheless, according to the press release by the Department of Government Information issued on December 12, 2020, Mihijaya Sevana was among the housing schemes, in which the isolation status was lifted on the aforementioned date. This decision has been based on the results of the PCR tests related to the relevant housing schemes (Department of Government Information 2020). During these periods, the inhabitants of the Mihijaya Sevana underwent severe problems and the isolation completely hindered their means of income. Since they were not allowed to leave the housing scheme, the breadwinners could not go to work. Most of the inhabitants in Mihijaya Sevana were doing blue-collar jobs—street vending, portering, Three-wheel driving, etc...., which meant that the government's work principles like work-from-home, under the 'New Normal' were not an option for these people. Furthermore, army officers were stationed at the entrances and exits to the high-rise to ensure that no one from the high-rise would leave the premises.

[10] The total no. of occupied houses only includes the houses where the relocated families are living. It does not include the housing units, which are used as UDA offices, community halls, pre-school, etc.

14.5 Findings

During the research, it was evident that the built environment of the high-rises, especially the architectural design of the floors and the vertical structure aided, or, more correctly, strengthened the surveillance methods followed by the government in order to control the behavior of the inhabitants. Two major surveillance or observation methods emerged from the analysis of interviews with inhabitants and based on these we argue that two disciplinary methods were followed/exercised by the government officials as well as by the inhabitants themselves during the isolation period:

1. The existence of an institutionalized network of control.
2. The existence of an internal surveillance.

These two disciplinary methods are elaborated in the next sections.

14.5.1 *Institutionalized Network of Control: The Medical Clinic and Police Post in Mihijaya Sevana*

Just as the system of permanent registration was used by town officials for surveillance during the plague as explained by Foucault, a similar system of registration was evident within the selected field location Mihijaya Sevana high-rise. As mentioned in the methodology, out of the four blocks of Mihijaya Sevana, research was carried out only among inhabitants of Block C. Two government institutions were situated within Mihijaya Sevana: a police post[11] and a medical clinic. By the time of the pandemic, both these institutions were situated within Block C and they had created a network

[11] One of our interviewees, Meena, explained to us how this high-rise ended up having its own police post. She is an active member of the *Mihijaya Sevana Kalamanaakarana Samithiya,* or Mihijaya Sevana Management Society, of her block. Usually each block has its own management society formed of representatives of the residents. The main purpose of this society is to assist the UDA manager in maintaining the environment of each block and also to represent inhabitants' issues. According to Meena, initially, her block and other adjacent blocks did not have a police post. The residents after forming the Mihijaya Sevana Management Society in July 2020 requested a police post. The aim was to deter violent encounters between groups of young men from Mihijaya Sevana and other nearby high-rises and maintain a peaceful environment within the high-rise. The idea of a police post was not favored by some male residents in the high-rise, claiming that it would cause unnecessary interference with their behavior. However, when the management society inquired about their request they were informed by the local police station that there were some pre-requisites to establish a police post within the premises. According to the information received from the police, all the inhabitants of the high-rise should agree and sign a formal request, and even after signing the request, they were told that it would take at least 4-5 years to establish a police post. This discouraged the inhabitants and they gave up on their idea. However, to the surprise of the inhabitants, in October 2020, within a week, a police post was established in a vacant housing apartment near the 'short-cut entrance' to the high-rise. During the isolation period, this police post was one of the surveillance points, ideally situated near the 'short-cut entrance' of Block C from which the police could observe the movements of the inhabitants. Currently this police post has been moved to Block D not far from its previous location and has been turned into a complete police

between themselves to observe and surveil the spread of the virus and the movement of the inhabitants within the high-rise. We term this constellation an 'institutionalized network of control'. These two institutions were continuously keeping records of COVID-19 patients or inhabitants who had influenza-like symptoms within the building and both institutions communicated with each other regularly. Through the records that were kept, both the medical clinic and the police post were able to surveil the inhabitants closely.

According to Meena,[12] during the lockdown period, the entrance and the exit to the high-rise premises were guarded by police personnel. There was a main entrance and a 'shortcut entrance' to the building. Furthermore, there was a main exit. In each block, there is one place to enter and one place to exit the particular block and another shortcut entrance. During the isolation period, policemen were stationed at the main entrance and the main exit, thus allowing no one to leave the premises without proper authorization from the police. Further, the police post was situated at the short-cut entrance to Block C. Thus, all the entry and exit points were monitored by the police.

As the inhabitants were aware that they were being monitored, they self-regulated themselves and did not leave the building for fear of being seen and arrested for breaking lockdown measures. Shironi, another interviewee resident of the high-rise, explained her experience of being monitored by the police as follows:

> This particular block has three elevators and one staircase. But if someone wants to leave the building there is only one main entrance, one main exit and one shortcut entrance. During the lockdown, the residents were given an announcement about the upcoming lockdown. And then the police closed the gates. We could not go out. When the lockdown was imposed for three months straight it felt like that we were like prisoners. As if that we were sentenced to prison not for three months but for three years (Shironi 2021, personal communication, 4 April).

When interviewees were inquired about their perceptions of the pandemic situation and the mobility patterns and potential in their previous areas of residence ('*watta*') which were spread horizontally, most of them mentioned a significant difference. For example, Krishna used the following expression:

> If we were still at the *watta* the situation might have been different. Even during the lockdown, we might have been able to get more chances to leave without being seen by the police and make money. The police could never cover all the entrance and existing points to the watta. (Krishna 2021, personal communication, 4 April)

The establishment of the police post near the shortcut entrance and placing officers at the entrance/exit of the premises created a sense of being watched, just as in the panopticon. Even if there had perhaps not been police officers at these places at all times, the inhabitants did not have the means to verify it. Therefore, they were not willing to take the risk of being caught and punished for violating the health

station with a complaint room and a holding cell. Now all the activities of the inhabitants related to law and order in the high-rise are under the jurisdiction of this police station.

[12] All names used in the quotations here are pseudonyms to ensure the anonymity of all personally given information of the interviewees. The only reference to the real living situation is the given name of the residential area.

guidelines. Thus, they practiced self-disciple during the pandemic. According to one of the interviewees, the nature of the built environment of the high-rise and the existence of a police post within the building, has also given them a sense of an "invisible authority" residing over them compelling them to be disciplined:

> Even people, who behave badly when they are outside, even try to be disciplined when they come inside (this building). Unlike in the *watta,* people are afraid to quarrel when they are here. Because they know that the police post is on the ground floor. Police have given its contact number to every house. So even if it is not someone from the floor where the people quarrel, someone from the upper or lower floor might call and inform the police about the quarrel. If this happened in the *watta,* it would take at least an hour for the police to come to the scene. By then everything has already happened. (Azafa 2021, personal communication, 26 October)

This situation crucially directly confirms what Foucault explained as Bentham's principle on power in relation to the panopticon:

> "(…) power should be visible and unverifiable. Visible: the inmate will constantly have before his eyes the tall outline of the central tower from which he is spied upon. Unverifiable: the inmate must never know whether he is being looked at any one moment; but he must be sure that he may always be so". (Foucault 1975, p. 201)

Considering what the inhabitants expressed on how these measures would have worked if they were still in the *watta*, it is evident that the built environment of the high-rise has strengthened the surveillance by the police, an institution Foucault identifies as using disciplinary measures in society.

During the lockdown period, no government officials or police officials came into the building, as they were not willing to take the risk of being infected. Thus, social exclusion is augmented by and justified through discriminatory social profiling. Instead, the medical clinic, which was brought inside the high-rise during the pandemic, maintained direct communication with the police post which was situated at the shortcut entrance, and together they created a perfect model of surveillance through their communication.

> During this time (isolation) we were so afraid to even go to get some medicine for a common cold. Because if we go, they will do the PCR, putting a stick into your nose. There is a clinic in the building. If we go there to buy some medicine for a cold or fever, the clinic would get our details and would immediately call the police post at this block. Then the police would come to our house and say "Somebody from this house has gone to get medicine for a cold and fever and we need to do PCR for all of you." We were so afraid of PCR that we never went to take any medicine. We drank boiled coriander and stayed at home. (Meena 2021, personal communication, 4 April)

This interconnectedness between government medical clinic and police post is what we term an '**institutionalized network of invisible authority**', which can also be elaborated using panopticism. In his explanation of how panopticism functions within society, Foucault states that panopticism exists in society not because existing methods are being replaced by it, but because these disciplinary modalities have managed to infiltrate the existing ones, link each other, etc. (Foucault 1975, p. 216). This interconnectedness between the government medical clinic and police post is a good example of how disciplinary modalities have linked different institutions that hide power structures of control behind institutionalized necessities to keep the pandemic in check. This can also be considered a very perfidious system of patronage.

14.5.2 The Corridor—A Modern Spatial Arrangement of Panopticism for Internal Surveillance

Spatialities are a powerful means to consolidate social control in built structures that lead to a perfidious reproduction of hidden power structures. As such, the corridor of the high-rise providing access to the apartments became an object of materialized surveillance structures, in its blurred role between being a public and a private space at the same time. During the interviews, it was evident that this space, the *corridor*, was used for different purposes by the inhabitants, and the inhabitants' prevailing understanding of the corridor was blurred and became conflicted during the period of isolation. Based on the interviews, it was clear that the inhabitants identified the corridor as a space that serves different purposes but also as a space that poses different threats for them. Those perceptions can be identified in four main categories: The corridor as:

1. a space to socialize
2. a neutral space of freedom (public space)
3. an unsafe space
4. a space of surveillance.

The first two understandings of the corridor by the inhabitants showcase a positive connotation while the third and fourth showcase a negative connotation of the corridor (Fig. 14.1).

14.5.2.1 The Corridor as a Space to Socialize

Socializing was one of the main purposes for which the inhabitants used the corridor. They viewed the corridor as a space to communicate with each other, especially across the floor or with a person living on a different floor.

> The corridor is the place where we talk with each other. If I need to talk with someone right across from my floor, I go to the corridor and call for them. When we talk with someone from the upper floor, we use hand signals for small chats. Like they might ask how are you doing? (shows a hand signal, while mouthing the words) and I would say I'm fine let's talk later (shows another hand signal while mouthing the words) But if I need to talk with them further, I'll be signaling them to come down to my floor to talk with me. (Meena 2021, personal communication, 26 October)

However, compared to the socializing experience they had at their place of origin this is somewhat different. When they were living horizontally in their former settlement, socializing was not something into which they had to put a special effort. In

Fig. 14.1 A photograph of a corridor in Block C, Mihijaya Sevana (*Source* Photograph by authors 2020)

most cases, they had their long-term neighbors, friends, or relatives living beside their houses, and chatting with each other was a part of their daily routine. Nevertheless, because of the procedure followed by the UDA in relocating these inhabitants, they have been relocated to different high-rises and different floors within the same high-rise. It is essential to explore the implication of these relocation arrangements combined with the spatial structure of the high-rise effect on the social relationships of the inhabitants. According to Meena, their new life at the high-rise has less social interaction with both former and new neighbors compared to their previous settlement, mainly due to the spatial arrangement of the building. Since it takes an extra effort for them to visit their friends, neighbors, and relatives (going to an upper or lower floor) she said that most of their interactions are now limited to the corridors.

In a way, it can be argued that in the first place the spatial arrangement of the high-rise has a negative impact on social interaction and the inhabitants have identified the corridor as a place to continue whatever social relationships they could salvage within the new environment.

14.5.2.2 The Corridor as a Space for Free Movement

During the isolation period, inhabitants were not allowed to leave the high-rise. As the policemen were stationed at every entrance/exit point, they did not leave the building for fear of getting caught by the police. This limited their movement to the house and to the corridor.

> This block was promised to have a pre-school. Now, we have it. Then a bank to do our banking. Then a community hall to have weddings and funerals. But, we don't have any park. The only play park is the one within the pre-school. During the lockdown, we were not even allowed to walk in the small area beside these buildings. The only place was the corridor. (Mohammed 2021, personal communication, 4 April)

Although the corridor was perceived as a space of free movement for the inhabitants, this freedom was curtailed when some severe measures were taken to control the spread of the virus. There had been instances when people were not even allowed to walk in the corridor. This type of restriction may have caused serious challenges to the mental health of the inhabitants of the high-rises. Dias et al. argue that during the COVID-19 people who owned households with private gardens or terrace space, a balcony or shared garden were comfortable and households who had no access to any sort of private open space were considerably less comfortable (2021, p. 625). In the case of Mihijaya Sevana, the housing units do not have private balconies so the only space they could use apart from their houses is in the corridor. However, at one point the free movement in the corridor was restricted, which put the inhabitants under extreme stress. The prolonged isolation caused stress within the high-rise inhabitants, resulting in massive protests from the neighboring high-rise 'Methsanda Sevana' against the isolation measures (Dias et al. 2021).

14.5.2.3 The Corridor as an Unsafe Space

Although the corridor was used by the inhabitants as a space to socialize, it was also a space that posed a threat to them. According to Anjula, a resident of Block C, the corridor is not a space to be considered 'safe'.

> During the daytime, we usually keep the doors shut, as the corridor is not safe. Strangers usually walk through, and we can't question their whereabouts. So we are afraid to let our children play there. If someone dragged them and went, we would not be able to find them. There are hundreds of houses here. We may not know; someone might drag them to an empty house on an upper floor and might abuse them. Recently, some lady from an upper floor was screaming that her child was lost. People from a few floors began to search for the child everywhere. Finally, the mother found the child playing at a neighboring house. Without her

noticing the child had gone to the neighboring house to play. So we have asked our children to play inside the house. (Anjula 2021, personal communication, 26 October)

Thus, the inhabitants saw the corridor as an unsafe place for their children. Since it is a public space, they cannot guarantee safety in the corridor as they have lost their authority over the space in which they live. The only space where they can ensure their children's safety is their own house. Anjula further explained how this would have differed if they had still been at the *watta.*

When we were at the *watta* our children played leisurely even until dark. We could always keep an eye on them while working from our homes because at the *watta* we keep our doors open. There is no chance for strangers to pass through without anyone from the *watta* noticing. (Anjula 2021, personal communication, 26 October)

Strangers were not the only thing that made the corridor a perceived unsafe space. Frequent thefts within the high-rise increased this sense of unsafety to another level. Rajika explained an experience of one of her neighbors as follows:

Even though the corridor is right in front of your house, you cannot keep anything in the corridor. There are lots of thefts happening in the high-rise. One of my neighbors had kept their bicycle in the corridor in front of their house. It was parked there for a few weeks. There were brief power-cuts at night for a few days and we never suspected anything. One morning, after such a power-cut, my neighbor realized that someone had stolen their bicycle. They even made a complaint to the police post, but the police have said that they can't help them find it because they can't find any suspect and have advised them to protect their own belongings. There are some drug addicts in this block. I suspect that it is them who have made the power-cuts and committed the thefts. The electricity control room is on the ground floor. They must have found some way to create brief power cuts. You can't even dry your clothes on cloth racks in the corridor. These drug addicts would even steal them. (Rajika 2021, personal communication, 4 April)

Thus, the corridors in front of their houses are not considered as a safe space, where the inhabitants can store any personal belongings. However, Meena claimed that this was entirely different compared to their previous houses at the *watta.*

Such kinds of thefts cannot happen at the *watta.* Even if someone steals from a house at the *watta,* there will always be someone who saw the thief. When you ask around and spread the word about the theft, the ones who saw such an unusual scene or unfamiliar person would inform you saying "Akka I saw someone taking your gas cylinder and walking in that direction". (Meena 2021, personal communication, 4 April)

Therefore, it became evident that a sense of security was among the inhabitants when they were living horizontally with their community. They knew their community very well as they were living in the same location for decades or generations. However, it should also be noted that topics like 'safety' are highly emotional. This issue could also trigger criticism from the inhabitants on the architectural design and would lead them especially to consider these 'new spaces' as unsafe when compared to their former settlements. As they have emotional attachments to their former settlements, since it is where they grew up and were raised over four generations, emotional bonds based on years of personal experience can form a 'place attachment'[13] to their former settlement, something which might take years to form within their new settlement.

14.5.2.4 The Corridor as a Space of Surveillance

During the isolation period, the corridor was also used as a space of surveillance. It is interesting that this surveillance was implemented both by government agents like the police and by the inhabitants themselves.

According to Shironi, although they were allowed to walk in the corridor during isolation, at one point they were placed under home confinement. During this period, they were not allowed to walk in the corridor, not even in the corridor space in front of their houses. "During the second corona wave, we were prohibited even from gathering in the corridors. Occasionally police officers would roam to check whether there are any unnecessary gatherings in the corridors". (Shironi 2021, personal communication, 4 April).

This indicates how police officers were using the corridor as a space for their observation. Since all doors faced the corridor, it was easy to spot anyone who had left their house. If the police officer is on the top floor, s/he can observe all the floors and the doors, just by walking around the particular floor, using the corridor, thus giving a good vantage point for the police officer (Fig. 14.2).

On the other hand, the same space was used by the inhabitants themselves for social control of people who leave their houses or loiter in the corridors. According to Meena, the inhabitants themselves would keep surveilling the neighbors."There are some people, who do not care about the health guidelines and loiter on the corridors. When we see them, we would say "*Ayyo why are you waiting in the corridor? Please go inside your house and stay safe, without spreading the virus*" (Meena 2021, personal communication, 4 April).

It can be argued that these self-regulatory behaviors associated with the corridor such as keeping the doors closed during the day and the internal surveillance created by the inhabitants to keep their own neighbors inside their houses are unintended outcomes of the built environment. Identifying the corridor as a public space is the main reason behind this unintended outcome. It is essential to distinguish what the inhabitants consider to be public and private within the high-rises in this regard. Based on the ideas expressed during the interviews, the inhabitants consider the corridor to be a space over which they have no control. The only space that they can control and manipulate is their own house. The inhabitants consider the corridor to be an unsafe space where their children can be harmed, and their belongings stolen.

Also, more importantly, they view the corridor as a space used as one of the observation points of the police during the pandemic time to monitor them. This understanding of the corridor as an unsafe public space rather than a safe private

[13] Place attachment is an emotional "bonding that occurs between individuals and their meaningful environment" (Scannell and Gifford 2010, p. 1).

Fig. 14.2 The building structure of Block C, Mihijaya Sevana (*Source* Photograph by authors 2020)

or even a communal space has allowed government authorities to use it for their surveillance. Furthermore, the fact that the corridor gives an outsider (or the authority in this case) the chance to observe the inhabitants from just one point of observation has created a feeling of being constantly watched. This confirms Foucault's claim that "visibility is a trap" (1975, p. 200). Because, even when there are no police officers observing from the corridor, inhabitants may be under the impression that they are being constantly watched and would self-regulate and adhere to the safety measures imposed by the health officials, especially during the home confinement period.

The ability of the police to use this kind of surveillance can again be linked to the built environment of the high-rises as a materialized manifestation of controlling power structures. Being built vertically and in a similar manner as a panopticon, the high-rise has given the police the advantage of initiating surveillance and thus applying an "invisible authority", without providing the inhabitants the chance to confirm whether they are being watched or not.

14.6 Conclusion

Governance in the modern world is much advanced compared to earlier modes of governance. The case study of Mihijaya Sevana during the COVID-19 pandemic substantiates this claim further. Using the nature of the built environment of the high-rise, the state proved that its existing disciplinary modalities can be strengthened and used to control the behavior of thousands of inhabitants while using very minimum human resources. The observation and surveillance exercised within the high-rise during the pandemic, which was quite visible, but not verifiable, managed to infiltrate inhabitants' social and material spaces and led them to discipline themselves. Therefore, the existence of panopticism was clearly evident during this period.

However, the most important question regarding such disciplinary practices in times of pandemics lies elsewhere. The COVID-19 pandemic was not something limited to one country. It was a global emergency and states all around the world took various measures to mitigate the spread of the virus. Most of these measures varied and changed according to the level of spread and were not aimed at certain social groups in society. In the case of Mihijaya Sevana, the disciplinary measures were so unique to its inhabitants and were not like what was witnessed in other parts of the country. The inhabitants of Mihijaya Sevana were treated with prejudice and were under some pre-forms of social profiling/stigmatization by the state similar to how people affected by the plague were profiled in France. In this case, these inhabitants were treated as if they were not to be trusted to follow the state's quarantine orders, thus they needed to be put under surveillance and controlled. It can be argued that the social stigmatization of the inhabitants of Mihijaya Sevana was based on their place of origin, i.e. slums, and shanties. These speculations of the state should be questioned on the grounds of whether the state followed the same kind of surveillance and control over other social groups such as middle-class citizens or upper-class citizens. Although there were roadblocks by police or army officials covering the entrances and exits to some areas, they were not placed to surveil the residents of those areas. Moreover, the houses of the residents were spread horizontally, so such surveillance was not practically possible. However, the special nature of the high-rises, which are vertically spread, and a larger population residing in one physical space assisted the state to keep a close observation of the inhabitants' movements. Although emergency situations like pandemics could lead to limitations on personal liberties and rights of the citizens, they cannot be used to justify unfair treatment towards certain social groups like high-rise inhabitants in this case.

Given the presented findings, we conclude that a critical analysis is essential to understand everyday life experiences of deprived urban dwellers better and to uncover hidden power aspects of governmentality. Qualitative data add significant insights and are an important tool to unveil subaltern voices to complement technical data emerging from quantitative surveys etc. Embedded in a critical theoretical context like the given framework of discussion these findings can help to formulate recommendations for practitioners and planners for a socially more just city planning and urban governance.

References

Abeyasekera A, Maqsood A, Perera I, Sajjad F, Spencer J (2019) Discipline in Sri Lanka, punish in Pakistan: neoliberalism, governance, and housing compared. J Br Acad 7(2):215–244. https://doi.org/10.5871/jba/007s2.215

AdaDerana (2020) Isolation status on six urban flats in Colombo lifted. Available at: http://www.adaderana.lk/news/69881/isolation-status-on-six-urban-flats-in-colombo-lifted. Accessed 28 June 2022

Alfano V, Ercolano S (2022) Stay at home! governance quality and effectiveness of lockdown. Soc Indic Res 159:101–123. https://doi.org/10.1007/s11205-021-02742-3

Brenner N, Schmid C (2015) Towards a new epistemology of the urban? City 19(2–3):151–182. https://doi.org/10.1080/13604813.2015.1014712

Collyer M, Amirthalingam K, Jayatilaka D (2017) The right to adequate housing following forced evictions in post-conflict Colombo, Sri Lanka. In: Brickell K, Fernandez M, Vasudevan A (eds) Geographies of forced eviction—dispossession, violence, resistance. Palgrave Macmillan, London, pp 47–69

Connolly C, Keil R, Ali SH (2021) Extended urbanisation and the spatialities of infectious disease: demographic change, infrastructure and governance. Urb Stud 58(2):245–263. https://doi.org/10.1177/0042098020910873

Corpuz JCG (2021) Adapting to the culture of 'new normal': an emerging response to COVID-19. J Public Health 43(2):344–345. https://doi.org/10.1093/pubmed/fdab057

Cyr J, Bianchi M, Gonzalez, Perini A (2021) Governing a pandemic: assessing the role of collaboration on Latin American responses to the COVID-19 crisis. J Polit Latin Am 13(3):290–327. https://doi.org/10.1177/1866802X211049250

Department of Government Information (2020) Press release. Press release. 18 Dec 2020. Available at: https://www.dgi.gov.lk/news/press-releases-sri-lanka/2542-press-release-2020-12-15

Development Planning Unit (2018) Muyadora Uyana—relocating quality of life in Colombo, Sri Lanka. University College London, London

Dias N, Jayakody C, Amaratunga D, Abeynayake C, Jayasinghe A (2021) Health resilient cities in a post Covid world. In: Senaratne R, Amaratunga D, Mendis S, Athukorala P (eds) COVID 19: impact mitigation, opportunities and. building resilience "from adversity to serendipity". Perspectives of global relevance based on research, experience and successes in combating COVID-19 in Sri Lanka, vol 1, pp 621–635. National Science Foundation, Colombo

Dobusch L, Kreissl K (2020) Privilege and burden of im/mobility governance: on the reinforcement of inequalities during a pandemic lockdown. Gender Work Organ 27:709–716. https://doi.org/10.1111/gwao.12462

Enright T, Ward K (2021) Governing urban infrastructures under pandemic conditions: some thoughts. Urban Geogr 42(7):1023–1032. https://doi.org/10.1080/02723638.2021.1893050

Foucault M (1975) Discipline and punish—the birth of the prison. Vantage Books, New York

Ghertner DA (2010) Calculating without numbers: aesthetic governmentality in Delhi's slums. Econ Soc 39(2):185–217. https://doi.org/10.1080/03085141003620147

Gunatilleke G, Nathaniel V (2014) The national involuntary resettlement policy. Verité Research, Colombo

Jayawardhana T (2010) Infrastructure challenges in post-conflict development. In: Sri Lanka—state of the economy report 2010. Institute of Policy Studies of Sri Lanka, Colombo, pp 1–15

Keil R (2020) The space and time a pandemic makes. Plan Rev 56(3):4–9. https://doi.org/10.1080/02513625.2020.1851895

Ministry of Health and Indigenous Medical Services (2020) Sri Lanka preparedness & response plan—COVID-19 April 2020. Ministry of Health and Indigenous Medical Services, Colombo, Sri Lanka

Nyashanu M, Simbanegavi P, Gibson L (2020) Exploring the impact of COVID-19 pandemic lockdown on informal settlements in Tshwane Gauteng Province, South Africa. Glob Public Health 15(10):1443–1453. https://doi.org/10.1080/17441692.2020.1805787

Onyishi CJ, Ejike-Alieji AU, Mbaegbu CC, Ezeibe CC, Onyebueke VU, Mbah PO, Nzeadibe TC (2021) COVID-19 pandemic and informal urban governance in Africa: a political economy perspective. J Asian Afr Stud 56(6):1226–1250. https://doi.org/10.1177/0021909620960163

Parnell S (2020) The enabling conditions of post-pandemic city government. Environ Plan b: Urb Anal City Sci 47(7):1143–1145. https://doi.org/10.1177/2399808320950041

Parnell S, Robinson J (2013) (Re)theorizing cities from the global south: looking beyond neoliberalism. Urban Geogr 33(4):593–617. https://doi.org/10.2747/0272-3638.33.4.593

Perera I (2015) Forced evictions in Colombo: high-rise living. Center for Policy Alternatives, Colombo

Perera I (2016) Living it down: life after relocation in Colombo's high-rises—a survey of households relocated by the Urban Regeneration Project. Center for Policy Alternatives, Colombo

Perera I (2020a) We can't feel the earth beneath our feet anymore": dispossession and high-rise living in Colombo. Polity 8(1&2):51–56

Perera I (2020b) The urban regeneration project. Bank Information Center

Perera I, Ganeshathasan L, Samaraarachchi S, Ruwanpathirana T (2014) Forced evictions in Colombo—the ugly price of beautification. Center for Policy Alternatives, Colombo

Reich MR (2020) Pandemic governance in Japan and the United States: the control-tower metaphor. Health Syst Reform 6(1):e1829314. https://doi.org/10.1080/23288604.2020.1829314

Roy A (2007) Urban informality: toward an epistemology of planning. J Am Plann Assoc 71(2):147–158. https://doi.org/10.1080/01944360508976689

Roy A (2015) What is urban about critical urban theory? Urb Geogr 37(6):810–823. https://doi.org/10.1080/02723638.2015.1105485

Scannell L, Gifford R (2010) Defining place attachment: a tripartite organizing framework. J Environ Psychol 30(1):1–10. https://doi.org/10.1016/j.jenvp.2009.09.006

Sharifi A, Khavarian-Garmsir AR (2020) The COVID-19 pandemic: impacts on cities and major lessons for urban planning, design, and management. Sci Total Environ 749:1–14. https://doi.org/10.1016/j.scitotenv.2020.142391

Sri Lanka National Involuntary Resettlement Policy (2001) [online] Available at: https://landportal.org/node/43813

Strub H (1989) The theory of panoptical control: Bentham's panopticon and Orwell's nineteen eighty-four. J Hist Behav Sci 25(1):40–59. https://doi.org/10.1002/1520-6696(198901)25:1%3c40::aid-jhbs2300250104%3e3.0.co;2-w

Urban Development Authority (2018) Urban regeneration programme [online]. https://www.uda.gov.lk/urban-regeneration-programme.html. Accessed 10 Apr 2022

Vanclay F (2017) Project-induced displacement and resettlement: from impoverishment risks to an opportunity for development? Impact Assess Project Appraisal 35(10):3–21. https://doi.org/10.1080/14615517.2017.1278671

World Health Organization (2020a) August feature countries—a monthly selection of case studies [online]. https://cdn.who.int/media/docs/default-source/sri-lanka-documents/5th-edition-august-2020a-covid-19-country-case-studies.pdf?sfvrsn=27d968d5_0. Accessed 10 Apr 2022

World Health Organization (2020b) Maintaining essential services: operational guidance for the COVID-19 context. World Health Organization

Chapter 15
Policy Recommendations for Built Environment Professional Bodies in Upgrading Their Professional Competencies to Address Displacement Contexts

Anuradha Senanayake, Chamindi Malalgoda, Chathuranganee Jayakody, Dilanthi Amaratunga, and Richard Haigh

Abstract Displacement due to extreme weather conditions and conflicts is a frequent phenomenon in the contemporary world. With the overall alteration of their day-to-day lives, the displaced communities require urgent and sustainable built environment interventions to rebuild their lives. It is vital that the needs of the displaced and host communities are satisfied by the built environment. However, in some instances, built environment professionals such as planners, designers, and architects are not aware of the cultural, social and economic needs of the displaced and host communities. This indeed has a grave impact on the overall satisfaction and impact of built environment interventions. Hence, this study is an investigation of the role of built environment professionals in relation to displacement. The study was based on evaluating the need for built environment professional bodies to upgrade their professional competencies to address the needs of host and displaced communities. The need was justified recognising existing gaps in built environment professional competencies and further formulating policy recommendations on how built environment professionals could accommodate the needs of the displaced and host communities in their interventions effectively. The methodology employed consisted of three stages including an integrative literature review, mapping the findings of the literature review against the competency framework developed by the research team, and eventually recognising the gaps of the existing competencies through a competency audit. To cover the existing lacunas in the built environment competencies, the research team have come up with 14 policy recommendations including considering contextual differences, the role of stakeholders, well-being, and living standards of

---

A. Senanayake (✉) · C. Jayakody · D. Amaratunga · R. Haigh
Global Disaster Resilience Centre, School of Applied Science, University of Huddersfield, Huddersfield, UK
e-mail: anuradha.senanayake@hud.ac.uk

C. Malalgoda
Global Disaster Resilience Centre, School of Arts and Humanities, University of Huddersfield, Huddersfield, UK

© The Author(s), under exclusive license to Springer Nature Switzerland AG 2023
M. Hamza et al. (eds.), *Rebuilding Communities After Displacement*,
https://doi.org/10.1007/978-3-031-21414-1_15

the displaced, enhancing social cohesion, and formulating legal and policy frameworks. The overall objective of these recommendations is to refine the boundaries of built environment professionals in empowering the displaced communities socially, culturally, and economically.

Keywords Displacement · Built environment interventions · Built environment professional competencies · Policy recommendations

15.1 Introduction

People being forcefully displaced from their habitats and countries is one of the main challenges the contemporary society is attempting to address. Within just the first half of the year 2021, it was reported that 84 million people had been displaced in comparison to the 82.4 million that were displaced upon completion of the year 2020 (UNHCR 2021b). Conflict and extreme weather events are recognised as two of the main trends in the annually reported cases of new displacements. The year 2020 marked the highest rate of new displacement for the decade with 40.5 million people being displaced (IDMC and NRC 2021). In a detailed investigation into the said displacement statistics, out of the said 40.5 million, a majority of 30.7 had been displaced due to extreme weather events (IDMC and NRC 2021). However, in the context of internal displacement in 2020, a majority of 48 million were internally displaced as a result of conflict situations or violence (IDMC and NRC 2021). This is a significant increase in comparison to the 10.8 million new displacements reported due to conflict situations in 2018 (IDMC and NRC 2019). Further, the United Nations Refugee Agency has reported 91,920,446 of persons of concern in relation to displacement by December 2020 (UNHCR 2021a). In terms of the global context of displacement in 2020, East Asia and the Pacific rank at the top (30.3%) while Europe and Central Asia report the least (0.8%) amount of displacement cases reported (IDMC and NRC 2021).

In terms of disasters, climate change has been recognised as a risk multiplier that contributes towards disaster risks (IFRC 2020). Despite the fact that the year 2020 is known for the COVID-19 pandemic, it has been recognised as a year that was dominated by other forms of extreme weather events such as flooding, landslides, droughts, and storms (UNHCR 2021b). In 2020, a total of 389 natural disasters were reported (CRED and UNDRR 2021). Further, more than 100 disasters which occurred in the first six months of the COVID-19 pandemic affected more than 50 million people in the year 2020 (IFRC 2020). The Asia Pacific region has been a vital geographical region in terms of disaster-induced displacement where between 2008 and 2018, 80% of global disaster-induced displacement was reported from the said region, estimated to be approximately 187 million cases (IDMC and NRC 2019).

In terms of conflict-induced displacement, with the prevailing conflict situations in countries such as DR Congo, Afghanistan, Ethiopia and Nigeria, a significant rise in internal displacement statistics from those countries in 2023 has been forecast

(DRC 2021). It has been further noted that the overall violence against country citizens has increased by 22% in 2021 in comparison to 2020 (DRC 2021). The conflict situation in the Ukrainian context adds a novel chapter to the conflict-induced displacement history in the world. It has been estimated that 18 million people will be affected and 12 million people will require humanitarian assistance due to the ongoing conflict in 2022 (Loft and Brien 2022). The conflict situations and violence in Yemen, Namibia, and Iraq have further been recognised as contributing factors for the increasing internally displaced people due to conflicts as well as increasing vulnerability in extreme weather events (IFRC 2021).

This escalation of forced displacement is a major concern as it tends to increase the level of the vulnerability of the displaced community (Hamza 2021). Such an increase in vulnerability due to displacement has been interpreted in the perspectives of social, cultural, and economic aspects (Cazabat 2018; Christensen and Harild 2009). It has further been pointed out that social capital, health, livelihood, and natural environmental resources of displaced communities tend to be reduced due to displacement (Cazabat 2018). Given that displacement contexts lead to a broader impact on the living fabric of the displaced communities, recommendations have been made on managing these broader impacts with proactive political recognition and investment in internal displacement, capacity development of individuals and organisations to reduce displacement and increase access to real-time data on displacement (IDMC and NRC 2020).

With this broader impact on the displaced community, intellectual scrutiny has taken place in terms of the role of built environment in disaster and conflict-induced displaced contexts (Amaratunga et al. 2022). It has been pointed out that the built environment plays a major role in the displacement contexts through providing adequate and appropriate housing, sufficient infrastructural facilities to both host and displaced communities, addressing their livelihood challenges, and consciously creating social cohesion between displaced and host communities (Jayakody et al. 2021). Built environment has been further acknowledged for enhancing social cohesion between the displaced and the host communities (Amaratunga et al. 2020; Jayakody et al. 2022) and enhancing resilience of the displaced community (León et al. 2019). The need of preserving the identity and culture in post displacement built environment contexts has been further emphasised (Mull 2016). Built environment has also been recognised as a vital component in child development in urban displacement contexts (Conti et al. 2020).

In post displacement, planned relocation plays a vital role as a recovery process (Senanayake et al. 2022). As an integral part of the post disaster recovery process, addressing physical needs by providing housing units for the displaced has been recognised as a favoured option (Sridarran 2018). However, when considering the experiences of the Global South, deteriorating housing conditions have been recognised as one of the main reasons for housing dissatisfaction among the displaced communities post relocation (Fernando and Punchihewa 2013; Sridarran et al. 2016). Further, it has been mentioned as beneficial for the authorities to work with built environment professionals at times of crisis to reflect on lessons learned in the previous interventions and to recognise key areas that need to be addressed in future risks (Max Lock Centre 2009).

Therefore, the role of the built environment in addressing the needs of the displaced as well as the urgency of stepping up their competencies cannot be disregarded (Amaratunga et al. 2022), especially because there is research which reveals that people usually end in worse and more vulnerable conditions after displacement and resettlement (Vanclay 2017). While acknowledging the important roles of the built environment and the progress made so far in addressing the needs of the displaced and the host communities, it is critical for built environment professionals to upgrade their professional competencies to address the various needs of the displaced and the host communities.

This study was a result of a three-year-long research collaborative titled "REGARD (REbuildinGAfteR Displacement)". This is a project initiative co-funded by an EU Erasmus + programme grant, led by the University of Huddersfield's Global Disaster Resilience Centre, United Kingdom. The project lead is joined by a consortium of five higher education institutions from four countries in Europe and Asia. This study is dedicated to presenting the outputs of one of the project's objectives of proposing policy recommendations to built environment professional bodies in order to upgrade their professional competencies to address the needs of the host and displaced communities.

Hence, this study is an investigation of the research problem of 'what is the need for built environment professional bodies to upgrade their professional competencies to address the needs of the host and displaced communities?'. In order to investigate the said research problem, this study utilises two sub objectives of investigating the competency gaps of the built environment professional bodies in addressing the needs of displaced and host communities and making recommendations to upgrade built environment competencies to accommodate needs of the displaced and host communities efficiently.

15.2 Methodology

The methodology of this study can be explained in several stages. Mainly, the methodology could be interpreted in three stages of managing secondary and primary data. Primary data collection was based on the partner country contexts of Estonia, Sri Lanka, Sweden, and the United Kingdom.

15.2.1 Stage 1—Mapping of the Initial Literature Review Findings Against the REGARD Competencies

The initial stage dealt with secondary data where an integrative literature review was conducted to recognise issues and challenges faced by host and displaced communities in displacement contexts. The review included literature representing both the Global South and the Global North to represent the country contexts of all the project partners.

The review revealed a range of issues and challenges faced by host and displaced communities raising the concern of addressing them to be a timely requirement. Afterwards, the thematic data analysis method was utilised in terms of recognising the prevailing challenges and issues in built environment interventions in displacement contexts. The summary of the themes recognised from the literature review could be presented as follows and these issues and challenges will be discussed in detail under the findings and discussion section (Fig. 15.1).

Next, these recognised challenges and issues were mapped against the competency framework developed as a major output of the project. The competency framework (hereinafter referred to as the REGARD competencies) was developed in a three-step process using a Delphi technique with an international panel of 19 experts and 3 rounds of questioning.

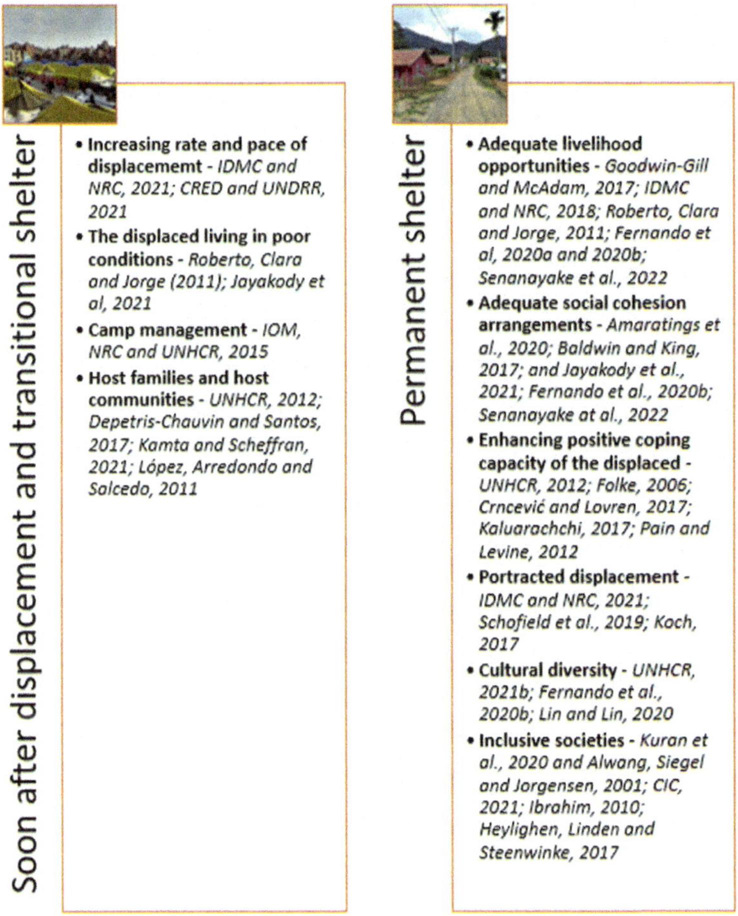

Fig. 15.1 Thematic map of the literature review conducted (*Source* Prepared by the authors)

The final competency framework developed (Fig. 15.2) covers three layers of competency with broader foundational competencies relevant to every profession and practitioner interested in displacement. The second and the third layers of the competencies are more specific levels of competencies relevant to built environment professionals and practitioners in the context of mass displacement.

Fig. 15.2 REGARD competency framework for built environment professionals (*Source* Amaratunga et al. 2022, p: 38)

15.2.2 Stage 2—Recognising Competency Gaps by Evaluating the Outcomes of the Stage 1 Mapping Exercise Based on the Findings of the Competency Audit

The competency audit was another major output of the project which was initiated by evaluating the professional competency documents of the partner countries in search of built environment professional competencies. Next, the built environment professional competencies identified in professional bodies' documentation were reviewed with respect to REGARD competencies developed in the aforementioned competency framework of the project. This was done as a mapping exercise to examine which REGARD competencies are currently covered (or not) in professional documentation, and the extent of to which such competencies are covered.

Further, it was investigated whether these competencies are key/partially key to the general built environment professional competencies. Four possible outcomes of each compared point were determined:

K: REGARD Key Competency is Covered in the professional documentation.
P: REGARD Key Competency is Partially Covered in the professional documentation.
R: A relevant Key Competency but Not Covered in the professional documentation.
N: Competency not relevant in REGARD context.

The second stage of this study was dedicated to validating the outcomes of the mapping exercise of stage one along with the competency audit outcomes. The validation process initially evaluated each tier of the competencies along with the overall possible outcomes of the competency audits and the summary of the said task is in Table 15.1.

Eventually, the outcomes of the competency audit were evaluated against the outcome of the mapping exercise of stage 1 to recognise the competency gaps and to justify the need for a policy upgrade in the field of built environment addressing matters of mass displacement.

Table 15.1 Competency audit outcome percentage as per each REGARD competency tiers

Outcome	TIER 1—foundational competencies (%)	TIER 2—built environment competencies (%)	TIER 3—occupational competencies	
			3a: planning and design (%)	3b: construction and facilities management (%)
K	15	31	32	27
P	82	65	60	54
R	82	65	60	54
N	3	4	1	20

Source Prepared by the authors

15.2.3 *Stage 3—Developing Policy Recommendations Based on the Results of Stage 2*

Finally, following the recognition of existing gaps in built environment competencies in stage 2, policy recommendations were developed to address these lacunas.

15.3 Findings and Discussion

This section will present the findings in terms of the recommendations developed by the research team in addressing the main gaps recognised in terms of the built environment professional competencies in displacement contexts. During the verification process, the research team recognised that the following competencies are not recognised as key competencies in any of the built environment professions in the partner country contexts (Fig. 15.3).

Based on the above-recognised gaps, the following policy recommendations were formulated to highlight the need for upgrading the existing built environment competencies to address the needs of the displaced communities effectively.

15.3.1 *Understanding the Contextual Differences and Dynamics of Displacement*

The rate of forced displacement in recent times requires urgent and more strategic action to address the needs of the displaced and their host communities. For instance, the reported statistics on displacement notes that conflict, violence, and climate

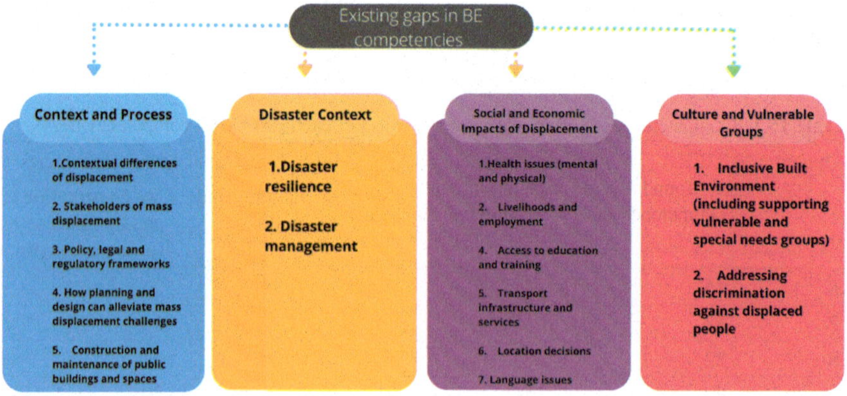

Fig. 15.3 Existing gaps in the built environment competencies (*Source* Prepared by the authors)

change-driven displacement was higher in the first half of 2021 (IDMC and NRC 2021). It has further been highlighted that with the introduction of the biological hazards of the COVID-19 pandemic, the displacement discourse requires further scrutiny (CRED and UNDRR 2021). This highlights the dynamic nature of displacement contexts and the problem in built environment professionals continuing to use the same competencies in unprecedented periods of forced displacement. With the increasing rate and pace of forced displacement, built environment interventions should consider contextual differences of displacement and further consider how planning and design alleviate mass displacement challenges. This requires timely update of the competencies to handle the dynamic nature of displacement contexts and temporary camps and resettlements more effectively.

Hence, built environment professionals should understand the contextual differences and dynamics of displacement. However, it is notable that the findings of the competency audit claim that this has not been considered as a key competency despite its acknowledgement as a REGARD competency under tiers of 1 and 2.

As it is evident that the nature, trends, causes and scales of displacement are rapidly changing, it is vital that these features are captured in built environment interventions in the contexts of displacement. Further, it is important to recognise the contextual dynamics in the Global South and the Global North in terms of the geographical contexts of the prevailing displacement discourse. Therefore, it is important to endorse policy strategies such as sustainable development and green environment to both alleviate displacement and assist the displaced communities with built environment interventions.

15.3.2 Considering Several Stages of Displacement

The process of displacement can be perceived through the lenses of temporary shelter, soon after displacement, transitional shelter, and permanent shelter in terms of relocation or resettlement (Fernando and Punchihewa 2013). Camp management is vital in the stage of soon after displacement and temporary shelter. For those who have lost property, lived through traumatic events, and are suddenly stranded or displaced outside the safeguards of their own homes and communities, camps can offer a safe haven in terms of receiving medical treatment, food, shelter, and other basic services. While camps cannot provide permanent sustainable solutions, if they are well-managed, they can temporarily meet the human rights of displaced populations and provide them with temporary refuge (IOM et al. 2015).

In terms of the long-term aspect of displacement, elongated conflicts in countries increase the number of people living in protracted displacement (IDMC and NRC 2021). In terms of disaster-induced displacement, it has been reported that displaced communities are often forced to live in informal temporary settlements which does not comply with building regulations, and with insecure land tenure and limited access to essential basic needs (Schofield et al. 2019). Hence, built environment professionals are required to update their professional competencies to include strategies to address the long-term needs of the displaced and the host communities in protracted situations. This is important because living in protracted displacement without adequate access to needs enhances the vulnerability of the displaced and the host communities (Koch 2015).

Addressing needs of the displaced in several stages of the displacement process requires a solidified legal and a policy framework. In this context, there is a need for having an overall strategy to recognise contextual differences of displacement and to strategise a legal and a policy framework in terms of the role of built environment professionals in mass displacement. However, it has not been recognised as a key competency in any of the built environment professions in the partner countries. Therefore, it is vital to recognise several stages of displacement (soon after, sometime after, and long time after) and to consider suitable built environment interventions accordingly. It is vital that such built environment interventions accommodate the needs and requirements of several stages of displacement.

15.3.3 Mainstreaming Disaster Management and Resilience in Built Environment Initiatives

It is critical to note that displacement deprives people from their employment, education, and other social and cultural activities. As a consequence, dependency on others and the inability to adequately cater for their families significantly affect the self-esteem of the displaced (UNHCR 2012). Hence, it is important to pay attention to providing the displaced and the host communities opportunities that can enhance their coping capacity, self-sufficiency, and self-esteem. In this context, the concept of resilience is important, and this concept could be explained in three views: stability (Buffer capacity), recovery (Bouncing back), and transformation (Creativity) (Folke 2006).

Studies based on Bangladesh, Sri Lanka, and Serbia have emphasized the importance of building community resilience amongst displaced communities (Crncević and Lovren 2017; Kaluarachchi 2018). Further, specifically in conflict settings, in order to build resilience of communities one should analyse power dynamics to ensure that the benefits reach the most vulnerable (Pain and Levine 2012). Therefore, it is important that built environment professionals consider measures to address this by upgrading their competencies to cover the factors that can enhance the positive coping capacity of the displaced and the host communities.

The prevailing built environment competencies do not give adequate attention to disaster management and resilience even though they are recognised as competencies under the REGARD competency tiers 1 and 2. In developing resilience in built environment interventions to address displacement contexts, it is vital to incorporate fundamentals of disaster management and disaster resilience in terms of rebuilding the communities better.

Therefore, it is important to mainstream disaster management and resilience in built environment initiatives in the context of mass displacement. Such initiatives will be beneficial in preventing further disasters and creating a built environment which accommodates the positive coping capacity of the displaced.

15.3.4 Prioritising the Well-Being and the Living Standards of the Displaced

One of the main needs of the displaced and the host communities is a quality and livable environment. Recalling what happened during the many years of conflict in Colombia, Roberto et al. (2011) state that upon arrival, the displaced persons typically resided in poor areas of urban centres. They report that the displaced and host communities experienced harsh living conditions. The need for adequate and appropriate housing facilities for the displaced and their hosts has been recognised as one of the roles of built environment professionals (Jayakody et al. 2021).

The findings of the competency audit reveal that health issues (mental and physical), livelihoods and employment (including access to means, land, etc.), access to education and training, transport infrastructure and services under the REGARD competency tier 1, and the competency of location decisions under REGARD competency tier 3a are not considered as key competencies in the partner country contexts. Hence, built environment professionals need to upgrade their competencies making novel housing codes that are of quality and a livable nature for the displaced and the host.

This requires sustainable built environmental interventions which provide necessary facilities to enhance the well-being and the living standards of the displaced. Such intervention should concentrate on infrastructure and other services that would fulfil the main needs of the displaced (water, electricity, and waste management) in addition to providing adequate shelter. The facilities should also cover transportation to make sure that there is no disruption in their adequate access to healthcare, education, and employment. Further, built environment interventions should prioritise the well-being and living standards of the displaced in deciding locations for such interventions. Location of various relevant built environment interventions should be decided based on the overall needs of the displaced. Further, built environment interventions should accommodate strategies to enhance social cohesion amongst the displaced to ensure that their social capital is not disrupted.

15.3.5 Mainstreaming the Concept of Inclusive Built Environment to Accommodate Vulnerable Groups

In terms of the concept of inclusivity, vulnerable groups need to be discussed. Such commonly recognised vulnerable groups are the elderly, children, people with disabilities (mental and physical), people with chronic diseases, and single-headed households (Kuran et al. 2010; Alwang et al. 2001). Kuran et al. (2010) however, claim the need for recognising the intersectionality of this concept of vulnerable groups. In this context, an inclusive environment recognises and accommodates various needs of several above-mentioned vulnerable groups. It facilitates dignified, equal and intuitive use by everyone. It does not physically or socially separate, discriminate or isolate (CIC 2021). In the context of built environment, as per Heylighen et al. (2017), inclusive built environment creates buildings, places, and spaces that can be used easily, safely and with dignity by all, regardless of age, disability or gender. In the context of Sri Lanka, studies have cited disaster resilient housing that includes features such as shrine rooms, wheelchair ramps, lowered electricity sockets, and shop fronts to accommodate people with disabilities (Ibrahim 2010).

However, the existing built environment competencies do not recognise addressing discrimination against displaced people under the REGARD competency tier 1 and inclusive built environment (including supporting vulnerable and special needs groups) under the REGARD competency tier 2 as key competencies. Therefore, this recommendation has two important aspects to be addressed. First of all, it is vital that built environment interventions recognise the importance of vulnerable groups and the needs specific to said groups in their interventions. The displaced communities may include women, children, people with disabilities, elders, and people with chronic diseases. Conducting a needs assessment and a social impact assessment prior to planning the interventions would allow built environment professionals to recognise the context of vulnerable groups in their setting. Next, built environment interventions should not be affected by prejudicial and stereotypical attitudes regarding the displaced. Such interventions should prioritise including the displaced communities into the mainstream society. Hence, built environment interventions should be pioneers of mitigating discrimination against the displaced.

15.3.6 Incorporating Cultural Sensitivity into Built Environment Interventions

In terms of mass displacement trends, it has been reported that people from various cultural backgrounds become internally displaced due to conflicts and disasters (UNHCR 2021b). It is important to understand the cultural backgrounds of displaced communities. For example, in the Sri Lankan context, there is a relocated community who are dissatisfied with their relocated setting due to the fact that the settlement was built on a cemetery without conducting cultural rituals that purify a land (Fernando et al. 2020b). A study conducted in Taiwan claims culture as a vital force that caters towards the livelihood needs and community resilience (Lin and Lin 2020).

The existing built environment competencies do not recognise cultural sensitivity in housing and sociology of housing as key competencies. These competencies are however recognised under the REGARD competencies framework's tier 3a. Therefore, to bridge this lacuna, it is vital that built environment professionals familiarise themselves with the diverse cultural backgrounds of displaced communities to ensure that their cultural needs are sufficiently fulfilled.

Hence, it is vital that the cultural values, norms and practices of the displaced are respected in terms of built environment interventions. The displaced may have religious and cultural interpretations in terms of their built environmental set-ups and their aspirations attached to them. Hence, conducting a socio-cultural survey would allow built environment professionals to recognise the cultural set-up of the displaced. Further, elements of the sociology of housing should be incorporated in order to make these initiatives more culturally sensitive.

15.3.7 Integrating Stakeholders into Built Environment Intervention

It is evident that a built environment intervention in the displacement process includes various stakeholders. Understanding the perspective of such stakeholders is vital for built environment intervention to be long lasting. Host families and host communities stand as vital stakeholders in this process.

It has been noted that host families are salient in the displacement process which has not received adequate attention (UNHCR 2012). This is an important part in the study of forced displacement that built environment professionals need to take into consideration, especially due to the fact that humanitarian actors are increasingly advocating for greater assistance for the displaced living outside camps and with host families (UNHCR 2012). The essence of this is for them to receive equal services with the rest in camps or host communities. It is also noted that upon displacement, there is a need to understand the preferences of the displaced as some prefer to live with host families (UNHCR 2012).

Stakeholders of mass displacement and their characteristics are not a key competency that has been recognised as per the outputs of the competency audit even though it has been recognised as a REGARD competency under tier 1. Therefore, the built environment professionals need to upgrade its competencies to include this and address matters accordingly.

15.3.8 Recognising the Needs of the Host Community

Built environment intervention in mass displacement contexts must consider the needs and the role of host communities, even though it has not been considered as a key competency. A study conducted in the Colombian internal displacement context claims that the hot community was impacted in terms of their rental prices, food prices, and poverty index (Depetris-Chauvin and Santos 2017). It has been further added that humanitarian aid should be equally distributed between both the displaced and host communities to make sure that the host community is not feeling neglected (Kamta and Scheffran 2021). Therefore, studies have reiterated that the effect on the host community also should be considered in displacement contexts (López et al. 2011). Therefore, it is vital that the needs and role of host community are not overseen in built environment interventions post displacement.

15.3.9 Creating a Built Environment Which Accommodates Livelihood Initiatives

Having access to livelihood opportunities is one of the main needs of the displaced community that needs to be addressed (Goodwin-Gill and McAdam 2017). However, it has been recognised that this need has not been sufficiently met (Roberto et al. 2011). The gravity of the impact on livelihood has been recognised in interlinked themes of disrupting business networks and reducing income opportunities due to displacement, losing future career prospects due to lack of education, over exploitation of natural resources leading to reduced crop yield, state of mental health affecting the employability of the displaced communities, and the lack of documentation preventing displaced communities working (IDMC and NRC 2018).

This has been further scrutinised with special reference to the built environment where it has been mentioned that displaced communities living in camps may not have access to livelihood opportunities and loss of livelihood can lead displaced communities to live in informal settlements (IDMC 2018). A study conducted in the Sri Lankan context provides an example from the Global South on the need for considering the impact on livelihoods when planning post disaster relocation projects (Fernando et al. 2020a, b; Senanayake et al. 2022). This indeed reflects that there is need for built environment professionals to upgrade their competencies to include what constitutes adequate economic opportunities for the displaced and their host communities and start creating such opportunities as a matter of urgency.

The competency of employment and livelihood opportunities in construction and maintenance under the REGARD competency tier 1 is not recognised as a key competency. In terms of livelihood strengthening of the displaced, the built environment can play a major role in this regard by aligning all interventions such as housing space and public facilities in such way that they accommodate livelihood initiatives for the displaced.

15.3.10 Recognising the Overall Societal Impacts of Displacement

The creation of an environment that promotes social cohesion between the displaced and the host communities is an essential role of the built environment (Jayakody et al. 2021). However, it has been noted that social cohesion and relations between the displaced and the host communities are complicated (Roberto et al. 2011). According to scholars, built environment plays a significant role in creating social capital and social cohesion (Amaratings et al. 2020; Baldwin and King 2017; Jayakody et al. 2021). However, it is vital to note that if there is no strategic planning in terms of social cohesion in the context of the built environment, the effects could be detrimental. A study conducted in one of Sri Lanka's landslide-induced relocation settings claims that there was a concern that the host community had been neglected when providing facilities and donations, developing misunderstandings between the two communities (Fernando et al. 2020b).

Therefore, it is vital that these overall societal impacts of the displaced are considered in built environment interventions. There is also one such competency under the tier 1 that has not been recognised as a key competency in the partner countries. Hence, it is vital that built environment interventions be mindful of the health impacts (physical and mental) as well as the impact the displacement would cause on their social capital. Further, the interventions should prioritise other fundamental needs such as access to education and employment.

15.3.11 Development of Public Spaces and Buildings for Recreational Facilities

Construction and maintenance of public buildings and spaces in mass displacement contexts is a competency in the REGARD competency tier 3b that has not been recognised as a key competency as per the competency audit. However, recreational and public facilities in relocation settings are considered to be opportunities to enhance the social networking amongst the displaced (Senanayake et al. 2022). A study conducted in the Sri Lankan context noted that the community organisation meetings held in the community centre at the relocation setting had positive effects in bringing communities together (Fernando et al. 2020b). Hence, built environment interventions in mass displacement contexts should address the recreational needs of the displaced. They should consider the development of public spaces and buildings to address these needs. The scope and nature of such facilities should be decided as per the requirements and composition of the displaced communities. Such initiatives should accommodate the access of the displaced to fulfil their basic needs and recreational needs.

15.3.12 Participation of Displaced in the Construction and Maintenance

Infrastructure provision and management in mass displacement contexts is another competency in the REGARD competency tier 3b that has not been recognised as a key competency as per the competency audit.

Built environment interventions should have an overall strategy to implement and monitor the interventions to make sure that both short-term and long-term goals are met. One of the most sustainable mechanisms is to have a strategy in place to incorporate the displaced in the built environment intervention process. They could ideally be involved in the construction and maintenance process which will eventually give them a sense of ownership over the built environment intervention. The involvement of the displaced in the construction process has been proven to increase housing satisfaction of the community (Vithanagama et al. 2015). In the field of relocation, informed participation of the displaced throughout resettlement processes has been claimed to be vital (Reddy et al. 2015; Vanclay 2017). This will provide the opportunity for the built environment professionals to maintain the effectiveness of their interventions implemented and to recognize further developments needed.

15.3.13 Addressing Language Issues

Language plays a major role in the context of displacement, specifically in terms of refugee integration and resettlement (Jayakody et al. 2022). It has been reported that language barriers affect refugees in terms of their access to educational needs, health and well-being needs, and economic needs (Hamza 2021). Even though the REGARD competency tier 1 recognises language issues as a required competency, the prevailing built environment competencies of the partner countries do not consider it to be a key competency. In this context, it is notable that the success of integration and resettlement is based on the language compliance of the displaced and the host society. Hence, built environment interventions in relation to integration and resettlement processes should be sensitive towards language issues.

15.3.14 Having a Legal Framework for Built Environment Initiatives in Mass Displacement

The progress built environment professionals are making in the field of mass displacement is commendable. However, it has been noted that there is no general policy and a legal background to address built environment interventions in the contexts of displacement (Amaratunga et al. 2022). Further, it has already been discussed under

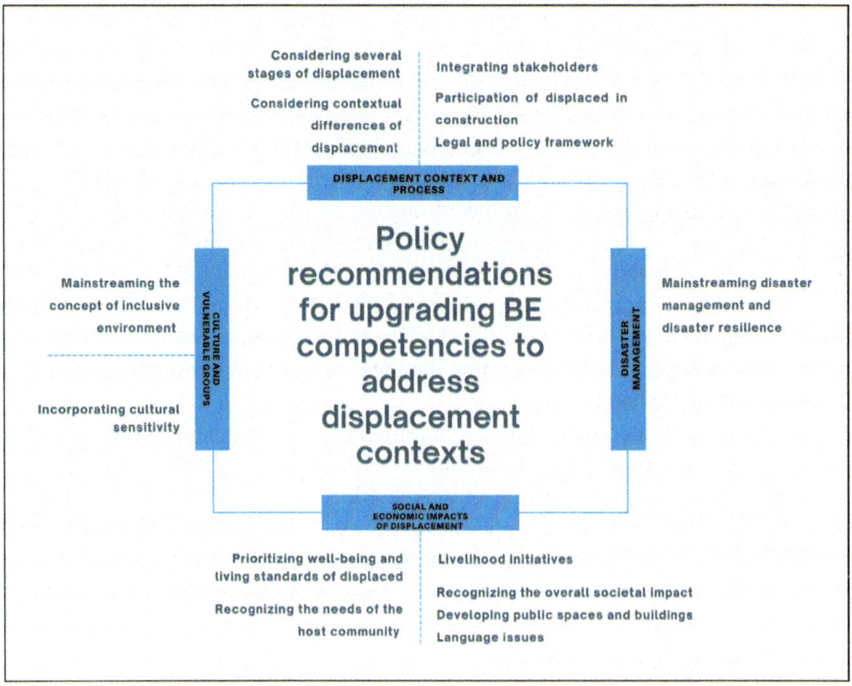

Fig. 15.4 Summary of policy recommendations (*Source* Prepared by the authors)

sub Sect. 15.3.2 that the existence of policy, legal, and regulatory frameworks relevant to the built environment is not considered as a key competency. Hence, this calls for urgent action towards making such an arrangement to accommodate a legal framework for built environment initiatives.

Incorporating the above-mentioned policy recommendations into the existing competencies will make a long-lasting impact on the displaced. Hence, an overall legal framework should be developed to regularise built environment interventions in mass displacement.

The aforementioned policy recommendations could be summarised under the initially recognised gaps in the built environment competencies as follows (Fig. 15.4).

15.4 Conclusion

Displacement is a common phenomenon in the contemporary world that alters the living fabric of the displaced community. With major disruptions to their basic needs and access to resources, building back better requires extensive planning, execution, and monitoring. This study emphasised the vital role the built environment could play in empowering the displaced through their interventions. The study was focused on investigating the need for upgrading the built environmental professional competencies to accommodate the needs of the displaced and host communities effectively. The findings revealed that the existing built environment competencies do not prioritise disaster management, disaster resilience, livelihood opportunities, societal impact of displacement, cultural sensitivity, and vulnerable layers of the displaced. Therefore, the findings of the study revolved around the 14 policy recommendations made by the research team in upgrading the built environment competencies to accommodate the needs of the displaced and host communities. Even though the findings were prepared based on data collected from Estonia, Sri Lanka, Sweden, and the United Kingdom, the contexts cover both perspectives of the Global South and the Global North. The recommendations could be utilised by built environment professionals, practitioners, and policymakers in redefining the focus of built environment interventions in displacement contexts.

The 14 policy recommendations require built environment professionals to consider the contextual differences and various stages of displacement. Recommendations also consider the role of stakeholders such as host families and host communities. Further, the need for prioritising the well-being of the displaced through considering their living standards, cultural aspects, vulnerable groups, language issues, livelihood opportunities, and societal impacts of displacement is highlighted. Moreover, it has also been recommended to create an environment to enhance social cohesion among the displaced and further between the host and displaced communities through public and recreational facilities. The need for formulating an overall policy and a legal framework to accommodate these required upgrades in the built environment competencies is also highlighted.

The future direction of the present study lies in the action of exploring mechanisms to implement the above suggested policy implementations. When evaluating the scope of the existing gaps and the suggested recommendations to address those gaps, conducting continuous appropriate trainings and collaborating with subject experts in the fields of Sociology and Economics could be considered as foundational steps towards implementing the said recommendations. Capacity building of built environment professionals in the fields of disaster management and disaster resilience could be achieved by introducing massive open online courses (MOOCs) which built environment professionals could attend based on their professional commitments. Therefore, it could be concluded that the existing built environment competencies require an upgrade in order to accommodate the needs of host and displaced communities effectively. It is high time that the role of built environment in displacement contexts is redefined. An expansion in their role beyond just providing a housing unit and infrastructure to be the link that connects host and displaced communities and to be a pioneer that empowers the displaced socially, culturally, and economically is vital.

Acknowledgements The authors would like to thank the Erasmus+ Programme of the European Commission for funding the REGARD project. The authors would like to acknowledge the significant contribution to the research process and co-development of the project outputs by all partner institutions, especially those of Emlyn Witt and Irene Lill, Tallinn University of Technology (leading partner of the output that developed the REGARD competency framework); Mo Hamza, Lund University; Champika Liyanage, University of Central Lancashire (leading partner of the output that conducted the competency audit); Nishara Fernando and Maduri Wasana Fernando, University of Colombo.

References

Alwang J, Siegel PB, Jorgensen SL (2001) Vulnerability: a view from different disciplines. The World Bank

Amaratunga D, Haigh R, Jayakody C, Malalgoda C (2020) Rebuilding after forced displacement. Built Environ J. RICS J. Available online at: https://ww3.rics.org/uk/en/journals/built-environment-journal/rebuilding-after-forced-displacement-.html. Accessed 17 Feb 2022

Amaratunga D, Malalgoda C, Haigh R, Jayakody C, Senanayake A, Fernando N, Hamza M, Liyanage C, Lill I, Witt E (2022) A built environment perspective on post-disaster and conflict-induced displacement. A report of the REGARD project: rebuilding after displacement. ISBN: 978-1-86218-208-0

Baldwin C, King R (2017) What about the people? The socially sustainable, resilient community and urban development. Available online at: https://www.anthro.ox.ac.uk/sites/default/files/anthro/documents/media/what_about_the_peopl_report_baldwin_king_2017-brookes.pdf. Accessed 17 Feb 2022

Cazabat C (2018) Thematic series. The ripple effect: economic impacts of internal displacement. IDMC

Christensen A, Harild N (2009) Forced displacement: the development challenge. The World Bank, Washington D.C.

CIC (2021) Essential principles for creating an accessible and inclusive environment for clients, developers, and contractors. Available online at: https://cic.org.uk/admin/resources/cic-essential-principlesguide-for-clients-contractors-and-developers-2018.pdf. Accessed 17 Feb 2022

Conti R, Dabaj J, Galli G, Rigon A (2020) Child development and participation in urban displacement, DeCID Thematic Brief #3. University College London, London

CRED and UNDRR (2021) 2020 the non COVID year in disasters: global trends and perspectives. CRED, Brussels

Crnčević T, Lovren VO (2017) Displacement and climate change: improving planning policy and increasing community resilience. Int J Clim Change Strateg Manage 59(1):105–120

Danish Refugee Council (DRC) (2021) Global displacement: forecast 2021. DRC, Copenhagen

Depetris-Chauvin E, Santos RJ (2017) KNOMAD working paper 27: the impacts of internal displacement inflows on host communities in Colombia. KNOMAD

Fernando N, Punchihewa AG (2013) Relocating the displaced strategies for sustainable relocation. Friedrich Ebert Stiftung, Colombo

Fernando N, Senanayake A, Amaratunga D, Haigh R, Malalgoda C, Jayakody RRJC (2020a) Disaster, displacement and relocation: an analysis of the needs and policy implications on a displaced community in Sri Lanka. In: 10th international conference on structural engineering and construction management: special session on disaster risk reduction, pp 128–145. University of Peradeniya

Fernando N, Senanayake A, Amaratunga D, Haigh R, Malalgoda C, Jayakody RRJC (2020b) Impact of the disaster-induced relocation process on the displaced communities in Kegalle district, Sri Lanka. In: European Scientific Journal special edition: contemporary Sri Lankan Society and Politics: felicitation volume in honour of renowned Sri Lankan Sociologist Professor ST Hettige. ESJ ISSN: 1857-7881 (Print) e-ISSN 1857-7431

Folke C (2006) Resilience: the emergence of a perspective for social-ecological systems analyses. Glob Environ Change 16(3):253–267

Goodwin-Gill GS, McAdam J (2017) Climate change, disasters and displacement. UNHCR, Geneva
Hamza M (2021) Refugees' integration in the built environment: the Sweden case. Sustainability 13:12812. https://doi.org/10.3390/su132212812
Heylighen A, Linden VV, Steenwinke IV (2017) Ten questions concerning inclusive design of the built environment. Available online at: https://www.researchgate.net/publication/311556315_Ten_questions_concerning_inclusive_design_of_the_built_environment. Accessed 25 Jan 2022
Ibrahim M (2010) Post-disaster housing reconstruction in a conflict affected district, Batticaloa, Sri Lanka: reflecting on the climate smart disaster risk management approach: SCR discussion paper 6. Institute of Development Studies, Brighton
IDMC and Norwegian Refugee Council (NRC) (2021) Internal displacement in a changing climate. [Online] Available online at: https://www.internal-displacement.org/sites/default/files/publications/documents/grid2021_idmc.pdf. Accessed 17 Feb 2022
IDMC and NRC (2018) 'Nowhere to return to, Iraqis' search for durable solutions continues: November 2018. IDMC, Geneva
IDMC and NRC (2019) Disaster displacement: a global review, 2008–2018. IDMC, Geneva
IDMC and NRC (2020) GRID 2020: global report on internal displacement. IDMC, Geneva
International Federation of Red Cross and Red Crescent Societies (2020) World disasters report 2020: come heat or high water—tackling the humanitarian impacts of the climate crisis together. IFRC, Geneva
International Federation of Red Cross and Red Crescent Societies (2021) Displacement in a changing climate: localized humanitarian action at the forefront of the climate crisis. IFRC, Geneva
IOM, NRC, UNHCR (2015) Camp management toolkit. IOM, NRC and UNHCR
Jayakody C, Malalgoda C, Amaratunga D, Haigh R, Liyanage C, Witt E, Hamza M, Fernando N (2022) Approaches to strengthen the social cohesion between displaced and host communities. Sustainability 14:3413. Available online at https://doi.org/10.3390/su14063413. Accessed 17 Feb 2022
Jayakody RRJC, Malalgoda C, Amaratunga D, Haigh R, Liyanage CL, Witt E, Fernando N (2021) Role of the built environment in rebuilding displaced and host communities. In: Multi-hazard early warning and disaster risks, pp 69–92. Springer, Cham
Kaluarachchi Y (2018) Building community resilience in the re-settlement of displaced communities. Procedia Eng 212:443–450
Kamta FN, Scheffran J (2021) A social network analysis of internally displaced communities in northeast Nigeria: potential conflicts with host communities in the Lake Chad region. Geo J. https://doi.org/10.1007/s10708-021-10500-8. Accessed 17 Feb 2022
Koch EJH (2015) Protracted displacement in Georgia: structural vulnerability and "existing not living". Hum Organ 74(2):135–143
Kuran CHA et al (2010) Vulnerability and vulnerable groups from an intersectionality perspective. Int J Disast Risk Reduct 50:1–8
León J, Mokrani C, Catalán P, Cienfuegos R, Femenías C (2019) The role of built environment's physical urban form in supporting rapid tsunami evacuations: Using computer-based models and real-world data as examination tools. Front Built Environ 4:89. https://doi.org/10.3389/fbuil.2018.00089
Lin PS, Lin W (2020) Rebuilding relocated tribal communities better via culture: livelihood and social resilience for disaster risk reduction. Sustainability 12(11):4538. Available online at: https://doi.org/10.3390/su12114538. Accessed 17 Jan 2022
Loft P, Brien P (2022) Ukraine crisis 2022: Aid and refugees. House of Commons Library
López RCD, Arredondo CA, Salcedo J (2011) The effects of internal displacement on host communities a case study of Suba and Ciudad Bolívar localities in Bogotá, Colombia. School of Economics Project on Internal Displacement, Brookings Institution, London
Malalgoda C, Amaratunga D, Haigh R (2014) Challenges in creating a disaster resilient built environment. Procedia Econ Finan 18:736–744

Max Lock Centre (2009) The built environment professions in disaster risk reduction and response, a guide for humanitarian agencies. Max Lock Centre, London. Available online at: https://www.ifrc.org/PageFiles/95743/B.a.07.Built%20Environment%20Professions%20in%20DRR%20and%20ResponseGuide%20for%20humanitarian%20agencies_DFDN%20and%20RICS.pdf. Accessed 17 Feb 2022

Mull R (2016) The architecture of displacement: a collaborative architectural investigation of place-making with displaced communities. University of Brighton. Available online at https://cris.brighton.ac.uk/ws/portalfiles/portal/23477350/MULL.pdf. Accessed 17 Feb 2022

Pain A, Levine S (2012) A conceptual analysis of livelihoods and resilience: addressing the insecurity of agency—HPG working paper. ODI, London

Reddy G, Smyth E, Steyn M (2015). Land access and resettlement: a guide to best practice. Greenleaf, Sheffield

Roberto CVL, Clara IAA, Jorge S (2011) The effects of internal displacement on host communities. A case study of Suba and Ciudad Bolívar Localities in Bogotá, Colombia. School of Economics Project on Internal Displacement, Brookings Institution, London. Available online at: https://www.brookings.edu/wp-content/uploads/2016/06/Host-Communities-Colombia-English.pdf. Accessed 6 Dec 2021

Schofield H, Lovell E, Flinn B, Twigg J (2019) Barriers to urban shelter self-recovery in Philippines and Nepal: lessons for humanitarian policy and practice. J Br Acad 7(s2):83–107

Senanayake A, Fernando N, Wasana M, Amaratunga D, Haigh R, Malalgoda C, Jayakody C (2022) Landslide induced displacement and relocation options: a case study of owner driven settings in Sri Lanka. Sustainability 14:1906. Available online at: https://doi.org/10.3390/su14031906. Accessed 17 Feb 2022

Sridarran P (2018) A framework to enhance post-disaster resettlement process through adaptable built environment in Sri Lanka. Available online at: http://eprints.hud.ac.uk/id/eprint/34573/1/Pournima%20Sridarran%20FINAL%20THESIS.PDF. Accessed 17 Jan 2022. Accessed 17 Feb 2022

Sridarran P, Keraminiyagel K, Amaratunga D (2016) Community integration and participation to improve the built environment of the post-disaster involuntary relocations. Available online at: http://eprints.hud.ac.uk/id/eprint/29257/1/Pournima.pdf. Accessed 17 Jan 2022

UNHCR (2021a) Global report 2020. UNHCR

UNHCR (2021b) Mid year report on forced displacement 2021b; The mid year trends. Available online at: file:///C:/Users/smart/Documents/MID%20YEAR%20REPORT%20ON%20FORCED%20DISPLACEMENT%202021b.pdf. Accessed 17 Feb 2022

UNHCR (2012) IDPs in host families and host communities: assistance for hosting arrangements. Division of International Protection, Protection Operational Support April 2012. Available online at: https://www.refworld.org/pdfid/4fe8732c2.pdf. Accessed 17 Jan 2022

Vanclay F (2017) Project-induced displacement and resettlement: from impoverishment risks to an opportunity for development? Impact Asses Project Apprais 35:1, 3–21, Available online at https://doi.org/10.1080/14615517.2017.1278671. Accessed 17 Feb 2022

Vithanagama R, Mohideen A, Jayathilaka D, Lakshman R (2015) Planned relocation in the contexts of natural disaster: the case of Sri Lanka. Centre for Migration Research and Development Colombo, Colombo, Sri Lanka

Part IV
Socio-economic Interventions for Sustainable Resettlement

Chapter 16
Socio-Economic Effects of War Against Terror-Induced Displacement on Host Communities in District Kohat, Pakistan

Shakeel Mahmood, Attia Atiq, Samiullah, and Ahmad Raza Khan

Abstract This study is an effort to highlight the socio-economic effects that residents of Kohat had to face due to temporary settlement of temporarily dislocated persons (TDPs) during war on terror. Huge population has been temporarily settled in the study area leading to social and economic issues particularly rising crimes, house rents and expenditure, and decline in daily wages and income. Therefore, the need of this study was felt to investigate the socio-economic impacts of TDPs on host communities. Both primary and secondary data was utilized to achieve objectives of this study. For this purpose, four sample sites were selected using stratified purposive sampling techniques to conduct questionnaire survey. Questionnaire survey of 280 households was conducted for detail micro-level analysis. Descriptive statistical analysis was used to visualize the temporal variations in selected variables. The relation between monthly income, expenditure and crime rate was determined by applying linear regression model. ArcGIS 10.2 is used to geo-visualize the spatial distribution of TDPs camps and sample sites. Analysis revealed that the study area was hosting large number of TDPs since 2007. Monthly income and expenditure of the hosting community remained same while, contradictory to that, the income of TDPs decreased by 50 percent and expenditure increased. A sharp increase was seen in daily wages from 2005 to 2009 but after the arrival of dislocated people the increase in wages was very slow. It slowed down the pace increment in wages. Similarly, increase in crime incidences was also very sharp. In the year 2011, crime incidents decreased because of start of return TDPs to their native areas but during 2012–14, crime rate increased again as more dislocated families from Federally Administered Tribal Area (FATA) came to Kohat.

S. Mahmood (✉) · A. Atiq
Department of Geography, Government College University, Lahore, Pakistan
e-mail: shakeelmahmoodkhan@gmail.com

Samiullah
Department of Geography, University of Peshawar, Peshawar, Pakistan

A. R. Khan
Department of Political Science, Government College University, Lahore, Pakistan

© The Author(s), under exclusive license to Springer Nature Switzerland AG 2023

M. Hamza et al. (eds.), *Rebuilding Communities After Displacement*,
https://doi.org/10.1007/978-3-031-21414-1_16

Keywords Forced displacement · IDPs · Insurgency · TDPs · Impacts · Income · Crime · Wages · Pakistan

16.1 Introduction

Twenty-first century, despite all civilization evolutions and developments, is still marred by internally displaced persons (IDPs). Worldwide, the number of forcibly displaced persons from their homes exceeded 80 million in 2020 (Council 2015; Depetris-Chauvin and Santos 2017; UNHCR 2022). Out of the total, some 46 million (58 percent) were IDPs caused by armed conflicts or disasters. The native areas of IDPs are thinned out because of devastation (IDMC 2022). Even with the deployment of humanitarian support, overloaded population and pressure on services and facilities overwhelms host communities (George and Adelaja 2021). They are the world's most vulnerable people who are physically, socially, economically and culturally cut off from their native environment (Stepputat and Sorensen 2001; Laliberte et al. 2003; Malhotra 2007). The internal displacement occurs not only in developing countries but also in western industrialized nations (Koyama and Dunn 2011). The civil and international wars have dislocated about six million people in Asia, twelve million in Africa, three million in Europe and nine million in Latin America (Lischer 2008; Ghimire 2012). In contrast to refugees, IDPs—unfortunately—have no guaranteed protection by any legal mechanism (Leus et al. 2001; Roberts et al. 2008). They are dependent victims but with passive recipients for assistance (Harrell-Bond and Voutria 1992) and have faced problems of humanitarian rights, violence, armed conflicts, military operations, civil and international wars (Makhoul and Ghanem 2009; Ghimire 2012). In short, 260million people are displaced in fifty-two countries worldwide (Holmes and De Pineres 2011).

The World Food Program (WFP) and World Health Organization (WHO) have been working for the betterment of IDPs (Weiss 1999). United Nation Commission on Human Rights (UNCHR) developed a framework for protection and assistance of IDPs. In 1993, the UN developed guidelines and principles for IDPs in collaboration with international legal experts and UN Agencies (Nyborg and Nawab 2017). The UN guiding Principles on Internal Displacement (1998) define IDPs as "Persons or groups of persons who have been forced to flee or obliged to flee or leave their homes or places of habitual residence, in particular as a result of or in order to avoid the effects of armed conflict, situations of generalized violence, violations of human rights or natural or manmade disasters, who have not crossed an internationally recognized State border."

In Pakistan, natural disasters like floods and earthquakes have displaced largest number of people from their native areas (Amir-ud-Din and Malik 2016; Mahmood et al. 2016a). For instance, the devastating 2005 earthquake displaced 3.5million people (Mohsin 2013) and 2010 super flood displaced over 7million from their homes (Mahmood et al. 2016b). Likewise, human-induced hazards, particularly conflicts, terrorism and military operations, have pushed the country into number of problems in the last two decades (Jadoon, 2017). The war against terror has also resulted in internal displacement of more than three million people from Federally Administered Tribal Areas (FATA); currently Khyber Pakhtunkhwa, and some districts of Malakand division, bordering Afghanistan, including Dir Upper, Dir Lower, Swat and Malakand in 2009 (Mustafa 2013; Sayed and González 2014). More than 2 million people have voluntarily returned to their native areas (Sayed and González 2014). The war against terror is still continued and the country has lost a large number of precious human lives and tens of billions US$ economic damage. The problem of IDPs was instigated in 2007 but properly given the name of temporarily dislocated persons (TDPs) by the Foreign Office of Pakistan in 2014 (Ali 2014). Over the years, Pakistan has been facing insurgency in FATA and remained under military operations for quite long time. Continuously increasing number of dissident activities beyond FATA resulted in socio-economic, political and security problems. These conditions led the Pakistan Army to start offensive strikes in FATA and other parts of Khyber Pakhtunkhwa to eliminate perpetuating neo-existential terrorist threats (Javaid 2016). The chronological details of these military operations to eliminate militants are; Rah-i-Haq-I, launched in district Swat in (2007), Rah-i-Haq-II, launched in districts of Swat and Shanglain (2008), Rah-i-Haq-III, launched again in districts of Swat and Shangla in (2009), Sirat-i-Mustaqeem, launched in Khyber Agencyin (2008), Sherdil, jointly operated with Frontier Corps in Bajaur Agencyin (2008), Thor Thandar, launched in districts of Buner, Lower Dir and Shanglain (2009), Brekhna was launched in Mohmand Agency (2009), Rah-i-Rast, generally recognized as Swat Operation (2009), Rah-i-Nijat was launched in South Waziristan (2009), Zarb-i-Azb in North Waziristan (2014) and (2017), Radd-ul-Fasaad was launched throughout the country to control the activities of militants and destroyed their networks (Samiullah et al. 2017).

Pakistan is playing a crucial role in combating insurgency and war against terror since the episode of 9/11 (Orakzai 2012). The military operation in 2009 against militants in Malakand Division and FATA, bordering Afghanistan, resulted internal displacement of about 2.3millionand 2.7million people, respectively, from their homes (Mustafa 2013). Millions of people have been dislocated from all seven agencies formerly FATA (Khyber, Mohmand, Bajaur, South Waziristan, North Waziristan, Kurram, and Orakzai) and six Frontier Regions (FR Bannu, FR Lakki Marwat, FR D.I.Khan, FR Peshawar, FR Kohat and FR Tank) and a lot of them have been dislocated more than once (Ahmad et al., 2018). This was the largest human exodus in Pakistan (Akhunzada et al. 2015). In the year 2011, 0.35milion TDPs from Bajuar Agency, 0.273million from Waziristan Agency, 0.245million from Mohmand Agency were registered and huge number of unregistered TDPs (Orakzai 2012). By the end of 2015, the number of dislocated persons reduced to 1.459 million. They were settled in TDPs camps located in districts of Kohat, Bannu, D.I. Khanand Mardan located in Khyber Pakhtunkhwa (Amir-ud-Din and Malik 2016).Very few (10%) TDPs were living in the camps while remaining (90%) were residing in the host communities causing burden on them (Mohsin 2013). As consequence number of socio-economic issues came to surface affecting the host communities (Jan et al. 2011). This paper is an attempt to consider impacts of TDPs on host communities of district Kohat Khyber Pakhtunkhwa, Pakistan. In particular, selected variables of this study to highlights the impacts are the change in population, daily wages, house rent, monthly income, expenditure and registered crimes after the arrival of TDPs.

Pakistan is the serious victim of (Orakzai 2012) war on terror (Orakzai 2012; Ahmad et al. 2018). In the last two decades, war against terror has been drastically affecting the economy, exports, physical infrastructure, tax collection, investment and social fabric of the country (GoP 2017). Since the insurgency started in Pakistan following the US invasion of Afghanistan, more than 75,000 Pakistani civilians, security forces personnel, women and children have been killed in gun, bomb and suicide attacks. Besides, thousands of others have been seriously injured or handicapped and are unable to win bread for their children which ultimately also created other social issues in the society. Similarly, economy of such a developing country suffered a massive cost of more than 120billion US$ on account of loss of lives, economic opportunities and damage to the country's infrastructure (The NEWS 26/05/2017). Terrorism in Pakistan has attracted substantial local and international attention, since the war on terror started at international level in 2001 and at domestic level in 2006. Therefore, enormous research work has been published mainly concentrating on causes of terrorism in Pakistan. Some recent studies including Bird et al. (2008), Gul et al. (2010), Irshad (2011), Nasir et al. (2011), Shahbaz (2013), Shahbaz et al. (2013) have explained causes and effects of terrorism on local and regional level in detail. Contrary to this, the current study concentrates on analyzing the impacts of forced displacement on host communities. War on terrorism has not only affected the areas where it was carried out rather, it also affected the areas where temporarily dislocated persons were stationed. These dislocations took place in several areas. There is lack of studies regarding the impact of TDPs on host communities is in context of the district Kohat, Khyber Pakhtunkhwa. Therefore, it is important to understand the socio-economic impacts of TDPs on host communities. This study has specifically targeted the socio-economic effects of war on terror on host communities; through displacement.

16.2 Material and Methods

16.2.1 Study Area

District Kohat is situated in province Khyber Pakhtunkhwa (Pakistan). Geographically, it is extended from 33° 09′N latitude to 33° 74′N latitude and from 71°E longitude to 72°E longitude (Fig. 16.1). Relatively, the study area shares boundary with Orakzai Agency in the north and north–west. Administratively, the study area comprised of 31 Union Councils (UC; smallest electoral unit) with total geographical area of 2545 km^2 out of which 25 percent is cultivated. Population of the study area was 0.56 million with density of 221 persons/km^2 (GoP 2000). Climatically study area is hot. June is the hottest month with maximum temperature of 40 °C and

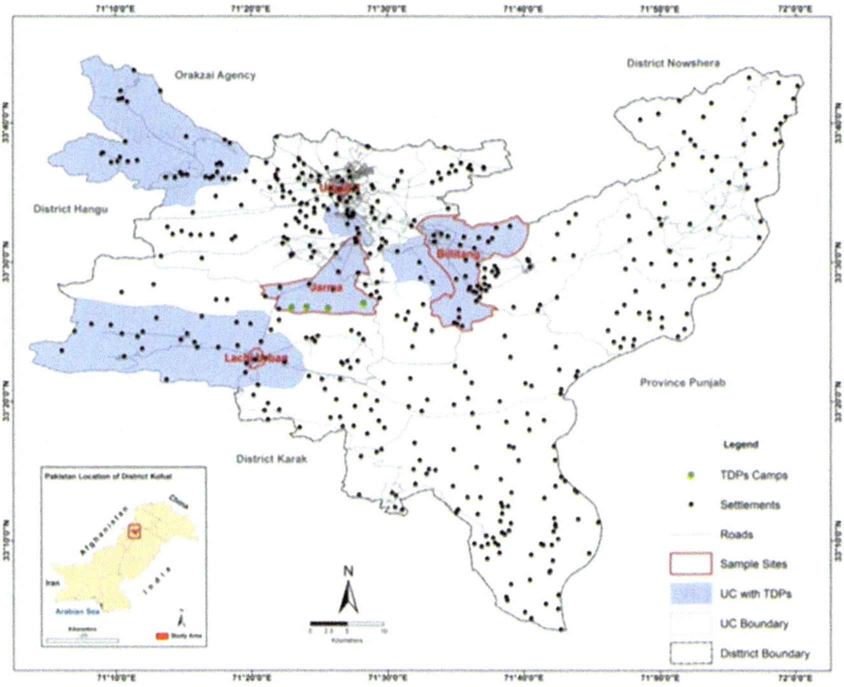

Fig. 16.1 Location of the study area and distribution of TDPs

minimum temperature of 27 °C. January is coldest month with maximum temperature 18 °C and minimum temperature of 6 °C. A large number of internally dislocated families from the war affected tribal areas of Pakistan during 2003–2014 came to the study area. The concentrated UC were Jerma, Lacchi and Billitang.

16.2.2 Research Problem

In Pakistan, war on terror has forcibly displaced millions of people from their homes. These displaced persons carry some small savings because of the sudden nature of war against terror. This has also affected the areas where forcibly displaced persons were stationed. The host communities were facing rapidly growing socio-economic issues. Therefore, this study has specifically targeted the social and economic effects of war against terror-induced displacement on host communities to assist decision makers.

16.2.3 Objectives of the Study

- To highlight changes in population of district Kohat.
- To determine the socio-economic impacts of TDPs on host community of district Kohat.

The selected variables of this study are the change in population, daily wages, house rent, monthly income, expenditure and registered crimes after the arrival of TDPs.

16.2.4 Data Acquisition and Analysis

This study is based on primary and secondary data. Primary data was collected through questionnaires, interviews and personal observation while secondary data was collected from concerned government departments. TDPs were distributed in nine UCs out of which four were selected using stratified purposive sampling techniques making sample size of 44 percent. The selected sample sites were Lachi Urban, Urban-6, Jerma and Billitang (Fig. 16.1). The latter are rural areas. Semi-structured questionnaire was designed for local residents and TDPs. Questionnaire survey was conducted in the sample sites using random sampling technique. A total of 280 households were surveyed in all sample sites, out of which 50 percent surveyed households were local residents. The local residents were asked regarding the house rent, daily wages and crime occurrences before and after displacement. On the other hand, TDPs were also asked regarding their family size, monthly income and expenditure before and after displacement. Global positioning system (GPS) is used to acquire the spatial location of TDPs camps.

Secondary data was acquired from concerned government departments. Population data (2005–2014), registered TDPs and crime records (2005–2014) were collected from the District Management Office, District Kohat. Base map of the study area was acquired from Survey of Pakistan (SOP), Islamabad. Descriptive statistical analyses were used to determine the variation in selected variables. Similarly, the relation between monthly income, expenditure and crime rate is determined by applying linear regression model. ArcGIS10.2 is used to develop the map of study area depicting administrative boundaries, roads, settlements, TDPs camps and sample sites.

16.3 Results and Discussion

The war against terror has changed situation in FATA and resulted in internal displacement of enormous number of families to the adjacent districts of Khyber Pakhtunkhwa. The TDPs belong to Orakzai Agency who migrated to the two adjacent districts of Hangu and Kohat. They were settled in camps and provided with basic facilities. The study area is hosting a large number of TDPs. They were spatially distributed in nine UCs, out of which three were urban (Fig. 16.1). The concentrated UCs were Jerma, Billitang and Lachi. The camps of Jerma were completely occupied. There were four food distribution points namely Kohat-1, Kohat-2, Kohat-3 and Kohat-4 for TDPs. At the same time, a large number of families were also living in residential areas on rent. This study is carried out in four sample sites comprising of two urban areas. The major effects of TDPs on study area are elaborated in the following sections.

The studied population in sample sites was 2697 out of which 1204 (595 male and 609 female) were local residents and 1493 (704 male and female 789) were TDPs. The number of TDPs was more in the studied population because of large family size (average 9 persons) as compare to local residents (average 7 persons; Table 16.1). In the year 2009, approximately, 36,513 families comprising of 328,000 individuals and 5242 families comprises of 31,430 individuals in 2010 shifted to study area from Orakzai Agency. This raised population of the study area from 0.703 to 1.06 million. In the year 2011, a large number (36,160) of TDPs went back to their native areas due to clearing the area from terrorists and militants. This out flux from the study area has dropped the total population to 1.028 million. But in the year 2012, displacement occurs again because of the military operation in FATA, and nearly, 28,000 families entered the study area and stayed there till 2014. In the year 2014, the population of Kohat reached 1.139 million (Fig. 16.2) with total number of 28,735 of TDPs' families. The presence of large number of TDPs in the study area has given birth to number of socio-economic and environmental effects. This study focuses on effects of TDPs on income, expenditure, daily wages, house rent and crimes.

The rent of125 m^2 (5marla) and 250 m^2 (10marla) houses was 3500 and 5000 PKR*, respectively, till 2008. But due to the influx of TDPs house rent increased dramatically to 10,000 and 20,000 PKR (Fig. 16.3). The increase is about 1.85 and 3 times respectively which is a big change and host communities were not able to pay the increased rent. More than 80 percent of the TDPs were living in residential colonies of host communities. That was the foremost reason of rise in house rent. Local community showed concerns about this increase in house rent as majority

Table 16.1 Composition of studied population in district Kohat

Sample site	Local residents		TDPs	
	Male	Female	Male	Female
Jerma	144	147	160	184
Billitang	156	152	170	179
Lachi Urban	159	168	192	224
Urban-6	136	142	182	202
Total	**595**	**609**	**704**	**789**

Source Field Survey 2014

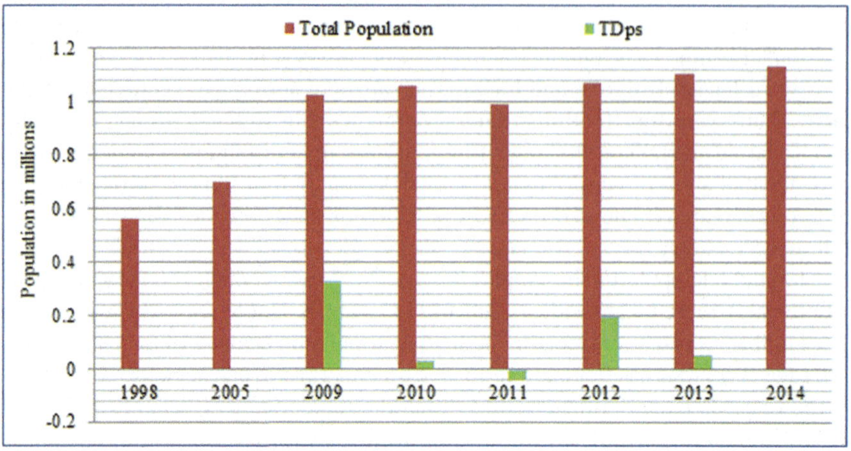

Fig. 16.2 Change in population caused by influx and out flux of TDPs in district Kohat (1998–2014)

of the local respondents said that they could not afford such high rent because the monthly income hardly fulfilled their household spending.

The impact of TDPs on wages was different from that on house rent. Daily wages of the labor and masons were higher till 2009. They were earning 300 and 600 PKR daily. The influx of TDPs resulted in sharp decrease in wages by 100 PKR in each cadre till 2012. The main reason was that TDPs were jobless and they had to earn their livelihood, so they were working on low wages in order to exploit work opportunities. This caused decrease in overall wages which also caused problem for the local wagers. From 2012 onward an increase has taken place due to inflation and return of TDPs. The wages have now attained the position of 2009 (Fig. 16.4). Analysis reveals that the supply of daily wagers increased while demand remained the same. Consequently,

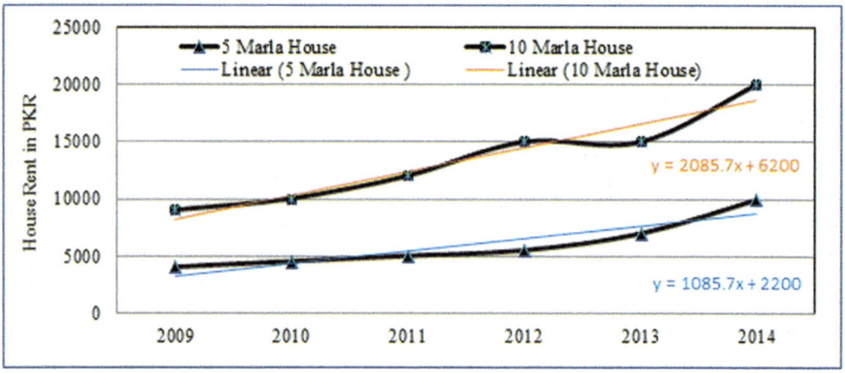

Fig. 16.3 Change in house rent from 2009 to 2014 after field survey 2014

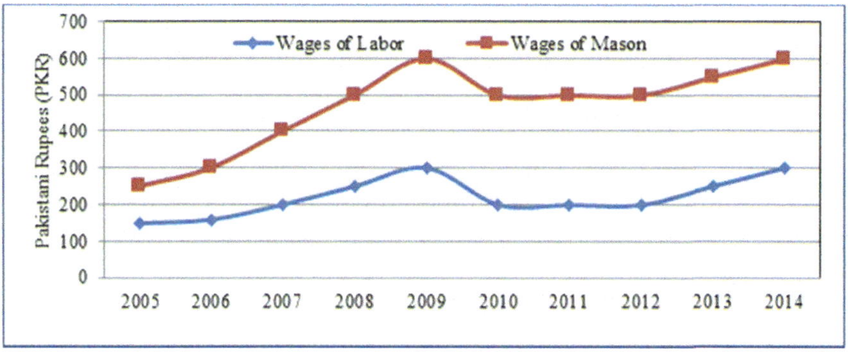

Fig. 16.4 Change in wages of labor and Mason from 2005 to 2014 after field survey 2014. *1 US$ = 208 PKR, 2022

daily wages dropped during the stay of TDPs in the area. Similarly, both the TDPs and host communities were also facing difficulty in securing regular income.

Worldwide, crime is growing problem of many cities in developed and developing countries (Haider and Ali 2015). Crime rate is climbing in developing countries (Broadhurst et al. 2008). The crime incidents increase where people from diverse geographic areas and economic background interact to each other. The study area has been hosting large number of TDPs families migrated from FATA with diverse economic background. They were also facing problem in earning their livelihood. Analysis reveals that, earlier the influx of TDPs, the annul registered crime events were less than 20 per annum but after the influx of TDPs, a sharp increase in criminal activities like hurt, attempted murder and murder have occurred and reach to its maximum in 2013 and then decrease. The number of murder cases has uplifted from 5 in 2005 to more than 30 in 2014 (Fig. 16.5). Crime rate has shown a dramatic change from 2005 to 2014, the heinous crime like murder, attempted murder and hurt showed a tremendous change before and after TDPs.

The textual analysis of the interviews concerned officials revealed that the influx of TDPs resulted in an increase in crimes and about 3000 TDPs were registered in criminal activities from 2005 to 2014 in study area. Because of having an easy access to weapons, these people felt easy to commit crimes like murder, kidnapping and attempted murder for sake of money. According to the local community, there was a group of approximately 40 people which was a cause of constant fear, among the locals for years, because they were allegedly involved in many crimes like kidnapping, house robbery, car snatching and even murder. They also occupied the property of local for the purpose of hiding themselves from being arrested. Many respondents (12%) also told about their own family member being murdered.

The analysis reveals that war on terror caused internal displacement which negatively affected monthly income of the TDPs. They were earning more in their native areas due to established economic activities. But after displacement their income decreased by 50 percent. Similarly, after the displacement expenditure also increased; and even doubled in some cases. In the host communities, crime rate also increased.

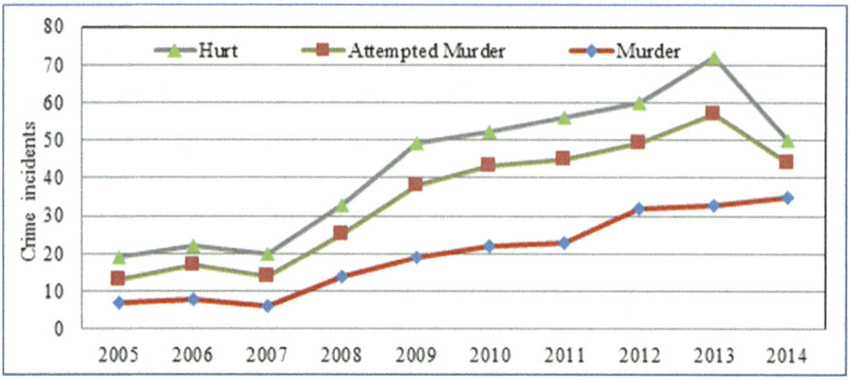

Fig. 16.5 Crime rate from 2005 to 2014 after GoP 2014

There was a slight increase in crime rate from 2005 to 2008. In the year 2009, large number of families entered the study area as TDPs and crime rate incidences and jumped from 13,000 to 19,000 per annum and then dropped to about 15,000 in 2011 because large number (36,160) TDPs went back to their native areas of Orakzai Agency. The crime rate increased very sharply from 2011 and attained maximum incidences of more than 40,000 by the year 2013 (Fig. 6a, b). There is a need of proper management of TDPs camps by providing the basic amenities and services. Similarly, proper mechanism of distribution of food items and commodities of daily needs was also needed. If the dislocated persons have the basic needs in their camps then they will not go outside the camp to earn livelihood. It was also found from the analysis that the per month rent of 05 Marla and 10 Marla houses increased from 4000 and 8000 PKR to 10,000 PKR and 20000 PKR, correspondingly. In any community, how the household head will manage living on rent, where the average per month income is less than 25,000 PKR with family size of more than 8 persons? This is the main economic impact of stationed TDPs on host community. Similarly, it was hard for the TDPs and they were struggling to find work opportunities even at low wages because they were supposed to earn their livelihood and to feed their family. It is highly recommended that the TDPs camps should include all basic facilities particularly food, shelter and health.

The analysis further indicates that magnitude and incidence of temporary displacement in Pakistan has exceeded, in the recent past. Yet, after 2016, dislocated communities started going back to their native areas—particularly (FATA). Pakistan Army has cleared the area from terrorists and has taken control. Similarly, Pakistan Army is playing a leading role in rehabilitation and resettlement of displaced communities (Javaid 2016).

Fig. 16.6 **a**, **b** Change in income and expenditure of TDPs per month and crime occurrence in district Kohat after field survey 2014; GoKP 2014

16.4 Conclusions and Recommendations

The study concludes that the hosting communities have been accommodating large number of TDPs since 2009 not only in camps but also in urban and rural areas. The presence of large number of people from diverse geographic units—with different economic backgrounds—has resulted in certain effects on hosting communities. The pressure on the recourses of host community was high and population was overloaded. The study concludes that district Kohat was hosting large number of TDPs. This increased the population of the study, area from 0.703 to 1.139 million. The rents of 125 m^2 and 250 m^2 houses increased by 1.85 and 3 times respectively. This created problems for the local people because they could not afford such high rents.

Daily wages were on rise from 2005 to 2009, but after the arrival of TDPs, wages dropped because the dislocated persons were struggling to find work opportunities, even at lower wages. Similarly, the crime rate also climbed very rapidly. In 2011, crime incidents were started reducing because TDPs started going back but after 2012–14 it started increasing again because of the influx of new badge of TDPs. Most of the displaced families were resettled by Government of Pakistan, through the unforgettable services of Pakistan Army, in their native areas but still twenty to twenty-five thousand families are residing as TPDs in the districts of Bannu, D. I. Khan and Kohat. There is a need to establish planned TDPs camps which could not only host a considerable number of migrating families but could also provide basic facilities, to ensure retaining of displaced people in camps. The impact of temporary displacement is often most severe on the most marginalized and vulnerable people, both in the host and dislocated community.

It is recommended that proactive measures must be taken to prevent the occurrence of temporary displacement by early identification of human induced or natural disasters. Persistent assessment and monitoring of risks generated by disasters is need of the time and preparedness for appropriate and timely responses is what Pakistan needs to plan and invest in. Resettlement in native locations is highly recommended because it will further strengthen the ties with their own community and area.

It is also recommended that in case of such situations, there is a need to develop strong ties and social cohesion between the host communities and dislocated persons. Similarly the capacity and skills of the dislocated persons should be identified to establish some centers for their work and provide jobs on temporary basis. Similarly, schools and Basic Health Units (BHUs) should be maintained in TDPs camps where eligible persons from displaced communities can work. In case of unavailability of such capable persons then they must be trained and deployed to work and serve their own community in the BHUs and schools. Comprehensive planning can not only decrease suffering of the TDPs but can also lessen the adverse effects of such unfortunate incidents on host communities as well.

References

Achvarina V, Reich SF (2006) No place to hide: refugees, displaced persons, and the recruitment of child soldiers. Int Secur 31(1):127–164

Ahmad S, Mubeen M, Kalim I (2018) Internal displacement and migration issues in FATA (2007–2016). Glob Polit Rev 3(1):147–155. https://doi.org/10.31703/gpr.2018(III-I).16

Akhunzada WA, Qadir A, Maqsood N, Rasool G, UR Rahman R (2015) Internally displaced persons (IDPS); Anxiety and depression. Prof Med J 22(3)

Ali Z (2014, September 13) Call IDPs temporarily dislocated persons, govt agencies told. DAWN.COM. https://www.dawn.com/news/1131711

Amir-Ud-Din R, Malik S (2016) Protecting the vulnerable: the case of IDPs in Pakistan. Eur Online J Nat Soc Sci 5(1):82–92

Bird G, Blomberg B, Hess G (2008) International terrorism: causes, consequences and cures. World Econ 255–274

Broadhurst R, Lee KW, Chan CY (2008) Crime trends. Understanding criminal justice in Hong Kong, pp 45–68
Council NR (2015) Internal Displacement Monitoring Centre. 2012. "Global overview 2011: people internally displaced by conflict and violence–India". RefWorld.org
Depetris-Chauvin E, Santos RJ (2017) The impacts of internal displacement inflows on host communities in Colombia. KNOMAD, August, 2017–08
George J, Adelaja A (2021) Forced displacement and agriculture: implications for host communities. Sustainability 13(10):5728
Ghimire A (2012) Placing internal displacement in the sustainable development agenda: the case of Nepal. Peace and Sustainable Development in South Asia: The Way Forward, pp 23–46
GoKP (Government of Khyber Pakhtunkhwa) (2010) Government of KP Environmental profile of Khyber Pakhtunkhwa. Environmental Protection Agency, a, Peshawar
GoP (Government of Pakistan) (2000) District census report of Kohat, 2000. Population census organization, Islamabad
GoP (Government of Pakistan). Economic Survey of Pakistan (2017). pp 279. www.finance.gov.pk/survey/chapters_17/Pakistan_ES_2016_17_pdf.pdf
Gul TG, Hussain AH, Bangash SB, Khattak SWK (2010) Impact of terrorism on financial markets of Pakistan (2006–2008). Eur J Soc Sci 18(1):98–108
Haider A, Ali A (2015) Socio-economic determinants of crimes: a cross-sectional study of punjab districts. Int J Econ Empirical Res 3(11):550–560
Harrell-Bond BE, Voutira E (1992) Anthropology and the study of refugees. Anthropol Today 8(4):6–10
Holmes JS, De Piñeres SAG (2011) Conflict-induced displacement and violence in Colombia. Stud Confl Terrorism 34(7):572–586
IDMC Internal Displacement Monitoring Centre. Available online http://www.internaldisplacement.org/countries/nigeria. Accessed on 22 June 2022
Irshad M (2011) Terrorism in Pakistan: causes & remedies. Dialogue (Pakistan) 6(3):224–241
Jadoon A (2017) Persuasion and predation: the effects of US military aid and international development aid on civilian killings. Stud Conflict Terrorism 1–25
Jan M, Paracha SA, Anwar M, Imran M (2011). Military operation in Wazirstan: public perception in Pakistan on terra. Gomal Univ J Res (27):93–109
Javaid U (2016) Evaluating rehabilitation process in FATA: challenges and prospects. J Polit Stud 23:323–339
Koyama D, Dunn PA (2011) Internal displacement: a study of homelessness in the city of Toronto. J Intern Displacement 1(2):59–80
Laliberté D, Laplante B, Piché V (2003) The impact of forced migration on marital life in Chad. Eur J Population 19(4):413–435
Leus X, Wallace J, Loretti A (2001) Internally displaced persons. Prehosp Disaster Med 16(3):116–123
Lischer SK (2008) Security and displacement in Iraq: responding to the forced migration crisis. Int Secur 33(2):95–119
Mahmood S, Khan A, Mayo SM (2016a) Exploring underlying causes and assessing damages of 2010 flash flood in the upper zone of Panjkora River. Nat Hazards 83(2):1213–1227
Mahmood S, Khan A, Ullah S (2016b) Assessment of 2010 flash flood causes and associated damages in Dir Valley, Khyber Pakhtunkhwa Pakistan. Int J Disaster Risk Reduction 16:215–223
Makhoul J, Ghanem M (2009) Displaced Arab families mothers' voices on living and coping in postwar Beirut. J Middle East Women's Stud 5(3):54–73
Malhotra C (2007) Internally displaced people from Kashmir: some observations. Indian Anthropologist 71–80
Mohsin ZR (2013). The crisis of internally displaced persons (IDPs) in the federally administered tribal areas of Pakistan and their impact on Pashtun women. Tigah: A J Peace Dev 3(2):92–117
Mustafa U (2013) Determinants and implications of major conflicts in Pakistan. Millennial Asia 4(1):67–86

Nasir M, Ali A, Rehman FU (2011) Determinants of terrorism: a panel data analysis of selected South Asian countries. Singap Econ Rev 56(2):175–187

Nyborg I, Nawab B (2017) Social vulnerability and local adaptation in humanitarian response: the case of Pakistan. IDS Bull 48(4)

Orakzai SB (2012) The war against terror and the issue of internally displaced people in Pakistan In the post 9/1 period. J Intern Displacement 2(1):76–78

Roberts B, Ocaka KF, Browne J, Oyok T, Sondorp E (2008) Factors associated with post-traumatic stress disorder and depression amongst internally displaced persons in northern Uganda. BMC Psychiatry 8(1):38

Samiullah RG, Gul S, Mahmood S (2017) Impacts of operation Zarb-e-Azb on spatio-temporal distribution of terrorist incidents in Khyber Pakhtunkhwa. Abasyn J Soc Sci 10:137–148

Sayed SA, González PA (2014) Flood disaster profile of Pakistan: a review. Sci J Pub Health 2(3):144–149

Shahbaz M (2013) Linkages between inflation, economic growth and terrorism in Pakistan. Econ Model 32:496–506

Shahbaz M, Shabbir MS, Malik MN, Wolters ME (2013) An analysis of a causal relationship between economic growth and terrorism in Pakistan. Econ Model 35:21–29

Stepputat F, Nyberg Sørensen N (2001) The rise and fall of 'internally displaced people' in the central Peruvian Andes. Dev Change 32(4):769–791

The NEWS (26/05/2017). https://www.thenews.com.pk/print/206709-Pakistan-suffers-1231-bn-losses-in-terror-war

UNHCR UNHCR-UNHCR Global Trends 2019. 2020. Available online https://www.unhcr.org/globaltrends2019/. Accessed on 22 June 2022

United Nations (UN) Guiding principles on internal displacement, 1998, Para 2,3 (1) and 4

Weiss TG (1999) Whither International efforts for internally displaced persons? J Peace Res 36(3):363–373

Chapter 17
Social Capital and Community Organizing in Community-Based COVID-19 Management in Two Resettlement Sites in the Philippines

Melissa Quetulio-Navarra and Anke Niehof

Abstract The study yields insights on the community-based actions for pandemic management in two urban resettlement sites in the Philippines. Studies that interrogate the origins of community-based actions in government-built resettlement sites during the pandemic have been scant. We investigated in two communities how social capital, community organizing, and collective action intersected in wielding the community leaders' combined agency in building their own COVID-19 management system. The data were collected through phone interviews with community leaders and residents during the lockdown period. In a context of low local government support, the communities proved their ability to build their pandemic management system. Nonetheless, one community was much quicker in setting up the crisis management system and was even able to adjust it to the shifting quarantine policies and corresponding needs of the residents. The notable differences between the two communities can be attributed to their different community organizing experience and different stocks of social capital.

Keywords Community organizing · COVID-19 pandemic · Social capital · Resettlement

17.1 Background

By 13 April 2022, 501 million people globally have now been infected by COVID-19 with over a million number of confirmed deaths (Worldometers 2022). According to the World Bank (2020), the COVID-19 crisis is exacting a massive toll on the poor and vulnerable. Millions of people will fall into extreme poverty, while millions of existing poor will experience even deeper deprivation—the first increase in global poverty since 1998. This will mean an estimated additional 18 million extremely poor

M. Quetulio-Navarra (✉)
Ateneo de Manila University, Quezon City, Philippines
e-mail: mnavarra@ateneo.edu

A. Niehof
Wageningen University, Wageningen, The Netherlands

© The Author(s), under exclusive license to Springer Nature Switzerland AG 2023

M. Hamza et al. (eds.), *Rebuilding Communities After Displacement*,
https://doi.org/10.1007/978-3-031-21414-1_17

people in fragile and conflict-affected situations. The pandemic deepens existing sources of fragility, enlarges instability in settings tainted by violence and conflict, and exacerbates the risk of women to become a victim of domestic and gender-based violence, as evidenced by a substantial increase in emergency calls for domestic violence cases.

In the Philippines, the confirmed cases number 3, 681, 851 and deaths 59,778 (Worldometers 2022). On 16 March 2020, the Philippine government imposed a total lockdown in Luzon, known as the enhanced community quarantine (ECQ), as a preventive measure to curtail the COVID-19 outbreak. All modes of domestic travel, including ground, air, and sea, were suspended. People were not allowed to leave their homes except in case of an emergency. Border closures and entry bans were enforced. Police officers and military personnel were deployed at checkpoints to ensure that people complied. The administration also implemented strict social distancing (Duddu 2020; CSIS 2020) and educated communities on healthy lifestyles. Different kinds of lockdown continued throughout the month of April 2020. By May 15, there was a moderate easing in certain areas, depending on the number of the cases (Hapal 2021). The government aimed to recover the failing economy, which had shrunken by 0.2% during the 45 days of lockdown. On June 1, the National Capital Region's enhanced community quarantine was switched to general community quarantine, allowing some businesses and activities to resume and some public transportation to operate, the rest of the country being in modified general community quarantine, the loosest quarantine phase. This allowed businesses such as malls and restaurants to open and, with limitations, physical social gatherings to take place.

17.2 The Community Perspective

The lockdown received an excessive amount of backlash from the Filipinos. Some touted the Philippine government's pandemic response as a "tragedy of errors," leaving the population impoverished and plummeting in rising hunger (Beltran 2020). In the last week of March, the government released an economic relief package of nearly $4 billion for primarily low-income families under the Social Amelioration Program (SAP). However, the program—implemented by the Department of Social Welfare and Development—was fraught with procedural problems. There were constant complaints that the assistance never reached its intended beneficiaries and that the allocations per town were far too small. Especially, the residents of poor urban communities who depend on their daily wages in the service industry to survive were severely hit by the pandemic and the lockdown. A leader from such a community said: "COVID-19 has changed the way we live. In our urban poor community, we face uncertainty every day of our lives. The new demands placed on us by the enhanced community quarantine (ECQ) have made life even more precarious" (Calinaya et al. 2020). A leader from another poor urban locality in Metro Manila expressed his frustration as follows: "When the authorities realized they had to call for a lockdown of the city, they should have immediately assembled key officials, business people, and civil society representatives, including POs at local levels, to plan the future scenarios. That would have meant infinitely better preparation. How? Through onsite pre-positioning of sufficient food stocks, water, medicine, health equipment, fuel, and other priority items including cash. Communities could have been mobilized to cooperate in that effort—from planning through implementation and evaluation" (Tenolete and Racelis 2020).

While most poor populations in the global south floundered in misery because of the pandemic and their government's poor response to it, there were few urban poor organizations who braved the circumstances and initiated actions to mitigate impacts on their communities. One case is the Federation of the Urban and Rural Poor (FEDURP) in Sierra Leone, Africa, led the fight against COVID in Freetown Municipality, the epicentre of the pandemic. FEDURP's actions included the development of a case monitoring app, producing sensitizing messages, setting up a community kitchen for the vulnerable, and engagement with state and local authorities to the enhance the government's response to needs. Alternatively, the Malawi SDI Alliance case focused on information dissemination about COVID-19 in 35 federation groups located in three informal settlements (SDI 2020). In the Philippines, DAMPA, a federation of 230 women grassroot organizations collaborated with local governments on improving the delivery and implementation of assistance regarding masks, sensitization on most social distancing, and ensuring food security to their grassroots communities (UN Habitat 2020). The Philippines Homeless People's Federation (HPFPI) also responded to the immediate pandemic-related needs of its more than 9000 members in106 communities across the country. It did an information drive on containing the virus, distributed soap and alcohol cleanser, built communal washing facilities, and distributed food packs to families containing 3-5 kg of rice, canned sardines, instant noodles, biscuits and coffee. In some cases, the packages included baby milk, medicines, and vitamins (Carampatana and Tuazon 2020).

Collective action is a common activity among these grassroot organizations. Braving the health risks, they quickly implemented actions and delivered assistance to their poor constituencies. Collective action aims to address common interests through joint action of groups (Samper 2017), in which factors such as shared interest, organizational structure, group mobilization, and the opportunity for action are essential (Zhao and Wu 2020). Studies on collective actions to address a crisis like the COVID-19 assistance have linked the organized interventions to shared dissatisfaction with the state's response towards the crisis (Edelenbos et al. 2018) and the moral imperative among citizens to help in the pandemic response (Villadiego 2020).

Putnam (1993, 2000) cites social capital as a potent element that facilitates collective action and conquers Olson's "dilemma of collective action." According to Olson (1971), collective action for community and societal goals is difficult to achieve because of the natural tendency for individuals to free-ride on others' participation. However, Putnam (1993, 2000) contends that when people have history of civic engagement, they develop norms of reciprocity, trust, as well as networks of social relations that cultivate in them shared goals and galvanize corresponding collective efforts towards community well-being. These three norms of reciprocity, trust, and social networks are referred to by Putnam as social capital. In urban poor settlements, erosion of civic engagement and sheer social capital are being addressed by community organizing led by professional community organizers of non-government organizations (Gittell and Vidal 1998; Ohmer and DeMasi 2009). Community organizing is mobilizing disempowered people through leadership capacitation to confront power and resolve community-shared issues (Pyles 2021). While the relationship of social capital and community organizing is beyond argument, their role in managing the COVID-19 pandemic at grassroots level has been understudied.

The present study focuses on two urban resettlement sites in the Philippines, exploring how the community stocks of social capital and community organizing history influenced the development of a leaders-led, community-based COVID-19 management system. The two cases were examined to reveal the intersection of social capital, community organizing, and collective action in wielding the leaders' combined agency in building and implementing an endogenous pandemic management system during the critical periods of quarantine impositions-from enhanced community quarantine (ECQ) until the transition to modified enhanced community quarantine (MECQ), then finally the shift to general community quarantine (GCQ).

17.3 Social Capital and Community Organization

Social capital is a widely investigated concept that has been viewed to explain several development-related outcomes ranging from economic growth (Fukuyama 1995;

Knack and Keefer 1997), local governance (Putnam 1993), to community development (Kay 2006; Putnam 2001). It is also seen as a substitute for other capital, like physical and financial capital, which poor families such as those in the resettlement sites lack (cf. Putnam 1993). Social capital is a multidimensional concept comprising various forms such as ties or networks, norms of reciprocity, and trust that facilitate coordination and cooperation towards joint benefit (Coleman 1990; Putnam 1995, 2000). Social capital can be examined at the individual and contextual level (Carpiano and Moore 2020). Moreover, social capital is also the source of civic activism (Portes 1998; Portes and Landholt 1996). Woolcock and Narayan (2000) distinguished three different types of ties or structural social capital: bonding, bridging and linking social capital. While poor families survive using their strong bonding ties with family and close friends, getting ahead in life can only be realized with bridging social capital, referring to ties beyond one's family and close friends, and with linking social capital or one's vertical ties with the powerful and the decision makers. By the same token, Briggs (1998) deems bridging and linking social capital as critical in attaining development and growth.

Social capital is seen to have a positive association with health and other community outcomes (Putnam 1993; 2001). In a similar vein, support from individuals and cognitive social capital (i.e. trust, social harmony) has been found to be associated with child nutritional status across four countries (DeSilva and Harpham 2007). During the COVID-19 pandemic, studies (Bai et al. 2020; Barrios et al. 2020; Bartscher et al. 2020; Borgonovi and Andrieu 2020; Ding et al. 2020; Fraser and

Aldrich 2020) have pointed out the correlation between high stocks of social capital in various places (state, county, prefecture) with compliance to COVID-19 pandemic safeguards measures, fewer confirmed cases, and a slower growth rate of infection over time.

In a study done in resettlement sites in the Philippines and Indonesia, the number of people who supported the resettled families as well as the norms of reciprocity and trust had positive effects for the families concerned in both countries (Quetulio-Navarra 2014). Additionally, the Philippine resettlement community benefitted significantly from the networks of women who started organizing themselves and subsequently found solutions to the lack of basic services like power and electricity (Quetulio-Navarra et al. 2017).

Consensus-based community organizing refers to establishing strong and participatory people's organizations through which the oppressed can join in decision making, even if only partially, in their common interest (Maglaya 2014). Pyles (2021) encapsulates the goal of community organizing as empowerment and social change. Empowerment is increasing the levels of understanding, engagement, and personal power of an individual citizens through several activities like consciousness raising, participation in social action, and taking on leadership roles. Empowerment is also shifting power to the side of the organized group through increasing their membership, training leaders, gaining a reputation for canny strategy, raising money to fund infrastructure and staff, and demonstrating the ability to engage large numbers of people in public action (Schutz and Sandy 2011). Empowerment is also seen to sustain social change such that the more people feel empowered the more sustainable organizing becomes over time (Pyles 2021). Organizing for social change refers to "the process through which a group of individuals orchestrate their skills, resources, and human potential to gain control of their future" (Papa et al. 2006, p. 31).

Social change is also the aim of community building which is a type of community organizing that concentrates on strengthening the community's social and economic fabric through facilitating access to outside resources like basic services or livelihood programs (Smock 2004). This type of organizing is the usual approach of international non-government organizations and of the government in implementing development programs or projects in poor communities. The types of community organizing may differ in how they generate social capital and networks and the forms they take. However, they all aim at developing social capital and networks to stimulate civic engagement and to provide a mechanism for ordinary citizens to impact public decision making that is relevant to the improvement of their social and economic conditions (Ohmer and DeMasi 2009).

Community organizing was indispensable in building self-reliance among the urban poor homeowners' associations in the provinces of Rizal, Iloilo and Albay, and in Quezon City and Muntinlupa municipalities, who all were acquiring land for their housing projects. The process, led by the Philippine Action for Community-led Shelter Initiatives, included community consultations, socio-economic surveys, orientations, organizational development, savings orientation and book-keeping (Teodoro and Co 2009). The alternative approach to housing for informal settlers relies heavily on community organizing inputs in urban poor communities. In the so-called People's Plan model, the community's people's organizations are the project beneficiaries, who should be properly consulted and meaningfully participate in the project planning and management (Bonagua 2020).

17.4 Study Design

17.4.1 The Two Communities and Their Community Organizing History

Two resettlement communities in the Philippines were chosen for this qualitative study conducted during the COVID-19 pandemic. One is a distant resettlement site located in Laguna Province, while the other case is an in-city site located in Pasig, Metro Manila. Both sites were built by the National Housing Authority (NHA), a government agency mandated to generate housing for the country's poor informal settler families. As presented below, the two locations have functioning homeowner's associations, have undergone community organizing by non-government organizations, and have experienced implementing community development projects. Thus, the two sites have a comparable background and social structure.

17.4.1.1 Southville 7 (SV7)

The Southville 7 resettlement community in Calauan, Laguna, was constructed in 2000 but families only started resettling in 2009. The community is divided into three sites. Site 1 houses families relocated from the Pasig River and those affected by government projects in Metro Manila such as road widening. Site 2 accommodates households from the Pasig River and identified danger zones in Calauan. Families displaced by Typhoon Ondoy reside in Site 3. Together, 3744 families currently reside in SV7.[1] In this study, we focused on Site 3, where 1962 families live. Site 3 has four homeowner's associations (henceforth HOA).

During 2012–2015, under the ADB project "Promoting Partnerships and Innovation in Poor and Underserved Communities," the entire community of Southville 7 was assisted by the ABS-CBN Lingkod Kapamilya Foundation (ALKFI), a non-profit arm of one of the Philippines' biggest media networks. With the assistance of ALKFI, the ADB grant was invested in community building, in the form of multisector partnerships, community development and innovation, and social entrepreneurship. HOA Officers received training in management and leadership, project management, financial management, and community organizing and development. In addition, they participated in training on Grassroots Entrepreneurship and Management, gender

[1] Data based on Joly Homes Foundation 2019 Socio-Economic Census in Southville 7.

sensitivity, savings and investment mobilization, community development planning, and participatory strategic planning for local leaders. Such capacity building should include both hard and soft skills. It is clear that trainings on financial management, community development planning, and leadership development were effective in confidence building. However, because of the low level of trust among residents the officers also needed strong interpersonal skills to improve relations with their constituents and other associations. An Asian Development Bank (ADB) evaluation report (2018) cited issues with the community organizer and the community organizing program. Currently, Site 3 is being assisted by the Joly Homes Foundation, a non-government organization (NGO) focusing on issue-based community organizing in resettlement sites and informal settlements. Since 2019, the Foundation has guided the Site 3 HOAs in organizing for social enterprise and in establishing a HOA-led cooperative that offers micro-credit services to its members. Moreover, HOA leaders are coached in the resolution of community issues like water supply, communal gardening, and keeping the playground.

17.4.1.2 Manggahan Residences (MR)

Manggahan Residences in Pasig City is an in-city resettlement site that presently accommodates 495 households previously residing just across the site, along the east bank floodway area. The construction of MR started in the early 2000s and resettlement started in 2015. The community is being managed by the leaders of the Alliance of People's Organization along Manggahan Floodway (APOAMF).

When the informal settlements along the Manggahan Floodway were threatened with demolition by the government in 2009, the CO-Multiversity (COM), an NGO specializing in consensus-based community organizing, started assisting the affected families in their quest for in-city, people-led housing (instead of the usual poorly-built distant relocation sites). The organizing efforts united the involved ten urban poor local organizations, with members totalling around 2800 informal settler families, together forming the Alliance of People's Organization along Manggahan Floodway (APOAMF). According to the leaders, over the years, the community organizing done by the COM brought the families newfound dignity and agency in resolving their housing issue. In view of the aim of the APOAMF to implement people-led, in-city socialized housing, the leaders received training in capacity building, leadership, and disaster-preparedness. They were also coached on how to engage with different stakeholders, how to build consensus among their members, and how to draft simple reports and letters. After years of struggle, rallies, and negotiations, APOAMF got the approval of the government for their People's Plan Housing Project, which was built just across their original slum residence. The project can accommodate 900 households in 15 five-storey buildings with 60 housing units per building. Since 2015, 495 households have relocated to their new units. The new HOA, led by a female President, has retained the name APOAMF and COM has continued to guide the residents' social integration in their new housing community. The HOA led the community in the creation of an Estate Management Policy to guide the leaders and households in managing their buildings and other community facilities and has formulated resolutions to establish order, peace and security in the Manggahan Residences. The HOA leaders also maintained their ties with relevant organizations and politicians.

17.4.2 The Respondents and the Interview

Due to the pandemic and the lockdowns imposed by the government in Metro Manila and in nearby provinces, phone interview was the only feasible mode of data collection for this research. Eight respondents, four from each site, were interviewed via phone call in July 2020. For each site, two HOA leaders and two members, female and male for each respondent type, were chosen for the in-depth interviews that would last from one hour to one hour and a half. Prior to the conduct of every interview, the researchers elicited the verbal consent of each interviewee, which was recorded for safekeeping.

The in-depth interviews were guided by a semi-structured questionnaire, which aimed at eliciting the respondents' stories from the day they learned about the government's lockdown order until they transitioned to different types of quarantine arrangements. The questions focused on the following four subject areas:

1. First reaction of the families towards the lockdown and preparations at household and community level
2. Families means of managing their lives under lockdown
3. HOA leaders' role in managing the community under this new circumstance
4. Level of participation and motivations of the families and the community as a whole in the new community-level COVID-19 management system.

17.4.3 Data Analysis

The interviews were recorded then entered into Atlas ti version 8.4.5 for thematic analysis to identify key themes in the four interview domains. A coding frame based on the four domains was prepared. During coding, codes were added to categorize the contents of the narratives and memos were attached to every recording, which served as a summary of the recording. This was done separately for the two sites. Subsequently, the codes were organized into one big network to analyse the logical connections of the codes and to check for codes that needed to be merged or separated. The networks were interpreted and analysed, then patterns and themes were identified. Finally, the two sites were compared in view of the networks and findings were noted.

17.5 Results and Discussion

The comparison of Southville 7 and Manggahan Residences communities centred around the individual, household, and community experience in tackling the COVID-19 pandemic at three critical junctures: the onset of lockdown, then the establishment of the community-based pandemic management system under ECQ, and finally when the communities transitioned from MECQ to GCQ. In the analysis of their pandemic management journey, the following areas were explored to establish how much these are implicated in both the process and outcome of building a COVID management system that matters hugely to the community residents.

1. Communal practices pre-COVID
2. Community organizing strategies pre-COVID
3. Community's social capital stocks
4. Leaders' and members' cooperation in the community COVID management system
5. The role of local government leaders in the community COVID management system.

17.5.1 Immediate Reaction Towards the Lockdown

The news about the nationwide "lockdown" settled similarly among the residents and HOA leaders of the two sites. All were in shock and in disbelief that the government would proceed in implementing the national lockdown. They thought that it was just government propaganda, to divert the people's attention from real social issues and the health crisis. They were very much aware of their own vulnerability. They were worried about how their families could survive the lockdown without enough food supplies. Some were doubtful about the feasibility of the lockdown since poor families have to work every day for their daily wages. They claimed that all families in their community could not adequately stock food supplies and essential items since they could not afford advance purchases and lacked refrigerators to keep perishable items. However, they thought that with enough rice supplies they could survive the lockdown, even if they would not have dishes to go with the rice. The money allocated for their monthly housing amortization was used for buying rice. Residents who received the news while at work elsewhere, wanted to be with their families as soon as possible.

According to the Manggahan Floodway community leaders, the lack of knowledge about the disease and its preparation requirements, in combination with poverty, created panic and heightening fear about the pandemic and the lockdown among the residents. One leader mentioned:

> *When we [the leaders] realized that COVID-19 is indeed real and people are dying from it, we got quite scared. We started asking ourselves, what will happen to us, to our community when COVID infection has reached our place? How are we going to survive?*

Both communities reveal how the top-down approach of the government in handling the pandemic struck vulnerable communities of low-income households and limited access to information. It is also apparent that while the HOA members are concerned about their family's' survival during the pandemic, the leaders have the same concern, but extended it to the community.

17.5.2 Setting up Community's COVID-19 Management During ECQ (March–May)

Both the Southville 7 and the Manggahan Residences considered setting up the COVID-19 management system in their community during the onset of the enhanced community quarantine (ECQ) as the toughest job they ever did. Figure 17.1 illustrates how the two communities started in setting it up, who were involved, and what the priorities were. The succeeding discussion explains how the leaders in each site went about establishing their pandemic management system in view of the social, political, and economic context and how this interfaced with their stocks of social capital and community management practice.

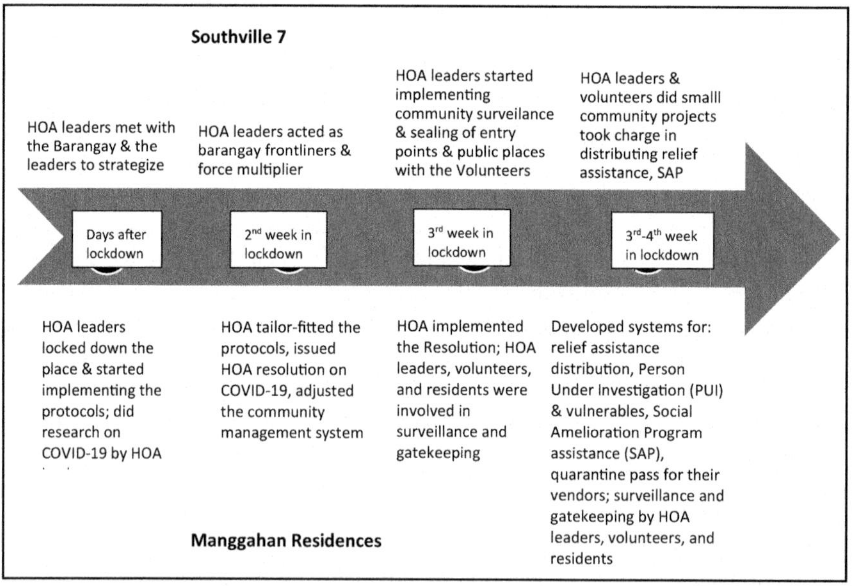

Fig. 17.1 Timeline of setting up COVID-19 management system under ECQ. *Source* Authors' interviews with respondents

17.5.2.1 Southville 7 (SV7)

Few days after the lockdown declaration, the four female presidents of the four HOAs and few of their top officers (also called leaders) were called in by the Barangay officials (village officials) to orient them on the government-issued safety protocols for COVID-19. These were: social distancing, hand washing, wearing of face mask, quarantine procedures, curfew, sealing of entry points in the community, and assigning checkpoints.

The top officers then convened a meeting with their HOA officers to echo the orientation done by the Barangay. Those who did not attend were visited in their homes. The leaders were also constantly updating their Facebook chatgroup.

Their next mission was to implement the instruction of the Barangay to seal all the exit and entry points in Site 3 and not let non-residents enter the site. Only residents of Site 3 were allowed to enter the site.

The narratives from the leaders convey a strong bonding capital among the HOA officers and few blocks in Site 3. Leaders shared that in the first few weeks after the enhanced community quarantine or "lockdown," they would meet almost every day with most of the HOA officers and block leaders close to the ever-present officers. The meetings would revolve around creating a system to have the residents comply with the COVID-19 protocols. The stocks of social capital that resulted in trust and cooperation among these leaders had been nurtured by the almost monthly HOA meetings and few community projects done together prior to COVID-19, like the *Bayanihan* on Sundays, when they would do cleaning and planting in the site. *Bayanihan* was designed as a community activity, but in fact the leaders were the main participants.

The management of COVID-19 within Site 3 started with the homeowner's association (HOA) leaders as well as block leaders,[2] essentially acting as a force multiplier of the local government (barangay) in the surveillance of the community in terms of compliance to the protocols. The barangay set up a tarpaulin in public places for information about the health protocols and distribution of COVID-19 leaflets, but leaders would do house-to-house orientation on the protocols and posted updates on the HOA Facebook page. The barangay requested the leaders to perform frontline jobs in the community, such as acting as marshals at the designated gate and assisting the health personnel in monitoring identified "person under investigation for COVID-19" (PUI) during quarantine procedures.

In the following weeks, the leaders started to customize the barangay directives according to the needs of the community. Site 3 has an area size of around 27 hectares and almost 2000 residents, which made it impossible for the leaders of the four HOAs (about 60 persons) to do effective surveillance in the entire site around the clock. They started recruiting volunteers among their members. Ten men

[2] Two layers of community leaders are present in Southville 7-the Homeowners' Associations (HOA) leaders who are elected to different positions by the HOA members and the block leaders who represent the residents in each block. Unlike the HOA leaders, the block leaders have been chosen to represent their own block by majority of their block residents and report to the HOA leaders.

joined the volunteer surveillance team. Of these 10 men, one was a HOA officer. The others were mostly husbands of HOA officers. They had lost their jobs at that time, were ready for a new role, and were concerned about their community. They wanted their wives to stop doing the roving in the site, hence volunteered as replacements. As shared by a respondent who also acted as a volunteer, the volunteers were not instructed on how to do the surveillance, neither by the HOA nor by the barangay. They based their surveillance approach from what they saw on TV and social media. They were not also given a regular allowance or support for food and gas for their motorcycles. Food assistance from concerned residents and HOA officers was sporadic. As mentioned by one volunteer:

> *We encountered problems in volunteering because we didn't have budget and we were not affiliated with the barangay although we had an ID as designated roving officers in Site 3. We didn't have budget for coffee, even face masks. We would use our own money for food, face masks, and even gasoline.*

However, they would do their job religiously, reminding everyone in Site 3 to stay at home, observe social distancing, wear a face mask, and to wash their hands every time. The leaders and volunteers relayed that enforcing the protocols was considered an uphill climb during the first few weeks of ECQ. Residents, male and female, whom they found incorrigible after several reminders, would be turned over to the barangay for punishment. This could take the form of a heavy scolding from the barangay personnel, doing squats for hours, or for some even overnight imprisonment. Roving personnel and leaders acknowledged the need for surveillance and punishment to achieve protocol compliance of the residents. In the weeks that followed, the volunteers were joined intermittently by barangay staff, not to assist them in surveillance but to wait for violators whom they would arrest and report to their superiors. The volunteer who disclosed this added that instead of boosting their morale, the presence of the barangay staff made them question the volunteer work they were doing; the situation struck them as unfair. Whereas the barangay staff are paid by the government, here they were doing volunteer work and exposing themselves to the risk of getting infected. Why could the barangay not do the roving alone? Or, when assisting the barangay, why were they not receiving any financial support for their gas or even snacks from the local government? These were the questions in their minds, which never got any response. Part of the surveillance work of the leaders and the volunteers was to make public places off limits to the residents, teenagers in particular, by putting up fences and sealing entry and exit points. In their free time, the volunteers and leaders did other community work, like fixing the solar streetlights, adding road humps, and clean-up drives.

Another area that the leaders had to agree on was the distribution of the relief assistance (from different agencies) to the families in the community. The relief assistance consisted of food supplies enough to feed a family of three or four for three days or hygiene products. Usually, the process of distribution would be activated by the barangay upon receiving news about an upcoming delivery. The barangay would then call in the HOA presidents and inform them about the assistance, how many families it could cover, and when it would exactly come. Then the HOA scenario on assistance distribution would be followed:

The HOA President would announce about the coming relief assistance. She would call the leaders to discuss about the strategy for distribution. Then we would gather the assistance, then call the block leaders who would distribute the assistance to their block members. Block leaders would line up while observing social distance when waiting for the assistance being handed over by the HOA leaders. […] Some Sites (referring to Site 1 and Site 2) in Southville 7 have problems with relief assistance distribution, but not in our HOA, Phase 5, in Site 3.

If the relief assistance would not be enough for all the families in Site 3, they would prioritize families already experiencing food shortage and personally deliver the assistance to their houses, to avoid jealousy among those who did not receive any. There were also times that they would distribute the limited assistance to their volunteers who were not getting any compensation for their work. It is interesting to mention that while HOA leaders would take the credit for the orderly distribution of the relief assistance, the volunteers claimed otherwise. According to a volunteer, it would always be the HOA President and the volunteers who were heavily involved in the distribution and leaders were rarely around during this activity. The four female presidents and few HOA officials were seen taking an active role in the release of the assistance to the identified beneficiaries in Site 3 by coordinating with the local government the selection of eligible families. The assistance consists of cash under the Special Amelioration Program (SAP) of the government in view of the pandemic. The leaders attributed the relatively quick release of assistance to their good working relationship with the local government.

In comparison with their other areas of pandemic management, there was no policy or agreed practice for the selection of families who were most in need of assistance. HOA presidents appeared to be primarily responsible for the interventions needed. When asked how she knew about cases of hunger among her members, a President replied that the household head would let her know of a family's emergency situation. As a response, the President would either use her own money to help or approach the barangay for rice provision. She would also prioritize this household in rice distribution and would seek the understanding of those who would not receive. She further shared that the HOA did not have funds for such needs.

Whereas the strong bonding ties among the HOA leaders were crucial for the coordination with their linking tie, the local government of Calauan, and for subsequently implementing new norms in Site 3, their social capital did not expand to include bridging ties in the community. Such ties might have been substantially helpful in enforcing and tailoring welfare-centric pandemic protocols. In the same vein, social cohesion is very much present among the leaders, but not in the Site 3 community as a whole. This may explain the marginal support to the roving activities, the poor observance of the protocols, the cold treatment towards violators of the protocols, and the limited action towards families in peril. One HOA member said:

HOA members were not so active. During COVID I recognized the value of a HOA board member. I volunteered in place of my wife.

17.5.2.2 Manggahan Residences (MR)

Few days after the declaration of the lockdown, the APOAMF female President and the rest of the leaders still did not have a clear understanding about COVID-19. They had many questions. How lethal is it? How does one get infected? How to avoid it and keep it at bay? One leader shared that although they had a sort of an orientation on the government protocols from the barangay and information on social media, they were still confused and extremely worried about how to protect their families and the community. The leaders, particularly the President, started doing research about COVID-19, and in their Facebook chatgroup, they discussed their findings and ways to protect the community from the impact of the pandemic. One leader said:

> *We gathered and studied data and information about COVID from the provided guidelines and those provided by social media. Following this, we tailor-fitted the regulations and protocols according to our community situation. We had to do this ourselves because no authority could explain to us about COVID.*

In the middle of this crisis and without adequate information, the solidarity among the leaders (majority women)—from the Council level down to the Building Representatives and Floor leaders in each of the eight buildings in MR—was instrumental in quickly implementing the emergency protocols in the community. They started even before the Barangay Resolution was issued about not letting non-residents enter the premises, wearing of face masks, and hand washing. Moreover, the established community management system in the site hugely helped in cascading the protocols to the residents. The existing HOA communication system comprises three layers, and for each layer there is a Facebook chat group. The lowest level is the residents and floor leader, the middle level the floor leaders and the building leaders, and the highest level are the building representatives and the HOA Council leaders. Previously, they would activate these levels through chat or face-to-face interaction. During the pandemic, the floor leaders were expected to check on the health conditions and general status of their members either through Facebook chat or meeting with social distancing. Subsequently, a report about their residents would be submitted to the Council for information or action. The Council also added an announcement system that they installed in their open space to announce the emergency protocols three times a day. One leader related:

> *The HOA installed large speakers to communicate with the entire residents from outside to their homes. We used this to relay information, announcements, safety precautions, reminders.*

After conducting adequate research and having gained an understanding on the nature of the virus and how to avoid it from penetrating MR, the leaders customized the regulations according to their community characteristics and issued a Board Resolution that covered all the applicable protocols and regulations for all the residents in the community. The Board Resolution, posted on each building, was about not to wear sleeveless shirts outside their homes, hand washing, wearing a face mask, social distancing, prohibited social activities, rules for continuation of home-based businesses, and the restriction of entry into the community to residents only. In order to enforce these new behaviour-changing regulations, the HOA saw the urgent need for additional volunteers to man the gate. Prior to the pandemic only one person was assigned at the gate. During the pandemic, they added volunteers to work in shifts. The volunteers would receive a small honorarium from the HOA budget. However, when the HOA leaders saw that there were always arguments between the gatekeeper volunteers and the residents, they decided to assign also a HOA leader at the gate to assist the volunteers for four hours per day. The arguments would always be about a resident insisting to allow the entry of his or her non-resident family member or friends. With a leader at the gate, those who complained could be given an explanation from an authority in the community. This resulted in a noticeable decrease in the number of fights or arguments at the gate. Simultaneously, the HOA Council female leaders took charge of doing roving activities in the community twice a day—day and night—to check the compliance of the residents with the HOA Resolution. They checked whether teenagers were staying outside and whether those going out would only buy essentials. Those who violated the protocols, most of them teenagers, would be reprimanded by the HOA Council while explaining patiently the need to follow the protocols. Leaders would repeatedly inform and warn the members of the ECQ policies. The leaders would practice maximum tolerance in handling violations, arguing that their approach in surveillance was to raise awareness and support to the protocols, not to wreak fear among the residents. Sometimes, the President would talk to the parents of a teenager who had violated the protocols. Incorrigible violators were handed over to the barangay for punishment.

Notwithstanding, the leaders expressed disappointment on the little support they were getting from the barangay. They thought that during the first few weeks of the lockdown enforcing the protocols would have been much easier if there had been a visible barangay-HOA partnership. They further said that the first two weeks were the toughest in enforcing the protocols, since—as they clearly understood—changing behaviour is not easy. One resident disclosed that the about ten per cent of the residents who did not observe the protocols were persons who had already been unsupportive of the HOA leadership in their previous residence. After two weeks, the residents generally started complying with the protocols and the number of violations dropped dramatically. Furthermore, the residents started assisting the leaders in their surveillance activities. They would now remind fellow residents to follow the protocols and would report the violators and those who refused to be corrected to the leaders. Then, the leaders would come and deal with these persons.

The HOA agreed on a policy and arranged a set of procedures for the distribution of relief assistance that frequently came to their community. The HOA leaders with the concurrence of the residents agreed that they would observe equity and fairness in the distribution, taking into careful consideration a household's vulnerabilities and size. Before the distribution, the leaders had to ensure first that the assistance would be enough to cover all households. If inadequate, they would combine it with another batch of relief assistance or use their HOA funds in buying items for topping up the insufficient assistance. The following statement of a resident shows that the residents were proud of the way the relief assistance distribution was organized:

We have a good system in relief distribution. Council leaders would inform the building representatives, then they will give instructions to the floor leaders, who will now request us to line up per group while observing social distancing. Then our names will be called one by one. Our leaders are really hardworking and without them we don't know how we would have survived. They never got tired of looking for relief assistance for the community.

The member households considered vulnerable, like those experiencing hunger, were looked after by the leaders. The HOA had profiled all households, so that they knew who at a time of crisis would need help most—like households with sick people or an unemployed household head, women-headed households. When the government implemented the program for cash assistance to poor families, the leaders were proactive by giving the local government in advance their list of eligible families. Thus, the processing and distribution of the assistance could be done quicker than in other communities.

Except for the computer shop within the building, all home-based small *sari-sari* or sundry stores were allowed to operate, provided they observed the protocols and the 6 pm curfew. Further, to avoid financial collapse among residents who were previously vending fruits and vegetables on carts outside the gate, the HOA allowed them to continue vending at the open space inside the community with the condition that they would strictly follow the protocols. The basketball court was converted into parking space for the Jeepneys and tricycles of the residents who were not allowed to operate. Lastly, the quarantine pass[3] policy was adjusted by the HOA according to the requests of the residents. Household heads who got a quarantine pass to go to work asked for their sons or daughters aged 18 and above to be given a pass as well, so they could also go out to buy essential items for the family in the event that the household head forgot. When three cases of a suspected "person under investigation" (PUI) emerged in the community, the HOA also established quick protocols, including quarantine for the residents of the entire floor where the PIU was residing. For faster contact tracing and to limit the number of leaders exposed to COVID-19, the President volunteered to assist all the families needing essential items or food supply from outside.

The community also carried on with their weekly clean-up drive—the *Bayanihan*—and the daily cleaning of hallways, for which every household appoints one member in the pool of cleaners. Absence in the activity is fined. The daily cleaning-up activity covers the hallway of every building floor, the Sunday *Bayanihan* program has the immediate building surroundings as the area of assignment. To avoid COVID, the leaders decided to intensify their cleaning activity by consistently adding a disinfectant agent.

One resident expressed her view on how the HOA leadership dealt with the pandemic that affected their entire community as follows:

We were very scared in the beginning of the lockdown, but our fear was dissipated by the constant presence and action of our leaders. They never abandoned us.

[3] Under the ECQ, the Barangay issues the quarantine pass to an adult member preferably head of the household (except for 60 and above) and authorizes the holder to leave his/her residence to buy essentials or go to work.

The strong bonding social capital found among all the HOA leaders of MR (from the Council leaders down to the building leaders and the floor leaders) enabled the smooth and quick setting up of the COVID-19 management system, from introducing emergency protocols to Board Resolution-based protocols. The bonding social capital made it possible for all the leaders to instantly come together (virtually) and carefully plan their way forward and determine their roles in managing the pandemic given the community's social and physical conditions. The functioning of the communication system, the observance of the protocols only after two weeks of implementation, the maximum tolerance approach of the leaders, the involvement of the residents in surveillance, and the collaboration between the residents and the leaders are illustrations that the community is in like manner rife with bridging social capital and that connections are not just concentrated among the leaders but are also seen in the entire population of the Manggahan Residences. However, also here the barangay did not play a proper role in supporting the community COVID management system. Articulated assistance for surveillance was asked for from the barangay, but no help came.

17.5.2.3 Comparing the Leadership in the Two Sites During The First Phase/ECQ

Although the leadership in two communities acted immediately on addressing the pandemic to days after the lockdown announcement of the government, the leaders in MR were faster in understanding the potential catastrophic impact of COVID-19 if not controlled. Setting up as well as adjusting the relevant health safeguards measures in the community as quickly as possible was critical. The prompt action of the Manggahan leaders can be explained by their leadership ability that was groomed by years of empowerment and capacitation training by their partner NGO, the rich bonding ties that are present among the leaders as nurtured by long history of community organizing, and the bridging ties between the leaders and their members. Alternatively, Southville 7 leaders were able to tap immediately their linking ties with the local government at the onset of the ECQ and was able to gain access to crucial information about the health crisis. However, they were unable to harness this particular type of social capital as they were assembling their community system for pandemic management, and this may be explained by the absence of well-functioning leadership structures that they could build on. Thus, they were rendered dependent briefly on the local government leaders on crisis management directions.

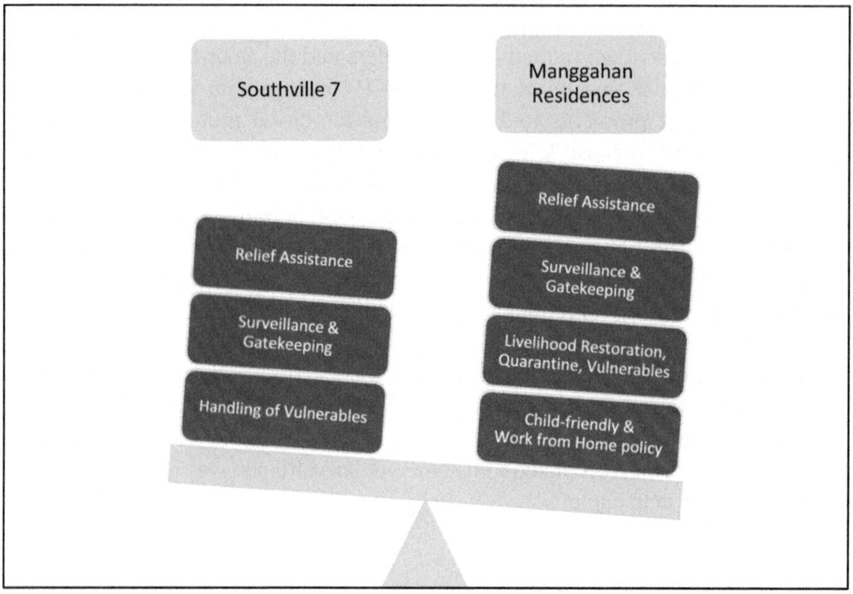

Fig. 17.2 COVID-19 management system in the two sites after the lockdown. *Source* Authors' interviews with respondents

17.5.3 The COVID-19 Management System During the MECQ to GCQ (May–July)

The two communities continued in the enhancing and implementing their community-based pandemic management system when Metro Manila and Laguna shifted to modified enhanced community quarantine (MECQ) in the second half of May and then to general community quarantine (GCQ) since June, adding more areas of intervention as shown in Fig. 17.2.

17.5.3.1 Southville 7 (SV7)

The HOAs carried on with the implementation of their pandemic management system until the community shifted to MECQ and then later on to GCQ. Under the new quarantine model, the leaders faced another spate of challenges that threatened the implementation of their COVID-19 management system. The leaders and volunteers observed that the gradual lifting of restrictions in physical mobility somehow undermined the new pandemic norms developed during the ECQ. Some people started ignoring the protocols again, the wearing of masks and observing social distancing in particular. The community's exposure to the virus increased since most of the household heads who went back to work would now freely go outside the community and return to Site 3. The number of volunteers for community roving dwindled as some of them resumed working since they could not rely on government assistance. Some did not want to continue because they no longer wanted to work with the barangay staff, as shared by one volunteer:

I don't want to work with the barangay anymore. I don't agree with the unfair sharing of roles. We will do the surveillance then they will just wait for our report on the violators whom they will arrest for punishment. They're the ones who are paid by the government to do the job we are doing. I still volunteer but not for COVID anymore, I still do rove at night in Site 3 to ensure that residents or outsiders will not throw their garbage in undesignated places.

17.5.3.2 Manggahan Residences (MR)

The Manggahan HOA continued enforcing their protocols and policies in the community, even when they transitioned to MECQ and GCQ. The new quarantine modality created new problems for the COVID-19 management system as household heads wanted to go out and earn income for the family, prompting the HOA leaders to adjust and refine the protocols according to the needs of the families and the community as a whole. One leader said:

Covid cases are rising in our barangay and therefore the Council refuses to shift to GCQ less strict policies. We still enforce ECQ policy.

Firstly, the HOA retained the 6 pm curfew for everyone except for residents who showed proof of work outside that did not allow them to be home at 6 pm. Fruits and vegetable sellers started selling outside again but only until 6 pm. Secondly, in response to the requests of the parents and children who were stuck in their 22-m^2 housing unit during ECQ, they developed a policy on the children's playtime in their floor's hallway. Children would be allowed to go out and play, but only in the hallway. After some time, those who were working from home (WFH) started complaining about the noise of the playing children. The HOA collected the schedule of the WFH residents so they could tailor fit the playing times of the children and released a play schedule. In the morning, with a child minder around, the children could play from 8 to10 and in the afternoon from 4 to 6. They also advised the WFH residents to try sleeping with a headset. Lastly, HOA leaders resumed the meetings with the residents per building, still observing social distancing. One leader admits that the pandemic brought the HOA leaders closer since they had to constantly communicate with one another regarding protocols and community issues. It was also surmised that throughout ECQ and GCQ the Manggahan community remained organized, except for the few who kept opposing the community-based response to the pandemic.

Manggahan Residences remains more advanced and resilient than SV7 in pandemic management even after the quarantine arrangement had shifted to less restrictive ones (MECQ to GCQ) laying bare new supervision challenges to the community leaders. The deficiencies that had afflicted the COVID-19 community-based Management in SV7 during the early part of the pandemic lingered even in the following quarantine phases and slowly eroding the norms that were built at the start of the lockdown. In contrast, MR kept on improving their pandemic control system and remained abreast with the changing demands of the context.

17.6 Conclusions

The cases of SV7 and MR show the impact of the COVID-19 crisis on a resettlement community comprised of poor families and reveal the roles of social capital and community organizing in the HOA's response to the pandemic. The two communities shared similar sentiments about the effects of the crisis on their families and on the community as a whole. The HOAs in both settings were scared and felt greatly deficient in terms of managing the crisis. Nonetheless, the HOAs were aware of their responsibility towards their residents. Hence, with or without the adequate support of the local government, they proceeded in leading their communities in setting up and implementing their COVID-19 management system.

However, whereas in both communities, the systems are community-based and HOA-led, they differ starkly in their preparation and implementation strategies. From the narratives, it transpires that these differences can be attributed to the community organizing experience, shared development victories, and the characteristics of their social capital.

Community organizing built on consensus and bottom-up issue resolution rather than community building as a goal, seems to leave more lasting positive effects for the community's social capital, development aspirations, and for the accumulation of human capital of HOA leaders. This confirms the assertion of Pyles (2021) that the lasting impact of community organizing in the pursuit of social change in a marginalized community can be attributed to the embeddedness of empowerment all throughout the issue resolution process. Moreover, it was illustrated in the two communities the positive correlation and interface of community organizing, social capital, and development. The community organizing for people-led housing guided by COM, spanning the period of 2009–2019 from the Manggahan slums to MR and the tangible development victories (i.e. preparation of the community-led housing project plan, funding for housing project) facilitated the forging of strong ties among the leaders and within the community, as well as linking ties with the other stakeholders in the government connected to socialized housing and with different NGOs. Trust and reciprocity within the locality are high, as illustrated by the stories of the respondents. In the process of organizing the Manggahan community, the APOAMF leaders also had enriched their human capital with the formal and informal trainings they received from COM. These were translated in the formation of social institutions in the community such as the HOA and its Committees necessary in community management. This capacity building enabled the leaders to strategize collectively, even when experiencing the economic and social shocks of the COVID crisis. In contrast, the seven-year community building program by the ABS-CBN Lingkod Kapamilya Foundation in SV7, perhaps due to its top-down nature, yielded bonding ties among the HOA leaders but few connections across the residents and vertical ties like with the barangay. Trust and reciprocity in the community were low. Similarly, human capital enrichment of the HOA leaders focused on confidence building, rather than on how to relate with their members and engage in participatory decision making and issue resolution.

In MR, the level of civic action and support by the APOAMF leaders and the majority of their residents in effectuating their community's pandemic management system affirms that such action springs from social capital and that its presence can be predicted by levels of social capital (Portes 1998; Portes and Landholt 1996). This relationship also explains the little support the HOAs in SV7 received from their residents. In SV7, the overall level of social capital is low. Leaders did not have established connections with their members, cross cutting ties among neighbours were few, and a sense of obligation to help the community during the pandemic crisis was virtually lacking. Decades ago, Temkin and Rohe (1998) surmised that over time, neighbourhood stability and the sense of attachment and loyalty among residents are determined significantly by the social capital they share. This has remained true to this day in the two resettlement sites when they were confronted with an unprecedented crisis. Further, the data show that a community can sustain transformative change under shifting conditions when its social capital stock is rich at multiple dimensions of social capital (ties, networks, trust, reciprocity, etc.). Relatedly, individual and community levels of social capital are viewed as linked and interdependent. Investing in both is therefore worthwhile.

The stories of the two sites also point out the limitations of the community-based pandemic management system, because its comprehensive implementation is partly contingent on an effective and supportive local government. The HOA leaders in both sites claimed that they have good relationships with their local government and some politicians. But during the COVID crisis, these linking ties were only useful in two cycles of relief assistance and in fast-tracking the SAP distribution. Overall, leaders and members from both sites shared the feeling of being exploited and stretched too far by the barangay representatives. They were disappointed for not receiving the forms of assistance that only the local government can provide, like readily available ambulances or transportation for those who needed to be hospitalized, quarantine places, and sufficient assistance of food and hygiene kits during lockdown and quarantine. Their collective action might have made it possible to cover these needs, but their locally raised resources gathered from the poor residents could only go so far. Contrastingly, their vertical ties with NGOs resulted into different forms of relief assistance for the entire community that helped them get by during the lockdown.

The female-dominated HOA leadership in the two settings also highlights the indispensable role of women in crisis management, both at household level and community level. Such is telling of female leaders' genuine concern for the entire community, their integrity in leading, as well as their courage even in the face of serious health risks. Women in the sites are in charge of the HOA on top of their familial roles, because most of the men are not keen on leading and being accountable for the welfare of their locality, as illustrated by the husband who only volunteered to take over his wife's duty as a leader in the roving assignment. Lastly, the study impresses on us how consensus-based community organizing led by professionals can be the key to achieve sustainable community development and to build community resilience in dealing with economic, social, and physical shocks in a resettlement context.

Acknowledgements The authors are grateful to Ateneo de Manila University's Research and Creative Work (RCW) program that funded the research. The authors are thankful for the assistance and participation of the leaders and members of the APOAMF HOA in Manggahan, Pasig, and the leaders and members of HOAs in Site 3, Calauan Laguna namely, Legacy of Light HOA, SSP1 HOA, USSICA HOA, and UPMC HOA in this study and the Joly Homes Foundation and Kem Tanaleon.

References

ADB (2018) Philippines: promoting partnerships and innovation in poor and underserved communities. Accessed November 15, 2020. https://www.adb.org/projects/documents/phi-45320-001-icm

Bai JJ, Jin W, Wan C (2020) The impact of social capital on individual responses to COVID-19 pandemic: evidence from social distancing. SSRN paper. https://papers.ssrn.com/sol3/papers.cfm?abstract_id=3609001

Barrios JM, Benmelech E, Hochberg YV, Sapienza P, Zingales L (2020) Civic capital and social distancing during the Covid-19 pandemic (No. w27320). National Bureau of Economic Research. https://www.nber.org/papers/w27320

Bartscher AK, Seitz S, Slotwinski M, Siegloch S, Wehrhofer N (2020) Social capital and the spread of Covid-19: insights from European countries. https://www.iza.org/publications/dp/13310/social-capital-and-the-spread-ofcovid-19-insights-from-european-countries

Beltran M (2020) "The Philippines' pandemic response: a tragedy of errors. The Diplomat. https://thediplomat.com/2020/05/the-philippines-pandemic-response-a-tragedy-of-errors/

Bonagua K (2020) A people's plan to house the poor. Philippine Daily Inquirer. https://opinion.inquirer.net/103959/peoples-plan-house-poor#ixzz6bKiMaenN

Borgonovi F, Andrieu E (2020) Bowling together by bowling alone: social capital and Covid-19. Covid Econ 17:73–96

Briggs XDS (1998) Brown kids in white suburbs: housing mobility and the many faces of social capital. Hous Policy Debate 9(1):177–221

Calinaya R, Igup I, Posadas JJ (2020) [FIRST PERSON] Dilemmas for the urban poor in Barangay Doña Imelda, QC. Rappler. https://r3.rappler.com/move-ph/ispeak/261682-first-person-dilemmas-urban-poor-dona-imelda-coronavirus

Carampatana T, Tuazon R (2020) Community-led COVID-19 response: the work of the Philippines Homeless People's Federation. IIED. Accessed 9 Oct 2020. https://www.iied.org/community-led-covid-19-response-work-philippines-homeless-peoples-federation

Carpiano RM, Moore S (2020) So what's next? Closing thoughts for this special issue and future steps for social capital and public health. Soc Sci Med 257:113013. https://doi.org/10.1016/j.socscimed.2020.113013

Coleman JS (1990) Foundations of social theory. Harvard University Press, Cambridge

CSIS (2020) Southeast Asia Covid-19 Tracker. Accessed 10 Oct 2020. https://www.csis.org/programs/southeast-asia-program/southeast-asia-covid-19-tracker-0

DeSilva MJ, Harpham T (2007) Maternal social capital and child nutritional status in four developing countries. Health Place 13(2007):341–355

Ding W, Levine R, Lin C, Xie W (2020) Social distancing and social capital: why US counties respond differently to COVID-19. Available at SSRN 3624495. https://papers.ssrn.com/sol3/papers.cfm?abstract_id=3632620
Duddu P (2020) Coronavirus in Philippines: how the COVID-19 affects Philippines. Airport Technol. Accessed 10 Oct 2020. https://www.airporttechnology.com/features/coronavirus-affected-countries-philippines-measures-impact-tourism-economy/
Edelenbos J, van Meerkerk I, Schenk T (2018) The evolution of community self-organization in interaction with government institutions: cross-case insights from three countries. Am Rev Publ Adm 48(1):52–66
Fraser T, Aldrich DP (2020) Social ties, mobility, and COVID-19 spread in Japan. https://papers.ssrn.com/sol3/papers.cfm?abstract_id=3544373
Fukuyama F (1995) Trust: the social virtues and the creation of prosperity, vol 99. Free Press, New York
Gittell R, Vidal A (1998) Community organizing: building social capital as a development strategy. Sage Publications, Thousand Oaks, CA
Hapal K (2021) The Philippines' COVID-19 response: securitising the pandemic and disciplining the pasaway. J Curr Southeast Asian Aff 40(2):224–244
Kay A (2006) Social capital, the social economy and community development. Community Dev J 41(2):160–173
Knack S, Keefer P (1997) Does Social capital have an economic payoff? A cross-country investigation. Q J Econ 112(4):1251–1288. https://doi.org/10.1162/003355300555475
Maglaya F (2014) Organizing people for power: a manual for organizers. Education for Life Foundation, Quezon City
Ohmer M, DeMasi K (2009) Chapter 1: approaches to community organizing and their relationship to consensus organizing. In: Consensus organizing: a community development workbook: a comprehensive guide to designing, implementing, and evaluating community change initiatives. Sage Publications, United States
Olson M (1971) The logic of collective action. Public goods and the theory of groups, 2nd edn. Cambridge, Massachusetts
Papa MJ, Singhal A, Papa WH (2006) Organizing for social change: a dialectical journey of theory and praxis. Sage Publications, New Delhi
Portes A (1998) Social capital: its origins and applications in modern sociology. Ann Rev Sociol 24:1–24
Portes A, Landholt P (1996) The downside of social capital. Am Prospect 26:18–23
Putnam R (1993) Making democracy work: civic traditions in modern Italy. Princeton University Press, New Jersey
Putnam RD (1995) Bowling alone: America's declining social capital. J Democr 6(1):65–78
Putnam RD (2000) Bowling alone: the collapse and revival of American community. Simon and Schuster, New York
Putnam RD (2001) Social capital: measurement and consequences. Isuma Can J Policy Res 2(Spring 2001):41–51
Pyles L (2021) Progressive community organizing transformative practice in a globalizing world. Routledge, New York
Quetulio-Navarra M (2014) Mending new communities after involuntary resettlement in the Philippines and Indonesia. Wageningen University, Netherlands
Quetulio Navarra M, Znidarsic A, Niehof A (2017) Gender perspective on the social networks of household heads and community leaders after involuntary resettlement. Gender Place Cult 24(2): 225–246. https://doi.org/10.1080/0966369X.2016.1277185
Samper J (2017) Eroded resilience, informal settlements predictable urban growth implications for self-governance in the context of urban violence in Medellin, Colombia. UPLanD J Urban Planning Landscape Environ Des 2(2):183–206
Schutz A, Sandy MG (2011) Collective action for social change an introduction to community organizing. Palgrave MacMillan, New York

SDI. 2020. Community Based Organisations are Key to Covid-19 Response. Accessed 9 Oct 2020. http://knowyourcity.info/2020/06/community-based-organisations-key-covid-19-response/

Smock K (2004) Democracy in action: community organizing and urban change. Columbia University Press, New York

Temkin K, Rohe W (1998) Social capital and neighborhood stability: an empirical investigation. Hous Policy Debate 9(1):61–88

Tenolete J, Racelis M (2020) Coronavirus in Baseco: a community leader's assessment. Rappler. Accessed 9 Oct 2020. https://amp.rappler.com/move-ph/ispeak/256081-opinion-coronavirus-baseco-community-leader-assessment

Teodoro and Co. (2009) Community-driven land tenure strategies: the experiences of the Homeless People's Federation of the Philippines. Environ Urban 21(2):415–441 [International Institute for Environment and Development (IIED)]

United Nations Habitat (2020) Community organizations are indispensable partners in the COVID-19 crisis. Accessed 2 Oct 2020. https://unhabitat.org/community-organizations-are-indispensable-partners-in-the-covid-19-crisis

Villadiego L (2020) Spaniards find beauty in helping each other amid COVID-19 crisis. Al Jazeera. Accessed 30 Apr 2020. https://www.aljazeera.com/indepth/features/spaniards-find-beauty-helping-covid-19-crisis-200319105933362html

Woolcock M, Narayan D (2000) Social capital: implications for development theory, research, and policy. World Bank Res Obs 15(2):225–249

World Bank (2020) Saving lives, scaling-up impact and getting back on track, World Bank Group COVID-19 crisis response approach paper. Accessed 10 Oct 2020. http://documents1.worldbank.org/curated/en/136631594937150795/pdf/World-Bank-Group-COVID-19-Crisis-Response-Approach-Paper-Saving-Lives-Scaling-up-Impact-and-Getting-Back-on-Track.pdf. October 9, 2020

Worldometers (2022) Coronavirus cases. Accessed 13 Apr 2022. https://www.worldometers.info/coronavirus/

Zhao T, Wu Z (2020) Citizen–State collaboration in combating COVID-19 in China: experiences and lessons from the perspective of co-production. Am Rev Publ Adm 1–7

Chapter 18
A Guideline for Host Communities in Selecting Effective Livelihood's Interventions for Refugees in an Informal Refugee Resettlement: A Case Study of Chiang Mai Province, Thailand

Watchara Pechdin and Mokbul Morshed Ahmad

Abstract Promoting refugees' livelihoods have emerged as a critical challenge for their host communities, noting that ineffective interventions could endanger refugee assistance regimes, and risk escalating the conflicts between host and the refugee community. Efforts toward them from relevant stakeholders are required specific, cautious, and careful interventions. Keeping this concept in mind, this paper sets out to investigate the intervention characteristics in order to support strategic planning and management on livelihood interventions for informal refugee camp, applying a case study from the Koung Jor refugee camp, Thailand. Strengths, weaknesses, opportunities, threats (SWOT) analysis was adopted. Documentary survey and semi-structured interviews with 10 key informants were also held. Key findings suggest that a participatory method is suitable for addressing a specific issue from the bottom up. Meanwhile, the desire to intervene holistically, employing relevant institutions' strengths, should be considered. When cross-disciplinary collaboration is required, the partnership method is advised.

Keywords Intervention characteristics · Livelihood strategy · Refugee livelihoods · Informal camp · Informal refugee · Livelihoods enhancement · SWOT

18.1 Introduction

The implementation of initiatives to enhance the living conditions of refugees has become a challenge for the communities that are hosting them. It is important to keep in mind that inefficient implementations could place the refugee assistance

---

W. Pechdin (✉)
Disaster Preparedness, Mitigation and Management (DPMM), Asian Institute of Technology, Khlong Luang District, Pathum Thani, Thailand
e-mail: w.pechdin@gmail.com

M. M. Ahmad
Development Planning Management and Innovation (DPMI), Asian Institute of Technology, Khlong Luang District, Pathum Thani, Thailand

© The Author(s), under exclusive license to Springer Nature Switzerland AG 2023
M. Hamza et al. (eds.), *Rebuilding Communities After Displacement*,
https://doi.org/10.1007/978-3-031-21414-1_18

regime in danger and run the risk of increasing tensions between the host and the refugee communities. Unaware assistance might encourage refugees to be dependent, inducing refugees to present themselves as needy and try to receive what they can (Vriese 2006). Assistance for refugees must be then based on careful analysis taking into account availability of their resources (Guarnieri 2003). This suggests policy-makers rigorously look at a distinct refugee discourse, such as migration history, characteristics of hos communities, physical and mental conditions, or economic status, which necessitates specific livelihood interventions (Jacobsen and Fratzke 2016).

Generally, refugees are permitted to establish shelters in restricted areas where their livelihoods are conditional upon their receiving of assistances or basic needs from external parties or humanitarian organizations (Jacobsen and Fratzke 2016). This makes their livelihoods dependent on external interventions which create a layer of vulnerability to them. Vulnerability subsequently creates their life insecurity, especially when their existence in the host country or community is not acknowledged formally (UNICEF 2014).

One of the informalities that contributes to their vulnerability discourse is their residence in an informal settlement area. Informal settlement is residential areas or housing units occupied by people who do not have a legal claim to the land. It may accommodate displaced refugees or refugee-like persons; persons who are in similar conditions to refugees (Huq and Miraftab 2020). The informality situation may be a direct consequence of an insufficient domestic legal framework and ineffective policies addressing refugee plight, limiting refugee rights to apply for international protection and access to a variety of basic services—such as healthcare, education, and social services—as well as the labor market (Bhuyan 2013). As with camp inhabitants, informal settlement occupants are frequently moved several times across and within national borders. It is viewed as de facto camps, devoid of the social and substantive safeguards that individuals with rights are meant to enjoy (Huq and Miraftab 2020).

Roles of the host country toward the informal refugee resettlement become complicated due to its unique context and requirement. Implementing without good practice might create tensions between surrounding communities and the refugees (Vriese 2006). Moreover, overlooking the importance to provide assistance to them would deteriorate image of the country from a humanitarian standpoint in international forums, which also affects the host country's recognition.

As a matter of fact, most of government interventions on refugee livelihoods are encouraged to focus on their own influencing factors. Due to different locations and host community's socioeconomic structure including nationality, culture and traditions, and local governance, certain scholars have confirmed that each refugee community requires its own specific intervention (e.g., Bartlett et al. 2020; Ellis et al. 2007; Keygnaert et al. 2008; Martin 2005; McCarthy and Marks 2010; Van Der Velde et al. 2009). One effective intervention in a particular refugee community might not be effectively applicable for other refugee communities, thereby requiring investigation on their own determined factors.

One unique context is referred to international conflict-affected area. This context is extremely faced various challenges as it has involved with both local-to-local and government-to-government level. Unaware interventions might impact not only local and refugee communities, but it might create a huge inter-government tension.

Keeping this concept in mind, this chapter sets out to investigate interventions of each characteristics toward informal refugee resettlement in the international conflicted affected area. This study is in housed in a border area of Chiang Mai province, Thailand which hosts the refugees from Shan state, Myanmar. In search of this context, it is expected to provide effective recommendations to guide the community on selecting appropriate strategies of refugee interventions, in particular deep insight with regard to key success factors and barriers.

18.2 Informal Refugee Resettlement in Thailand

Thailand shares border with four Countries; Lao PDR in the east and northeast, Cambodia in the southeast, Malaysia in south, and Myanmar in the west and northwest where is only one border that still has refugee camps. The camps host displaced people from ongoing ethnic conflicts and have faced several forms of tensions, and some recommendations have been sent to international organizations and donors to carefully support activities working related to Myanmar's ethnic conflicts (Pechdin and Ahmad 2022; Ahmad 2017; Siddique and Ahmad 2012; South et al. 2010).

The Myanmar refugee camps are located in the inner part of Thailand in placed in Chiang Mai, Mae Hong Son, Tak, Kanchanaburi and Ratchaburi Province. Since Thailand does not sign the Refugee Convention 1951, people who had fled to settle in shelters along Thailand–Myanmar border as a result of civil war are only acknowledged as "displaced persons". Humanitarian reasons permitted them to have their resettlement under the supervision of Ministry of Interior and Ministry of Defense. Refugees in their camps are limited in some activities such as unable to go outside the refugee camp, or to have their own plantation area outside the camps. This is mostly driven by national security concerns from the national policymakers. As a result, their basic needs are only allocated by authorized agencies which the allocation is just sufficient for living. This single dependence has become an essential risk to them as they are sensitive to external shocks made by those allocators such as an unpredictability in providing humanitarian aids, delaying allocations of basic needs as well as frustrating social protection system.

Currently, there are ten active camps in Thailand, but only nine are recognized by United Nations High Commissioner for Refugees (UNHCR), while only one camp, Koung Jor Camp, has not been recognized yet. The location of the Koung Jor Refugee Camp is shown in Fig. 18.1.

The term of informal refugee resettlement might be adopted for the Koung Jor refugee camp. The camp was established in 2002 and has been being a residence of refugees fled from Shan state, Myanmar, resulted from intense armed conflict with the Myanmar military. Its status is close to the definition of an informal refugee

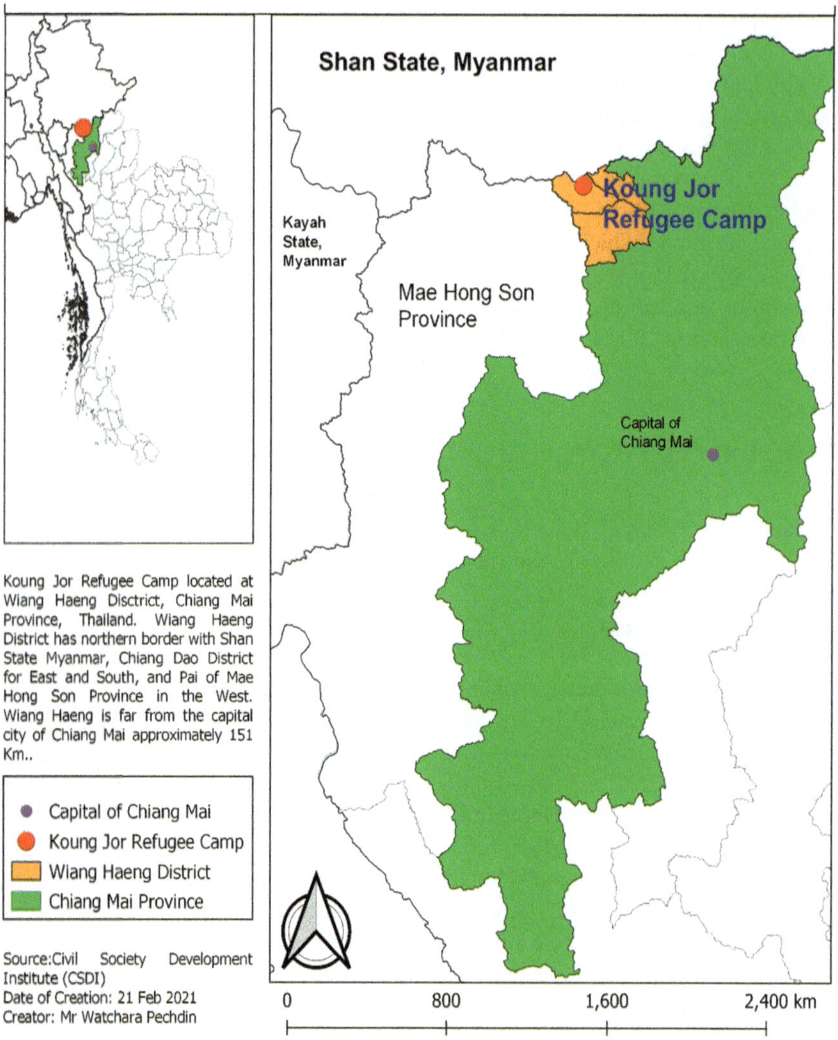

Fig. 18.1 Location of Koung Jor refugee camp. *Source* Developed by Author

resettlement due to its lack of recognition, not allowing them to register in or asking for assistance program under UNHCR activities, particularly regarding the third country resettlement program (Shan State Committee (Thai Border) 2018). Thus, those residing in the Koung Jor refugee camp is more vulnerable to the external shocks and cannot claim for any legal protection.

Currently, the Koung Jor refugee camp is inhabited 281 refugees consisted of 143 males and 145 females, and percentage of age under 18 years old is 27%.

18.3 An Overview of Refugee Livelihood Interventions

An overview of intervention strategies put forward to a government to achieve a successful interventions of refugee livelihoods is previously discussed in various perspectives. Factors determining effectiveness of the interventions are varied in types of intervention, legal frameworks, right designs, or implementation programs. This section overviews three mainstream frameworks for the intervention strategies, detailing as follows.

18.3.1 Participatory Approach

Participatory approach follows the concept of a bottom-up strategy. It encourages policymakers to ensure and incorporate people's voice and participation into development planning, as to reveal all kind of risks of livelihood activities, especially on issues found in relation to gender, resource allocation, or employment opportunities (Chopra and Hohe 2004; Vriese 2006). In addition, some studies such as suggested by Lam et al. (2003) point that a participatory approach plays as essential roles to motivate refugees for changes. This is a result of the requested changes satisfying the context of refugees. However, there are debates on adopting participatory approach. Dilemmas of this approach have been raised by many authors, particularly in the divergence from the existing evidences (Wood and Kallestrup 2021). One of the divergence is attributed to the differences on community characteristics (Schafft and Greenwood 2003). Distinct characteristics of the recruited participants might reveal a preference differing from the prior evidence, and this would compromise the adoption of such implementation.

In addition to the debates, it has been discussed on ethical consideration for developing participatory program. Since refugees are frequently considered as vulnerable individuals owing to their past circumstances; some of them may be traumatized as a result of violations, executions, or other severe treatments, this approach is then sensitive to their feelings. According to Ellis et al. (2007), the study advised that the refugee research must adhere to the ethical standards which set to protect the participants' feeling. By fostering a more welcoming environment for participation, the participatory intervention's efficacy can be enhanced.

Overall, we observed that this sort of intervention is useful when policymakers desire to concentrate their efforts on developing certain communities with specific goals or interventions. The participatory method would identify the particular factors and constraints affecting the intervention's implementation. Together with undertaking rigorous ethical consideration, this approach will offer the effective outcome as a result.

18.3.2 Holistic Approach

As refugee livelihoods are influenced by various factors including with economic, social, political, and environmental factors, a holistic approach is developed to address this complexity. The approach is mostly utilized for interventions related to availability of food and natural resources, access to market, legal framework, health and health-related issues, and living conditions of host communities (Ehntholt and Yule 2006; Hughes and Beirens 2007; Kale 2019; Pacione et al. 2013; Vriese 2006). Particularly in refugees' living conditions connected to the local communities, adopting a holistic approach is recommended (Vriese 2006). Some studies support that a holistic intervention is important approach to fulfill building and creating good environment for refugees, especially when the context of people–place relationship is prominent. Evidence from Kale (2019), adoption of this approach provides critical insight on the role of location in well-being and the importance of considering how everyday multisensory settings affect refugees' emotions of safety and happiness. This contributes to policy developer to establish the effective intervention for promoting the refugee livelihoods.

More to the points, a holistic approach benefits to deliver an implementation with the utmost care for vulnerable individuals in society. Due to the combination of relevant methodologies, a holistic approach can incorporate approaches that are appropriate for the study's environment, particularly when sensitive issues or vulnerable participants are involved. Evidence from Ehntholt and Yule (2006), adoption of this approach is advised in order to provide appropriate treatments to young refugees who have multiple trauma. This adoption would solve complex system of the difficulties in mental health issues.

However, refugee integration has been identified as a critical challenge in adopting the holistic approach (Dragonas 2021). This was influenced by the efficacy of the multiple collaborations of the stakeholders, seeing that effective integration can only take place if the host society provides services, access to jobs, and acceptance of migrants in social interaction. Therefore, in the case that there is ineffective social integration, a struggle in implementing holistic approach might be a result.

In a nutshell, it can be summarized that the holistic approach benefits to solve complex problems as it involves with various stakeholders. Nonetheless, integration challenges can become essential factors to discourage its success. Therefore, those challenges have to be well documented and addressed while launching the holistic intervention.

18.3.3 Partnership Approach

It is worth noting that the resolutions for effective refugee management could not be implemented effectively by the host country or communities alone, and the formation of permanent and sustainable partnerships is required (Vriese 2006). This approach highlights the importance on effective collaborations among the key stakeholders in achieving refugee-related development.

Since development of any activities supporting refugee livelihoods requires effective collaborations of key persons, specifically humanitarian agencies, government authorities in both local and national level, development actors, domestic and international NGOs, community-based organization, donors as well as beneficiaries are encouraged to be partnered. Some evidence indicated that effectiveness of this approach is motivated by an engagement (Abdi et al. 2022; Demirsu and Cihangir-Tetik 2019; George et al. 2020; Moreno-Serna et al. 2020; Veronis 2019). In Jacquez et al. (2019) who took into account on stress reduction in Latino immigrants, pre- to post-intervention results revealed significant improvements in social support and stress management. One reason was due to an effective engagement, having significant impacts for continuously monitoring the interventions.

On the other essential point, this approach could allow each organization to do their own activities with their specialization. Evidence from George et al. (2020) applying to Syrian refugee resettlement in Toronto and Mississauga, the study pointed that when each institution conducted its tasks efficiently and effectively in accordance with its specialization, it can speed the process for forthcoming refugees seeking work opportunities. This implies that in environments requiring specialization, partnering actions would be more productive.

However, there are growing challenges toward the partnership approach as issues related to the decision-making process keep arising. Decision-making process becomes an expensive cost for implementing the partnership approach when involved partners demand to make their own decisions. It consequences to delay decision in some activities, especially there is an activity requiring prompt response. Ineffective implementation might be a result. Due to this, there is advice to address a clear governance system within the partnership to dampen this struggle.

18.3.4 Refugee Livelihoods Intervention Challenges

Challenges on intervening refugee livelihoods have changed dramatically. Findings addressed by United Nations High Commissioner for Refugees (UNHCR) and Mae Fah Luang Foundation (2014) suggest that sustainable solutions beyond refugee livelihoods should include the concerns on returning to their origin, future livelihood desired particularly issues related to work opportunities, kind of support such as land, health services, employment as well as basic education.

However, a research or policy implementation often disregards the context of local communities surrounding refugee camps. Vriese (2006) argues that it is crucial to understand the living conditions of host communities and their relationship with refugees. This is to ensure that the quality of life experienced by refugees is solely perceived by the local population. Another example given by López et al. (2011) shown that there are frequent conflicts between internally displaced persons (IDPs) and host communities due to economic issues. Host communities could not well understand the state assistance provided to those refugees. This could potentially lead to hostility toward IDPs/refugees.

Moreover, Sebba (2006) points that the contention between refugees and host communities might appear in the form of land resources due to an increasing population in the refugee's camp and land value. Local communities possibly reverse their good perception toward refugee at the beginning as an asylum seeker seemed to be a threat of their living.

In addition, the negative perception of host communities could be extended to labor market. Whitaker (1999) was of the view that an increasing arrival of refugees in the host communities could change the local context of labor market. Increasing refugee population in society could increase competition for local civil. Additionally, in some occupations, particularly labor-intensive occupations, refugees may be employed for a low wage, either voluntarily or via exploitation. It would have the effect of lowering the average pay in society. However, some arguments point that the local communities could gain from refugee employments when there is effective management of labor demand and supply (Bakewell 2002).

In this connection, it is suggested that the investigation on refugee livelihoods should extend to host communities. Vriese (2006) viewed that factors that positively influence the relationship with the host communities are a shared ethnic background, existing economic interaction before the influx, and sharing linguistic and cultural attributes. These factors help for a creating peaceful co-existence and or local integration. Therefore, concurrent investigation on livelihoods of host communities is able to decrease tensions between the displaced and the local communities.

In summary, to overcome these challenges, intervention characteristics must be tailored to the societal context, with a specific emphasis on cohabitation between refugee populations and their neighbors. This is to ensure the interventions satisfy the needs of both communities.

18.4 Methodology

18.4.1 Analytical Framework

This study follows the concept of exploratory research design which is an effort to explore appropriate livelihood interventions toward refugees in the informal refugee resettlement, applying a case study to the Koung Jor refugee camp, Thailand. The intervention strategies including with participatory, holistic, and partnership are examined by the SWOT analysis (strengths, weakness, opportunities, threat) in accordance with the capital asset management, vulnerability context, and roles of policies and institution faced by host communities. This effort is to persuade host communities to foresee the potential impacts of each intervention characteristics, having insights on its determiners and become a guideline for a community on selecting an appropriate intervention. As a result, they would provide effective interventions with specific and in accordance with the refugee and host communities in the context of informal refugee resettlement.

18.4.2 Participants and Data Collection

Primary and secondary sources of data were used for this research. During June 2021–January 2022, primary data were gathered through non-participant observation at the Koung Jor refugee camp together with gathering through semi-structured interviews with 10 key informants who have direct experience with and knowledge of the Shan refugees. Selected residents of the Koung Jor refugee camp, NGOs involved in refugee activities in this area, academics, and experts, other Shan displacement in Thailand, and local Thai government leaders were all covered. In terms of the interview conclusions, information was returned to informants to be reviewed again, or questions may be repeated as well. This is to ensure that the data collected is accurate, and that respondents have consented to the recording of their responses. After that, all data collected will be brought forward to analyze its determinant factors by the SWOT analysis.

18.4.3 Ethical Consideration

In this research, ethical issues were seriously taken into consideration. As refugees in Koung Jor refugee camp are induced by the conflicts, most people have an experience on extrajudicial killing, torture, and sexual violence. In addition, most people have an extreme concern on non-voluntary repatriation since they are not eligible for third country resettlement undertaking by UNHCR (Shan State Committee (Thai Border) 2018). Therefore, those refugees are recognized as a vulnerable person.

Hence, the trigger questions of this study were first forwarded to the respective agencies before organizing key informant interviews. After that, all of the matters were considered by authorized persons/agencies in the study area particularly the head of refugee community and head of neighboring communities.

During interviewing process, video or sound was not recorded, taking notes were taken place. Important information regarding to the objectives and background of the study were informed prior starting the interview. Moreover, the respondents were informed that they have right to reject answering uncomfortable question, or right to withdraw consent from this study when they feel discomfort and insecure at any time.

In the reporting stage, no information in the written report was attributed to their identity or personality prior to the permission given by them.

18.5 Key Findings

Living in the informal refugee resettlement consequences to create a layer of vulnerability to its habitants, differing from recognized refugees. We have drawn findings from the field observations and summarized key points from semi-structured interview with the participants who were involved and engaged with the Koung Jor refugee camp. Profile of the participants is described in Sect. 18.7. Each intervention characteristics is investigated in accordance with SWOT framework in relation to refugee livelihoods.

18.5.1 Participatory Interventions

Any interventions will be useful when it represents an actuality. The participatory method is one kind of them. The approach is recognized as the strategy to express what in the bottom up to the top. We have observed and analyzed in accordance with its characteristics, essential results are showed as follows.

18.5.1.1 Strengths

Applying to our case study, an intervention designed from participatory approach has resulted in support to develop the livelihoods of refugee in the informal refugee resettlement significantly. As informed by *Participant 1, 2, and 3,* in the context of informal refugee resettlement, in particular induced by forced displacement, those habitants are in extreme concern over their life securities issues; forced repatriation, legal enforcement, for example. In Thailand, where the country has not participated in the refugee convention 1951, unrecognized refugee status limited them to express what are their conditions. Therefore, this approach has strong point as it alters them to express their real issues and existing conditions arisen in their livelihoods, specifically as an unrecognized population which far beyond adopting conceptualization from recognized refugees in other recognized refugee camps in Thailand.

18.5.1.2 Weaknesses

Its weaknesses are apparent when the organizer's personality contrasted with that of the personality of refugees. This brings critical concerns to engaged with the refugees. With this concern in mind, it may put restrictions on the gathered information's quality. Our main refugee key informants *(Participant 2 and 3)* reported that they had some experiences with these issues. When they attended an event hosted by some organizations, particularly from an international organization, they felt discouraged to express their true objectives, concerning that their informal status might be endangered by an unknown personality.

18.5.1.3 Opportunities

Our findings indicated that designing interventions from the participation commenced an opportunity by establishing negotiation platform to the refugees *(Participant 1)*. This empowers refugees to negotiate their critical benefits during the participation. Some evidence was reported that when the refugees have the right to negotiate, it pushes them to provide high-quality collaboration and fosters attitudes of shared responsibilities for the policy executions *(Participant 1)*. This shapes policy to fit the context of informal refugee resettlement, resulting in increasing effectiveness of the interventions.

18.5.1.4 Threats

Participatory intervention could create circumstances for micro-politics within refugee communities *(Participant 7 and 8)*. This is partly because some of them might report only things that are essential for them, less to develop for the whole refugee community. This can foresee as a potential risk to endanger benefits sharing within the refugees.

18.5.2 Holistic Interventions

Owing to the fact that factors affecting livelihoods are not separable, a holistic intervention could provide the effective solutions for a complex circumstance. The holistic interventions emphasize the relevance of institutional roles by outlining the institutional framework and connecting the micro to the macro. As a result, it necessitates a shift in policy assessment away from local viewpoints replaced by the institutional policy frameworks. Undertaking the SWOT analysis, it is identified as follows.

18.5.2.1 Strengths

The livelihood interventions drawing from the holistic approach have the key strengths from an outstanding work allocations. It is suitable for the context of Myanmar ethnic conflicts which its issues have involved with various concerns *(Participant 6)*. As we knew that this approach includes various stakeholders to design and work together for implementing the interventions, in the context of informal refugee resettlement, we found that this strength stands an essential role to solve complicated issues and enhancing effectiveness of the livelihoods intervention. Such evidence was observed with the spread of COVID-19 in 2020, the lack recognitions on refugee status deprived the refugees in the Koung Jor refugee camp for external help *(Participant 2 and 3)*. By adopting holistic approach, the relevant stakeholders including local government agencies, health sector, school, and civil society have completed their missions and efforts in accordance with their specialization to aid the refugees in the camp in time and effectively *(Participant 1–5, 9–10)*.

Additionally, we discovered that the holistic interventions could reduce difficulty arising from the cost-bearing issue *(Participant 1 and 10)*. Since works and responsibilities are allocated to each stakeholder, the costs of operation are also allocated accordingly.

18.5.2.2 Weaknesses

The weaknesses of the holistic intervention have emerged in mismatching skills, especially contributed by the insufficient knowledge and skills between the local practitioners and respective organizations *(Participant 1 and 10)*. Due to the fact that the intervention's work duties and obligations are assigned by a specific agency or organization in a central planning, certain assigned works require specific abilities that are not found in the local area. This results in hanging teamwork capabilities, which delays in executing program implementation *(Participant 1 and 10)*.

More importantly, we found drawback from this approach on engagement issues *(Participant 1, 6 and 7)*. Besides the majority of relevant public organizations are not officially engaged due to recognition of the camp as an informal status, they were unable to perform well collaboration, imposing by country's laws and regulations. This circumstance may increase the possibility of misallocating tasks to unrelated organizations, resulting in less productive performance. Our key informants *(Participant 1)* stated that they were afflicted by this onerous situation since some stakeholders would not serve the expertise effectively. In occasion, work has to be repeated as assigned stakeholders are unfamiliar with the context, aftereffect to draw incompetent conclusion. This leads to redundancy in the work.

18.5.2.3 Opportunities

Opportunities arising from the holistic intervention are described as experience sharing among stakeholders *(Participant 1)*. This is critical for improving the system and mechanisms for governing the refugee affairs. Sharing experience gained from the interventions, particularly in an informal refugee camp where unique solutions are necessary and various circumstances exist, would improve the foundation and competent capacity for future development.

18.5.2.4 Threats

Challenges have arisen while implementing holistic interventions without discontinuity. Participants *(Participant 1,7, and 10)* addressed that when the context of the organization has influenced the effectiveness of the livelihood's development. Most of focal points or government official frequently relocated their work, having results on discontinuing intervention implementation. Such case contributed to livelihoods development would not reach its goal effectively.

18.5.3 Partnership Interventions

The partnership concept itself makes use of each stakeholder's field of expertise. Thus, this sort of intervention would help narrow the capability gap for livelihood intervention by establishing partnership. The following key challenges are in accordance with the SWOT analysis.

18.5.3.1 Strengths

The strongest point of this approach is it emphasizes collaborations across specializations. Especially when humanitarian assistance is sponsored by an international organization or authorized government agencies, *Participant 1,7, and 10* pointed that this strength complements one another's operations by giving expertise to local organizations who are knowledgeable with the local environment to be tasked with operative interventions. This creates more benefits to the implementation than having remote organizations undertake it on their own. Additionally, establishing partnerships with local organizations results in cost savings associated with operations, logistics, and translation, for example.

In the context of an informal refugee camp, being partner is critical for planning and managing livelihood intervention schemes since certain information is not publicly available and, in some cases, is out of date. Collaborations with either local government or community-based non-governmental organizations help reduce this difficulty by regular sharing information *(Participant 1 and 10).*

18.5.3.2 Weaknesses

Common weakness of the partnership approach is a decision-making issue, likewise holistic approach. This potentially influences the effectiveness of the implementation. In the context of informal resettlement, our key informants *(Participant 1 and 10)* claimed that they were often delayed in executing activities as a result of waiting on decision-making by its partner. With the lack of flexibility of decision-making, sometimes even in minor justification making, they have to wait for the partners to deliberate for the conclusion. This is significantly challenging while working dealing with the unrecognized refugee where the issue arises dynamically, essentially when the issues connected to financial implication that demands prompt response.

18.5.3.3 Opportunity

Implementing livelihoods interventions in a collaborative manner would result in increased human capabilities by experience sharing among practitioners both in local and national/international level. According to the local organization *(Participant 10)*, forming partners with external or outside groups would expose them to the context of comprehensive refugee involvement. They are capable of preparing themselves in response to external requirements. On the other side, being in partnership will allow them for exchanging experience between local and external organizations *(Participant 1)*. It will enable them to have a thorough understanding of the existing conditions and challenges in the informal refugee camp.

18.5.3.4 Threats

Being a partner is mostly influenced by political sanction, much more so when the relationship is launched across the country. This becomes the threat for intervention with partnership approach. In our case study, this difficulty kept on rising when the Myanmar government sought for international collaboration on the country's peacebuilding process, less supporting to the local NGOs *(Participant 1–5)*. This discourages international donor or organization to the support livelihoods development.

18.6 Concluding Remarks

The strengths, weaknesses, opportunities, and threats of three interventions, namely participatory, holistic, and partnership approach were done to develop a guideline for communities in selecting appropriate interventions for the refugee in the informal refugee resettlement. The key findings reflect their own key success factors and barriers.

As discussed above, we found common internal challenges, i.e., strengths and weaknesses of the three interventions are associated with sharing information and expertise. Beginning with the theme of information sharing, it particularly affected the effectiveness of the participatory approach. We were of the view that this factor is influenced by concerns over forced repatriations of the refugees. Having recognition that forced refugees are mostly caused by violent, extrajudicial, torture, or unjust experiences from their original place, this induced those refugees to limit sharing their information. Then, it becomes the critical challenge for participatory approach to extract in-depth information in order to develop livelihood intervention strategies. Therefore, in terms of the strengths, it can be a strong point in the case that participation with refugee is efficiently placed. It could provide fruitful information for the policymaker in designing intervention program specific for the context of informal refugee resettlement. On the other hand, in terms of weaknesses, it can turn to be a weak point when ineffective participation is occurred. This approach might face inconsistent information as the obtained information varied from the existing studies, leading policymakers to confuse with the information. Ineffective planning might be a result.

Taking expertise into account, this theme particularly determines the success of implementing the holistic and partnership approach. Work is allocated according to the stakeholders' expertise. However, we observed that these factors are easily influenced by the decontextualization. Some experts are assigned from central agencies/government might not truly understand the local context, especially the informal refugee resettlement context where it is full of various complexities. Making a decision or taking actions may not reflect the real context of the informal refugee resettlement. Ineffectiveness of implementing the holistic and partnership approach might be a result. This is a challenge for communities in selecting the holistic and partnership interventions.

Another issue is the external challenges, i.e., opportunities and threats. As we observed, political circumstances particularly caused by the local situation and government play a critical role in making opportunities and threats for the host communities for all kind of interventions. When the circumstances become friendly, this situation could stimulate the opportunities for the interventions such as motivate participants to exchange their views or experiences according to intervention strategies implemented. In contrast, when situation becomes insecure, those participants will restrain from their actions or opinions in order to protect themselves from threats or any potential accusations.

In a nutshell, our key findings highlight the key success factors and barriers toward implementing livelihood strategies in different kinds of interventions. Host communities on developing and selecting any interventions are recommended to ensure that the selected interventions contextualize the existing conditions at the local level. In case of future studies, we suggest the scholars to seek opportunities to research on social integration process in the context of informal refugee resettlement as well as an integration process of unrecognized refugees into the host communities. As we noted that, characteristics of unrecognized refugees or informal refugee resettlement have faced distinct vulnerabilities. Hence, there is a need for an integration process. Investigation on factors determining the success and barriers of social integration process would address the existing conditions of informal refugee resettlement challenges.

18.7 Profile of Participants

The study recruited 10 key informants for organizing the semi-structured interviews. Those who participated are as follows:

(a) Participant 1 is a representative from the local NGO, which the organization's responsibility is to provide humanitarian assistance to Shan people.
(b) Participant 2 and 3 are resident in the Koung Jor refugee camp.
(c) Participant 4 is a Shan citizen who has been living in Thailand since 2020 and his/her family has experienced from unjust treatment from the conflicts in the Shan state.
(d) Participant 5 is Shan citizen who has been working in Thailand since 1998.
(e) Participant 6 is an academic expert, who has been doing research in Thailand–Myanmar border more than 40 years.
(f) Participant 7 is Myanmar expert, who retired from the Myanmar news report.
(g) Participant 8 is a representative from border community in Chiang Mai, where community is affected by consequences of conflicts in the Shan state.
(h) Participant 9 is border trader whose business operated in the Chiang Mia border more than 20 years.
(i) Participant 10 is a representative from the local government organization in the Chiang Mai's border community.

References

Abdi SM, Miller AB, Agalab NY, Ellis BH (2022) Partnering with refugee communities to improve mental health access: going from "why are they not coming" to "what can I (we) do differently?" Cult Divers Ethnic Minor Psychol 28(3): 370–378. https://doi.org/10.1037/cdp0000476

Ahmad MM (2017). NGO Field Workers in Bangladesh. https://doi.org/10.4324/9781315195254

Bakewell O (2002) Review of CORD community services for Angolan refugees in Western Province, Zambia

Bartlett R, Robinson T, Anand J, Negussie F, Simons Smith J, Boyle JA (2020) Empathy and journey mapping the healthcare experience: a community-based participatory approach to exploring women's access to primary health services within Melbourne's Arabic-speaking refugee communities. Ethn Health 1–17. https://doi.org/10.1080/13557858.2020.1734780

Bhuyan, R. (2013). Unprotected and unrecognized.

Chopra J, Hohe T (2004) Participatory intervention. Glob Gov 10:289

Demirsu I, Cihangir-Tetik D (2019) Constructing the partnership with Turkey on the refugee crisis: EU perceptions and expectations. J Balkan near East Stud 21(6):625–642

Dragonas T (2021) Toward a holistic approach to refugee integration. Int J Learner Divers Identities 28(2):1–14

Ehntholt KA, Yule W (2006) Practitioner review: assessment and treatment of refugee children and adolescents who have experienced war-related trauma. J Child Psychol Psychiatry 47(12):1197–1210. https://doi.org/10.1111/j.1469-7610.2006.01638.x

Ellis BH, Kia-Keating M, Yusuf SA, Lincoln A, Nur A (2007) Ethical research in refugee communities and the use of community participatory methods. Transcult Psychiatry 44(3):459–481. https://doi.org/10.1177/1363461507081642

George U, Benayoune A, McDermott T (2020) A partnership approach to Syrian refugee resettlement in Toronto and Mississauga: preliminary findings

Guarnieri V (2003) Food aid and Livelihoods: challenges and opportunities in complex emergencies food security in complex emergencies: building policy frameworks to address longer-term programming challenges, Tivoli, Italy

Hughes N, Beirens H (2007) Enhancing educational support: towards holistic, responsive and strength-based services for young refugees and asylum-seekers. Child Soc 21(4):261–272. https://doi.org/10.1111/j.1099-0860.2007.00098.x

Huq E, Miraftab F (2020) "We are all refugees": camps and informal settlements as converging spaces of global displacements. Plan Theor Pract 21(3):351–370

Jacobsen K, Fratzke S (2016) Building livelihood opportunities for refugee populations: lessons from past practice, Migration policy Institute Washington, DC. https://www.migrationpolicy.org/sites/default/files/publications/TCM-Dev-Jacobsen-FINAL.pdf

Jacquez F, Vaughn LM, Suarez-Cano G (2019) Implementation of a stress intervention with Latino Immigrants in a non-traditional Migration City. J Immigr Minor Health 21(2):372–382. https://doi.org/10.1007/s10903-018-0732-7

Kale A (2019) Building attachments to places of settlement: a holistic approach to refugee wellbeing in Nelson, Aotearoa New Zealand. J Environ Psychol 65:101315

Keygnaert I, Wilson R, Dedoncker K, Bakker H, Van Petegem M, Wassie N, Temmerman M (2008) Hidden violence is a silent rape: prevention of sexual & gender-based violence against refugees & asylum seekers in Europe: a participatory approach report. Ghent University. International Centre for Reproductive Health (ICRH)

Lam TK, McPhee SJ, Mock J, Wong C, Doan HT, Nguyen T, Lai KQ, Ha-Iaconis T, Luong T-N (2003) Encouraging Vietnamese-American women to obtain Pap tests through lay health worker outreach and media education. J Gen Intern Med 18(7):516–524. https://doi.org/10.1046/j.1525-1497.2003.21043.x

López RCV, Arredondo CIA, Salcedo J (2011) The effects of internal displacement on host communities a case study of Suba and Ciudad Bolívar localities in Bogotá, Colombia

Martin A (2005) Environmental conflict between refugee and host communities. J Peace Res 42(3):329–346. https://doi.org/10.1177/0022343305052015

McCarthy C, Marks DF (2010) Exploring the health and well-being of refugee and asylum seeking children. J Health Psychol 15(4):586–595. https://doi.org/10.1177/1359105309353644

Moreno-Serna J, Sánchez-Chaparro T, Mazorra J, Arzamendi A, Stott L, Mataix C (2020) Transformational collaboration for the SDGs: The Alianza Shire's work to provide energy access in refugee camps and host communities. Sustainability 12(2):539. https://doi.org/10.3390/su12020539

Pacione L, Measham T, Rousseau C (2013) Refugee children: mental health and effective interventions. Curr Psychiatry Rep 15(2). https://doi.org/10.1007/s11920-012-0341-4

Pechdin W, Ahmad MM (2022) Revisiting critical consequences from armed conflicts in the Shan State, Myanmar: implications for border development of Chiang Mai province, Thailand. Kasetsart. J Soc Sci 43(4):1067–1074

Schafft KA, Greenwood DJ (2003) Promises and Dilemmas of participation: action research, search conference methodology, and community development and community development society. Journal 34(1):18–35. https://doi.org/10.1080/15575330309490101

Sebba KR (2006) Land conflicts and their impact on refugee Women's Livelihoods in Southwestern Uganda (New Issues in Refugee Research: Research Paper No. 127, Issue)

Shan State Committee (Thai Border) (2018) As conflict escalates in Shan State, aid must not be cut off to Shan-Thai border refugees. https://shanrefugees.org/2018/07/23/as-conflict-escalates-in-shan-state-aid-must-not-be-cut-off-to-shan-thai-border-refugees/

Siddique MH, Ahmad MM (2012) Variables affecting fieldworkers of NGOs in Pakistan. Dev Pract 22(2):216–228. https://doi.org/10.1080/09614524.2012.642341

South A, Perhult M, Carstensen N (2010) Conflict and survival: self-protection in south-east Burma. Asia Programme Paper: ASP, 2010

UNICEF (2014) Syrian refugees staying in informal tented settlements in Jordan. UNICEF, New York

United Nations High Commissioner for Refugees (UNHCR), Mae Fah Luang Foundation (2014) Displaced persons in the temporary shelters along the Thai-Myanmar border: futures hopes and aspirations. https://reliefweb.int/report/thailand/displaced-persons-temporary-shelters-along-thai-myanmar-border-future-hopes-and

Van Der Velde J, Williamson DL, Ogilvie LD (2009) Participatory action research: practical strategies for actively engaging and maintaining participation in immigrant and refugee communities. Qual Health Res 19(9):1293–1302. https://doi.org/10.1177/1049732309344207

Veronis L (2019) Building intersectoral partnerships as place-based strategy for immigrant and refugee (re)settlement: The Ottawa local immigration partnership. Can Geogr/le Géographe Can 63(3):391–404. https://doi.org/10.1111/cag.12559

Vriese MD (2006) Refugee livelihoods: a review of the evidence

Whitaker BE (1999) Changing opportunities: refugees and host communities in Western Tanzania (New Issues in Refugee Research: Research Paper No. 11, Issue)

Wood B, Kallestrup P (2021) Benefits and challenges of using a participatory approach with community-based mental health and psychosocial support interventions in displaced populations. Transcult Psychiatry 58(2):283–292. https://doi.org/10.1177/1363461520983626

Chapter 19
Changes in Social Capital After the 2011 Great East Japan Earthquake—Results of an Awareness Survey in Taro District, Miyako City, Iwate Prefecture

Shinsuke Tanimoto and Masaaki Minami

Abstract Ten years after the Great East Japan Earthquake, reconstruction projects are progressing in the affected areas and new towns are being formed. However, the recent spread of COVID-19 infection is thought to have affected interactions among residents. In this study, we conducted a multiple-choice questionnaire survey in the Taro district of Miyako City, Iwate Prefecture, which was severely damaged by the Great East Japan Earthquake, to investigate residents' social capital. Cross-tabulations, chi-square tests, and residual analysis revealed the following trends. First, most respondents indicated that the frequency of interaction with local residents and community activities decreased or remained the same after the earthquake. Second, Changes in residents' social capital compared to pre-disaster levels are significantly related to their social capital now, 10 years after the earthquake. Finally, respondents with no roommates and female respondents showed a higher rate of decline in social capital compared to pre-disaster levels. In the disaster-affected areas, the importance of social capital is becoming increasingly recognized for promoting future reconstruction community planning and inhibiting solitary death. It is necessary to respond to residents whose social capital is likely to decline more than before the disaster, such as women and those who live alone.

Keywords Social capital · Questionnaire survey · Great East Japan earthquake

19.1 Introduction

Eleven years have passed since the Great East Japan Earthquake struck, and urban development projects for reconstruction are underway in affected areas. Projects to rebuild seawalls destroyed by the tsunami, raise urban areas, and clear mountains to create new urban areas are being implemented, and the physical layout of affected towns will continue to change dramatically from their predisaster state as a result.

S. Tanimoto (✉) · M. Minami
Faculty of Science and Engineering, Iwate University, 4-3-5 Ueda, Morioka, Iwate, Japan
e-mail: stani@iwate-u.ac.jp

© The Author(s), under exclusive license to Springer Nature Switzerland AG 2023

M. Hamza et al. (eds.), *Rebuilding Communities After Displacement*,
https://doi.org/10.1007/978-3-031-21414-1_19

Immediately after the earthquake, many residents were forced to live in evacuation centers. Residents then moved to temporary housing and are now living in rebuilt houses in newly constructed urban areas. Within this new environment, new social communities are being formed. In addition to infrastructure reconstruction, disaster recovery requires fostering social capital among residents, and various studies have shown the importance of such efforts. Using Kobe City, Hyogo Prefecture, which was severely damaged in the Great Hanshin-Awaji Earthquake, as an example, Tachiki et al. examine the relationship between (a) the degree of self-rated trust for other residents and (b) the frequency of participation in residents' activities and the reconstruction of their lives after the disaster (Tachiki et al. 2002). Moreover, Evans discusses a case study in which residents in an area with active citizen participation cooperated with each other to promote reconstruction activities in the aftermath of the Great Hanshin-Awaji Earthquake (Evans 2002). Several more examples follow. Weil examines the relationship between post-disaster population recovery and involvement in organizational activities in Hurricane Katrina-affected areas (Weil 2010). Chandra and Acosta also use the example of recovery from Hurricane Katrina, noting that rebuilding communities does not stop with rebuilding infrastructure but includes regenerating social networks (Chandra et al. 2009). By examining interviews with survivors of the disaster, Aldrich also points out the importance of social networks among victims of the Great East Japan Earthquake (Aldrich 2019).

Thus, the importance of social capital in the process of disaster recovery has been the subject of numerous studies. It is thought that changes in the social capital of residents due to the disaster also affect the social capital of residents after the reconstruction project. However, there is not enough knowledge on this subject. Hsueh examines the impact of social capital on aid and recovery satisfaction using the example of recovery from the damage caused by Typhoon No. 23 that hit Izu Oshima, Japan (Hsuch 2019). Karunarathne also uses the example of past flood disasters in Sri Lanka to identify changes in social capital before, during, and after flooding (Karunarathne 2021). Both studies track the status of social capital over the post-rehabilitation project period. However, they do not capture the situation after a relatively long period of time, 10 years after the disaster.

By the time 10 years have passed since the Great East Japan Earthquake, some infrastructure reconstruction projects have been completed in the affected areas. Regarding the social capital of residents, which had been highlighted even before the Great East Japan Earthquake, efforts to maintain and improve local communities are still underway in many areas. However, due to the recent spread of COVID-19, there are concerns about the impact on interactions that took place prior to the spread of the disease in the affected areas. Therefore, this study aims to conduct an awareness survey of residents in the Taro-area of Miyako City, Iwate Prefecture, an area affected by the Great East Japan Earthquake, to understand the current status of social capital of residents during the COVID-19 disaster, and to understand changes in social capital of residents in the affected area before and after the Great East Japan Earthquake. From the results obtained, we aim to analyze the relationship between the current social capital of Taro district residents and the changes before and after the disaster.

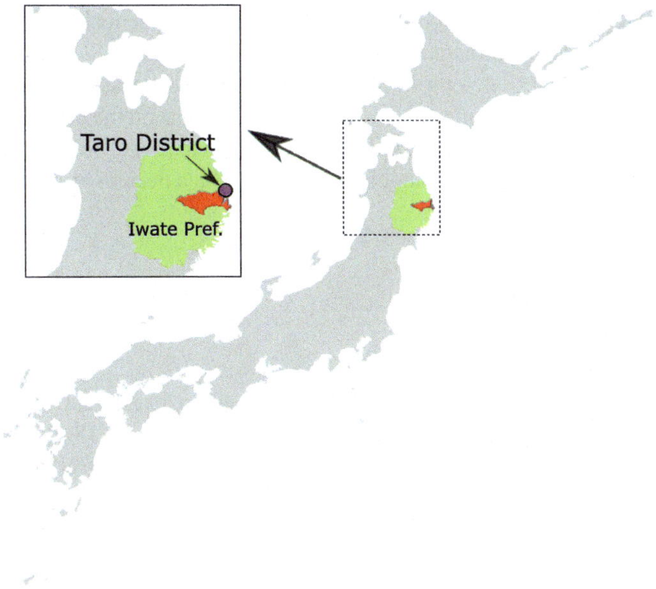

Fig. 19.1 Survey area

19.2 Survey Area

This study focuses on the Taro district of Miyako City, Iwate Prefecture, which was severely damaged by the Great East Japan Earthquake and tsunami (Fig. 19.1).

The Taro district has been hit by large tsunamis many times in the past, and each time, the area has been severely damaged. During the 1896 Meiji Sanriku Earthquake, a 15-m tsunami hit the Taro area, killing 1850 people. In 1933, the Showa Sanriku Earthquake hit Taro, bringing with it a 10-m tsunami that killed 574 people. In response to these disasters, the Taro area has continued to employ efforts to mitigate tsunami damage, including the construction of a 10-m-high tide embankment and annual evacuation drills. However, in 2011, the Great East Japan Earthquake hit the area with a 17-m tsunami, higher than previous tsunamis, and 181 people were killed.

Subsequently, the Taro area was rebuilt to make the community more resilient to tsunami disasters. New urban areas were constructed on higher ground by grading and leveling mountains, and some parts of the city inundated by the tsunami were raised. Between the time of the earthquake and the completion of the new urban area, the physical nature of Taro's residential areas and the daily realities of its residents changed several times. Immediately after the disaster, residents who lost their homes were forced to live in evacuation centers and were later moved to temporary housing, where they stayed for several years. When relocating to temporary housing, the government took care to maintain the community that had been formed prior to the disaster. For example, people who lived close to each other before the earthquake were assigned proximal rooms in their temporary housing as well. After higher ground was prepared and public disaster housing was constructed, residents left their temporary housing and moved to permanent housing in new locations.

Population decline has long been a problem in rural areas of Japan, and even before the earthquake, the population in the Taro area was on the decline. According to the census, the population in 2010, before the earthquake, was 4302, while the population in 2020 (the latest available data) was 2729, a decrease of about 37% over the past decade.

19.3 Research Method

A multiple-choice questionnaire was used in this study to investigate the social capital of Taro residents. The survey was conducted between March 18 and 31, 2021. A total of 1848 questionnaires were distributed, and 524 responses were received. Questionnaires were dropped directly into residents' mailboxes and collected via enclosed return envelopes.

19.4 Survey Items

The survey asks about the current state of residents' social capital (SC) and how it has changed from predisaster to post-disaster.

Various indicators have been proposed to measure social capital. Using publicly available data, Fraser et al. developed a new set of social capital indices, including coupled, bridging, and connected social capital indices, as well as a social vulnerability index, and made them available to 1741 municipalities in Japan (Fraser 2021). They attempt to index social capital based on indicators such as voter turnout, number of community centers per capita, and equal employment opportunities for men and women. Han et al. on the other hand, analyze social capital and health in urban and rural China (Han 2022). They measure social capital based on indicators such as frequency of socializing with neighbors and participation in neighborhood volunteer activities as structural aspects of social capital. In addition, Saito and colleagues have developed an index to measure community-level social capital related to health (Saito et al. 2017). From the results of a survey of older adults in 30 municipalities in Japan, they extracted 11 indicators, including frequency of participation in volunteer activities and trust in community members. The results have been used in other studies such as Watanabe et al.'s (2022), which was conducted in a Japanese city with an aging population and low population density.

The area covered in this study is a relatively small area within a city and an area with an aging population. For this reason, we decided to measure social capital in terms of frequency of daily activities and socializing, as adopted by Han et al.

19.4.1 *Status of SC at the Time of the Survey*

Seven questions asked about the current state of SC (Q1 to Q7). The questions asked about the degree to which people know and interact with others in their community on a daily basis and the degree to which they participate in community activities. In response to these questions, respondents were asked to choose the most personally relevant answer from four options. Option 1 reveals the highest degree of SC and Option 4 indicates the lowest SC. Table 19.1 shows the survey questions and response options.

Table 19.1 Social capital questionnaire and choices

	Questionnaire	Choices 1	Choices 2	Choices 3	Choices 4
Q1	Socializing with local people	Most of the people in the community	Half of the people in the community	A few people in the community	Very little socializing
Q2	Frequency of greetings and conversations with neighbors	Often	At times	Not much	Not at all
Q3	Frequency of neighborhood association and community association activities	Often	At times	Not much	Not at all
Q4	Frequency of participation in local disaster drills	Often	At times	Not much	Not at all
Q5	Frequency of activities of Firefighting Group and voluntary disaster prevention organizations	Often	At times	Not much	Not at all
Q6	Frequency of volunteer activities	Often	At times	Not much	Not at all
Q7	Opportunities to meet people from other districts	Often	At times	Not much	Not at all

Table 19.2 Respondent attributes

(*n* = 524)		(%)
Sex	Male	41.4
	Female	43.7
	Unknown	14.9
Someone living with you	No	14.9
	Yes	84.5
	Unknown	0.6
Frequency of going out	More than 4 times a week	44.5
	At least 2 to 3 times a week	26.0
	At least 2 to 4 times a month	22.3
	Less than once a month	6.3
	Unknown	1.0
Living in disaster public housing	No	88.2
	Yes	11.8

19.4.2 Pre- and Post-Disaster Comparison of SC

For each of the aforementioned questions, we asked about changes in SC before and after the earthquake. Three options were given: "increased from before the earthquake," "same as before the earthquake," and "decreased from before the earthquake."

19.4.3 Personal Attributes

Respondents were asked about their gender, whether they live with a roommate, and how often they go out the home. The frequency of going out was defined as "4 or more times a week," "2–3 times a week," "2–3 times a month," or "less than once a month." Respondents were asked to select the item that best fit their situation from the four options (Table 19.2).

19.5 Analysis Method

In this study, we first conducted a simple tabulation of the responses to understand the current status of SC and the changes before and after the earthquake. Next, we cross-tabulated the results between the current status of SC and the changes before and after the earthquake to understand the trend of the relationship between the two. Furthermore, tests of independence were conducted to quantitatively analyze the significance of the relationship between the two. Finally, a residual analysis was conducted to analyze responses with statistically significant characteristics.

Based on these results, we quantitatively clarified how the SC of residents in the affected areas changed as a result of the earthquake and how this change is relates to the current status of residents' SC. The results obtained are shown in Table 19.2.

19.6 Current Status of Social Capital

Figure 19.2 shows the results obtained from the Taro area residents' awareness survey. We will describe the results here in detail.

The results of Q1 ("Number of people in the community that I know or interact with") show that half of the respondents answered with "most people in the community" or "half of the people in the community." Next, the results of Q2 ("Frequency of greetings and conversations with neighbors") reveal that about 80% of respondents chose "often" or "sometimes." A similar trend, about 80% of respondents chose "often" or "sometimes," can be seen in Q3 ("Frequency of participating in neighborhood association or neighborhood association activities") and Q7 ("Opportunities to meet people from other neighborhoods"). These questions investigate relatively routine social interactions or those that occur within the respondents' immediate sphere of influence. As such, there was a tendency for responses to these questions to cluster toward options indicating a relatively high degree of SC.

However, less than 10% of respondents answered "almost none" to Q1 ("How many people in the community do you know or interact with") and "none at all" to Q2 ("How often do you greet and talk with your neighbors"), indicating that some respondents are at relatively high risk of social isolation. More than half of the respondents chose "not much" or "not at all" for Q4 ("frequency of participation in

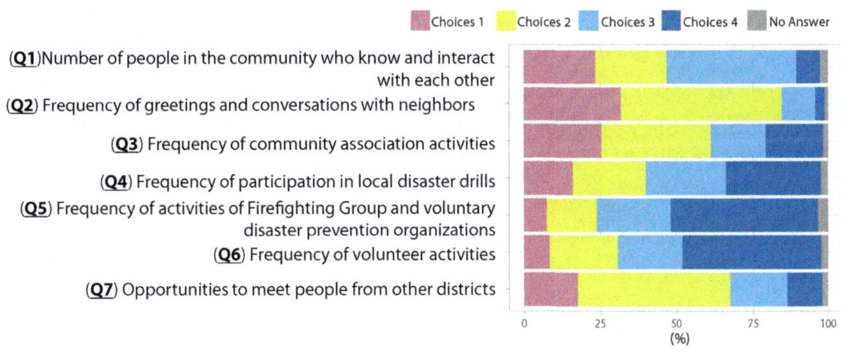

Fig. 19.2 Current state of social capital

local disaster drills "), Q5 ("frequency of activities in local fire brigades and voluntary disaster prevention organizations"), and Q6 ("frequency of participation in volunteer activities").

Q1 and Q2, which asked about interaction with people in the community, showed different trends depending on the depth of the interaction, with Q2, which is considered to be relatively shallow, showing a tendency for the frequency of interaction to increase.

A similar trend was observed in Q3 through Q6, which asked about the frequency of community activities, with Q3 and Q4, which are considered relatively familiar activities for residents, showing a tendency for the frequency to increase.

Furthermore, Q7, which asked about interaction with people from other areas, showed the same tendency as Q2.

19.7 Changes in Social Capital Before and After the Earthquake

Figure 19.3 shows the results of a simple tabulation of changes in the frequency of interaction with people in the community before and after the earthquake. More than half of the respondents answered "unchanged" for all questions. The figure also shows that 20–30% of respondents answered that the frequency had "decreased." Notably, less than 10% of all respondents answered that their frequency of interaction increased after the earthquake.

These results show that many respondents' interactions with local people remained unchanged before and after the earthquake, while about 20–40% of respondents indicated that their interactions in the local community had decreased since before the earthquake.

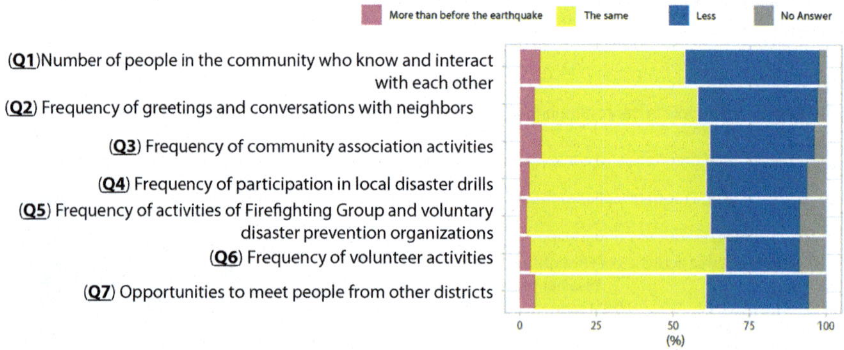

Fig. 19.3 Comparison of social capital before and after earthquake

The percentage of "Less" tends to be relatively higher in Q1, Q2, and Q7, which asked about interaction with others, than in Q3, Q4, Q5, and Q6, which asked about community activities.

19.8 Relationship Between Current Social Capital and Changes Before and After the Earthquake

This section describes the current state of SC for respondents and changes in SC before and after the earthquake. First, for each question item from Q1 to Q7, we conducted a cross-tabulation of the current state of SC and changes in SC before and after the earthquake to understand the response trends of both. From the results of the cross tabulation, a chi-square test was conducted to quantitatively analyze the relationship between the two. In addition, a residual analysis was conducted from the results of the cross-tabulation to analyze whether there were any statistically significant characteristics of the responses.

Figure 19.4 shows the results of a cross-tabulation of the number of people who have met or interacted with Q1, "People in the community with whom I have met or interacted," 10 years after the earthquake and the changes before and after the earthquake. Respondents were classified according to the results of their responses to the changes before and after the earthquake, and the ratio of responses to the number of people they currently know or interact with was calculated for each group. From Fig. 19.4, it can be read that the more respondents reported that they had more acquaintances and contacts before and after the earthquake, the more likely they were to choose the option with more current acquaintances and contacts.

The *p*-value was much lower than 0.01 when a chi-square test was conducted, indicating that the number of people who are acquainted with and interact with people in the community 10 years after the earthquake is significantly related to the changes before and after the earthquake.

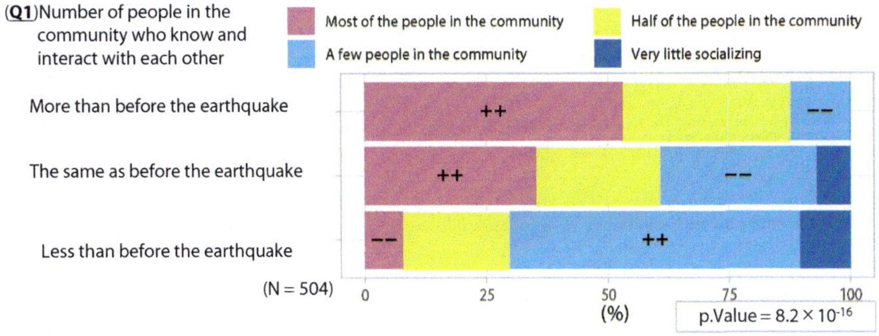

Fig. 19.4 Comparison of current state of social capital with changes before and after earthquake (Q1)

The + and – marks in Fig. 19.4 indicate the results of the residual analysis. Categories with significantly higher percentages of responses than the other respondent groups are marked with +, while categories with significantly lower percentages are marked with –. A ++ or – mark indicates a 1% probability of significance, and a + or – mark indicates a 5% probability of significance. In the "Most people in the region" category, the "Increased from before the earthquake" and "Same as before the earthquake" categories were marked with a + sign. This indicates that the percentage of respondents who chose "know/interact with most people in the community" was significantly higher among those who chose "increased from before the earthquake" and "same as before the earthquake." On the other hand, "Almost everyone in the community" was marked with a – sign in the group of respondents who chose "decreased from before the earthquake," indicating that the percentage of respondents who chose "know and interact with most people in the community" was significantly lower among those who chose "less than before the earthquake." When looking at the responses to "few people in the community," those who chose "increased from before the earthquake" and "same as before the earthquake" were marked with –, while those who chose "decreased from before the earthquake" was marked with +. This is the opposite of the result for "most people in the region." The results indicate that those who decreased their acquaintance and interaction with local residents before and after the earthquake were significantly less likely than other respondents to select the option of many people who acquaint and interact with them now, 10 years after the disaster.

Figure 19.5 shows the results of the cross-tabulation in Q2, "Frequency of greetings and conversations with neighbors." The format of the graph is the same as in Fig. 19.4, and the trend of the responses is the same as in Q1. Respondents who said they greeted and talked with their neighbors more often than before the earthquake read a trend toward greeting and talking with their neighbors relatively more often now, 10 years after the earthquake.

The results of the chi-square test showed that the *p*-value was well below 0.01, indicating that there is a significant relationship between the frequency of greeting

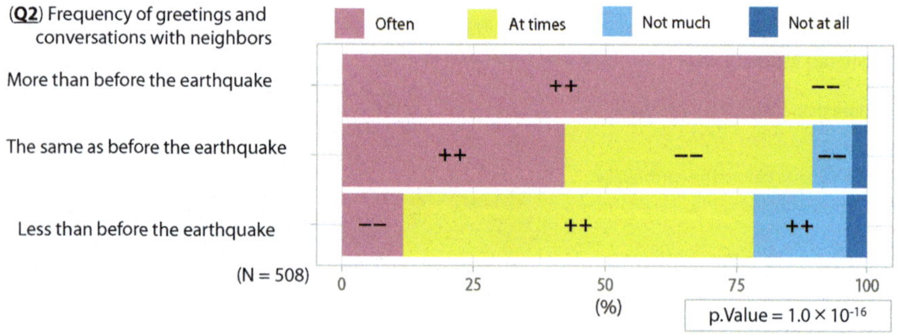

Fig. 19.5 Comparison of current state of social capital with changes before and after earthquake (Q2)

and conversing with people in the community 10 years after the earthquake and the change before and after the earthquake.

The results of the residual analysis show that most of the responses indicate – or +. The percentage of respondents who responded "frequently" was marked with ++ for "increased from before the earthquake" and "unchanged from before and after the earthquake," while "decreased from before the earthquake" was marked with –. The tendency for respondents to greet and talk with people in the community more often 10 years after the earthquake was statistically indicated by those who responded with "more than before the earthquake" or "the same as before and after the earthquake." On the other hand, those who chose "sometimes" showed the opposite tendency to those who chose "often." Respondents who chose "not often" marked "unchanged before and after the earthquake" with a – and those who answered "decreased from before the earthquake" with a ++. The tendency can be confirmed that those who greeted and talked with their neighbors less frequently after the earthquake tended to greet and talk with their neighbors relatively less frequently even now, 10 years after the earthquake.

Figure 19.6 shows the cross-tabulation results for Q3, "Frequency of Activities in Neighborhood Associations and Neighborhood Councils." The tendency of the responses is similar to that of Q1 and Q2, with respondents whose frequency of activities increased from before the earthquake tending to engage even more frequently now, 10 years after the earthquake.

The results of the chi-square test showed that the *p*-value was well below 0.01 for Q3 as well. It can be understood that there is a significant relationship between the frequency of activities in neighborhood councils and neighborhood associations 10 years after the earthquake and the changes before and after the earthquake.

The results of the residual analysis show that the "frequently do" response was marked with ++ for the "increased from before the earthquake" and "unchanged from before and after the earthquake" groups of respondents. On the other hand, those who responded with "less than before the earthquake" marked –. This indicates that respondents who were more active than before the earthquake tended to be

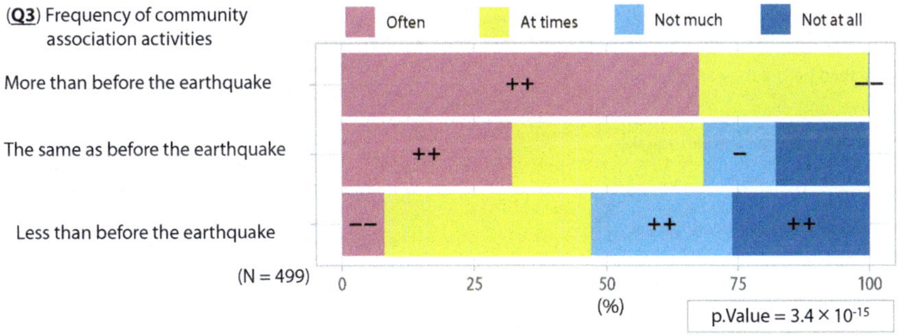

Fig. 19.6 Comparison of current state of social capital with changes before and after earthquake (Q3)

significantly more active now, 10 years after the earthquake. On the other hand, the respondents who indicated "I don't participate much" were marked with – or – in the groups of "increased from before the earthquake" and "unchanged before and after the earthquake." Those who chose "less than before the earthquake" were marked with ++, and the group of respondents who chose "less than before the earthquake" also marked "no participation at all" with a ++ mark. This indicates that the respondents whose frequency of activities decreased from before the earthquake tended to be significantly less likely to engage in activities 10 years after the earthquake.

Figure 19.7 shows the cross-tabulation results for Q4, "Frequency of Participation in Local Disaster Prevention Drills." The trend of responses is similar to that of Q1 through Q3, where respondents whose frequency of participation in local disaster drills increased after the earthquake tended to have a higher proportion of relatively frequent responses even now, 10 years after the earthquake.

The results of the chi-square test showed that the *p*-value was well below 0.01 for Q4 as well. The frequency of participation in local disaster drills also shows a significant relationship between the frequency of participation 10 years after the earthquake and the change in frequency before and after the earthquake.

Looking at the results of the residual analysis, a ++ mark was assigned to the group of respondents who answered, "I do it often," with "increased from before the earthquake" and "unchanged before and after the earthquake" being marked with a ++. On the other hand, answers of "less than before the earthquake" were marked with a –. In addition, "Never" was marked with – for those who answered, "Increased from before the earthquake" and "Remained the same before and after the earthquake." Answers indicating "less than before the earthquake" were marked with a ++. This indicates that the proportion of respondents who answered that the frequency of their activities had increased from before the earthquake was significantly higher than that of those who answered that the frequency of their activities was relatively high, even 10 years after the earthquake. On the other hand, respondents who were less active than before the earthquake tended to be significantly more likely to be less active now.

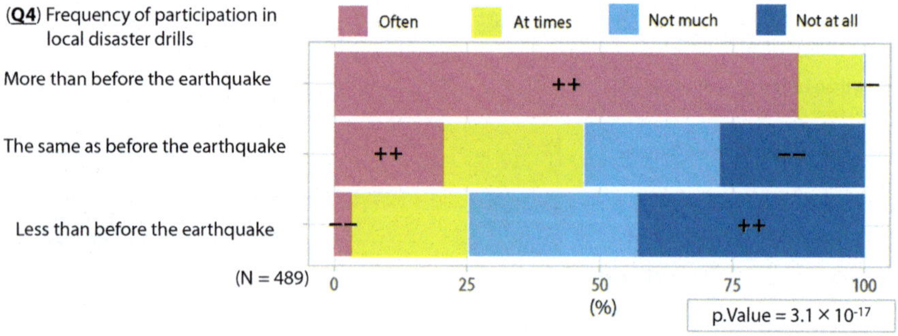

Fig. 19.7 Comparison of current state of social capital with changes before and after earthquake (Q4)

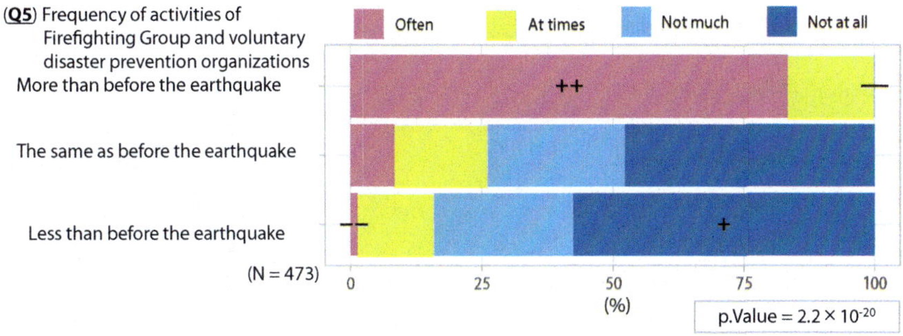

Fig. 19.8 Comparison of current state of social capital with changes before and after earthquake (Q5)

Figure 19.8 shows the cross-tabulation results for Q5, "Frequency of activities of fire brigades and voluntary disaster prevention organizations." The "increased from before the earthquake" group tends to have a relatively high percentage of respondents who answered that they "often participate" in firefighting groups. The respondents whose frequency of activities increased from before the earthquake tended to be more frequent even now, 10 years after the earthquake.

The results of the chi-square test showed that the *p*-value was well below 0.01 for Q5 as well. In terms of the frequency of activities of fire brigades and voluntary disaster prevention organizations, we can understand that there is a significant relationship between the frequency of activities 10 years after the earthquake and the changes before and after the earthquake.

The results of the residual analysis show that a ++ mark was placed in the "increased from before the earthquake" group of respondents for the "frequently do" response. Those who answered with "less than before the earthquake" were marked with a −. For those who answered, "not at all," a − mark was assigned to the "increased from before the earthquake" group, while a + mark was assigned to the "decreased from before the earthquake" group. No + or − marks were given to the group of respondents who answered that their participation had "remained the same before and after the earthquake." This indicates that the respondents whose frequency of activities increased from before the earthquake tended to participate in such activities significantly more frequently now, 10 years after the earthquake.

Figure 19.9 shows the cross-tabulation results for Q6, "Frequency of Participation in Volunteer Activities." The group of respondents whose frequency of volunteer activities increased from before the earthquake tends to have a higher percentage of respondents who are still active in volunteer activities 10 years after the earthquake.

The results of the chi-square test showed that the *p*-value was well below 0.01 in Q6. It can be understood that there is a significant relationship between the frequency of volunteer activities 10 years after the earthquake and the changes before and after the earthquake.

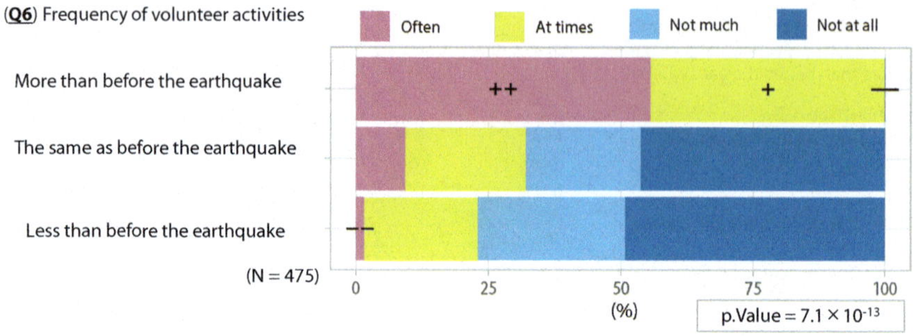

Fig. 19.9 Comparison of current state of social capital with changes before and after earthquake (Q6)

The results of the residual analysis show that the group of respondents who answered "frequently" were marked with ++ for "increased from before the earthquake." On the other hand, those who answered, "less than before the earthquake" were marked with −. In addition, for the group of respondents who answered, "increased from before the earthquake," a + mark was assigned to those who answered, "sometimes participate," while a − mark was assigned to those who answered, "not often participate." This indicates that the respondents whose frequency of activities increased since before the earthquake tended to have a significantly higher percentage of responses indicating a relatively high frequency of activities even 10 years after the earthquake.

Figure 19.10 shows the results of the cross tabulation of Q7, "Opportunities to meet with people from other areas." The group of respondents who met more frequently with people in other areas than before the earthquake were more likely to choose the option of meeting more frequently with people in other areas 10 years after the earthquake.

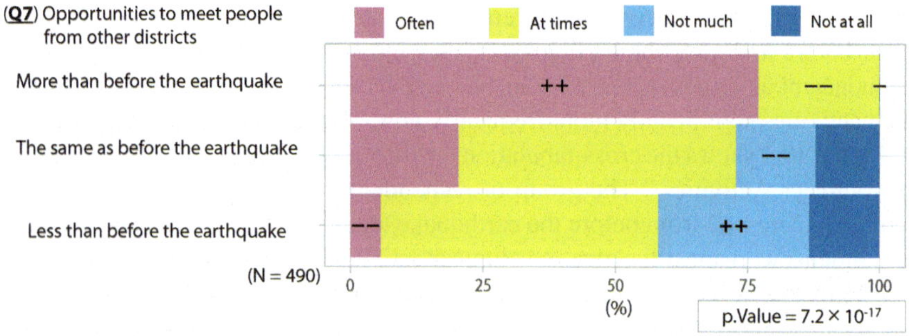

Fig. 19.10 Comparison of current state of social capital with changes before and after earthquake (Q7)

The results of the chi-square test showed that the *p*-value was well below 0.01 for Q7 as well. It can be understood from this that there is a significant relationship between the frequency of meeting with people in other areas 10 years after the earthquake and the changes before and after the earthquake.

The results of the residual analysis show that a ++ mark was assigned to the group of respondents who responded that their frequency "increased from before the earthquake" in response to the "more often" response. On the other hand, those who answered "less than before the earthquake" were marked with a −. The "Sometimes" response was marked with a − in the "Increased from before the earthquake" group. For those who selected "not often," a − or −− mark was assigned to the group of respondents who selected "more than before the earthquake" or "the same before and after the earthquake." Conversely, those who chose "decreased from before the earthquake" were marked with ++. This indicates that respondents who had more opportunities to meet with people in other areas after the earthquake tended to have significantly more opportunities 10 years later as well.

All items showed a significant relationship between the current status of social capital and changes before and after the disaster, but there were some differences in the tendency among the items. Q2, which is a relatively shallow level of interaction, and Q3 and Q4, which are considered relatively familiar activities, tended to receive more + and − marks, suggesting that relatively shallow interactions and familiar activities have a relatively high sensitivity to the impact of a large-scale disaster. Although + and − marks were also indicated in the other items, the sensitivity to the impact of disasters is considered to be relatively low.

19.9 Attributes of Respondents Who Decreased Their Involvement with the Community

The above results confirm the following trends for all SC items in the survey.

(1) There is a statistically significant relationship between the frequency and number of people involved in the community 10 years after the earthquake and the changes before and after the earthquake.
(2) The frequency and number of people involved in the community tended to increase 10 years after the disaster.
(3) The respondents who were involved with the community less frequently immediately after the earthquake tended to also be involved with the community less frequently now, 10 years after the earthquake.

These results indicate that the SC of residents in the disaster-affected areas may be polarized due to the effects of the earthquake and COVID-19. As an overall trend, the number and frequency of people involved with the community after the disaster tended to be lower among "women," people with "no roommates," people who participated in "infrequent outings," and "residents of public housing for disaster

Table 19.3 Percentage of respondents whose Involvement with the community decreased in number and frequency since preearthquake status

	(%)	Q1	Q2	Q3	Q4	Q5	Q6	Q7
Sex	Male	39.7	35.4	28.0	29.3	30.9	20.8	30.4
	Female	**51.4**	**46.2**	**42.5**	**40.8**	**35.3**	**31.3**	**41.6**
Someone living with you	No	**52.0**	**47.3**	**47.3**	**52.8**	**48.5**	**39.4**	**43.1**
	Yes	44.0	39.6	33.7	32.5	29.4	24.4	34.6
Frequency of going out	More than 4 times a week	39.2	35.2	28.2	30.8	26.3	20.0	27.8
	At least 2 to 3 times a week	43.3	37.9	38.0	39.7	33.9	31.5	38.0
	At least 2 to 4 times a month	53.9	50.4	42.6	37.2	37.2	31.8	**45.9**
	Less than once a month	**60.6**	**51.5**	**54.8**	**43.3**	**45.2**	**37.0**	45.2
Living in disaster public housing	No	44.0	38.9	34.1	33.8	31.4	25.2	35.3
	Yes	**51.7**	**52.5**	**47.4**	**46.4**	**36.4**	**36.4**	**37.9**

(colored cells) p <0.05

victims." In particular, gender and lack of roommates revealed four items with significant results in the chi-square test. In addition, the percentage of respondents who had decreased their frequency of greeting and conversing with people in the community tended to be higher among residents of public housing for disaster relief.

Table 19.3 shows the results of the percentage of respondents who reported a decrease in the number and frequency of their involvement with the community by individual attribute, and the colored cells indicate items for which significant results were obtained by the test of independence.

Therefore, we consider it necessary to closely monitor residents with these attributes in future community planning to prevent social isolation challenges.

19.10 Conclusions

Based on a survey of residents' attitudes conducted in the Taro district of Miyako City, Iwate Prefecture, which had been damaged by multiple tsunamis including the Great East Japan Earthquake tsunami, this study sought to understand how residents interacted with the community 10 years after the earthquake and how their activities changed before and after the disaster. In the midst of the social unrest caused by the COVID-19 pandemic, we analyzed how respondents interacted with their communities and how the SCs had changed since before the disaster.

The results show that amidst the spread of social unrest directly after the Great East Japan Earthquake tsunami, the number and frequency of social interactions as reported by respondents tended to decrease or remain the same as predisaster levels. A small number of people increased the number and frequency of their social interactions post-disaster. Furthermore, the results of the chi-square and residual analyses showed that, amid the spread of social unrest caused by COVID-19, residents' SC 10 years after the disaster showed a significant relationship with the change in SC before and after the disaster. The number and frequency of social interactions 10 years post-disaster tended to be lower for respondents who decreased their number and frequency of social interactions immediately after the earthquake. Finally, the number of respondents who decreased their interaction with the community after the earthquake tended to be particularly high among women and those who did not live with others. These results suggest that changes in social capital due to large-scale disasters may have long-lasting effects thereafter.

Because the Taro district is a regular tsunami hit area, consideration of the SC of residents, as well as the development of disaster prevention-related infrastructure, will play a major role in the recovery process during and after future disasters. In the areas affected by the Great East Japan Earthquake, the problem of solitary deaths has been pointed out, and the importance of SC has been reaffirmed here as well. Therefore, it is necessary to address the needs of residents who are likely to decrease the number and frequency of interactions with people in their community after a disaster, such as women and those living alone. In the future, we plan to conduct similar surveys in other disaster-affected areas to make comparisons among districts. We also plan to conduct another survey in the Taro area to understand changes in trends over time.

References

Aldrich PD (2019) Black wave: how networks and governance shaped Japan's 3/11 disasters. University of Chicago Press, Chicago

Chandra A, Acosta J (2009) The role of nongovernmental organizations in long-term human recovery after disaster-reflections from Louisiana four years after, Hurri-cane Katrina. RAND Occasional Paper. https://doi.org/10.1080/10705420802475050

Evans N (2002) Machi-zukuri as a new paradigm in Japanese urban planning: reality or myth? Jpn Forum 14(3):443–464. https://doi.org/10.1080/0955580022000008745

Fraser T (2021) Japanese social capital and social vulnerability indices: measuring drivers of community resilience 2000–2017. Int J Disaster Risk Reduction 52. https://doi.org/10.1016/j.ijdrr.2020.101965

Han Y, Chung YR (2022) The role of neighborhood social capital on health and health inequality in rural and urban China. Prev Med 156. https://doi.org/10.1016/j.ypmed.2022.106989

Hsuch YH (2019) The role of household social capital in post-disaster recovery: an empir-ical study in Japan. Int J Disaster Risk Reduction 39. https://doi.org/10.1016/j.ijdrr.2019.101199

Karunarathne YA (2021) Geographies of the evolution of social capital legacies in response to flood disasters in rural and urban areas in Sri Lanka. Int J Disaster Risk Reduction 62. https://doi.org/10.1016/j.ijdrr.2021.102359

Saito M, Kondo N, Aida J, Kawachi I, Koyama S, Ojima T, Kondo K (2017) Development of an instrument for community-level health related social capital among Japanese older people: the JAGES project. J Epidemiol 27:221–227. https://doi.org/10.1016/j.je.2016.06.005

Tatsuki S, Hayashi H (2002) Seven critical element model of life recovery: general linear model analysis of the 2001 Kobe panel survey data. In: Paper presented at the second workshop for comparative study on urban earthquake disaster management, pp 14–15

Watanabe J, Kimura T, Nakamura T, Suzuki D, Takemoto T, Takekosshi A (2022) As-sociations of social capital and health at a city with high aging rate and low population density. SSM Population Health 17. https://doi.org/10.1016/j.ssmph.2021.100981

Weil F (2010) The rise of community engagement after Katrina, In the New Orleans Index at Five, pp 1–20

Chapter 20
The Role of Social Capital as a Post-relocation Coping Mechanism: A Case Study of Kegalle, Sri Lanka

Anuradha Senanayake, Maduri Fernando, Nishara Fernando, Dilanthi Amaratunga, Richard Haigh, Chamindi Malalgoda, and Chathuranganee Jayakody

Abstract In the systematic approach toward disaster risks, planned relocation is considered to be the last resort when providing a durable solution for relocates. This study is an investigation of the role of "social capital" in rebuilding lives in the aftermath of relocation caused by post-disaster induced displacement. This study is based on field research conducted in Kegalle, Sri Lanka. The study employs primary data that was collected using a sample survey and in-depth interviews as techniques. A systematic sample of 60 households was selected from both donor and owner driven relocation settings to conduct the sample survey. 28 in-depth interviews were conducted with relocates, host community members, and project officials based on purposive samples. The findings of the study have been analyzed using the main thematic orientation of the main types of social capital, i.e., bonding, bridging, and linking. The findings suggest that social capital plays a major role in the coping strategies developed by relocates to face the alterations in their living fabric caused due to displacement and relocation. A few reasons for the significance given to social capital are the homogeneity of the community, living through the relocation, public facilities, community-based initiatives, and economic cooperation. Hence, the study concludes that integrating social capital is a vital coping mechanism for planned relocation.

Keywords Coping mechanism · Disaster-induced displacement · Planned relocation · Social capital

---

A. Senanayake (✉) · D. Amaratunga · R. Haigh · C. Jayakody
Global Disaster Resilience Centre, School of Applied Sciences, University of Huddersfield, Huddersfield, UK
e-mail: anuradha.senanayake@hud.ac.uk

M. Fernando
Social Policy Analysis and Research Centre, University of Colombo, Colombo, Sri Lanka

N. Fernando
Department of Sociology, University of Colombo, Colombo, Sri Lanka

C. Malalgoda
Global Disaster Resilience Centre, School of Arts and Humanities, University of Huddersfield, Huddersfield, UK

© The Author(s), under exclusive license to Springer Nature Switzerland AG 2023

M. Hamza et al. (eds.), *Rebuilding Communities After Displacement*,
https://doi.org/10.1007/978-3-031-21414-1_20

20.1 Background

Millions of men, women, and children around the world are displaced from their homes each year by sudden onset disasters caused by natural hazard events (IDMC 2010). Displacement is a situation during which inhabitants are forced to leave their own shelters due to sudden shocks or stresses mainly because of armed conflicts, civil unrest, or manmade or natural hazards (Amaratunga et al. 2022).

Three types of internal displacement are common around the globe: Displacements caused by development initiatives, natural disasters, and conflict (Muggah 2008). Apart from the above categorization, climate induced displacement is a new typology that has gained much attention with the rise of climate change in the world (Amaratunga et al. 2022). Among these, natural disaster induced displacement is caused by natural or environmental disasters, with people losing their habitats due to natural or environmental disasters. These disasters are said to be the leading cause of displacement worldwide, and those displaced are sometimes referred to as environmental or climate refugees (Muggah 2008).

In 2020, compared to the previous year, disasters caused more than three-quarters of the new internal displacements worldwide. Around 30.7 million people had to move due to disasters. According to the United Nations Development Program (UNDP), it occurred mainly due to storms and floods (UNDP 2021). A statement delivered by Asako Okai at the launch of the 2021 Global Report on Internal Displacement further notes that the recognition of climate and disaster induced displacement in international frameworks is evolving. In addition, the mitigating risk of displacement is factored in frameworks not only for sustainable development, disaster risk reduction, and climate change but also for social protection and human rights protection (UNDP 2021).

However, most disaster displacement in 2020 took place in East Asia and the Pacific region. Among various types of disasters, typhoons, floods, earthquakes, and volcanic eruptions caused nearly 12.1 million new displacements. Moreover, a third of the world's new disaster induced displacement took place in South Asia in the same year (IDMC 2021).

Sri Lanka is a country that is known to be affected by various hazards such as flooding, landslides, cyclones, and droughts (ADPC 2020). During the period 1990–2018, Sri Lanka recorded the highest proportion (58%) of flooding incidents (CRED 2019). As mentioned in Internal Displacement Index (2021) by the International Displacement Monitoring Center (IDMC), disasters were the sole cause of displacement in 2020, triggering 19,000 new displacements in Sri Lanka. The majority of these displacements occurred as a result of Tropical Cyclone Burevi, which hit the country in November (IDMC 2021). The context of landslide hazards constantly evolved during the period of 1952–2020 in Sri Lanka with a significant increase (Mahani 2013). In 2020, 101,778 families and 363,399 persons were recorded to be affected by disasters alongside 62 deaths caused by the same (National Disaster Relief Services Center 2021).

Relocation principally is about "moving away of people's lives," implying moving away from places where they would have been used to living and working among many others (Fernando and Punchihewa 2013). As mentioned in the "disaster displacement: How to reduce risk, address impacts, and strengthen resilience" guidelines by UNISDR (2018), measures to help displaced people achieve durable solutions should be integrated in recovery, rehabilitation, and reconstruction plans, recognizing them as disproportionally affected groups with specific needs in accessing services and the benefits of development and reconstruction programs (UNISDR 2018). Internally displaced persons (IDPs) have three options for durable solutions in terms of location based on their place of origin, their place of displacement or another part of the country (UNISDR 2018).

Social capital is an essential element in recovery from and resilience to disasters (Panday et al. 2021). Panday et al. (2021) further explain how literature has focused on the strength of social networks and leadership in disaster recovery, gender roles in disaster response, the simultaneously positive and negative influences of social capital in disaster response, and the need for context-specific cultural understandings of disaster response. This is because social capital in disasters seems to have different effects for different people, depending on specific contexts (Panday et al. 2021).

Social capital being the network of relationships plays a significant role in managing risk and crisis situations saving human lives from adverse effects and impacts of disasters (Behera 2021). As mentioned earlier, authors and scholars across many countries through their disaster studies show that social capital in the form of bonding, bridging, and linking rescued victims of disasters, evacuated the most vulnerable people to safer places, provided food and drinking water and other basic amenities to the victims of disasters, provided homes to the homeless, and provided solace to those who were psychologically depressed (Behera 2021). Thus, it must be noted that studies of social capital should take into account both the exclusionary practices and their negative consequences, as well as their advantages in various contexts (Panday et al. 2021).

Considering the above-mentioned facts, this study is an attempt to address the following research problem: What is the role of social capital as a post-relocation coping mechanism in the selected research field? This research revolves around the main objective of understanding the role of social capital as a post-relocation mechanism in the selected field of Sri Lanka. The following sub-objectives are utilized to attain the said main objective:

To investigate the impact of relocation on social capital.

To inquire of the social capital-based coping mechanisms developed by relocates.

20.2 Conceptual Framework

As this study mainly revolves around the main concept of "social capital," this section will introduce the said concept along with the various forms of social capital that have been utilized in the analysis.

20.2.1 Social Capital

Hanifan (1916) was the first social scientist who promoted social capital as goodwill, fellowship, mutual sympathy, and social intercourse among a group of individuals and families who make up a social unit (Behera 2021). Social capital as a theoretical concept is further explored and publicized in the works of Bourdieu (1986) Coleman (1988, 1990), Woolcock (1998), and Putnam (1994, 1995). Social capital is defined as *exchange of information, trust, norms, and values between and among the individuals, groups, families, and communities and with an existing network, while physical capital is fully tangible, and the social capital merely exists invisibly in the form of relationship between and among individuals* (Coleman 1988, p. 98).

Broadly, social capital is the ability of people to work together for common purposes in groups and organizations with a set of informal values or norms that motivates cooperation between and among groups (Behera 2021). Social scientists have defined social capital as the networks which connect individuals, families, groups with each other either through weak or strong ties (Behera 2021). Woolcock and Szreter showed a path in 2004 in which social capital can be categorized into horizontal and vertical networks such as bonding, bridging, and linking social capital, which can be widely used to save lives in emergencies (Behera 2021; Szreter and Woolcock 2004).

20.2.2 Types of Social Capital

Having introduced the concept of social capital, it is important to discuss the various types of social capital that have been analyzed in the present study.

The discussion of social capital is credited to three main theorists: Pierre Bourdieu, James Coleman, and Robert Putnam. Bourdieu stated that there are three kinds of capital named economic, social, and cultural. However in this paper, we only consider social capital. Bourdieu provides a concise definition of social capital as follows: "Social capital is the aggregate of the actual or potential resource which is linked to the possession of a durable network of more of less institutionalized relationships of mutual acquaintance or recognition" (Woolcock 2001, p. 67).

20.2.2.1 Bonding

Bonding social capital connects individuals who are similar in terms of their socio-financial position and demographic characteristics (Panday et al. 2021). Bonding social capital is often described as horizontal ties between individuals within the same social group (Claridge 2004). Bonding social capital refers to networks with a high density of relationships between members; individuals belonging to the network are interconnected because they know each other (Claridge 2004). These groups have a relatively high degree of homogeneity, and bonding has been argued to facilitate immediate rescue and early recovery of in-group members in disasters. Yet, bonding has been criticized for conveying advantages only to in-group members or increasing the likelihood of group members being held back and prevented from forming relationships with sources outside the group (Panday et al. 2021).

20.2.2.2 Bridging

Bridging social capital refers to connections between people who are not family, relatives, or kin, but who have similar financial status and political influence (Panday et al. 2021). This is also described as vertical ties often operating through formal hierarchical structures (Claridge 2004). As a result of bridging, people from across social divisions within a community or neighborhood come together, irrespective of their ethnicities, geographical, or occupational backgrounds (Behera 2021). Bridging is crucial for solving community problems by helping people to know each other, building relationships, sharing information, and mobilizing community resources (Bhandari and Yasunobu 2009).

20.2.2.3 Linking

Linking social capital refers to connections between groups and people in positions of power and includes vertical links to formal institutions, such as governmental organizations (Panday et al. 2021). Thus, linking social capital seems to reinforce existing systems of discrimination, allowing the (relatively) privileged to access resources while harming, or excluding, those on the edges of society (Behera 2021). In many ways, linking social capital is not much different from bridging social capital (Claridge 2004). It reaches out to unlikely people in dissimilar situations. Linking social capital may involve networks and ties of a particular community with states or other agencies (Bhandari and Yasunobu 2009).

Although social capital includes bonding, bridging, and linking as the main three types of social capital, this study only focuses on bonding and bridging. Therefore, a summary of the analytical framework of this present study is as follows (Fig. 20.1).

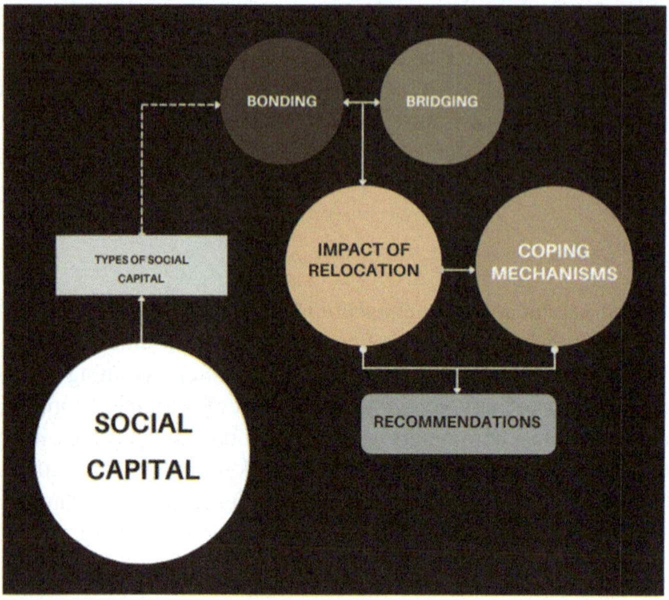

Fig. 20.1 Analytical framework of the present study. *Source* Prepared by the authors

20.3 Methodology

Following a detailed explanation of the aim of this paper and the conceptual framework, this section explains the methodology of the present study. This study was conducted in Kegalle, one of the most disaster-prone districts in Sri Lanka. At the end of 2020, Kegalle stood in the top 10 districts in Sri Lanka with the highest number of people affected with six deaths reported due to extreme weather events (National Disaster Relief Services Center 2021).

The year 2016 marks a juncture in the extreme weather events records of the Kegalle district when the landslide of the Samasara Mountain took place in May. This landslide incident estimated 301,602 people being affected, and approximately, 21,484 people being displaced (OCHA 2016). By the end of May 2016, Kegalle district had reported 50 deaths due to flooding and landslides (International Organization for Migration (IOM) Sri Lanka 2016).

In response to this disaster, the District Secretariat of Kegalle along with other state and non-governmental organizations took the initiative of providing temporary and permanent shelter for the displaced. This study was conducted on the permanent shelter stage. The permanent shelter process was initiated as a planned relocation initiative. Giving relocation options is recognized as one of the best practices integral in the planned relocation process (Karunasena and Rameezdeen 2010). Sri Lanka has a history of utilizing both owner-driven and donor-driven relocation options in planned relocation in displacement contexts (Fernando and Punchihewa 2013;

Senanayake et al. 2022). The present study's selected planned relocation initiative has utilized both donor-driven and owner-driven approaches. While donor-driven settlements were made of 211 housing units built by several donors, the owner-driven settlements had two elements. The donor-driven approach which was utilized in this project had the following strategies:

Government Resettlement Site (GRS): 10 perches of land allocated by the government and relocates were given LKR 1.2 million in five installments for housing construction. 949 housing units were built in GRS settings.

Individual Resettlement Site (IRS): If the relocates wanted to find their own land, this was the option they were asked to adopt. Relocates were given LKR 4 million for the land and LKR 1.2 million in installments to construct the housing units. Under this option, 569 housing units have been completed.

Based on the above background, the present study focuses on three main locations representing all three relocation options discussed above. The selected locations could be summarized as follows:

20.3.1 The Donor-Driven Setting

A settlement that was donated by an international donor was selected as the donor-driven location. The housing structure of this setting is a high rise housing structure which is a novel initiative in the disaster induced relocation context of Sri Lanka (Fernando et al. 2020a, b). Given the special nature of the housing structures, this location was selected as the donor-driven setting. This location has 234 families and 60 housing units. However, only 59 housing units are occupied by relocates, and the remaining housing units have been converted into the office of primary level government officers (Gramaniladhari, Development Officer, and Samurdhi Social Protection Grant Officer). The setting also has a community hall.

20.3.2 The GRS Setting

This location is in the W Division of the X District Secretariat Division. The relocates who were living in various settings of the Kegalle District and displaced from the 2016 landslide were relocated into this setting. There are 25 housing units in this location.

20.3.3 The IRS Setting

This is location is in the Y Gramaniladhari Division of the Z District Secretariat Division. This location has 23 relocates, and all were living in the same location prior to relocation.

Data collection of the present study was conducted in two phases. In the first stage, a random sample was collected from each setting to conduct a sample survey to investigate the impact on social capital due to relocation in the selected settings. As the second stage, a purposive sample of 10 each from the settings was selected to conduct in-depth interviews to inquire the social capital-based coping mechanisms developed by the relocates to tackle the said impact on social capital. Further, 8 host community members were interviewed and selected purposively to represent all three settings.

The data collection was conducted by a team of three researchers for two months. It was conducted in Sinhala and later translated into English for the purposes of analysis. Thematic analysis was utilized to analyze the collected data and revolved around the main theme of thc impact of relocation on social capital. Further, sub-themes were developed as per the typology of social capital that has already been discussed under the section on conceptual background.

20.4 Research Findings

The findings are presented as per the sub-objectives of the study. First, the overall impact on the social capital at the selected locations will be discussed. Later, the overall coping mechanisms incorporated by the relocates will be evaluated. A summary of the overall findings is in the following Figs. 20.2 and 20.3.

20.4.1 Impact of the Relocation on Social Capital

The communities at the selected locations were Sinhala Buddhist homogeneous communities. Due to the cultural similarities, they had a positive perception of the post-relocation community. It was evident that these settings have strong social capital. However, despite such a strong social fabric, there were the following signs of social impoverishment risks of social disarticulation as pointed out by Cernea (2000). The impact will be presented as per the selected locations.

20.4.1.1 Donor-Driven Setting

In terms of the donor-driven setting, the following key impacts were visible:

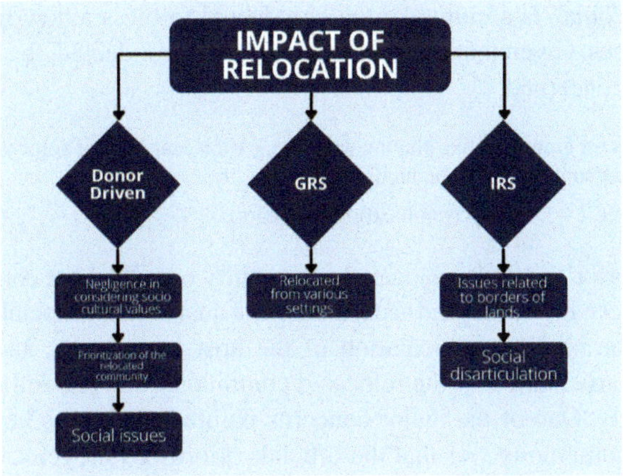

Fig. 20.2 Impact of relocation on social capital. *Source* Prepared by the authors

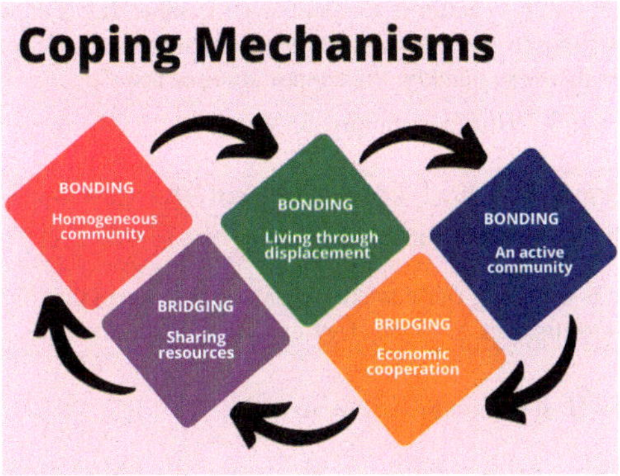

Fig. 20.3 Coping mechanisms developed by the relocates in terms of social capital. *Source* Prepared by the authors

(A) Negligence of socio-cultural values
Even though the community was homogeneous, there was a pressing issue at the donor-driven settlement. The major concern was related to the resettled land. The community was made aware that the houses were built on land that used to be the cemetery of the area. They showed concerns in this regard as officials had not taken necessary steps to conduct cultural purification rituals, and this had a huge impact on their cultural needs. In this context, the community believes that the community is cursed due to the lack of cultural rituals being done before the resettlement.

"We cannot build houses on lands just as we please. They should have conducted a Bahirava Pooja and chanted Pirith before starting construction."

(Interviewee 1 – Donor-driven location, 45 years)

"Bahirava Pooja" is a cultural ritual conducted to bless a new land against any evil influences. Given that such practices were not conducted, the respondents are very much concerned.

> "Taking spirits for granted is like playing with fire. We have taken their home without sending them to another safe place by conducting a pooja."
>
> (Interviewee 1 – Donor-driven location, 53 years)

(B) Issue of prioritizing the relocated community over the host community

This aspect can be analyzed according to the dual nature of social disarticulation. First is the social disarticulation of the host community, and second is the social disarticulation of the relocated community due to conflict with the host community. One of the major concerns pointed out by the key informants of the host community was that the officials prioritized the relocated community over them, and they were being overlooked in terms of dire needs.

> "They still get aid from different groups. For example, a foreign monk gave aid for them, and last time, a group gave milk powder for their children. We have poor children in our village, too. It is better if we also have the same aid as them. They can at least think of our children and consider giving them some sort of an aid. They got their water supply also very quickly. We cannot approve this."
>
> (Interviewee 1 – Host community, 53 years)

On the other hand, the host community of the donor-driven setting is also concerned about them losing their cemetery.

> "The biggest issue is us losing our cemetery; that is a big harm that happened to us. Now, we have to go a 1km distance to a cemetery or we have to bury. They also face this same problem."
>
> (Interviewee 3 – Host community, 42 years)

With these issues of prioritization, there was a visible slight disconnect between the host and resettled communities resulting in social disarticulation of the relocated community.

(C) Social issues

The host community key informants further pointed out that there is a threat of narcotics due to the relocated community. This was also reflected as a major point of social disarticulation of the relocated community.

> "They have drug issues. So, I think it is better to have an awareness program especially for the youth to prevent this issue. We are worried about our children. This matter was covered in the news as well."
>
> (Interviewee 2 – Host community, 34 years)

20.4.1.2 Owner-Driven Setting

Specifically, in the context of the owner-driven setting, despite the basic infrastructural facilities such as roads and drainage systems being present, there were no additional public facilities. While the community center at the donor-driven setting accommodated the relocated community to gather and enhance their network, the owner-driven setting had no such arrangements. Further, the specific concerns of the GRS and IRS locations could be summarized as follows:

GRS Setting

The impact of relocation in terms of social capital was different in the GRS setting. This can be explained as follows:

(A) Relocated from various settings
In comparison with the donor-driven setting, the relocates in the GRS setting were relocated from various administrative divisions of the Kegalle district. This made the relocates to be less close knit in comparison with the donor-driven setting.

IRS Setting

The impact of relocation in terms of social capital in the IRS setting can be explained as follows:

(A) Issues related to the boarders of lands
As the relocates of the IRS setting were given the opportunity to find their own lands, as mentioned before, the relocates of the selected IRS setting used to live in the same neighborhood that later developed into an IRS. Despite the positive aspect of continuing the same community post-relocation, this setting has several issues due to the borders of lands not being properly demarcated.

> "I am an old woman who lives in this house. The borders of my land have not been properly marked, and my neighbor is trying to take advantage out of this."
>
> (Interviewee 3 – IRS Setting, 65 years)

This has been one of the focal matters creating issues among the neighbors.

(B) Social disarticulation
The IRS setting has a housing unit where a 65-year-old mother is looking after her 45-year-old Autistic son. This has made both the mother and the son face several incidents of discrimination in the settlement. We were introduced to the house as "PissuGedara" (crazy house). The mother said that the son brings them water and other things needed for their day-to-day lives. However, it appears that neighbors make jokes and bully him when he walks down the road.

Nevertheless, in an overall investigation of the sense of community as pointed out by Downing (1996), it could be found that this setting has a sense of community due to the homogeneous nature of the post-relocation communities. None of the donor-driven settlement's respondents were looking forward to moving out of the relocated location. In the owner-driven settings, only two wanted to move out of the setting and that, too, was for economic concerns rather than social capital related concerns.

20.4.2 Coping Mechanisms to Address the Impact on the Social Capital Post-relocation

Given that there is a considerable amount of impact on social capital due to relocation in the selected settings, the coping mechanisms that were common in all three settings will be discussed in line with the main typology of social capital which has already been discussed in the conceptual background section.

20.4.2.1 Bonding Social Capital

As already discussed, bonding could be defined as bringing people together who are alike in important aspects such as ethnicity, age, gender, and social class (Putnam and Gross 2002). During the fieldwork period, bonding social capital was evident in the relationship among relocated communities. Several thematic aspects were unveiled during the sample surveys as well as the in-depth interviews.

Having a Homogeneous Community

Regarding the ethnic fabric of the setting, it is a homogeneous community where the majority of them belong to the same ethno-religious group: Sinhalese Buddhists. Therefore, the respondents highlighted that there are no issues with the neighbors or with other community members. They live in harmony most of the time and help each other whenever it is needed.

> "Whenever there is an issue at the setting (funeral, injury, or an accident), all of us come together and support each other."
>
> (Interviewee 5 – GRS setting, 45 years)
>
> "It is very easy for us to blend together well because we belong to the similar religious and ethnic backgrounds."
>
> (Interviewee 4 – IRS setting, 45 years)

Living Through Displacement

One of the strongest factors on bonding social capital is the fact that all of them have gone through the displacement process together. Even though some settings have people coming from various neighborhoods, the bond between the communities seems to be strong. The main reason for the said factor is that they have gone through the displacement process together. They stayed in the transitional and temporary shelters together for approximately two years. As per the respondents, during their stay in the said camps, they worked as a team to rebuild their disrupted lifestyles.

> "We are coming from different original settings. But, we have a strong bond. The reason is that we stayed in the same camp for 2 years together. We faced everything together as one."
>
> (Interviewee 3 – Donor-driven setting, 56 years)

An Active Community

In terms of the status quo of bonding social capital, community-based initiatives emerged as another vital element. In these settings, a majority of the respondents are members of various community-based organizations.

They have several community-based organizations and societies to strengthen existing bonds and develop more. The community has formed a death donation society ("Maranadhara Samithiya"), a development society ("Sanwardhana Samithiya"), a youth society ("Yowun Samithiya"), and a children's society ("Lama Samithiya").

It was also pointed out that the said community-based organizations organize activities such as New Year celebrations, carols, and Buddhist religious activities. The community of the donor-driven setting also invites monks to preach when the community conducts religious activities. Further, they invite the chief monks of the temples where they stayed during the temporary shelter stage. The Social Welfare Organization has also organized a sewing class for women in the settlement. Further, the community of the donor-driven setting has also got together in building a massive Buddha statue at the entrance. The Secretary of the Social Welfare Organization mentioned that they try their best to keep the community itself. On the other hand, in the donor-driven setting, their first ever community-based consignment was to organize the opening ceremony of the settlement.

20.4.2.2 Bridging Social Capital

As already discussed, bridging is bringing people together who are unlike one another (Putnam and Gross 2002). During the fieldwork period, the bridging social capital characteristics were visible between the relocated communities and the host communities. Given that both communities have different territorial relationships with the locations, it could be argued that the communities are different from one another. The recognized thematic orientations are as follows:

Economic Cooperation

During the in-depth interviews with the key informants from the host community, those who have their own businesses mentioned that they are getting economic benefits from the resettled communities.

> "Within two to three weeks, the host community started making mutual interactions with the resettled community. There is a small restaurant there (at the host community), and we go to that restaurant often. So, within a small time period, we made a connection with them."
>
> (Interviewee 6 – Donor-driven setting, 43 years)

Apart from the said restaurant, there are also retail shop owners and a respondent with a home-based salon of which resettled community members are customers.

> "They (host community) has a salon. They always come to me when they have any social function."
>
> (Interviewee 2 – Donor-driven setting, 45 years)

Sharing Resources

In-depth interviews with the key informants from the host community further unveiled that the resettled community share several resources that belong to them with the host community. Two of the main resources are the road across the donor-driven settlement and the community center at the same.

> "We have a community hall. We use it for our own purposes. But host community members also come to the community hall to attend some meetings such as the death donation society meetings and the meetings held by the Development Officers and the Village Level Officer."
>
> (Interviewee 6 - Donor-driven setting, 50 years)

The IRS setting has no water pipeline connections. To mitigate the issue of water supply, there is a public well at the setting. This was not provided by the authorities but by a businessman in the host community. The reason behind such supply shows hopes of development of strong social capital. The host community member provided this water well to the relocated community in memory of his wife who had passed away from a disease. Hence, the provision of the facility was purely a charitable purpose in his interpretation.

20.5 Discussion

The findings suggest that social capital in the selected settings has been one of the main ways of strategizing coping mechanisms. One of the strongest elements in making social capital one of the vital sources of coping mechanisms was the homogeneous nature of the community. Studies conducted on the refugee resettlement contexts in the Global North have recognized ethnicity being one of the main indicators that cater toward the building of a strong social capital (Bertrand 2000; Madjar et al. 2000). In the circumstances of the present study, even though it represents an internally displaced relocated community from the Global South, it could be observed that the homogeneous Sinhalese ethnic fabric has played a major role in curbing the impact of relocation on the social capital. Further, ethnicity-based formal and informal organizations have proven to be vital post-relocation (Wahlbeck 1998). Even though the settings did not have organizations based on the ethnicity per say, their community-based initiatives were very much connected with their ethnic identity. In the Sri Lankan context where ethnicity is widely equated with religion, Sinhala Buddhist is the strongest community. Even in the selected settings, the homogeneity of the community was based on the Sinhala Buddhist aspect. Therefore, initiatives such as building the Buddha statue at the entrance and inviting monks for religious rituals could be seen as indirect confirmation of the claim made by Wahlbeck (1998).

This context could further be evaluated under the concept of "homophilous social relationships" to explain that people who share similar lifestyles and other forms of similarities tend to develop strong ties in relocation settings (Quetulio-Navarra et al. 2013). Lin (2001) has explained this phenomenon in the context of development induced displacement. However, in terms of this study given that the community has lived through the displacement process, it could be seen that they share the similar life experiences that lead to a similar lifestyle. This eventually complements the development of a stronger social capital.

In relocation planning, it has been pointed out that as much as attention is being given to housing units, equal attention should be given to public spaces that accommodate communal activities such as libraries, communal shrines, and stores (Aldrich and Meyer 2015; Abe and Shaw 2014, Quetulio-Navarra et al. 2013). Further, the need for having recreational facilities at the relocation setting to enhance the relationships among the relocates has been pointed out (Senanayake et al. 2022). It was evident how the community center at the donor-driven setting was more beneficial toward strengthening the ties among the relocated community as well as between the host and the relocated communities. On the contrary, the owner-driven settings had no such facilities, and this was recognized as one of the major impacts of relocation on social capital that is common to both owner-driven settings.

Basque (2021) has pointed out that in a relocation initiative, it should ensure that the due honor is given to the land one is departing from and the one they are relocating to. The need of respecting the culture and norms of the relocating community has also been emphasized (Fernando et al. 2020a, b). In this context, the donor-driven setting shows a contradictory stance by not considering these factors. It is evident that overseeing these factors has a major impact on their post-relocation lives.

Maldonado et al. (2013), Yang (2014) have claimed that when an initial community is scattered due to relocation, and the number of strangers increases upon resettlement, it affects the trust within the relocated community. The IRS setting could be considered an ideal example of how the relocated community wanted to prevail this notion of trust by continuing the same neighborhood post-relocation.

Studies have further emphasized the importance of creating social cohesion among the host and the displaced communities (Amaratunga et al. 2020; Jayakody et al. 2022). It has been further noted that the built environment plays a major role in creating such cohesion (Jayakody et al. 2021). In the circumstances of the present study, economic cooperation and sharing resources (specifically water) through a charity initiative and the community center show positive aspects of development of cohesion in both communities, specifically in terms of the built environment.

20.6 Conclusion and Recommendations

In evaluation of the role of social capital as a coping mechanism in the selected post-relocation settings, it is visible that social capital plays a significant role in making the relocates resilient in the selected settings. The salient features in the selected locations were the relationship with the host communities and the active community initiatives that have been built in the post-relocation settings. Hence, it could be concluded that social capital plays a vital role in the selected settings. However, as explained above, the integration of social capital into the relocation process to enhance the resilience of the relocates could have been better in certain aspects. Therefore, the following general recommendations are made in terms of mainstreaming social capital as a coping mechanism in planned relocation. These recommendations will be useful for both built environment professionals as well as other relevant professionals involved in planned relocation.

(i) ***Considering the wider composition of the community in relocation***
As explained, homogeneity of the relocated community plays a vital role in forming bonding social capital in the context of relocation. Therefore, in planned relocation, it is vital that a thorough study is conducted in understanding the geographic, demographic, and cultural composition of the relocated community. This will allow the planned relocation initiative to make sure that the composition of the relocated community will not be disrupted. This was evident in a homogeneous community in the selected settings and further with the undisrupted neighborhood at the IRS setting.

(ii) ***Integrating cultural sensitivity into the relocation process***
It is important to consider the cultural beliefs, values, and norms of the relocating community. This will allow the planned relocation initiative to respect their beliefs in terms of land, housewarming, and community formation. This will eventually enhance the satisfaction and the resilience of the relocated community. This could prevent issues such as the dissatisfaction that has been developed at the donor-driven setting for being relocated to a land that used to be a cemetery. Further, this will assist the relocates in terms of developing their social capital among themselves (bonding) and with their host community (bridging).

(iii) ***Having public and recreational resources at the relocation sites***
In planned relocation initiatives, it vital for public and recreational facilities, apart from the basic infrastructural facilities, to be provided. This could be in the nature of a community center, community well, libraries, stores, parks, and sports facilities. This will allow enhancement of the interactions among the relocates and interactions between the relocates and the host community as was evident in the donor-driven setting through their community center. This will pave way toward the relocates developing their bonding and bridging social capital in the post-relocation setting. Further, it is notable that the settings still have the space to accommodate recreational facilities.

(iv) ***Accommodating community-led initiatives at relocation sites***
Relocation sites should have opportunities for communities to come together and work as a team, and this complements the contribution toward the development of bonding social capital. This could be seen in the donor-driven setting dominantly where the officials created an environment to encourage the relocated community to come together and make initiatives. Hence, creating such an accommodating environment is vital.

(v) ***Connecting host and relocated communities***
It is vital to interpret social capital as being not just among the relocates but also between the host and relocated communities. The findings of this study pointed out how the development of bridging capital takes place with positive connection between the relocated community and the host community. Hence, the planned relocation initiatives should recognize the commonalities between the two communities as well as opportunities for the two communities to be linked. As it was observed in the selected settings, the use of a community center, economic cooperation, and sharing the water resource could be recognized as such links. Therefore, accommodating and encouraging such links and opportunities are vital to ensuring cohesion between the both communities.

Acknowledgements The authors would like to thank the Erasmus+ Programme of the European Commission for funding the REGARD project. The European Commission support for the production of this publication does not constitute an endorsement of the contents which reflect the views only of the authors, and the Commission cannot be held responsible for any use which may be made of the information contained therein.

References

Abe M, Shaw R (2014) Community resilience after Chuetsu earthquake in 2004: extinction or relocation? In: Shaw R (ed) Community practices for disaster risk reduction in Japan, pp 191–208. [online] Available at https://link.springer.com/book/https://doi.org/10.1007/978-4-431-54246-9. Accessed 2 July 2022

ADPC (2020) Disaster risk reduction in Sri Lanka. UNDRR, Bangkok

Aldrich DP, Meyer MA (2015) Social capital and community resilience. Am Behav Sci [online] 59(2):254–269. https://doi.org/10.1177/0002764214550299

Amaratunga D, Haigh R, Jayakody C, Malalgoda C (2020) Rebuilding after forced displacement. Built Environ J RICS J. [online] Available at https://ww3.rics.org/uk/en/journals/built-environment-journal/rebuilding-after-forced-displacement-.html. Accessed 2 July 2022

Amaratunga D, Malalgoda C, Haigh R, Jayakody C, Senanayake AC, Fernando N, Witt E (2022) A built environment perspective on post-disaster and conflict-induced displacement: a report of the REGARD project: rebuilding after displacement. REbuildinGAfteRDisplacement

Basque NB (2021) Adapting together: exploring the relationship between social capital and planned relocation as a disaster risk reduction measure. Royal Roads University, Canada

Behera JK (2021) Role of social capital in disaster risk management: a theoretical perspective in special reference to Odisha, India. Int J Environ Sci Technol. [online] https://doi.org/10.1007/s13762-021-03735-y

Bertrand D (2000) Policies for the reception and integration of refugees: a comparative study of the UK and France (research update). Refugee Studies Centre, Oxford University, Oxford

Bhandari H, Yasunobu K (2009) What is social capital? A comprehensive review of the concept. Asian J Soc Sci 480–510. https://www.researchgate.net/publication/233546004

Bourdieu P (1986) The forms of capital. In: Richardson J (ed) Handbook of theory and research for the sociology of education. Westport, CT, Greenwood, pp 241–258

Cernea M (2000) Impoverishment risks, risk management and construction: A model of population displacement and resettlement. UN Symposium on Hydropower and Sustainable Development. Beijing

Claridge T (2004) Designing social capital sensitive participation methodologies. Soc Capital Res

Coleman JS (1988) Social capital in the creation of human capital. Am J Sociol (Special Issue on Organ Inst Sociol Econ Approaches Anal Soc Struc) S95–S120 [online]. Available at https://www.jstor.org/stable/2780243. Accessed 2 July 2022

Coleman JS (1990) Foundations of social theory. Cambridge, Harvard University Press

CRED (2019) EM-DAT: the emergency events database—Université catholique de Louvain (UCL)—CRED. Université catholique de Louvain, Brussels

Downing TE (1996) Mitigating social impoverishment when people are involuntarily displaced. In: Christopher Mcdowell (eds) Resisting impoverishment—tackling the consequences of development induced impoverishment. Oxford: Bergham Books

Fernando N, Punchihewa AG (2013) Relocating the displaced: strategies for sustainable relocation. Friedrich Ebert Stiftung, Colombo

Fernando N, Senanayake A, Amaratunga D, Haigh R, Malalgoda C, Jayakody RRJC (2020a, October) Disaster, displacement and relocation: an analysis of the needs and policy implications on a displaced community in Sri Lanka. In: 10th international conference on structural engineering and construction management: special session on disaster risk reduction. University of Peradeniya, pp 128–145

Fernando N, Senanayake A, Amaratunga D, Haigh R, Malalgoda C, Jayakody RRJC (2020b) Impact of the disaster-induced relocation process on the displaced communities in Kegalle district, Sri Lanka. Eur Sci J (Special Edition: Contemporary Sri Lankan Society and Politics: Felicitation volume in honour of renowned Sri Lankan Sociologist Professor ST Hettige). ESJ ISSN 1857-7881 (Print) e-ISSN 1857-7431

Hanifan LJ (1916) The rural school community centre. Ann Am Acad Political Soc Sci 67:130–138

IDMC (2010) Displacement due to natural hazard-induced disasters. Global estimates for 2009 and 2010

IDMC (2021) Global report on internal displacement 2021. Internal Displacement Monitoring Centre (IDMC), Geneva

International Organization for Migration (IOM) Sri Lanka (2016). Sri Lanka Floods and Landslides: Situation Report #2. IOM Sri Lanka

Jayakody RRJC, Malalgoda C, Amaratunga D, Haigh R, Liyanage CL, Witt E, Hamza M, Fernando N (2021) Role of the built environment in rebuilding displaced and host communities. In: Amaratunga D, Haigh R, Dias N (eds) Multi-hazard early warning and disaster risk reduction. Springer, London, pp 69–92

Jayakody C, Malalgoda C, Amaratunga D, Haigh R, Liyanage C, Witt E, Hamza M, Fernando N (2022) Approaches to strengthen the social cohesion between displaced and host communities. Sustainability [online] 14(6):3413. https://doi.org/10.3390/su14063413

Karunasena G, Rameezdeen R (2010) Post-disaster housing reconstruction: comparative study of donor vs owner-driven approaches. Int J Disaster Resilience Built Environ [online] 1(2):173–191. https://doi.org/10.1108/17595901011056631

Lin N (2001) Social capital: a theory of social structure and action. Cambridge University Press, Cambridge

Madjar V, Humpage L (2000) Refugees in New Zealand: the experiences of Bosnian and Somali refugees. School of Sociology and Women's Studies, Massey University (Albany), Auckland, New Zealand

Mahani RT (2013) Comparative study of donor driven vs. owner driven approach on the way to "Build Back Better" of Gaza Private demolished houses. Available online https://library.iugaza.edu.ps/thesis/110082.pdf

Maldonado JK, Shearer C, Bronen R, Peterson K, Lazrus H (2013) The impact of climate change on tribal communities in the US: displacement, relocation, and human rights. Clim Change 120(3):601–614 [online]. https://doi.org/10.1007/s10584-013-0746-z

Muggah R (2008) Relocation failures in Sri Lanka: a short history of internal displacement and resettlement. ZED Books, London and New York

National Disaster Relief Services Center (2021) Disaster situation report as (2020.01.01–2020.12.31) State Ministry of Internal Security, Home Affairs and Disaster Management

OCHA (2016) Sri Lanka: floods and landslides situation Report No. 2 (as of 26 May 2016). OCHA Resident Coordinator's Office, Sri Lanka

Panday S, Rushton S, Karki J, Balen J, Barnes A (2021) The role of social capital in disaster resilience in remote communities after the 2015 Nepal earthquake. Int J Disaster Risk Reduction 55:102112 [online]. https://doi.org/10.1016/j.ijdrr.2021.102112

Putnam RD (1994) Social capital and public affairs. Bull Am Acad Arts Sci 47(8):5–19

Putnam RD (1995) Bowling alone: America's declining social capital. J Democracy 6(1):65–78

Putnam R, Goss KA (2002) Introduction. In Putnam R (ed) Democracies in flux: the evolution of social capital in contemporary society. Oxford, Oxford University Press, pp 1–19

Quetulio-Navarra M Niehof A van der Vaart W (2013) Social capital in involuntary displacement and resettlement. Int J Soc Sci Humanity Stud 5(2):139–154

Senanayake A, Fernando N, Wasana M, Amaratunga D, Haigh R, Malalgoda C, Jayakody C (2022) Landslide induced displacement and relocation options: a case study of owner driven settings in Sri Lanka. Sustainability 14(3):1906 [online]. https://doi.org/10.3390/su14031906

Szreter S, Woolcock M (2004) Health by association? Social capital, social theory, and the political economy of public health. Int J Epidemiol 33(4):650–667. https://doi.org/10.1093/ije/dyh013

UNDP (2021) Displacement and climate change. Retrieved from United Nations Development Programme: https://www.undp.org/

UNISDR (2018) Disaster displacement: how to reduce risk, address impacts and strengthen resilience. UNSIDR

Wahlbeck O (1998) Community work and exile politics: Kurdish refugee associations in London. J Refug Stud 11(3):215–230

Woolcock M (1998) Social capital and economic development: toward a theoretical synthesis and policy framework. Theor Soc 27:151–208. https://doi.org/10.1023/A:1006884930135

Woolcock M (2001) The place of social capital in understanding social and economic outcomes. Contribution of human and social capital to sustained economic growth and well-being, pp 65–88

Yang H (2014) Assessing the evolvements and impacts of environmental migration in inner Mongolia. In Gemenne F, Brücker P, Ionesco D (eds) The state of environmental migration 2014: a review of 2013, pp 119–133 [online]. International Organization for Migration. https://publications.iom.int/system/files/pdf/state_environmental_migration_2014_0.pdf

Chapter 21
Relocated or Displaced? A Social Inquiry of Tsunami-Induced Relocation Programme in Southern Sri Lanka

Thushara Kamalrathne and Anuradha Senanayake

Abstract Sri Lanka has experienced several coastal hazards, and significant attention was given by the government and other humanitarian agencies after the Indian Ocean Tsunami in 2004. 1,009,474 people were affected by this tsunami in 2004, which was the most recent tsunami experience in Sri Lanka causing an economic loss of an estimate of US$208.2 billion to livelihoods and other infrastructure. Many victims of the tsunami in 2004 were resettled by the Sri Lankan government and several non-government agencies in various places known as tsunami resettlement schemes in the country. The objective of this paper is to highlight the social constraints faced by people resettled by these schemes. This study finds that the resettlement project has paid little attention to national and global safeguard policies such as the national involuntary resettlement policy and policies of aid agencies and donors. Livelihood restoration is a vital pillar in any resettlement project, but this resettlement project has a detrimental impact on livelihoods. Further, the absence of land ownership restricts children's access to government schools and impacts social respect and identities. Therefore, the study shows that the resettlement project lacks prior social planning and remains unsatisfactory as the constraints of the affected people have not been properly addressed. The field study for this research was carried out in the Dickwella divisional secretariat in the Matara district. Focus group discussions and key informant interviews were used to collect primary data. Michael Cernea's impoverishment risks and reconstruction model (IRR) (Cernea in, World Dev 25:1569–1587, 1997) was used as an analytical framework to study and analyze the impact of the resettlement.

Keywords Resettlement · Relocation · Displacement · Sri Lanka · IRR model

T. Kamalrathne (✉) · A. Senanayake
Global Disaster Resilience Centre, School of Applied Sciences, University of Huddersfield, Huddersfield, UK
e-mail: Thushara.Kamalrathne@hud.ac.uk

© The Author(s), under exclusive license to Springer Nature Switzerland AG 2023

M. Hamza et al. (eds.), *Rebuilding Communities After Displacement*,
https://doi.org/10.1007/978-3-031-21414-1_21

21.1 Background

> What is too little understood both by professionals and scholars alike is the social impact of displacement and relocation. When resident people are forced to move, certain general impacts can be expected. But the collective social impact on the community or other social organizations differs widely from case to case; to date, no model exists to predict the cumulative effect (Brenchin et al. 1991: 17).

Disasters and their grave impact on societies are one of the major considerations in contemporary society. Even though the year 2020 was predominantly known for the COVID-19 pandemic, it recorded 389 extreme weather events in general with a high percentage of flooding events (CRED and UNDRR 2021). The year 2021 recorded an increase of 432 extreme weather events with 223 flooding incidents (CRED 2022). Displacement of people following such extreme weather events is one of the most regular forms of displacement in the contemporary world (Turton 2006). In the year 2020, more than 40 million people were displaced out of which more than fifty per cent were displaced due to disaster events (IDMC and NRC 2021). Studies conducted in the global context on the disaster-induced displacement profiles claim East Asia and the Pacific as the regions with the highest number of displacements reported due to disaster events (IDMC and NRC 2021).

The year 2004 was a landmark in the discourse of extreme weather events in the global context. The earthquake that took place on the 26th of December 2004 in the Indian Ocean resulted in a tsunami that affected more than 200,000 people with 1.7 million displacement cases being reported (Doocy et al. 2007). Sri Lanka was one of the countries that was majorly affected due to the tsunami. The waves of the tsunami affected a considerable proportion of the coastline in Sri Lanka with an overall impact in 13 coastal districts (WFP 2005). This devastating event reported more than 30,000 deaths and approximately 100,000 tsunami-induced displacement incidents in Sri Lanka (Jayasuriya et al. 2006).

In the context of displacement, planned relocation has been considered a suitable disaster response strategy (Yonetani 2018). In terms of planned relocation, some studies have emphasized the importance of housing that is provided as an integral part of the said planned relocation process (Oo et al. 2018). These planned relocation strategies take several forms such as allowing the displaced to build their own housing units (owner-driven) and private and state parties donating housing units for third parties (Barenstein 2006; Fernando and Punchihewa 2013).

With the mass destruction caused by the tsunami to the coastal community, the Sri Lankan government declared that no construction could take place within 100 m to the Western and Southern coastal areas and 200 m to the Eastern and the Northern coastal areas (Buffer Zone). Further, the regulation made the authorities destroy the existing housing structures within the said zone, and more than 70,000 people were forcefully relocated (Fernando 2018). The planned relocation strategy that was utilized to reconstruct the tsunami affected community in Sri Lanka was donor-driven. In this circumstance, the Tsunami Housing Reconstruction Unit (THRU) under the guidance of the Urban Development Authority (UDA) was mainly in charge of permanent housing reconstruction projects. Donors built houses according to the UDA guidelines and site plans issued by the National Housing Development Authority (NHDA) after signing a Memorandum of Understanding (MoU) with THRU. Moreover, donors paid for the cost of basic amenities such as water, electricity, and sewerage for households and infrastructure in the settlement areas (Fernando 2012).

Disaster-induced displacement is such a phenomenon that changes the living fabric of communities, and a well-planned intervention is required to make sure that the lives of the displaced are built back better (Amaratunga et al. 2022). In this context, relocation itself has been interpreted as a vital intervention that requires the participants of the displaced to make sure that their aspirations are duly covered in the interventions (Fernando et al. 2009; Senanayake et al. 2022). Further, the Sendai Framework for disaster risk reduction emphasizes the importance of the participation of the community in managing a systematic approach to disaster risks (UNISDR 2015).

Relocation following displacement leads to a variety of impacts on the livelihoods of the relocates and causes several impoverishment risks (Satiroglu and Choi 2015; De Silva and Yamao 2007). The impacts have further been recognized as economic, physical, housing, cultural, and political (Sridarran et al. 2016). The economic impact has been interpreted around the issue of employment opportunities created post-relocation (Cao et al. 2012; Lakshman and Amirthalingam 2009). Relocation projects have further raised concerns in terms of post-relocation housing units in terms of the architecture and cultural aspects (Badri et al. 2006; Fernando 2012; Senanayake et al. 2022). Not considering cultural values and practices in the relocation process has been recognized as a major concern in terms of rebuilding the lives of the relocates post-relocation (Fernando et al. 2020b; Amaratunga et al. 2020). Ignorance of social issues is another fact that has been recognized as a fact that contributes towards the failure of relocation projects (Vanclay 2017). It has also been emphasized that the living conditions of the relocates are highly based on the status quo of the post-relocation neighbourhoods they were put into (Miltenburg et al. 2018). Studies have also emphasized the needs for prioritizing the needs and aspirations of both relocates and host communities and accommodating social cohesion amongst the said communities after displacement (Jayakody et al. 2022).

This makes it clear that a relocation initiative has a major impact on the relocates in terms of their living fabric. In the context of Sri Lanka, studies have been conducted on tsunami resettlement from the perspectives of vulnerability and relocation impact (Fernando 2012). This study is also based on a tsunami relocation settlement programme in Sri Lanka. However, the main objective of the study is investigating the social impacts of the tsunami-induced relocation programme in Sri Lanka.

21.2 Conceptual Framework

The conceptual framework of this study revolves around displacement, disaster, and relocation. This section elaborates these concepts and the relevant conceptual framework used in the present study. The said framework is the impoverishment risks and reconstruction model for resettling displaced population (IRR model) (Cernea 2000).

21.2.1 Introduction to the Main Concepts of the Study

The term 'displacement' was coined by Russian–American sociologist Eugene M. Kulischer with the introduction of the term 'displaced persons' (Kulischer and Jaffe 1962). Ever since concerns were raised regarding displaced communities, many initiatives related to best practices that should and can be utilized by state and non-governmental organizations emerged (Terminski 2013). The guiding principles on internal displacement define internally displacement persons as:

> Persons or groups of persons who have been forced or obliged to flee or to leave their homes or places of habitual residence in particular as a result of or in order to avoid the effects of armed conflict, situations of generalized violence, violations of human rights, or natural or human-made disasters and who have not crossed an internationally recognized State border. (United Nations 1999, p: 1).

Fernando et al. (2009) and Terminski (2013) identify three typologies of displacement, namely development induced displacement, conflict induced displacement, and natural disaster induced displacement. They also emphasize the fact that some people are permanently displaced whilst others are temporarily displaced due to disasters and conflicts. In this context, natural disaster induced displacement is where people lose their habitats due to natural or environmental disasters. The present study is based on a disaster-induced displaced community.

In this field of study, it is important to explain the difference between the concepts 'hazards' and 'disasters'. A hazard refers to the potential occurrence in a specific time period and geographic area of a natural phenomenon that may adversely affect human life, property, or activity to the extent of causing a disaster (International Federation of Red Cross and Red Crescent Societies 2000, p. 6). The occurrence of a hazard (earthquake, flood, or cyclone, for example) becomes a disaster when it results in injuries, loss of life and livelihoods, displacement, homelessness, and/or destruction and damage to infrastructure and property (International Federation of Red Cross and Red Crescent Societies 2000, p. 7). A cyclone that flows over an uninhabited island does not result in a disaster; however, it would be a disaster if it hit a populated area and caused extensive loss to life and property (International Labour Organization 2002).

In this context, it is clear that a hazard becomes a disaster when it affects the human population. Therefore, a disaster is extreme disruption to the functioning of a society that causes widespread human, material, or environmental losses that exceed the ability of an affected society to cope using its own resources (International Federation of Red Cross and Red Crescent Societies 2000, p. 6).

Planned relocation has been recognized as one of the main disaster response techniques post-disaster-induced displacement (UNHCR 2014). According to the Oxford Modern English Dictionary (1995), resettlement is defined as *'settle[ing] again'* (p. 920), and relocation is defined as *'locate[ing] in a new place'* (p. 912). When considering the definition of resettlement, settling again could be done in the original setting or in a new setting. The present study considers affected people settling in a new setting, and therefore, one may argue that the most suitable definition for the present study is the definition of relocation. Fernando et al. (2009) has pointed out that the definitions of these terms should be considered in the context of physical movement of people to a new place to live other than the previous place.

The term 'resettlement' refers to physical, pre-planned relocation, combined with appropriate support mechanisms, including social support in the new location (Terminski 2013, p. 13). This was earlier elaborated by Chambers in 1969, and according to him, resettlement is characterized by two main features: a movement of population and an element of planning and control. Given that planning is an important element in resettlement, and planned relocation has been one of the most important concepts discussed in displacement studies. The UNHCR (2014) explains resettlement as a component of planned relocation. Turton (2006) recognized two other aspects in relation to resettlement which are involuntary and voluntary resettlement. He distinguishes these two terms based on the consent of the relocates. Hence, involuntary resettlement is a process which is done without the consent of the relocates.

In terms of planned relocation options, the 'owner-driven' approach is known as an approach where the displaced community constructs their own post-displaced housing units with external financial aid. On the other hand, 'donor driven' is an approach where external donors or governments construct post-displaced housing units (Karunasena and Rameezdeen 2010).

21.2.2 Introduction to the Conceptual Frameworks of the Study

As the main objective of this study is to investigate the social impacts of the tsunami relocation programme that took place in Sri Lanka, the analysis will be based on the conceptual framework of impoverishment risks and reconstruction model for resettling displaced population by Cernea (2000).

The IRR model enables an understanding of the cumulative effects of planned resettlement and relocation and provides a rigorous approach to analyze of interconnectedness between the cascading social and economic effects that emerged within the system (Cernea 2000).

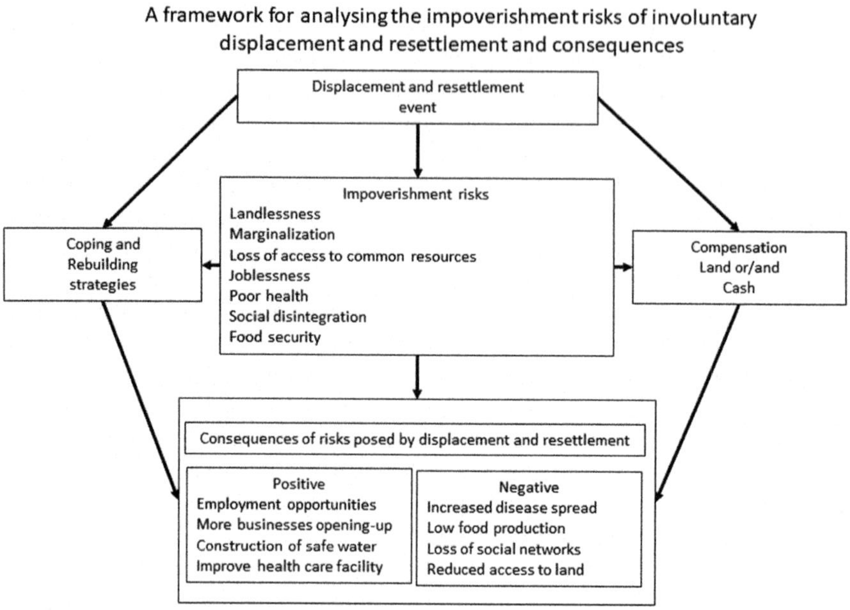

Fig. 21.1 Impoverishment risks and reconstruction model (IRR model). *Source* Cernea (2000)

This model discusses the social, cultural, and economic impoverishment risks the resettled go through. He discusses eight possible impoverishment risks that a relocate may face, and he further adds that the planned relocation should consider these risks in planning and execution (Cernea 1991). According to the aforementioned Fig. 21.1, the said impoverishment risks are landlessness, marginalization, loss of access to common resources, joblessness, poor health, social disintegration, and food security.

This model has been utilized in various studies to recognize the gravity of impoverishment risks resulted by involuntary relocation (Bang and Fewa 2012; Dhakal et al. 2011; Quetulio-Navarra et al. 2014). A study conducted in the contexts of Indonesia and Philippines has utilized the eight risks of the IRR model (excluding 'loss of access to common property' and 'increased morbidity and mortality' to develop a risk index to recognize the household level risk intensity post-relocation (Quetulio-Navarra et al. 2014). The loss of access to common property was highlighted in the nature of the lack of government services in studies conducted in Laos and Egypt (Adriansen 2009; Jacka 2008). The IRR model has also been applied in resettlement that takes place in the mining sector to recognize the impoverishment risks that arise from relocation (Vivoda et al. 2017). However, a study conducted in East Ethiopia recognized several risks such as cattlelessness, constrained community mobility, constrained access to education, loss of resiliency, and loss of human rights that are overlooked in the IRR model (Gizachew 2015).

Table 21.1 Summary of recognized risks

Impoverishment risk	Impact
Landlessness and homelessness	Land tenure Condition and quality of the housing units Infrastructure availability
Marginalization	Name of the relocation site Change of social class
Loss of access to common resources	Issues with public services Lack of eligibility for loans
Joblessness	Lack of livelihood restoration programmes
Poor health, food security, and social disintegration	Free health policy

Source Develop by the Authors

Therefore, this study utilizes this framework to recognize the impoverishment risks that have been emerged within the selected study locales post-relocation. The summary of the recognized risks in this study could be presented as follows (Table 21.1):

21.3 Methodology and Procedures

As explained before, the objective of this research is to explore the social impacts of the tsunami-induced relocation programme in Sri Lanka. It is aimed at understanding barriers and enablers encountered in terms of policymaking and practice in the process of disaster-induced relocations.

This research was carried out in two resettlement housing schemes used to resettle people who were affected by the 2004 Indian Ocean Tsunami in the Dickwella divisional secretariat in the Matara district of Sri Lanka. Both housing schemes were selected based on their spatial and social characteristics. One housing scheme (X) was selected from a remote area, whereas the other one (Y) was picked from a suburban area. Both schemes were built by the government with aid donated by several religious institutions and non-government organizations.

The research design is qualitative descriptive. The descriptive design aims to describe how things are experienced by people, with respect to the social setting or social phenomena under investigation (Leavy 2017). Bradshaw et al. (2017) highlight that qualitative descriptive research enables investigating lived experience as in phenomenology. In addition, qualitative description design is largely applied when research seeks to discover and understand a phenomenon, a process, or the worldviews of the people involved (Caelli et al. 2003). This research also investigated the social impacts of relocation from users' points of view. The four focus group discussions were administrated to collect primary data from settlers in both housing schemes and eight in-depth interviews were conducted with two Grama Niladaris (village officers), two government officials who had close official duties with settlers, and four key informants from the relocation schemes who had a strong connection with people in the particular area. Community members for the focus group discussions were selected purposively based on their gender, age, type of livelihood, and duration in the housing scheme. Two focus group discussions were conducted in each relocation setting and 8–10 members participated in the group discussions. The semi-structured interview method was applied in conducting focus group discussions, and the open-ended method was used to conduct in-depth interviews. Close observation was used to gather data on land, housing conditions, basic amenities, etc. The thematic method was administrated in qualitative data analysis and was designed in line with the analytical framework developed to comply with the IRR Model.

21.4 Social Impacts: Constraints Faced by the Settlers

In this section, the data analysis has been conducted according to the themes identified through the IRR model. The themes have been derived from the impoverishment risks component of the IRR model which analyzes the consequences of risks posed by forced relocation following displacement.

21.4.1 Landlessness

In both settings, housing units were given in 2006 and 2007 to the displaced people who lost their houses and basic amenities due to the 2004 tsunami, especially in the Southern province of Sri Lanka. The families were initially picked from several temporary camps including *Rathmale, Godauda*, and *Maaliyadda*. Housing schemes in the area were basically funded by different stakeholders including government agencies, faith-based organizations, and local and international non-government organizations. For example, scheme X was completely built by a faith-based organization and two international non-government agencies. Therefore, the conditions of the houses in the two relocation entities seemed to be uneven in size and structure. Scheme X has 90 houses that are single-storied but attached whilst scheme Y has 56 housing units including single houses that are two-storied. Amongst them, many households combined four housing units together.

Houses given to the relocates seemed to be uneven in terms of the type of the house, floor of the house (tile and cement floors), and the floor space. Similarly, the size of the lands is also not unique to all relocates as house space and land size have no rational relationship with the size of the family or other relevant criteria. Some families have 10 perches, whilst some others have 40–70 perch lands including the given housing units. The selection of the housing units and lands for families is seemingly asymmetrical which disappoints many relocates. For example, one family which has 7 members has a 10 perch land and house, whilst another family which has 2 members has a 70 perch land within the same relocation scheme.

The burning issue of the relocates in terms of property ownership is the lack of proof of ownership of their properties. Ownership of many houses and lands has been transferred by a government entitlement certificate which shows that they are not the absolute owners according to the land development ordinance of the country. Only 26 households in scheme X have deeds, and all other housing units have ownership by a letter known as a transfer order issued by the divisional secretariat mentioning that the particular land/house has been officially transferred to the name of the relocate. Therefore, the majority does not have proper official deeds for the land and further, almost all houses do not have a proper plan for their lands and have to be amended with the official deed showing land demarcations. However, it seems that many of the government relocation programmes have this issue, and issuing an official deed has become a missing political promise by several parties' propaganda in the country during the last decade (Fernando 2012).

The absence of absolute ownership of the house and lands has plenty of cascading social impacts in Sri Lanka. For example, as highlighted by many, illegibility for bank loans and mortgages, and illegibility for applying for admission to a school which is a major issue in Sri Lanka are a few key social problems that have emerged as a result of the lack of ownership of property. On the other hand, the lack of ownership is believed to be a factor in losing the sense of ownership amongst many relocates in relocation programmes. However, according to the field data, only 20 families out of 90 in scheme X are reported to be displaced by the 2004 tsunami, and all other households have been sold as a mutual agreement between the relocate and the buyer. Therefore, most of the housing units in both schemes have been sold and recolonized by outsiders who are looking for residence in the area. For instance, many house buyers in scheme X are involved in the fishing industry, especially fishermen and their families from *Kudawella* and *Nilwella* fishing harbours find that access from scheme X is considerably easy from the particular fishing harbours. Scheme Y is close to the Matara town area, and therefore, houses have been bought by residents who are looking for easy access to the main town area including those who are looking for access to public services such as education and other government offices. Interestingly, some members who responded to the field research had proper houses before the relocation even though they were affected by the tsunami, but severe damage had not happened to their properties in the same area.

21.4.2 Marginalization

As with many relocations and resettlements, a large number of relocation schemes are named something that ends with '*watta*' including housing schemes in the study area. From a cultural perspective, the term '*watta*' reflects the area where 'outsiders' settle with or without original settlers in an area. Sociologists emphasize that the term '*watta*' is recognized in urban settings as terminology that reflects marginalization or social exclusion (Silva and Atukorala 1991). Similarly, '*watta*' is used to describe settlements of the Indian Tamil population who were settled in upcountry plantations and several other places in Sri Lanka (Atukorala et al. 2010). In almost all cases, '*watta*' reflects a marginalized or excluded notion in relation to where a new group of people relocates or resettles.

Social marginalization experienced by the relocates has been caused by the new social identity that they have as relocates in the area after the tsunami. Some displaced people have re-built their own houses in areas where severe damage was not recorded, and some received financial aid from different sources. Some have completely migrated to their relatives' properties far from the impacted areas. In this context, many relocates being resettled lost all properties including housing and lands where the areas were severely damaged and could not be re-built for settling down. Ironically, some members who were deprived several of basic amenities during the relocation had significantly good economic stability before the tsunami. Some relocates they think that they are really 'poor' now even though they themselves identified as 'middle class' before the tsunami as many involved in the large-scale fishing industry. However, interestingly, a new social status and social hierarchy based on the current living conditions have emerged amongst relocates in the new social setting. This is very common in many relocation and resettlement programmes that have taken place in Sri Lanka since the 1960s. Several studies highlight the formation of new social structures, especially changes in 'social class' in resettlements as the new model of economy drastically changes the production relations amongst settlers (Brow 1981).

21.4.3 Loss of Access to Common Resources

Many relocates have been deprived of access to essential public services due to several underlying factors largely because of property ownership. According to the respondents, one of the key issues they face in public service access is the illegibility to apply to schools due to the unavailability of deeds which is a mandate in the Sri Lankan context to assess the residence of the applicant. It seems that many relocates use a letter or any other legal document provided by the divisional secretariat assuring the ownership which is a long process in terms of obtaining. Similarly, people need to show their official records to prove their residence and, in some cases, to confirm property ownership to obtain public services which are challenging for relocates.

The second issue as they revealed was illegibility to apply for any kind of loans issued by formal private or public banks in Sri Lanka. Candidates must present proof of their property ownership when they mortgage a house or land even to obtain a small-scale loan which is popular in terms of Small and Medium Enterprises. As a result of being excluded from formal financial networks, many relocates are trapped financially in informal sector loan providers including some merchants known as '*Mudalalies*' who provide informal loans with no documents by keeping high mortgages and charging a high rate of interest.

21.4.4 Joblessness

Failure to restore livelihoods or lack of policy measures for livelihood restorations is an important component in terms of understanding the huge job-related challenges encountered by relocates during the relocation process. As the Sri Lankan government declared the 200 m no-build zone from the coastal belt as a strategy to reduce the risk associated with future coastal hazards including tsunami or tidal waves, a number of tsunami-induced relocations have been established carried out in areas where the distance from the seashore is extended for the safety purposes. However, some studies have highlighted that this extension of distance from the coastal area has badly impacted the fishing industry as many people involved in fishing have been relocated too far from the coastal area where they had initially developed industry-based relations (Jayasuriya et al. 2006).

According to the field study, housing scheme X is located approximately 5 km distance from the coastal border which creates enormous problems in maintaining working relationships amongst people involved in fishing. For example, in a focus group discussion had with fishermen, they mentioned that they were largely deprived of short-term or casual job opportunities which are available at fishing harbours to fill within a shorter period. Many of those per-day vacancies seemed to be filled immediately within the proximate communities. Therefore, the likelihood of getting involved in temporary job opportunities is remarkably less for those living a few miles away from the harbours. Many of those casual vacancies come up due to the absence of a member or more in a permanent group or crew who are involved in long-range fishing in the deep sea for several days or months. Hence, those casual jobs sometimes let men earn a considerable amount of money depending on the number of dates they can engage in. Similarly, relocates of scheme X lost swift access to other short-term and daily-basis wage works in the fishing industry that is part of their livelihood even before the relocation had taken place.

Livelihood activities of women who have been involved either in the fishing industry or assisting fishing-related activities of their male counterparts have been badly damaged due to the distance between the new housing scheme and the fishing harbours just like many men. Many have become economically weaker and dependent. As women are largely deprived of the livelihood opportunities that they were engaged in in the original locations, a new set of income generation activities other than the traditional income ventures have been merged with the livelihood activities in the relocation setting. Amongst the most prevalent activities are sewing, packing, and selling spices. Almost all activities are domestic level and small scale. Apparently, such livelihood activities are neither well-established by connecting with a stable local market nor sustainable in terms of accessing the materials required, expansion of the market range, and developing stable buyers or a customer base.

Due to the uncertainty and inadequacy of their income ventures in terms of sustaining domestic level income activities, many women are economically vulnerable, especially in female-headed households. As a result of this instability, a large number of females have been trapped in loans provided by microfinance companies.

As revealed by research including government assessments, microfinance companies impose an unacceptable interest rate and high pressure to pay back loans. For example, some women entrepreneurs have taken several loans to recover the first loan issued by the same company whilst having no proper income. Eventually, this unforeseen loan process results in a large financial trap, and women who obtained loans without proper payback ability become highly vulnerable in terms of their social and economic resilience. Another notable fact regarding income activities amongst the relocates was the disruption of their livelihood network which they had largely developed based on fishing communities and other small vendors in their original settlements.

21.4.5 Poor Health, Food Security, and Social Disintegration

Food security is proportionately a cascading effect of joblessness. In this context, no one has lands to cultivate or develop any other livelihood. Therefore, food insecurity has a detrimental impact on the loss of livelihoods due to the damaged livelihood network in the area. As Sri Lanka has a free health policy, everyone of this group has the same, free access to health services. However, free access to health services is no longer a determinant of the health of an individual as affordability and susceptibility are largely underpinning health conditions. Therefore, poor health amongst the relocates is also a cumulative effect of joblessness and little of the loss of access to common resources. Social integration was a somewhat diluted discussion which has been triggered to understand disintegration between relocates in early resettlements and relocations where people were picked from several continents. People were selected for the relocation schemes from several temporary camps after the tsunami where people lived in proximate areas for a prolonged time. Hence, severe damage to the socio-cultural network happened as many of them belong to the same subcultural groups. Socio-economic homogeneity is the cause for the lack of heavy social disintegration between relocates.

21.5 Discussion and Conclusion

In terms of the policy background of planned relocation, globally, the phenomenon is governed by safeguard policies such as those of the World Bank (2004), Asian Development Bank (2009), and International Finance Corporation (2012). All these safeguard policies emphasize the need of considering relocation as the last resort, and if the ultimatum is to consider relocation, that the said should be planned with the participation of the relocates. Further, these policies claim that the relocates should be adequately compensated, fully informed of the relocation process, and economically empowered. The year 2001 was a turning point for the field of relocation in Sri Lanka with the introduction of the National Involuntary Resettlement Policy. The fundamental principles of the National Involuntary Resettlement Policy are that affected persons should be fully and promptly compensated, and if the persons affected have lost land, replacement land should be offered, and in the absence of replacement land, cash compensation should be offered (Gunatileke and Nathaniel 2014). Further, the existing safeguard policies of the World Bank and the Asian Development Bank urge to treat involuntary resettlement as a last resort, to fully compensate the relocates, and to make sure that the relocates are involved in the relocation process (Asian Development Bank 2014; and Road Development Authority 2016).

However, the findings of this study suggest that the resettlement project has paid inadequate attention to national and global safeguard policies such as the National Involuntary Resettlement Policy and policies of aid agencies and donors. Despite the existing policy framework requiring impact assessments, the resettlement programme lacks prior social planning and remains unsatisfactory as the social issues of the affected people have not been adequately addressed.

The mismatch between the policy and execution of the relocation project is largely visible due to the disintegration between the following three key elements of policy implementation (Fig. 21.2). In Sri Lanka, the interconnection between the research, policy or law-making, and implementation is largely lacking and is caused by ineffectiveness in the implementation of social development programmes.

Fernando et al. (2020a), Fernando et al. (2020b), and Senanayake et al. (2022) have emphasized the importance of empowering the livelihoods of the relocates in the Sri Lankan context. As suggested by some scholars, a livelihood network is a strong component of the social capital amongst communities which extends as a means of coping when an unexpected social or economic consequence has occurred (Amaratunga et al. 2022). Unfortunately, it has been pointed out that livelihood empowerment has not been considered a vital integral component of the relocation programme.

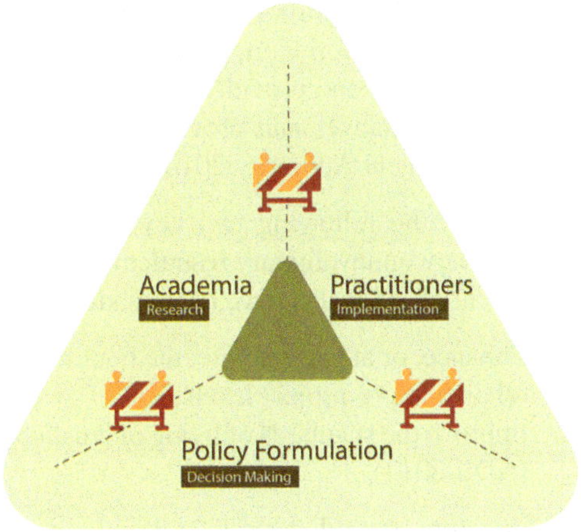

Fig. 21.2 Triangular model of policy and implementation. *Source* Author created

In the wider spectrum, the relocates have also lost their social capital which has been highlighted by a significant amount of research in the contexts of resettlement and relocation in Sri Lanka (Fernando et al. 2020a, 2020b). The development of social cohesion is vital to rebuilding the displaced post-relocation (Jayakody et al. 2022). Hence, given that the community is a homogeneous community, and this could be considered a strong point for rebuilding the lives of the relocates. This programme further lacks a proper assessment and monitoring process which is highlighted below in detail. This contradicts what Scudder (2005) has pointed out considering relocation as a long-term process which requires adequate monitoring and handing over mechanisms. The lack of a strong legal framework to address matters of relocation in the Sri Lankan context is another reason that makes relocation projects ineffective with no strong assessment mechanisms (Senanayake 2019). Commonly, poor prior planning and assessments have largely neglected the absolute need for relocation which is the key policy of the World Bank and Asian Development Bank in terms of involuntary resettlement. Those key points could be summarized as follows:

Avoid where feasible or minimize involuntary resettlement, exploring all viable alternative project designs (World Bank safeguard policy OP/BP 4.12);

Avoid involuntary resettlement wherever possible (ADB's Safeguard Policy Statement of 2009) (Road Development Authority 2016 and ADB 2014).

Overall, the study insists that key concerns and involuntary resettlement policies of several actors and donors have been largely neglected in this relocation programme. For example, inadequate attention to the following provisions of involuntary resettlement policies of the World Bank safeguard policy OP/BP 4.12 that is dedicated to involuntary resettlement has been largely identified through the findings:

1. Resettlement planning has the objective of providing displaced persons with a standard of living equal to, if not better than, their pre-displacement levels or to levels prevailing prior to the beginning of project implementation, whichever is higher;
2. The required measures to address the impacts resulting from involuntary taking of land a Relocation Policy Framework is required to ensure that the displaced persons are informed about their options and rights pertaining to resettlement, consulted on and provided with technically and economically feasible resettlement alternatives and provided compensation at full replacement cost. (Road Development Authority 2016).

Further, the following two key concerns of Asian Development Bank's safeguard policy on involuntary resettlement that is elaborated in their Safeguard Policy Statement of 2009 also have been inadequately addressed.

1. Enhance, or at least restore, the livelihoods of all displaced persons in real terms relative to pre-project levels; and
2. Improve the standards of living of the displaced poor and other vulnerable groups (ADB 2014).

Therefore, the relocates face double exposure to the impoverishment risks due to poorly understood social components, lack of prior planning, and missing links

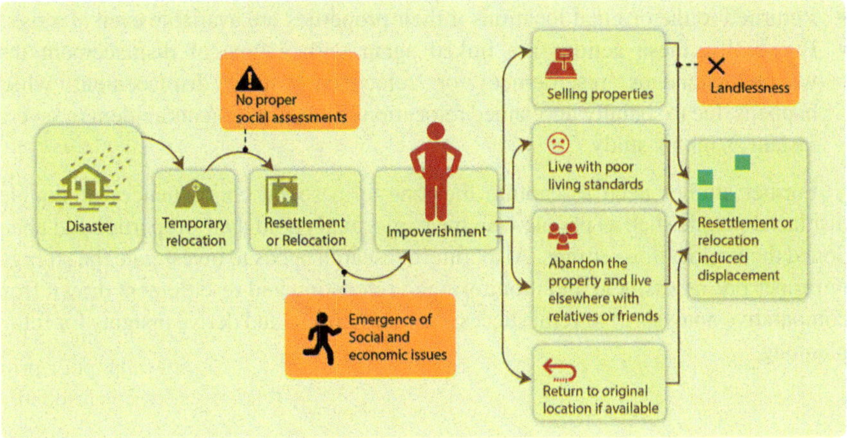

Fig. 21.3 Diagram on resettlement-induced displacement. *Source* Author created

between key elements of the relocation process which are highlighted in the following diagram (Fig. 21.3).

Figure 21.3 presents a diagram which elaborates how relocates become displaced again due to poorly planned relocation or resettlement programmes. Key elements in the diagram have been identified through the field study component. Accordingly, after a disaster, displaced people were kept in temporary settings such as camps and huts. Then, they are relocated or resettled in relocation or resettlement schemes on a permanent basis with the facilitation of government or non-government agencies. However, a serious gap in the process is the absence or lack of social assessments that have to be done prior to the resettlement or relocation, which should be the blueprint to decide the requirements of relocates in the new setting or rapidly conducted assessments to meet the donors' assessment requirement. This assessment process should include social, economic, and environmental feasibility, impact assessments that cover short-term and long-term social and economic impacts, and social management plans which are lacking in many relocation projects in Sri Lanka.

Ignorance of proper assessment causes an array of unforeseen social and economic issues amongst relocates. Many socio-economic grievances of people cannot be rectified due to the absence of project monitoring and evaluations, which is a fact derived from the study. These social issues are largely connected to impoverishment which is the main discussion that the paper intended to accomplish using the IRR model presented by Cernea (2000). As per the results of the study, impoverishment is caused by a complex of short-term and long-term socio-economic consequences that are interconnected and cannot be logically separated. Due to the range of cascading social impacts, many relocates have followed one of the following actions:

- Sold their property informally which caused landlessness again;
- Decided to live with poor living standards that increase the impoverishment risk;
- Abandoned the property and live with their relatives or friends;

- Returned to the original locations if their properties are available even at a risk.
- Eventually, these actions are linked again with a type of displacement that we understand as 'resettlement' or 'relocation-induced displacement' which happens due to socially unplanned relocations and resettlements evident in several instances in this study.

Further studies need to capture the long-term social consequences of double-displaced people such as people who have informally sold their properties and abandoned their relocation settings. Also, future research needs to investigate the strength of proper pre-assessment and monitoring in relocation and resettlement drawn from comparative analysis of several successful programmes and derive insights for future planning.

References

Adriansen HK (2009) Land reclamation in Egypt: a study of life in the new lands. Geoforum 40:664–674

Amaratunga D, Haigh R, Jayakody C, Malalgoda C (2020) Rebuilding after forced displacement. Built Environ J RICS J. (online). Available at: https://ww3.rics.org/uk/en/journals/built-environment-journal/rebuilding-after-forced-displacement-.html

Amaratunga D, Malalgoda C, Haigh R, Jayakody C, Senanayake A, Fernando N, Hamza M, Liyanage C, Lill I, Witt E (2022) A built environment perspective on post-disaster and conflict-induced displacement. A report of the REGARD project: rebuilding after displacement. ISBN: 978-1-86218-208-0

Asian Development Bank (2009) Safeguard policy statement. Manila

Asian Development Bank (2014) REG: mainstreaming land acquisition and resettlement safeguards in the central and West Asia Region. Asian Development Bank

Atukorala K, Pinnawala SK, Kamalarathne T, Seelagama P (eds) (2010) The plantation sector in transition research on the upcountry plantation sector in Sri Lanka, Department of Sociology, University of Peradeniya

Bang N, Fewa R (2012) Social risks and challenges in post-disaster resettlement: the case of Lake Nyos, Cameroon. J Risk Res 15:1141–1157

Badri SA, Asgary A, Eftekhari AR, Levy J (2006) Post-disaster resettlement, development and change: a case study of the 1990 Manjil earthquake in Iran. Disasters 30(4):451–468. https://doi.org/10.1111/j.0361-3666.2006.00332

Bradshaw C, Atkinson S, Doody O (2017) Employing a qualitative description approach in health care research. Glob Qual Nurs Res 4:2333393617742282. https://doi.org/10.1177/2333393617742282

Brenchin SR, West P, Harmon D, Kutay K (1991) Resident peoples and restricted areas: a framework for inquiry. In: West P, Brenchin SR (eds) Resident peoples and national parks, University of Arizona Press, Arizona, Tucson

Caelli K, Ray L, Mill J (2003) 'Clear as Mud': toward greater clarity in generic qualitative research. Int J Qual Methods 2(2):1–13. https://doi.org/10.1177/160940690300200201

Cao Y, Hwang S-S, Xi J (2012) Project-induced displacement, secondary stressors, and health. Soc Sci Med 74(7):1130–1138. https://doi.org/10.1016/j.socscimed

Cernea M (1991) The economics of involuntary resettlement: questions and challenges, World Bank, Washington

Cernea M (1997) The Risk and reconstruction model for resettling displaced populations. World Dev 25(10):1569–1587

Cernea M (2000) Risks, safeguard and reconstruction: a model for population displacement and resettlement. Econ Pol Wkly 35(41):3659–3678

CRED (2022) 2021 Disasters in numbers. CRED, Brussels

CRED, UNDRR (2021) 2020: the non COVID year in disasters: global trends and perspectives. CRED, Brussels

De Silva D, Yamao M (2007) Effects of the tsunami on fisheries and coastal livelihood: a case study of tsunami-ravaged southern Sri Lanka. Disasters 31:386–404

Dhakal N, Nelson K, Smith D (2011) Resident well-being in conservation resettlement: the case of Padampur in the Royal Chitwan National Park, Nepal. Soc Nat Resour 24:597–615

Doocy S, Gorokhovich Y, Burnham G, Balk D, Robinson C (2007) Tsunami mortality estimates and vulnerability mapping in Aceh, Indonesia. Am J Publ Health 97(S1):S146–S151

Fernando N (2012) Forced relocation after the Indian ocean Tsunami 2004: case study of vulnerable populations in three relocation settlements in Galle, Sri Lanka. UNU-EHS Graduate research series

Fernando N (2018) Lessons learnt from long term impact of 2004 tsunami relocation: a case study of selected relocation settlements in Akmeemana Divisional Secretary Division in Galle District, Sri Lanka

Fernando P, Fernando K, Kumarasiri M (2009) Forced to move: involuntary displacement and resettlement, policy and practice. Centre for Poverty Analysis

Fernando N, Punchihewa AG (2013) Relocating the displaced strategies for sustainable relocation, Friedrich Ebert Stiftung, Colombo

Fernando N, Senanayake A, Amaratunga D, Haigh R, Malalgoda C, Jayakody RRJC (2020a) Disaster, displacement and relocation: an analysis of the needs and policy implications on a displaced community in Sri Lanka. In: 10th International conference on structural engineering and construction management: special session on disaster risk reduction. University of Peradeniya, pp 128–145

Fernando N, Senanayake A, Amaratunga D, Haigh R, Malalgoda C, Jayakody RRJC (2020b) Impact of the disaster-induced relocation process on the displaced communities in Kegalle district, Sri Lanka. Euro Sci J Spec Edn. ESJ ISSN: 1857-7881 (Print) e-ISSN 1857-7431 (Contemporary Sri Lankan Society and Politics: Felicitation volume in honour of renowned Sri Lankan Sociologist Professor ST Hettige)

Gizachew A (2015) Refining the impoverishment risks and reconstruction (IRR) model: a study of the model's "overlooked" risks, evidences from the impacts of Tekeze Dam, North East Ethiopia. J Dev Agric Econ 9(4):66–79

Gunatileke G, Nathaniel V (2014) The national involuntary resettlement policy: dispelling misconceptions and assessing compliance. Verite Research, Sri Lanka

Internal Displacement Monitoring Centre and Norwegian Refugee Council (2021) Grid 2021: internal displacement in a changing climate. IDMC and NRC

International Federation of Red Cross and Red Crescent Societies (2000) Introduction to disaster preparedness: disaster preparedness training programme

International Finance Corporation (IFC) (2012) Performance standard 5—land acquisition and involuntary resettlement. In: Update of IFC's policy and performance standards on environmental and social sustainability, and access to information policy. Washington, DC, pp 58–65

International Labour Organization (2002) ILO generic crisis response modules. International Labour Organization

International Strategy for Disaster Reduction (UNISDR) (2015) Sendai framework for disaster risk reduction. United Nations

Jacka T (2008) The implications of aspirations: reconsidering resettlement in Laos. Crit Asian Stud 40:499–530

Jayakody C, Malalgoda C, Amaratunga D, Haigh R, Liyanage C, Witt E, Hamza M, Fernando N (2022) Approaches to strengthen the social cohesion between displaced and host communities. Sustainability 14(6):3413 (Online). https://doi.org/10.3390/su14063413

Jayasuriya S, Steele P, Weerakoon D (2006) Post-Tsunami recovery: issues and challenges in Sri Lanka. ABD Institute

Karunasena G, Rameezdeen R (2010) Post-disaster housing reconstruction: Comparative study of donor vs owner-driven approaches. Int J Disaster Resilience Built Environ 1(2):173–191

Kulischer EM, Jaffe AJ (1962) Notes on the Population Theory of Eugene M. Kulischer. Milbank Memorial Fund Q 40(2):187–206

Lakshman RWD, Amirthalingam K (2009) Displacement and livelihoods: a case study from Sri Lanka. In: Fernando P, Fernando K, Kumarasiri M (eds) Forced to move: involuntary displacement and resettlement—policy and practice. Centre for Poverty Analysis, Colombo, pp 57–82

Leavy P (2017) Research design: quantitative, qualitative, mixed methods, arts-based, and community-based participatory research approaches. Guilford Press, New York

Miltenburg EM, van de Werfhorst HG, Musterd S, Tieskens K (2018) Consequences of forced residential relocation: early impacts of urban renewal strategies on forced relocatees' housing opportunities and socioeconomic outcomes. Hous Policy Debate 28(4):609–634. https://doi.org/10.1080/10511482.2018.1424722

Oo BL, Sunindijo R, Lestari F (2018) Users' long-term satisfaction with post-disaster permanent housing programs: a conceptual model. Int J Innov Technol 9:28–32

Quetulio-Navarra M, Niehof A, Van der Horst H, Van der Vaart W (2014) Short-term risk experience of involuntary resettled households in the Philippines and Indonesia. Habitat Int 41:165–175

Road Development Authority (2016) Resettlement policy framework for transport connectivity and asset management project (TCAMP). Ministry of Higher Education and Highways

Satiroglu I, Choi N (2015) Development-induced displacement and resettlement: new perspectives on persisting problems. Routledge, Oxford

Senanayake A, Fernando N, Wasana M, Amaratunga D, Haigh R, Malalgoda C, Jayakody C (2022) Landslide induced displacement and relocation options: a case study of owner driven settings in Sri Lanka. Sustainability 14(3):1906 (online). https://doi.org/10.3390/su14031906

Senanayake A (2019) A review of existing laws and policies applicable to disaster induced resettlement in Sri Lanka. In: Fernando N, Siriwardana C (eds) Selected papers from the international conference on capacity building for research and innovation in disaster resilience. University of Colombo, Sri Lanka

Silva KT, Athukorala K (1991) The watta-dwellers: a sociological study of selected urban low-income communities in Sri Lanka, University Press of America. https://books.google.co.uk/books?id=8m3aAAAAMAAJ

Sridarran P, Keraminiyage K, Amaratunga D (2016) Consequences of involuntary relocations that affect the process of recovery: a literature review

Terminski B (2013) Development-induced displacement and resettlement: theoretical frameworks and current challenges. Development 10:101

Turton (2006) Who is the force migrant? In: Fernando N (ed) Forced relocation after the Indian ocean Tsunami 20014: case study of vulnerable populations in three relocation settlements in Galle, Sri Lanka. UNU-EHS Graduate research series

UNHCR (2014) Planned relocation, disasters and climate change: consolidating good practices and preparing for the future: report, Sanremo, Italy

United Nations (1999) Guiding principles on internal displacement, Office for the coordination of humanitarian affairs, New York

Vanclay F (2017) Project-induced displacement and resettlement: from impoverishment risks to an opportunity for development? Impact Assess Proj Appraisal 35(1):3–21. https://doi.org/10.1080/14615517.2017.1278671

Vivoda V, Owen J, Kemp D (2017) Applying the impoverishment risks and reconstruction (IRR) model to involuntary resettlement in the global mining sector. Centre for Social Responsibility in Mining, Australia

WFP (2005) Sri Lanka Tsunami WFP emergency needs assessment report: mission to Sri Lanka: 7–28 Jan 2005. WFP

World Bank (2004) Involuntary resettlement sourcebook: planning and implementation in development projects, World Bank, Washington

Yonetani M (2018) Mapping the baseline: to what extent are displacement and other forms of human mobility integrated in national and regional disaster risk reduction strategies? Platform Disaster Displacement

Chapter 22
Migrants and Resettlement: Mobilising Co-existence Through Social Cohesion

Shalini Attri

Abstract There have been debates on the vulnerability of human settlement due to environmental hazards, conflict zones and man-made disasters raising questions on the interface between the host and the displaced community categorised by cultural attitudes, norms and social structure. Social cohesion becomes an important strategy for rebuilding and resettlement fostering good relations and peaceful synchronicity. The paper has discussed government resettlement policies for migrants in context of North India, dimensions of social cohesion, co-existence of the diverse group based on collective efficacy and integration. The study is exploratory and observation based on displaced community due to environment, manmade disasters and pandemic in select North Indian states, and it has taken into account the existing academic and policy literature, reports, news, articles and documents from UN, ILO, *Nitiayog*. The research, while primarily focusing on internal displacement in North India, concluded that these shifts in the place and space caused fragmented identity, non-acceptance, non-assimilation, employment problematics and loss of basic rights. The need for social cohesion, relooking into migrant polices, mobilising international and national community towards co-existence through government and non-governmental organisations with action-based inclusive strategies, legal framework, social protection, employment opportunities and local integration thus can contribute to migrant's resettlement.

Keywords Migration · Environment · Policies · Social cohesion

22.1 Introduction

Migration is a pervasive global phenomenon. Several factors shape the decisions to migrate. The migrants share common heritage and ties with their homeland resulting

S. Attri (✉)
Department of English, BPS Women's University, Sonipat, Haryana, India
e-mail: shalini@bpswomenuniversity.ac.in

© The Author(s), under exclusive license to Springer Nature Switzerland AG 2023
M. Hamza et al. (eds.), *Rebuilding Communities After Displacement*,
https://doi.org/10.1007/978-3-031-21414-1_22

in diasporic identity, whereas migration as a concept is movement of people for settlement voluntarily or involuntarily. The Greek meaning defines diaspora as 'dispersion' (OED). Diasporas, the voluntary or forcible movement of people from their homelands into new regions, is a central historical fact (Ashcroft et al. 2013). Diasporic communities created by forced or unforced migration may all be affected by this process of dislocation and further regeneration as argued by critics of the diasporic and migrant discourses. Rapport and Dawson (1998) remark that 'home' for migrants is conceptualised in fluid terms as being neither here nor there …rather, itself, a hybrid, it is both here and there-an amalgam, a pastiche, a performance (Bhatt 2009, p. 87). The geopolitical experience of dislocated population explores identity shift closely associated with belongingness, inclusion and exclusion. On one hand, the migrants are seen as economic mediators while on the other they face economic scarcity aided with political and social discrimination. Critics like de Haas believe that, "it would be unrealistic to represent migrants as victims desperately fleeing situations of destitution, oppression and human misery as it would be to depict them as entirely rational and free actors who constantly make rational cost–benefit calculations" (de Haas 2021).

22.2 Migration and Environment

The effect of environment on human exodus has been illustrated through the Anthropocene that causes movement of humans. Some of the main reasons for migration are: (a) political instability and conflict, lack of economic opportunities, and lack of access to resources ('push' factors); (b) factors related to the region or country of destination, including the availability of employment and demand for workers, higher wages, political stability or access to resources ('pull' factors); (c) intervening factors that facilitate or restrict migration, including transportation, family or social networks, government immigration or emigration policies, economic ties such as trade and investment linkages, or social and cultural exchanges (Laczko and Aghazarm 2009, p. 48). Migration ranges from seasonal to temporary and permanent moves and can be local, national or international. Various adaptations to climate, adaptive capacity and social vulnerability are required for sustainable development and risk management.

Environmental change and disasters besides, other reasons like war, pandemic have always been major drivers of migration. It is predicted that the climate change in the twenty-first century will affect greater number of people likely to migrate because of weather-related events like floods, droughts, storms, etc. becoming recurrent and intense causing change in temperature patterns further impacting livelihoods and human safety. "Disasters triggered more than three quarters of new displacement recorded worldwide in 2020 and more than 98% were the result of weather-related hazards such as storms and floods" (IDMC GRID 2021). The climate change leading to environmental disaster affects the population at three levels: (a) decrease in agricultural production; (b) upsurging of weather events like drought, melting glaciers, cloud bursts; and (c) sea level rise destroying the life of people staying along the seaside, forcing dislodgement.

Many studies divulge that the "environmental migrants" exceeded the documented refugees which resulted from war and political persecution put together. "The United Nations High Commissioner for Refugees estimates that by 2030, there will be 30 million environmental refugees" (Ansohn and Pleskovic 2011, p. 23). Professor Myers similarly argues, "there could be as many as 200 million people overtaken by disruptions of monsoon systems and other rainfall regimes, by droughts of unprecedented severity and duration, and by sea-level rise and coastal flooding". Because of climate disasters, India will have approximately 45 million people compulsorily migrating from their native place by 2050, three times more than the present figures (Yashwant and Faleiro 2021, p. 2). Seven million people were internally displaced by disasters across 104 countries and territories as on 31 December 2020, whereas Afghanistan, India and Pakistan comprise the largest displacement. The economic impact of internal displacement was $20 billion in 2020. India saw 3.9 million displacements in 2020 and South Asia alone saw 9.2 million of disaster displacement in 2020 (IDMC GRID 2021). The Global Climate Risk Index 2021 places India among the top 10 countries affected by climate change. "India lost 87 billion dollars in 2020 due to natural disasters such as cyclones, floods and droughts, the World Meteorological Organisation (WMO) said in a report" (Nandi 2021). There will thus be a vertical increase in number of environmental refugees/migrants.

International Organisation for Migration (IOM) defines "environmental migrants as persons or groups of persons who, for compelling reasons of sudden or progressive change in the environment that adversely affects their lives or living conditions, are obliged to leave their habitual homes, or choose to do so, either temporarily or permanently, and move either within their country or abroad" (Laczko and Aghazarm 2009, p. 19). The migrants involuntarily move from rural to urban areas as a consequence of an environmental disaster causing destruction of their homes and lands. They often seek shelter in macro-cities because of the variety of opportunities available. These migrants often become non-existent due to under-representation, lack of residency rights or social entitlements and are placed into the category of illegal immigrants. According to 2011 Census of India, there were 450 million internal migrants in India by 'place of last residence' metric and based on this trend 600 million internal migrants are estimated in 2020 (Registrar General of India 2011). Surojit Gupta in his research states that "among this one-third are inter-state and inter-district migrants which accounts to 200 million, and within this 200 million, two-thirds are estimated to be migrant workers that roughly constitute 140 million" (Irudaya et al. 2020).

Looking at the large data on migrants in India, the question of climate induced internal displacement becomes a prominent point of discussion. The frequent natural calamities in North India including Kedarnath Tragedy following the catastrophic cloudburst in Uttarakhand in 2013, the Chamoli tragedy of February 2021, and new developmental projects taking place in Himachal Pradesh and Uttarakhand raise the question of resettlement of the migrants and the notion of internal displacement. No doubt, states like Orissa and Uttarakhand have been creating attempts to adjust and integrate the migrants, but ecological experts consider these measures as insufficient because of catastrophe at macro-level. Seasonal migration is another prevalent problem existing in North Indian states like Uttar Pradesh, Madhya Pradesh and Rajasthan as reported by researchers of International Institute of Environment and Development. Another major cause of increasing internal displacement is due to pandemic that led to reverse migration in North India as New Delhi being the center for displacement of labour migrants.

Miles DePaul, a political scientist suggests "a normative shift on the issue of climate-induced migration to ensure that those victimized by anthropogenic and natural climate change are met with a compassionate, coordinated global regime rather than strict national immigration policies" (Sharalaya 2018). Environmental hazards describe the spectrum of climate's relationship with human mobility and the circumstances of "trapped migrant population" for whom migration becomes the only possibility. Organisational flaws and improper planning additionally aggravate the misery of migrants.

The United Nations Guiding Principles on Internal Displacement explains internally displaced persons as those who "are forced to flee elsewhere within a state as a result of, or in order to avoid the effects of armed conflict, circumstances of generalized violence, violations of human rights or natural or man-made disasters" (UNHCR no date, p. 1). This causes: (a). fragmented identity, (b). loss of basic rights, (c). non-acceptance and non-assimilation and d). lack of employment.

The select North Indian states witnessed internal displacement due to climatic hazards and pandemic. The research was undertaken to understand the basis for uprootedness, displacement and the need for social cohesion that could create a constructive society. In this framework, the present research paper has analysed and addressed the problems of internally dislocated migrants by identifying and examining the levels of displacement in select North Indian states due to man-made/ environmental disasters/ pandemic. It has evaluated the government resettlement schemes, policies for migrants in India. The paper argued on the constructive role of integration through social cohesion of the migrants and the receiving community aiming at their inclusion economically, socially and politically.

22.3 Methodology

The research was based on an investigation of secondary data taken from the published sources and information available in libraries and adopted qualitative tools such as internet-mediated research and analysis. It was literature review based and a case study of internally displaced migrants in select states of North India. The information collection from extensive data sources included: national and subnational government authorities, civil society organisations, news media, newspapers, specific topic-related articles, literature, reports, published statistics from research organisations, existing academic and policy literature, and documents from UN, ILO, *Nitiayog*. The quantitative data was taken from World Bank, censusindia.gov.in,

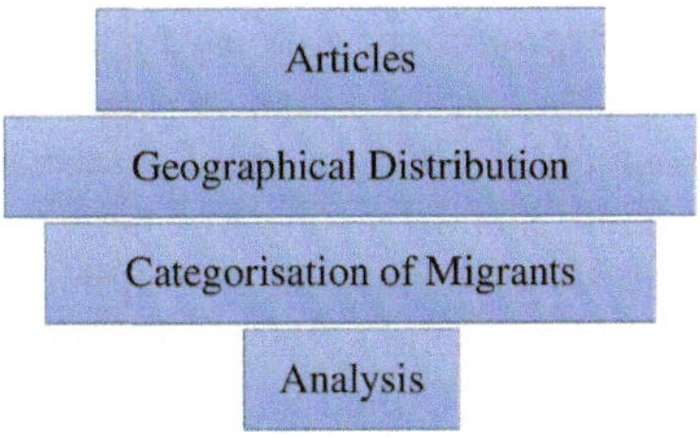

Fig. 22.1 Process of conducting research

international displacement monitoring centre.org, International Labour Organisation. The classification of migrants has taken into account the geographical division of India. The migrants taken up for the study were internally displaced community and were categorised with respect to North India including New Delhi, Uttarakhand and Himachal Pradesh. The group selected for study comprised of migrant labourers, villagers and residents displaced by environmental disaster or pandemic.

The results were based on the descriptive investigation of data, and the measurement was done by gathering, analysing and presenting the collected data. The research followed and adopted theoretical, deductive and prescriptive effort to define integration and social cohesion further exploring the connections that discussed the impact of migration and internal displacement. The emphasis was on issues faced by internally displaced persons especially the identity, policy, exclusion, voting right, mobility, integration and cohesion. The question addressed in the research was how macro-structural factors shaped micro-factors affecting social construct of the migrants and the reversing of subaltern status of the migrants through integration and social cohesion. The research was exploratory, observation based and document analysis grounded on subjective methodologies and decoding of the actions. The process of conducting research is as follows (Fig. 22.1).

22.4 Social Cohesion and Migration

It was significant to understand the emergence of migration as a glocal feature and the necessity of social cohesion for a sustainable development. The questions often raised in debates and discussions are: How can we mobilise people through social cohesion? Or how can policymakers integrate principles of social cohesion? India's geopolitical, economic and cultural construct make it mandatory to lay emphasis on the problems and issues of migrants. Migration, a continuous phenomenon, is required to be seen from the lens of social and economic inclusion. There were many acts and policies for the migrants in India, but as it is stated in many researches that by 2050, India will face large-scale displacements due to climatic disasters, it became a key point of discourse for understanding of India's existing scheme and its long-term plan for rehabilitation, reconstruction efforts and resettlement consideration for migrants. This can be achieved through integration and cohesion.

Integration is a two-way process, a kind of co-existence between migrants and host societies that makes a country or a nation cohesive. An interconnected society mobilises stronger citizenship with a focus on responsive governance. Such society stands in disagreement to polarisation thus creating resilience. A cohesive society works in two dimensions: vertical, i.e. relation between government and society, and horizontal, i.e. relation among people. Sustainable social cohesion according to United Nations norms focuses on human rights, dignity of person, human security, collective development and opportunities for individuals. Thus, trust, inclusion and shared identity become its focal point. Migration, an important enabler for sustainable development, has its dependency on cohesive society. The debates on migration and migration theory based on the idea of role played by migrants in the formation of nation states, political shifts, social structures, economic gains are quite prevalent in eastern and western discourse. The historian of migration studies, Nancy Foner opines about "the changing continuities in the migration patterns" (Brettell and Hollifield 2008, p. 44). These changing continuities exist in every civilisation and cause cultural change. In fact, German Jewish migrant anthropologist Franz Boas and his students shared strong interests in describing "migrants as culture-bearing agents engaged in the cultural transformation of the countries where they settled, rather than as marginalised people submitting to demands for their assimilation" (Brettell and Hollifield 2008, p. 44) thus giving a new dimension to migrants as active agents. Geertz's conceptualisation of global–local connection, distinction between culture and social meaning in local settings formed the basis of new productive approach to trans-localism. The social system and culture enable the host community to consider modes of societal integration and also to take into account discontinuity and change occurring in society. Therefore, migration also leads to cultural anxiety and social disjuncture where practices, beliefs and registers of social identity undergoes change contributing to a heterogenous society. Cultures and identities form the core component of social cohesion, and the local–global connection plays an important part in migrant cohesion.

Migration studies focused on the exclusion–inclusion process in twentieth century; however, Lowe (1996) opined that the immigrants are producers of culture (Brettell and Hollifield 2008, p. 48). Mixing of cultures caused assimilation and multiculturalism, a move towards social cohesion with the positioning of migrants in host country. Gordon (1964) stressed on the assimilation that results in pluralism and integration, and he gave three major alternatives: a melting pot tendency; an Anglo-conformity…on middle-class cultural patterns to the embracement of discredited theories of racial superiority; and a cultural pluralism emphasising the maintenance of native-country culture and institutions despite economic and local integration into the host country. Alba and Nee (2003) amended Gordon's account and proposed that in new assimilation theory, certain institutes, including those reinforced by civil rights law, play an important role in achieving assimilation (Brettell and Hollifield 2008, p. 75).

Another pointer of migration is that it causes many conflicts and fear among the natives against minority migrants because of economic competition which again becomes a vital point of contestation and the need of cohesive society to achieve sustainability. Thus, diversity, social resilience, creativity, development of interpersonal skills that results from coping up with differences, challenges and opportunities strengthen the bonds among diverse groups and is an answer to the fears existing among the hosts and the migrants. One major factor that enhances integration is the necessity of economic assimilation. Migrants are a social capital and economists examine the impact of migrants on the economies and labour markets where they move and to the areas that they leave behind. The migrants consequently have socioeconomic impact on the country, state or region where they settle.

Gonzalez (1961) gave a model of migratory wage labour and argued that migration would be "reflected in social organization in different ways depending on the nature of the sociocultural system affected as well as the type of migration itself" (Brettell and Hollifield 2008, p. 150). Further she identified "seasonal, temporary nonseasonal, recurrent, continuous, and permanent" as the five types of migration based on her research in the Caribbean region (Brettell and Hollifield 2008, p. 150). These models exist almost in every society and nation. Migration is formed and structured by glocal, i.e., global and international economies, and the interaction between the societies due to migration explains the connection between the family structure and the state, hence laying emphasis on how migration invents specified meanings in local cultural frameworks. Being ingrained characteristic of the economy, migration and development cannot be separated because of the growth and recognition of remittances on the native country and migrants' families provided by the migrants. The category of low-skilled migrant workers embodied as invisible to the economy forms the structures of the labour markets, states and countries further defining the public sphere and discourse. Therefore, migrant employment has been of importance for social cohesion and economic expansions.

The social setting alters an individual's identity that forms during the migratory state further causing a shift in hierarchies. Within the migrant spaces, immigrants engage in a host of community activities that becomes expression of their identity. The spatial patterns and the rapid latitudinal/longitudinal dispersion of immigrants within the host country elaborates the networks between their residential patterns, integration and changing identities. The individual and the state, social structure, cultural practices, institutional understanding and belongingness forms the process of inclusion and exclusion experienced by the migrants which is shaped by the host society. Dual process of subjectification and oppression, of production and submission places the displaced and dislocated at the boundaries. The mingling of migrants in the society becomes a vantage point of debate since they affect the making of a country, state or region.

The relationship between migration, social cohesion and diversity provides a mirror view of the 'host' country that stands for divergence but also [re]defines its existence through otherness. The process of assimilation happens when immigrants and natives are converging in some way with a frequency of constructive sociocultural structures. The regional hierarchies that constitute social entitlement and mobility, transnational affiliations, histories of multiculturalism and/or homogeneity are all integral to social cohesion. The idea of social cohesion is distinguished from normative and functionalist models entrenched in an essentialist understanding of what is called a 'a good society' or sustainable society. Hence, the challenge lies in making a society multi-cultural and diverse and that can be achieved through co-existence.

22.5 Cohesion and Co-existence

The research on social cohesion originated with the theory of Le Bon (1897) on collective behaviour; and Durkheim further defined social cohesion as interconnection between individuals focusing on the absence of hidden social conflict and the presence of strong social bonds. Freud (1921) formulated the definition of social cohesion on Le Bon's idea of the "unconscious identification of individuals, and illustrated that social cohesion identified individuals with similar characteristics, providing them strong emotional ties" (Fonseca et al. 2018, p. 3). Cohesion has an emotive feature that sustains the society. Beauvais defines it "as a continuous process taking place among similar groups having levels of agreement and collectiveness with a mechanism to resolve conflict" (Fonseca et al. 2018, p. 5). The individuals form diverse social identities by cataloguing with dominant groups or groups placed at periphery. This new social identity is again an indicator of cohesion.

The accepting of an influx of displaced group can lead to abrupt and extreme upsurge in the population which makes it difficult for the host country to become a service provider. These exogenous shocks can result in conflict (Mooney et al. 2021, p. 11). Many times, mixing of population can further increase the conflict through misinformation and hate speech on various platforms further amplifying tensions, divide and causing distress. This invites a dialogue on cohesion which becomes a critical tool to reduce tension while promoting a shared perspective. Cantle (2008) gave the community cohesion framework based on the four assumptions: "Migration is reality of life and diversity forms a part of migration; The problems are theorized in context of differences; Segregation has a strong negative impact on intercommunal relations because lack of communication between groups causes ignorance and illogical fears; Mutual understanding and a shared set of values emerges from interchange based on shared experiences" (Hickman et al. 2012, p. 36).

Being an important determinant of peaceful, democratic nation where multiple identity groups share geographic space, a cohesive society creates stronger bonds fostering trust in government. This is analogous to World Bank's explanation of five dimensions of social cohesion, i.e., "belonging, inclusion, participation, recognition, and legitimacy" (Berry and Roberts 2018, p. 4). Therefore "a cohesive society is one where all groups have a sense of being, participation recognition and legitimacy. Such societies are not homogeneous, and by respecting diversity, they harness the potential residing in their societal diversity" (United Nations 2020).

The intrinsic complexities and heterogeneities, plurality, cultural diversions have remained prominent in the contemporary society. Jacobs and Fincher remarked that "difference, unlike that of diversity, emphasize on the continuous power structures that unevenly shape lives and also complicates the picture by highlighting the ways in which such structures are shaped by the reliant circumstances of specific groups in specific settings" (Hickman et al. 2012, p. 4). On the contrary, multiculturalism as a signifier looks upon the differences and otherness as a constructive process. The reformulation of new definition and rethinking borders becomes a complex phenomenon. The integration of immigrants and their diasporic connections are neither incompatible nor binary opposites. Further, the assimilation in social relations makes us rethink on the spaces of possibility to understand the operative conditions of society. Ideas, practices, network of relationships and resources trace the dynamics of social cohesion within this social spaces. The intricacies of the socioeconomic, political and spatial factors make the concept of social cohesion to be examined differently from the way it is understood by policymakers, academicians, etc.

Social cohesion discusses two features of society: (a). 'Absence of latent conflict' and (b). 'Presence of strong social bonds' (Berry and Roberts 2018, p. 19). It is an enclosed system that identifies individual and group level phenomena, and their two-way effect and interface. In other words, it is about resolving conflict or absence of conflict. Gilchrist (2004) remarked that "the ability within the individuals and groups informing a social field to manage the inequalities, differentiations and tensions intervening within and between them in terms of what that they perceive as positive and successful" (Hickman et al. 2012, p. 13). Thus, a unified, interconnected society and nation mirrors stronger citizenship, responsive governance, resilience, and, also includes social capital, social contract, vertical and horizontal dimensions.

Social cohesion is influenced by structural factors, i.e. formal and informal dialogues, policymaking and public polices to address inequity (UNDP 2020, p. 37). The indicators of social cohesion include: (a). political indicators—government schemes, services, policies; (b). social indicators-exclusion and access to information; (c). economic indicators- development, living standards; (d). cultural indicators-prejudice, stereotyping, discrimination, inclusive activities, knowledge of other groups.

Community cohesion policies are intended to plug the gap existing in the policy for migrants; and this can be achieved by projecting shared experiences which produce attitudinal change and lessens the 'fear of difference'. For Harvey (1991), the late modern historical phase was characterised by reduction of the spatial distance, whereas Appadurai (2001) observes that this period also features increase in the international mobility of people, goods and cultural products (Hickman et al. 2012, p. 80). Self-representations and self-identities resulted from the multifaceted exchange between dominant power structures and their internalisation, intervention or refutation. This interplay characterised localised understanding of social cohesion, reproducing social categories and initiating or confining the capacity to solve conflicts in ways which are perceived constructively by the individuals and groups. Consequently, belongingness and fitting into the frame of society was influenced by socioeconomic transformations that impacts permanent residents' responses to the influx of immigrants. Social relations, narratives and experience of belonging exemplified social cohesion among the displaced. There was an explicit role of social interventions, the importance of public spaces and actions taken by the state for reselect. Migration as a discourse frame local/host culture and societal structures. The occurrence of localised discourses that theorised cohesion must be examined in context of macro-transformations encircling and influencing locality. Hickman, Mai and Crowley (2012) observe "the responsibility, direction, separateness in hampering/promoting of social cohesion and recognizing the historical role of migration in the emergence of localities" (p. 182).

The recognition of migrant's structural heterogeneity in society, pluralism, the need of social inclusion and effective interventions for the arrival of new groups can make them adapt to the environment. The paper suggested a background for promotion of resilient people, states, country; and it also offered the kind of interventions and restructuring of policies to be adopted to become a resilient nation. The application of appropriate frameworks on migration, internal displacement and resettlement plans when applied in Indian context exhibited India as a multicultural and multiethnic country where diversity, community dynamics, economic conditions extended horizontally as well as vertically within the groups, from people to leader and local to national level. The assessment of migration and their effect on future influx would contribute to sustainable development of India.

22.6 Existing Provisions for Migrant Resettlement in India

Before formulating the gaps present in migrant resettlements provisions in India, the existing provisions were conferred about. The foremost facility was mirrored in the foundational principles of free migration present in Article 15, 16 and 19 of the Constitution of India. Article 19(d) and 19(e) of the constitution of India guaranteed its citizens the freedom to move, work and reside in any part of the country. It has prohibited human trafficking, freedom from discrimination and equal opportunities (The Constitution of India 2019, pp. 25–27). Skill development and employment contributed to conflict deterrence and peace building, providing involved dialogues for social participation. Inculcating skills into the migrants by the Government led Skill India Initiative and the Pradhan Mantri Kaushal Vikas Yojana was an outcome-based skill training program for placement of skilled human resources. Similarly, rural housing schemes such as Prime Minister's Awas Yojana-Rural would help in mitigating pressures for push migration. Mahatma Gandhi National Rural Employment Guarantee Act (MGNREGA) was an Indian labour law and social security measure that safeguarded the 'right to work'. This national rural employment scheme ensured 100 days of work per year to every adult member of rural household. Lokhande and Gundimeda in their study concluded that 7.5 million seasonal migrant workers took refuge under MGNREGA during the lockdown (2021, p. 584). Likewise, Garib Kalyan Rozgar Scheme was launched in 2020 for migrant labourers.

The Interstate Migrant Workmen Act 1979 was amended in 2017 (Central Amendment Rules), and it was about employment, provisions, conditions of service and matters for interstate migrant workmen. Under this act, it became mandatory for the labour contractors to acquire a license, registering of the migrant workers with the government authorities, and arranging the issuance of passbook to authenticate the identity of the workers. The benefit of income, housing, health facilities, clothing, etc. were provided to the workers by the contractor (Kumar and Choudhury 2020, p. 3). National Food Security Act, 2013 similarly ensured food for two-thirds of India's population including migrants, transforming the 'welfare state' concept into 'rights-based' approach. *Sarva Siksha Abhyan* of Government of India (seasonal hostels) and Right to Education Act, 2009 stated that the migrant students can be a part of any school irrespective of where they reside. The act also provided space for child care services. Integrated child care development services for migrants and *Rastriya Swasthya Bima Yojana* (National Health Insurance Policy) provided benefits for health. Social cohesion is a possible adaptable determinant of health and wellbeing contributing to more sustainable development, whereas poor physical and mental health is deterrent to social cohesion.

There were other Indian laws for protecting the rights of the workers: Trade Union Act, 1926 safeguarded registration of trade unions; Payment of Wages (Amendment) Act, 2017 (Act No.01 of 2017) ensured that wages be paid in cash within the given time; Minimum Wages Act-1948 (Act No. 11 of 1948) offers fixing of lowest wage rates; Equal Remuneration Act, 1976 (Act No. 25 of 1976 amended Act No. 49 of 1987) suggested equality in payment to men and women; Unorganized Workers Social Security Act, 2008 (Act No. 33 of 2008) mandated social security to the unorganised workers (Chief Labour Commissioner no date). Similarly, Article 42 discussed about safe, unbiased and humane working conditions and maternity relief; Article 43 illustrated the right to a living wage and decent working settings with adequate standard of living, leisure and cultural opportunities for workers (The Constitution of India 2019, p. 35).

Although there were laws, acts and policy frameworks for the migrants especially for migrant labourers, there are still substantial gaps in the existing policies that do not entirely provide solutions to the resettlement problems and co-existence of the migrants.

Nitiayog's (Policy Commission) proposed draft on national migrant policy recommended an emphasis on institutional building and social cohesion. The movement of 10 million migrants (as per government estimates) from mega cities during the pandemic made *Nitiayog* formulates a draft national migrant labour policy. The policy envisaged migration as an essential component for growth of nation and the policies of government ought to work towards facilitating internal migration. It suggested that the Ministry of Labour and Employment should alter the 1979 Act for protection of migrants. The managing of migration resource centres in high migration zones, a national labour helpline, connecting of worker households to government schemes, and interstate migration management bodies for migrants recommended by the draft policy would manage the data as well as provided the migrants with the outreach for redressal of their problems. There was a need of constitution of a nodal agency for internal migration within Ministry of labour and employment which would ensure systematic gathering of data on migration influx, facilitating state coordination with a focus on food safety, assured health service, child care health and education. The establishment of national helpline for workers to share their grievances and violation of rights and inclusion of migrants in India's urban policy formulation is exigent.

Investing in education, high-impact skills and vocational training would certainly become a pathway for migrants and cohesion.

India's National Education Policy 2020 (NEP 2020) recognised migrants as socially and economically disadvantaged in their access to education. Lack of education has pushed them into exploitive works with no vertical socioeconomic mobility. Moreover, the inclusion of components of education in workers' rights and responsibilities for safeguarding of the interest of migrating youth necessitated their empowerment through information and knowledge. Similarly, legal remedy embedded in educational framework can become the powerful tools benefitting the migrants. The seasonal distress migration needs special attention where Panchayati Raj institution can play an effective role in providing decent work. Implementing of Sexual Harassment of Women at Workplace Act, 2013 is already providing safe space to migrant women. For children of migrant workers, interstate collaborations are needed for seamless adaptation and transition between the source and destination schools having instructional language continuity along with access to multilingual teachers. "The representation of the migrants can be through social dialogue and sensitizing of key stakeholders like political parties, media, government official and residents" (ILO 2020, pp. 26–37).

The contemplation expounded so far was on the concepts and theories of social cohesion, integration, and migration along with India's policies, acts and laws to safeguard the migrants. In the following data, select cases given are important to understand the internal displacement occurring because of environmental hazards or pandemic:

(a) **Displacement due to Environment**

India has abrupt and slow-onset natural and manmade environmental hazards like earthquakes, tsunamis, cyclones, storms, landslides and droughts that result in increasing risky high population exposure to social and economic vulnerability. According to Internal Displacement Monitoring Centre as on 31 December 2020, India has 3,856,000 new displacements due to such disasters and total number of IDPs due to disasters-929,318 (IDMC GRID 2021). The data regarding select North Indian states gives an overview of the catastrophe. The slope movement caused Chamoli (Uttarakhand) Tragedy in February 2021 that triggered landslides and the people of that region were displaced. In Himachal Pradesh, there were 146 displacements (destroyed housing) due to floods and rainfall; Uttar Pradesh had 3748 displacements according to disaster authorities due to heavy rainfall and flooding; Punjab had 24 displacements due to monsoon flooding (2021) (IDMC GRID 2021). Shifts in hydrological cycle, food security and water scarcity are the major cause of interstate migration. The 2018 report of Rural Development and Migration Commission points out that almost 70% of migration took place within the state Uttarakhand, further indicating the shifting of people from the state's high mountains to plains. The commission report revealed that between 2011 and 2017, 734 villages became entirely depopulated. Similarly, after 2011 the population in another 565 villages fell below 50% (Garg et al. 2021, p. 40). Kedarnath Tragedy (2013) in Uttarakhand occurred due to cloud burst causing flood which further affected 12–13 districts of the state culminating in large scale displacement. Many villages in Mandakani valley were wiped off, thousands were rendered homeless, and Guptakashi too became empty. Rajasthan in 2003 has 41,000 drought-affected villages forcing seasonal displacement, and in 2021, nearly 20 lakh hectares of Kharif crop was hit by flood and drought (The Times of India 2021).

(b) **Displacement due to Manmade Disaster**
Char Dham Project (2016) comprising of Gangotri, Yamnotri, Kedarnath and Badrinath in Uttarakhand was the reason for large-scale displacement making people homeless, and the compensation given according to residents and natives was meagre. Moreover, the zone was eco-sensitive. National Highway Authority of India's Kiratpur–Manali Highway (2013) in Himachal Pradesh displaced 10,000 people in the region. The project has led to the displacement of 350 families without giving them proper compensation. In Thalout, Aut, Dador and Panarsa villages, 90% of the market has been wiped out, and the natives and residents were compelled to abandon their homes and shops. Flood due to Tehri Dam water rise (2010) affected 1200 villages, and about 30,000 people from Haridwar, Nainital, Udham Singh Nagar and Almora districts of Uttarakhand became destitute after the collapse of their houses in landslides, cloudbursts and continuous rain (Kunwar and Tripathi 2010).

(c) **Displacement due to Other Reasons like Pandemic**
Millions of migrants have been forced to return home because of pandemic facing unemployment and poverty (ILO) at international level. India too saw reverse migration during pandemic catastrophe which became the main reason of the migrant labourers to seek refuge in their ancestral homes. According to a report, 807,032 migrant workers left Delhi during the first four weeks from 19 April to 14 May 2021, for their home states consequently causing reverse migration (The Hindu 2021).

22.7 Results

The few cases elaborated hitherto with reference to North India illustrated displacement triggered by environmental, man-made and pandemic hazards contributed to migration and were found to be the reason for internal dislocation. Displacement further resulted in fragmented identity, non-acceptance, non-assimilation, employment problematics and loss of basic rights. The migrants faced mental health crisis; discrimination, segregation, cultural denigration; loss of land and home. Language problems are faced by the children of migrants in schools, and they feel excluded. The components of social cohesion include three pillars, i.e., social capital, mobility and inclusion. OECD illustrates that "a cohesive society works towards the well-being of all its members, fights exclusion and marginalization, creates a sense of belonging, promotes trust, and offers its members the opportunity of upward mobility" (2011, p. 51). Thus, cohesion aims at assimilation of all, minimising disparities, avoiding marginalisation within groups enabling upward mobility. Inequality and hardship because of discrimination, intolerance and stereotypical thoughts can exclude certain sections of a people and drive them to the margins of society.

The 167-km Kiratpur–Manali, 90-km Parwanoo–Shimla four-lane project and other national highways approved for the Himachal Pradesh raised dissatisfaction among the affected families as remarked by a native and shop owner Bhagat Ram from Aut "This development is causing destruction and many are fighting their battle in the court. His family has lost 60 bighas of land, home and shop". In order to assert their representation people have thus formed "Four-Lane Sangharsh Samiti" (Chauhan no date). Right to Fair Compensation and Transparency in Land Acquisition, Rehabilitation and Re-settlement Act, 2013 states that the government should provide the affected people a market rate for their property acquired for a project or development work, along with solatium. However, Brig. Kushal Thakur heading the Four-Lane Sangharsh Samiti retorted that "the state government is determining land compensation rates that is still being calculated using an age-old method without considering the current value, productivity and commercial viability by the officials thus being unable to give displaced their due" (Manta no date). It is estimated that 40,000 families in the districts will lose their lands, homes and livelihoods because of the developmental work. Hence, actual data reflected was less, and compensation for resettlement was not in accordance with the Act.

Interstate Migrant Workmen Act, 1979 (Act No. 30 of 1979) covers only labourers migrating through contractor leaving out the independent migrants. Lack of registration under this act and non-maintenance of data by state labour department further demands framing of sound policies for interstate labour migration. Covid-19 pandemic triggered a large-scale movement of migrant labourer resulting in reverse migration of semi-skilled and unskilled labourers moving back from cities to villages. A total of 128 million workers' livelihood has been affected due to onslaught of pandemic. "A study and survey conducted on 11,000 migrant workers at varied location in India during pandemic discloses that 96% have not received rations, 70% didn't obtain any cooked food and 89% did not get any payment from their employers" (The Hindu Data Team 2020). Since the ration cards were issued under the Public Distribution System to the natives and not to the outsider migrant workers, the latter were, therefore, denied the benefits of this scheme. It was also found that distribution of ration was not enough and hard cash was required for milk, medicines, gas cylinders, etc. during severity of pandemic period which was the main reason for reverse migration. The migrant labourers during Covid-19 also made a complaint of discriminatory and apathetic attitude of the officials.

Many residents felt they became homeless because of the developmental projects. The impact of migration on women was more severe leading to "double marginalization" (the term borrowed from Gayatri Spivak). The MGNREGA in its existing form offered 100 days employment with no solution for other 250 days. It became the reason of conflict between the registered labourers of the villages and the migrants that further caused complications for the latter during the pandemic. There were reports of psychosocial problems like anxiety, fear and psychotic disorders. The cost of living being very high in the big cities, the migrants were willing to go back to their homes as they had no source of income which again became the major force for migrants to migrate back. The seasonal migrants are not permanent in host cities; therefore, they are unable to transfer their constituency and participate politically consequently becoming voiceless and devoid of representation. *Nitiayog* aptly pointed out that the state labour departments were less engaged with migration issues and emphasise on lack of administrative will to handle the exploitation of the migrants. "The local administration, given the usual constraints of manpower, is not in a position to monitor which becomes the breeding ground for middlemen to thrive on the situation and entrap migrants" (Mehrotra 2021). The contradiction/integration in institutional subsystems is not migrant inclusive. "System integration refers to the more or less functional or contradictory relationships between its institutional subsystems" (Hickman et al. 2012, p. 7). Hence, political, social and economic integration of migrants can be accomplished by emphasising on the sensitive issues like education, health, language, land, etc. to avoid patterned discrimination.

For the internal migrants, during pandemic the Government of India undertook initiatives for creating jobs at the local level for migrant workers through Pradhan Mantri Garib Kalyan Yojana, Aatmanirbhar Bharat Yojana and Pradhan Mantri Garib Kalyan Rozgar Abhiyaan. The objective was to make workers self-sufficient and helping them to recuperate their confidence by providing them livelihood and social protection. The states undertook identification of reskilling needs. For example, in Uttar Pradesh, the massive exercise of skill-mapping directed the creation of livelihood opportunities and launched portals and mobile-based apps for connecting workers with employment opportunities and providing access to social security and skill development. Its 'Sewa Mitra application' used a multi-dimensional approach control technology and brick-and-mortar institutions, and by collecting data on workers, it gives them employment. It has database of migrant returnees and registration mechanism for all workers. The state government through Krishi Vigyan Kendra, Rural Self-Employment Training Institute and the district administration conducted multiple trainings of women who were part of self-help groups. A fund of Rs 1.10 lakh was given to each Self-Help Group for internal lending and capital investment. Rajasthan government's Raj-Kaushal portal had a vast database of skilled, semi-skilled and unskilled workers, and there was empowering of women migrant workers. Through this portal, migrants in the age group of 18–60 were provided employment, after conducting surveys and skill-mapping (Skill Development and Employment Vertical 2020).

22.8 Discussion

The study illustrated that the programmes and policies in their current form will not produce desired results unless there is incorporation of social cohesion for inclusion in their design framework along with climate resilience as its outcomes. Redressal of underlying inefficiencies and the change in the attitudinal behaviour of the host and the migrants at migration destination locations would be a move towards sustainable goals. Although India's State Action Plans on climate change served as the primary policy document at the subnational level for matters related to climate change, yet the national- and state-level policies on coastal zone management focused majorly on climate-resilient infrastructure and overlooked the impact of human displacement.

Schiefer and Noll (2016, p. 579) mentioned the following indicators for social cohesion: Social relationship-Participation, Trust; Connectedness-Belonging, Identification; Orientation-Towards social responsibility, and Solidarity. Based on the indicators of social cohesion, there can be mobilising of coexistence for internally displaced migrants, ensuring their equitable access to health, education, decent work and legal identity. Having high-quality reliable data as 'disaggregation of data' is a way to ensure inclusiveness (UNSDG 2030). The credible data for identification of migrants becomes significant in mapping their number along with setting up of employment centres to facilitate labour market access. Inclusion of special requirements for low-income houses and provision for anticipatory wage employment will integrate the migrants. The unused land with irrigation facilities can be given to the migrated villagers where the land is occupied for developmental projects so as to protect the needs of the displaced people falling into the category of migrants. Housing being the key to inclusion forms a part of social protection policies. It promotes home ownership that generates sense of belonging.

Education is the significant to formation of cohesive society. Language and education being the key indicators for integration, there should be language interpreters or local language teachers, special schemes and trained person to educate the migrant children. By increasing the capacity in schools for the migrant children, the benefit of education can be given to the displaced migrant children.

Migrants suffer from various health problems, accordingly training of health and medical staff for being sensitive to migrant populations becomes a focal point. The need of counsellors and therapists for psychological care of the migrants becomes necessary for their wellbeing because of the displacement trauma.

Culture and tradition, being an important constituent of any society, provides a sense of belonging. Access to local practices and customs (traditions and festivals) and mentors for intercultural training and organisation of cultural activities enhances social cohesion. Similarly, the purpose of literature is to orient human beings towards social construction, culture and historical developments. Any text exhibits multiple meaning and functions as a realistic tool that is significant in the development of the society. Academic institutions bridge the links in society, and inclusion of migrant literature prepares the younger generation for a resilient and cohesive society.

There should be regular training programs for displaced persons/migrants through online/virtual learning platforms of government along with workforce intermediaries so that the migrants can be provided access to digital skill building. The administrative capacity needs to be enhanced by sensitising the administrators for identifying the potential of migrants for offering manpower to local governments. The involvement of the State Labour Departments for handling of exploitation is essentially required. Portability of social protection entitlements (IIED 2019), social security benefits and adoption of anti-discriminatory measures are perquisites for migrants' integration and align with the indicators of social cohesion. The migration driver indicators should categorise the population at places where migration might upsurge in future due to augmented environmental pressure and manmade disasters, and it can act as a vital tool for adopting various strategies to avoid displacement convulsions.

22.9 Conclusion

United Nations High Commissioner for Refugees resolution highlights the principle of burden and responsibility sharing, mobilising the international community and galvanising action for co-existence. 2030 Sustainable Development Agenda (out of 17 goals 10 contains indicators relevant to migrants) makes migrants as important enabler of sustainable development. The World Bank study suggests long-term strategy with a focus on responsive governance, by formal informal dialogues, policymaking and developing social relationships, connectedness and orientation-the indicators of social cohesion (Schiefer and Noll 2016). The adaptation assessments incorporating cultural, social, economic and psychological understandings of the contexts and reasons for migration reverse or minimise the lethal effects contributing to displacement. The research focused on policies and their role in social cohesion that can affect the societal change creating a stronger society. High-level cohesive society leads to high trust, social harmony and prosperity. The involvement of political and community participation and multicultural policies are the indicators for assimilation of migrants in the mainstream which would intertwine a cohesive society. Adequate compensation, voting rights for seasonal migrants, education for their children, housing facilities and employment are important components for building up a cohesive society. IOM views integration as an essential component of effective and comprehensive migration management to promote social inclusion and good relations across diverse groups, thereby contributing towards inclusive societies. It further recognises that successful integration is a dynamic and multi-directional process that involves mutual adaptation of migrants and the host society, based on principles of protection of fundamental rights, respect, tolerance and non-discrimination.

Cohesive society lessens disparities and polarisation which becomes a significant factor for successful integration of the displaced, acting as precondition for social cohesion which further elevates the economic progress aligning with the economic indicator of social cohesion. Employment is generally recognised as a significant factor for inclusion and eventual integration in society, and therefore, workplace integration becomes a major indicator for economic growth. Healthcare system is another important aspect that can work towards a constructive society. Thus, sensitivity towards the stress and problems migrants face build a viable society. For migrant families, education is an important means for skill development, knowledge gain, social mobility and intercultural skills.

An 'insider mediation' provides more inclusivity. Therefore, the addition of migrant voice becomes a necessary parameter for cohesion thereby necessitating voting rights for seasonal migrants or migrant labourers placing them in the mainstream.

The diasporic group is distinguished by gender, class, ethnicity, various forms (internal/international/permanent/temporary) of migration, and the effect of worldwide economic and political strategies. The WBG's Social Development Strategy of 2005 states that "cohesive societies enable women and men to work together to address common needs, diverse interests, overcome constraints and resolve differences in a civil, nonconfrontational way, promoting peace and security" (Berry and Roberts 2018, p. 15). This inclusive approach towards migrants can become a catalyst in integration. To conclude, policy revision should cover migrant's inclusion in economic, psychological, social, linguistic, navigational and civic spheres empowering the host communities and other local performers to receive and engage with migrants for an integrated and constructive society. Government and non-governmental organisations should adhere to action-based inclusive strategies, legal framework, social protection, employment and local integration creating functional structures for migrant's resettlement, co-existence, cohesion and integration.

References

Ansohn A, Pleskovic B (2011) Climate governance and development. The World Bank, Washington

Ashcroft B, Griffiths G, Tiffin H (2013) Postcolonial studies: the key concepts. Routledge, London

Berry PB, Roberts A (2018) Social cohesion and forced displacement: a desk review to inform programming and project design, p. 15. https://documents1.worldbank.org/curated/en/125521531981681035/pdf/128640-WP-P163402-PUBLIC-Social Cohesion and Forced Displacement.pdf. Accessed 5 Mar 2022

Bhatt W (2009) The gender dimension of migration in India: the politics of contemporary space in Orissa and Rajasthan. Dev Pract 19(1):87–93. https://www.jstor.org/stable/27752013

Brettell C, Hollifield JF (eds) (2008) Migration theory: talking across disciplines. Routledge, New York

Chauhan P (no date) Land acquisition brings misery to people, Tribune India News Service. https://www.tribuneindia.com/news/archive/himachaltribune/land-acquisition-brings-misery-to-people-627556. Accessed 10 Mar 2022

Chief Labour Commissioner (Central) (no date) Government of India. https://clc.gov.in/clc/acts-rules/acts-and-rules-0. Accessed 22 Feb 2022

de Haas H (2021) A theory of migration: the aspirations-capabilities framework. Comp Migr Stud 9:8. https://doi.org/10.1186/s40878-020-00210-4.Accessed25February2022

Fonseca X, Lukosch S, Brazier F (2018) Social cohesion revisited: a new definition and how to characterize it. Innov Euro J Soc Sci Res 32(2):1–23

Garg R et al (2021) Climate-induced displacement and migration in India. https://cansouthasia.net/wp-content/uploads/2021/02/Migration_India_20_02_2021.pdf. Accessed 22 Nov 2021

Hickman MJ, Mai N, Crowley H (2012) Place, belonging and social cohesion. Migration and social cohesion in the UK. Palgrave Macmillan, New York, pp 80–112

IDMC GRID (2021) Internal displacement in a changing climate. https://www.internal-displacement.org/sites/default/files/publications/documents/grid2021_idmc.pdf. Accessed 13 Jan 2022

IIED (2019) International Institute for Environment and Development. https://www.iied.org/. Accessed 22 Nov 2021

ILO (2020) Road map for developing a policy framework for the inclusion of internal migrant workers in India. International Labour Organization, Geneva

Irudaya RS, Sivakumar P, Srinivasan A (2020) The covid-19 pandemic and internal labour migration in India: a 'crisis of mobility.' Indian J Labour Econ 63:1021–1039. https://doi.org/10.1007/s41027-020-00293-8

Joseph J (2021) The times of India, 4 September. https://timesofindia.indiatimes.com/city/jaipur/nearly-20l-hectares-of-kharif-crops-hit-by-flood-and-drought/artileshow/85912853.cms. Accessed 12 June 2022

Kumar S, Choudhury S (2020) Migrant workers and human rights: a critical study on India's covid-19 lockdown policy. Soc Sci Hum Open Elsevier. https://doi.org/10.1016/j.ssaho.2021.100130. Accessed25March2022

Kunwar DS, Tripathi A (2010) Tehri sparks flood scare in 1200 villages downstream Ganga. The economic times. https://economictimes.indiatimes.com/news/politics-and-nation/tehri-sparks-flood-scare-in-1200-villages-downstream-ganga/articleshow/6597257.cms?from=mdr. Accessed 10 Mar 2022

Laczko F, Aghazarm C (2009) Migration, environment and climate change: assessing the evidence. International Organization for Migration, Geneva

Lokhande N, Gundimeda H (2021) MGNREGA: the guaranteed refuge for returning migrants during covid-19 lockdown in India. Indian Econ J 69(3):584–590. https://doi.org/10.1177/00194662211023848 (Top of Form Bottom of Form)

Manta D (no date) Inadequate compensation, NGT cuts rile residents. Tribune India news service. https://www.tribuneindia.com/news/archive/himachaltribune/inadequate-compensation-ngt-cuts-rile-residents-757754. Accessed 10 Mar 2022

Mehrotra K (2021) Explained: What is NITI Aayog's draft national policy on migrant workers? The Indian express. https://indianexpress.com/article/explained/niti-aayog-migrant-wo .Accessed 8 Mar 2022

Ministry of Law and Justice Government of India. https://legislative.gov.in/. Accessed 24 Feb 2022

Mooney L, Hasbun R, Mekker I (2021) Social cohesion and displacement toolkit international republican institute: advancing democracy worldwide. International Republican Institute, Washington. https://www.iri.org/wp-content/uploads/legacy/iri.org/wysiwyg/social-cohesion-toolkit-abridged_final.pdf. Accessed 6 Mar 2022

Nandi J (2021) India lost $87 billion to climate disasters in 2020: Report. Hindustan times. https://www.hindustantimes.com/india-news/india-lost-87-billion-to-climate-disasters-in-2020-report-101635272896946.html. Accessed 12 Jan 2022

OECD (2011) Social cohesion and development. Perspectives on global development 2012: social cohesion in a shifting world. OECD Publishing, Paris pp 51–63. https://doi.org/10.1787/persp_glob_dev-2012-6-en

Registrar General of India (2011) D series: office of the RGI and census commissioner. Government of India. https://censusindia.gov.in/census_and_you/migrations.aspx. Accessed 15 Feb 2022

Schiefer D, Van der Noll J (2016) The essentials of social cohesion: a literature review. Soc Indic Res 132(2):579–603

Sharalaya N (2018) Taking India's climate migrants seriously. https://thediplomat.com/2018/08/taking-indias-climate-migrants-seriously. Accessed 5 Mar 2022

Skill Development and Employment Vertical (2020) Empowering migrant workers through skill development and livelihood generation. Nitiayog, Delhi

The Constitution of India (2019) https://legislative.gov.in/sites/default/files/COI-updated.pdf. Accessed 5 Mar 2022

The Hindu (2021) Over 8 lakh migrant workers left Delhi on State-run buses, The Hindu, 22 May. https://www.thehindu.com/news/cities/Delhi/over-8-lakh-migrant-workers-left-delhi-on-state-run-buses/article34624216.ece. Accessed 11 Mar 2022

The Hindu Data Team (2020) Data 96% migrant workers did not get rations from the government, 90% did not receive wages during lockdown: survey, The Hindu, 20 April. https://www.thehindu.com/data/data-96-migrant-workers-did-not-get-rations-from-the-government-90-did-not-receive-wages-during-lockdown-survey/article31384413.ece. Accessed 1 Feb 2022

Un.org (2020) Social inclusion and poverty eradication. https://www.un.org/development/desa/socialperspectiveondevelopment/issues/social-integration.html. Accessed 13 Mar 2022

UNDP (2020) Strengthening social cohesion: conceptual framing and programming implications. https://www.undp.org/publications/strengthening-social-cohesion-conceptual-framing-and-programming-implications#modal-publication-download. Accessed 10 Mar 2022

UNHCR (no date) Emergency handbook. https://emergency.unhcr.org/entry/250553/idp-definition. Accessed 15 Mar 2022

Yashwant S, Faleiro J (eds) (2021) Climate-induced displacement and migration in India. https://cansouthasia.net/wp-content/uploads/2021/02/Migration_India_20_02_2021.pdf. Accessed 17 Mar 2022

The manufacturer's authorised representative in the EU is Springer Nature Customer Service Centre GmbH, Europaplatz 3, 69115 Heidelberg, Germany. If you have any concerns regarding our products, please contact ProductSafety@springernature.com

Printed and bound by CPI Group (UK) Ltd, Croydon, CR0 4YY
01/10/2026
02225872-0017